3rd International Conference on Competitive Materials and Technological Processes 2014 (IC-CMTP3)

IOP Conference Series: Materials Science and Engineering
Volume 123

Miskolc-Lillafured, Hungary
6 - 10 October 2014

ISBN: 978-1-5108-2278-8
ISSN: 1757-8981

Printed from e-media with permission by:

Curran Associates, Inc.
57 Morehouse Lane
Red Hook, NY 12571

CURRAN ASSOCIATES INC. proceedings.com

Some format issues inherent in the e-media version may also appear in this print version.

Printed by Curran Associates, Inc. (2016)

Institute of Physics
Dirac House, Temple Back
Bristol BS1 6BE UK

Phone: 44 1 17 929 7481
Fax: 44 1 17 920 0979

techtracking@iop.org

Additional copies of this publication are available from:

Curran Associates, Inc.
57 Morehouse Lane
Red Hook, NY 12571 USA
Phone: 845-758-0400
Fax: 845-758-2633
Email: curran@proceedings.com
Web: www.proceedings.com

Table of contents

Volume 123

3rd International Conference on Competitive Materials and Technology Processes (IC-CMTP3)
6–10 October 2014, Miskolc-Lillafüred, Hungary

Accepted papers received: 21 March 2016
Published online: 15 April 2016

Preface

Papers

Advanced Materials for Bio- and Medical Applications

Advanced Materials for Extreme Applications

Light-Weight Metals and Alloys

Membranes and Catalysts

Minerals for Environmental and Medical Application

Nanomaterials for Environment and Health

Novel Synthesis and Processing Technology

Phase Diagram as a Tool of Materials Science

Polymer Derived Ceramics

Processing and Properties of Silicate Ceramics

Testing and Characterization of Materials - Methods, Equipment and Errors

is-icbm1: The 1st International Symposium on Innovative Carbon Based Materials

ic-cmtp3

3RD INTERNATIONAL CONFERENCE ON COMPETITIVE MATERIALS AND TECHNOLOGY PROCESSES

October 6-10, 2014/ Miskolc-Lillafüred, Hungary

Organized by the International Scientific Advisory Board (**ISAB**),

International Organizing Committee (**IOC**) and

Session Chairs

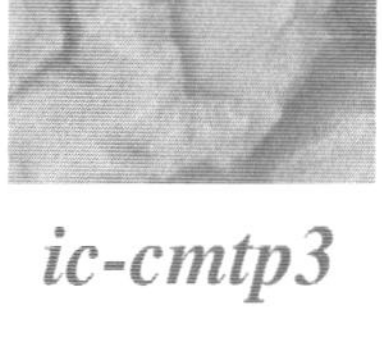

ic-cmtp3

IOP Conf. Series: Materials Science and Engineering **123** (2016) 011001 doi:10.1088/1757-899X/123/1/011001

Abstract

Competitiveness is one of the most important factors in our lives and it plays a key role in the efficiency both of organizations and societies. The more scientifically advanced and prepared organizations develop more competitive materials with better physical, chemical, and biological properties, and the leading companies apply more competitive equipment and technological processes.
The aims of the **3rd International Conference on Competitive Materials and Technology Processes** (ic-cmtp3), and the **1st International Symposium on Innovative Carbons and Carbon Based Materials** (is-icbm1) and the **1st International Symposium on Innovative Construction Materials** (is-icm1) organized alongside are the following:

- Promote new methods and results of scientific research in the fields of material, biological, environmental and technological sciences;
- Exchange information between the theoretical and applied sciences as well as technical and technological implementations;
- Promote communication and collaboration between the scientists, researchers and engineers of different nations, countries and continents.

Among the major fields of interest are advanced and innovative materials with competitive characteristics, including mechanical, physical, chemical, biological, medical and thermal, properties and extreme dynamic strength. Their crystalline, nano- and micro-structures, phase transformations as well as details of their technological processes, tests and measurements are also in the focus of the **ic-cmtp3** conference and the **is-scbm1** and **is-icm1** symposia. Multidisciplinary applications of material science and the technological problems encountered in sectors like ceramics, glasses, thin films, aerospace, automotive and marine industries, electronics, energy, construction materials, medicine, biosciences and environmental sciences are of particular interest.

In accordance with the program of the **ic-cmtp3** conference and **is-icbm1** and **is-icm1** symposia we have received more than **350** enquiries and registrations from different organizations. More than **240** abstracts were accepted for presentation. From them **12** were plenary lectures and **112** oral presentations. Researchers from **41** countries in Asia, Europe, Africa, North and South America travelled to Miskolc-Lillafüred (**Hungary**) and participated in the conference events. Including co-authors, the research work of more than **700** scientists were presented in the sessions and symposia of the **ic-cmtp3** conference.

Prof. Dr. László A. Gömze
chair, ic-cmtp3

THE INTERNATIONAL SCIENTIFIC ADVISORY BOARD (ISAB)

Dr. Priscila ANADÁO, University of São Paulo (Brazil)

Dr. Rafaella DEMICHELIS, Curtin University (Australia)

Prof. Dr. Cheng-Ji DENG, Wuhan University of Science and Technology (China)

Prof. Dr.Katherine T. FABER, Northwestern University (USA)

Prof. Dr. Saverio FIORE, CNR Institute of Methodologies (Italy)

Prof. Dr. László A. GÖMZE, University of Miskolc (Hungary)

Prof. Dr. Kiyoshi ITATANI, Sophia University (Japan)

Prof. Dr. Sergei N. KULKOV, Tomsk State University (Russia)

Prof. Dr. John M. PARKER, The University of Sheffield (UK)

Prof. Dr. Tohru SEKINO, The University of Osaka (Japan)

Prof. Dr. David S. SMITH, University of Limoges (France)

Prof. Dr. Bojja SREEDHAR, CSIR-Indian Institute of Chemical Technology (India)

THE INTERNATIONAL ORGANIZING COMMITTEE (IOC)

Dr. Priscila ANADÁO, University of São Paulo (Brazil)

Prof. Dr. Dong Sik BAE, Cgangwon National University (Korea)

Dr. Fares BEAINY, Volvo Construction Equipment (USA)

Prof. Dr. Philippe BLANCHART, ENSCI Limognes University (France)

Dr. Adorján BOROSNYÓI, Technical University of Budapest (Hungary)

Prof. Dr. Saveiro FIORE, National Research Council (Italy)

Prof. Dr. Elzbieta GODLEWSKA, AGH University (Poland)

Dr. Ludmila N. GÖMZE, IGREX Engineering Service Ltd. (Hungary)

Prof. Dr. Olga KOTOVA, RAS Institute of Geology (Russia)

Prof. Dr. Hiromitsu KOZUKA, Kansai University (Japan)

Prof. Dr. Sergey N. KULKOV, Tomsk State University (Russia)

Prof. Dr. Huey-Jinan LIN, National United University (Taiwan)

Prof. Dr. Vasily LUTSYK, Buryat State University (Russia)

Prof. Dr. Liviu MARSAVINA, Polytechnic University of Timisoara (Romania)

Prof. Dr. Fujio MIZUKAMI, National Institute ATST (Japan)

Prof. Dr. John M. PARKER, The University of Sheffield (UK)

Prof. Dr. Tomasz SADOWSKI, Lublin University of Technology (Poland)

IOP Conf. Series: Materials Science and Engineering **123** (2016) 011001 doi:10.1088/1757-899X/123/1/011001

Prof. Dr. Tsugio SATO, Tohoku University Sendai (Japan)

Prof. Dr. Michael SCHEFFLER, Otto-von-Guerike University (Germany)

Prof. Dr. Igor SHABALIN, The University of Salford (UK)

Prof. Dr. Bojja SREEDHAR, CSIR-Indian Institute of Chemical Technology (India)

Prof. Dr. Jean-Claude TEDENAC, University of Montpellier 2 (France)

Dr. Nicolas TESSIER-DOYEN, ENSCI, Limognes University (France)

Prof. Dr. Alexandr L. URAKOV, Izhevsk State Medical Academy (Russia)

Dr. Hao WANG, Wuhan University of Science and Technology (China)

THE CHAIRS OF THE SESSIONS AND SYMPOSIUMS

***Session 1.* Advanced Materials for Bio- and Medical Applications**

MD. Prof. Dr. Alexandr L. URAKOV,

Izhevsk State Medical Academy (Russia)

***Session 2.* Advanced Materials for Extreme Applications**

Dr. Ludmila N. GÖMZE,

IGREX Engineering Service Ltd. (Hungary)

***Session 3.* Advanced Nanomaterials with Predesigned Properties**

Prof. Dr. Elzbieta GODLEWSKA,

University of AGH (Poland)

***Session 4.* Biomaterials Derived Ceramics and Composites**

Prof. Dr. Bojja SREEDHAR,

CSIR-Indian University of Chemical Technology (India)

***Session 5.* Glasses, Coatings and Related Materials**

Prof. Dr. Nikolay NIKONOROW,

ITMO St. Petersburg (Russia)

Session 6. **Hetero-Modulus and Hybrid Materials**

Prof. Dr. Tohru SEKINO

The University of Osaka (Japan)

Session 7. **Light Weight Metals and Alloys**

Dr. Jean-Marie DREZET,

Ecole Polytechnique Federale Lausanne (Switzerland)

Session 8. **Materials with Extreme Dynamic Strength**

Prof. Dr. László A. GÖMZE,

University of Miskolc (Hungary)

Session 9. **Membranes and Catalysts**

Dr. Priscila ANADÁO,

University of São Paulo (Brazil)

Session **10. Minerals for Environmental and Medical Applications**

Prof. Dr. Saveiro FIORE,

National Research Council (Italy);

Prof. Dr. Olga KOTOVA,

RAS Institute of Geology (Russia)

Session 11. **Nanomaterials for Environment and Health**

Prof. Dr. Tsugio SATO,

Tohoku University Sendai (Japan)

Session 12. **Novel Synthesis and Processing Technology**

Prof. Dr. Sergey N. KULKOV,

Tomsk State University (Russia)

Session 13. **Phase Diagram as a Tool of Materials Science**

Prof. Dr. Vasily LUTSYK,

Institute of Physical Materials Science- SB RAS (Russia)

IOP Conf. Series: Materials Science and Engineering **123** (2016) 011001 doi:10.1088/1757-899X/123/1/011001

Session 14. **Polymer Derived Ceramics**
Prof. Dr. Michael SCHEFFLER,
Otto-von-Guericke University Magdeburg (Germany)

Session 15. **Processing and Properties of Silicate Ceramics**
Nicolas TESSIER-DOYEN,
University Limoges, Heterogeneous Materials Group G.E.M.H. (France)

Session 16. **Testing and Characterization of Materials – Methods, Equipment and Errors**
Dr. István KOCSERHA,
University of Miskolc (Hungary)

is-icbm1 – **The 1st International Symposium on Innovative Carbon Based Materials**
Prof. Dr. Igor SHABALIN,
The University of Salford (UK);
Dr. Hao WANG,
Wuhan University of Science and Technology (China)

is-icm1 – **The 1st International Symposium on Innovative Construction Materials**
Prof. Dr. Tomasz SADOWSKI,
Lublin University of Technology (Poland);
Dr. Adorján BOROSNYÓI,
Technical University of Budapest (Hungary);
Prof. Dr. Liviu MARSAVINA,
Polytechnic University of Timisoara (Romania)
Dr. Ludmila N. GÖMZE,
IGREX Engineering Service Ltd. (Hungary)

ACKNOWLEDGEMENTS

On behalf of the ic-cmtp3 conference and is-icbm1 and is-icm1 symposia organizers I would like acknowledge and thank the following organizations and societies for their support and help in press campaigns and media contributions:

AIPEA – Internatonal Association for the Study of Clays
Applied Rheology – international scientific journal on Rheology (Switzerland)
biomat.net – The Biomaterials Network (Portugal)
cfi – Ceramic Forum International (Germany)
Építőanyag-JSBCM – scientific journal of SZTE (Hungary)
GFC – Groupe Francais de la Céramique (France)
ICG – International Commission of Glass
IGREX Engineering Service Ltd. (Hungary)
KSSS – The Korean Society of Spine Surgery (Korea)
Landes Biosciences (USA)
ME – University of Miskolc (Hungary)
RAS – The Russian Academy of Sciences (Russia)
refractory WORLDFORUM – scientific journal (Germany)
SB-RAS – Institute of Strength Physics and Materials Science (Russia)
SBAO – Society for Biomaterials and Artificial Organs (India)
SZTE – Scientific Society on Silicate Materials (Hungary)
TSU – Tomsk State University (Russia)
TUS – Tohoku University Sendai (Japan)
UO – The University of Osaka (Japan)

I would like to thank colleagues of ISAB, IOC and the chairs of sessions and symposia for their support and help in organization and in successful transaction of ic-cmtp3, the 3rd International Conference on Competitive Materials and Technology Processes.

I would like to personally thank Prof. Dr. Tohru Sekino (The University of Osaka), Prof. Dr. Sergey N. Kulkov (Tomsk State University), Prof. Dr. Michael Scheffler (Otto-von-Guericke University, Magdeburg), Dr. Jean Marie Drezet (Ecole Polytechnique Federale Lausanne), Prof. Dr. Tomasz Sadowski (Lublin University of Technology) and to Dr. Adorján Borosnyói (Technical University of Budapest) for their excellent organization work in Japan, Russia, Asia and the European Union. Special thanks to Dipl. Ing. Rosemary Vocht-Mields for here strong support in scientific press and media.

Prof. Dr. László A. Gömze
chair of conference board

Peer review statement

All papers published in this volume of *IOP Conference Series: Materials Science and Engineering* have been peer reviewed through processes administered by the proceedings Editors. Reviews were conducted by expert referees to the professional and scientific standards expected of a proceedings journal published by IOP Publishing.

IOP Conf. Series: Materials Science and Engineering **123** (2016) 012001 doi:10.1088/1757-899X/123/1/012001

Modelling of the mechanical behavior of a polyurethane finger interphalangeal joint endoprosthesis after surface modification by ion implantation

A Beliaev[1], A Svistkov[1], R Iziumov[1], I Osorgina[2], A Kondyurin[3], M Bilek[3], D McKenzie[3]

[1]Institute of Continuous Media Mechanics UB RAS, Laboratory of Micromechanics of Structurally Inhomogeneous Media, Perm, Russia
[2]Perm State National Research University, Perm, Russia
[3]School of Physics, University of Sydney, Sydney, Australia

E-mail: beliaev@icmm.ru

Abstract. Production of biocompatible implants made of polyurethane treated with plasma is very perspective. During plasma treatment the surface of polyurethane acquires unique physic-chemical properties. However such treatment may change the mechanical properties of polyurethane which may adversely affect the deformation behaviour of the real implant. Therefore careful study of the mechanical properties of the plasma-modified polyurethane is needed. In this paper, experimental observations of the elastic characteristics of plasma treated polyurethane and modelling of the deformation behaviour of polyurethane bio-implants are reported.

1. Introduction

The problem of creation of biocompatible implants is raised in many papers [1-5]. Variety of materials is used to create them: metals, polymers and ceramics. All of these materials are suitable for various classes of prostheses. In this study will be considered the usage of polymers that are suitable for creating soft tissue implants inter alia hearing implants, breast implants, heart valves, interphalangeal prosthesis, etc. [6-12]. The most commonly used polymer in soft tissue implantology is silicone [13, 14]. However, this polymer has a number of disadvantages: not all polymer degradation products are removed from the body and some accumulate there. It may lead to adverse consequences for the patient. Also silicone is bioinert material and bioinertness is sometimes not enough for optimum functioning of the prosthesis in the body. The introduction of the implant into the body starts the process of encapsulation, which eventually leads to the formation of calcified shell around the implant. Polyurethane has unimportant advantage over the silicone: being bioinert it can be subjected by ion-plasma treatment [15, 16]. As a result, the treated surface becomes active and protein adheres to it [17, 18]. Correctly adsorbed protein not only becomes strongly bound on the surface but will retain its native conformation [19]. In this state protein is able to serve as a link between the foreign body (implant) and the living organism. In other words, the biocompatibility of the implant in the organism will be achieved.

2. Formulation of the problem and method of solution

At high nitrogen plasma fluence (approximately 10^{16} ions /cm^2) and 20 keV ion energy a carbon layer about 100 nm thickness is formed on the surface of the material [21]. Since the plasma treatment provides the surface of the material with unique physical and chemical properties, there is reason to believe that the treatment may also have an impact on the mechanical properties of the material.

Additionally, the presence of a carbonized layer on the surface presents a number of problems. The adhesion between the polyurethane and the carbon layer is extremely high because the layer is prepared by modification of the polyurethane surface layer. The carbonized layer in this case is very fragile. Studies using atomic-force and optical microscopy show that even a slight deformation of the polyurethane after treatment leads to cracking of the carbonized layer [22].Since the adhesion between the layer and the material is high the carbon layer does not peel off from the surface after cracking but continues to be there as a set of small fragments of varying size and shape. The problem is that during deformation of the material the space between the fragments is a strain concentrator.

This can lead to the formation of micro-cracks between the fragments that affect the material's macro strength. Furthermore, cracks open an area of unmodified polyurethane which is not covered by a protein and the cells of the immune system recognizes these areas as a "foreign body" and activate the foreign body reaction [20]. From this it follows that it is necessary to design such a shape of prosthesis which is close anatomically and in addition to minimize the deformation on the surface of the prosthesis.

This paper presents an experimental study of the mechanical properties of polyurethane-treated by plasma. Also the calculation of the stress-strain state of areal interphalangeal joint prosthesis is conducted and the estimation of efficiency of possible forms is held.

3. Experiment

3.1. Experimental study of treated polyurethanes

The study of the mechanical properties of polyurethane was carried out on the dynamic mechanical analyzer DMA / SDTA861e. The dynamic mechanical analysis (DMA) method has been chosen because it can be used to obtain information about not only the elastic, but also the dissipation properties of a material. Five groups of samples were studied: untreated (init) and 4 groups with different treatment fluences $5 * 10^{14}; 10^{15}; 5 * 10^{15}; 10^{16}$. Such fluence values were obtained by treating with nitrogen plasma whilst applying pulsed negative bias of 20 keV to a sample electrode and covering mesh for 40, 80, 400 and 800 seconds respectively. The samples were rectangular strips of dimensions10x4x2 mm. Measurements were made at 1, 4 and 7 Hz frequencies. For each frequency, the amplitude loading was varied from 1 to 10% with1.5% increments. For each amplitude E` (storage modulus) and E``(loss modulus) values were determined.

Analysis showed that characteristics of the material remain practically unchanged with different frequencies (therefore all the results will be given only for a frequency of 4 Hz). This is certainly a positive result since the viscosity is a negative characteristic of the material used for producing implants.

Figure 1 shows the values of storage modulus for samples treated with different fluences. *E`* values were found to depend on the deformation as is typical for polymers. *E`* dependence on the fluence of the ion treatment is very weak. The results of the measurement differ only within the instrumental error. On the basis of the measurements it can be concluded that the elastic properties of the material do not change after the plasma treatment.

Figure 2 shows the values of the dynamic loss modulus for samples treated with different fluences. It is observed that E` is about 5 times greater than E``. This confirms the fact that the behavior of the polymer can be described as elastic and we can disregard its viscosity when modeling an object made of this polymer.

Taking into account the results obtained using DMA analysis it was decided to consider polyurethane as a hyperelastic material. The properties of untreated polyurethane are used in the calculation due to the minimal influence of the ion surface modification on the elastic behavior of material.

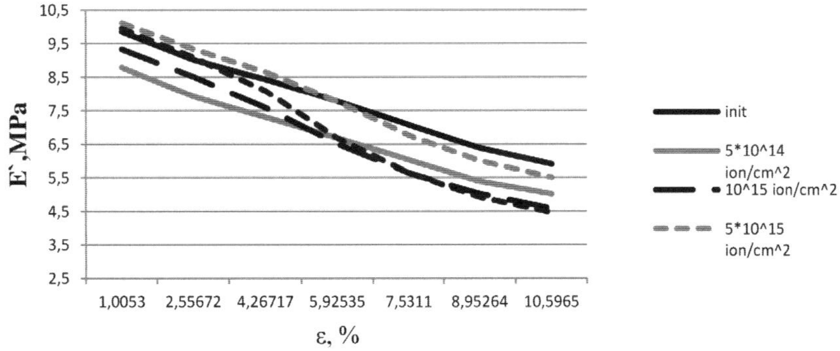

Figure 1. The values of the storage modulus $E`$ for polyurethane samples that were ion-treated with various fluencies

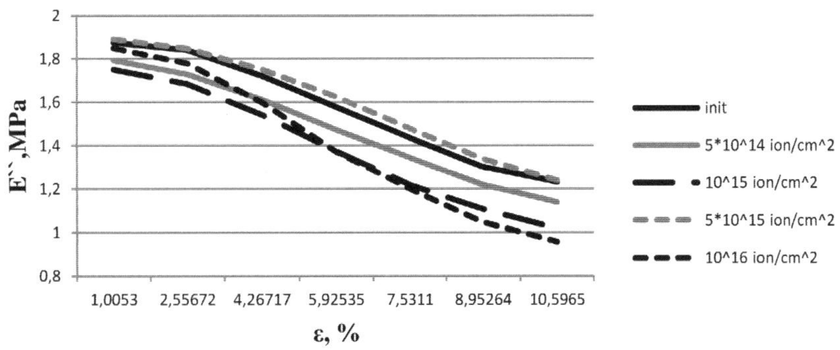

Figure 2. The values of the loss modulus $E``$ for polyurethane samples that were ion-treated with various fluencies

3.2. Interphalangeal joint prosthesis modeling

For interphalangeal prosthesis modelling in a nonlinear-elastic formulation it is necessary to have a curve of the deformation behaviour of the material. To obtain this data an experiment where uniaxial tension was applied to samples of polyurethane was carried out. Ring-shaped samples were used for this testing (as ring-shaped samples eliminate the error in the measurement of deformation associated with slippage of the sample in the grips). The diameter of the ring is 7 cm, and it has a 6 mm width and 2 mm thickness. The loading was carried out at a rate of 0.2%/min. As mentioned above, the effects associated with the viscosity of this material are negligible. However, such a low strain rate was chosen to completely eliminate the impact of temporary effects on the deformation curve. The results are shown in Figure 3.

Stretching was performed to 180% strain (the sample was not destroyed). The initial modulus corresponding to the linear portion of the curve is $E_{initial}$=0.93 MPa. The obtained curve was used in finite element calculations software ABAQUS. On the basis of the deformation behavior of the polyurethane sample, the stress-strain state of an interphalangeal joint prosthesis was calculated.

Figures 4a and 4b show photographs of the interphalangeal prosthesis prototype.

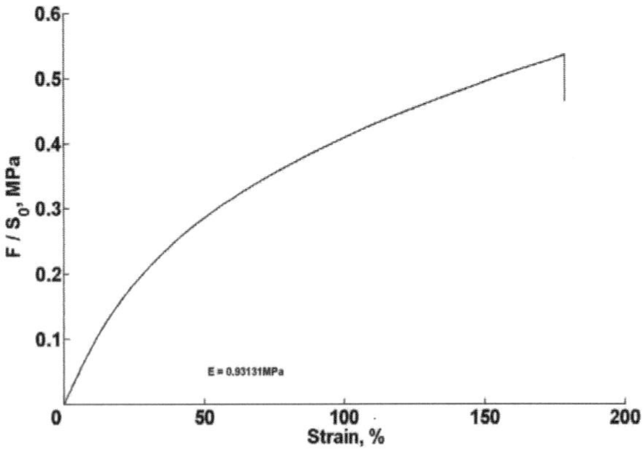

Figure 3. Loading curve of the polyurethane sample under uniaxial tension

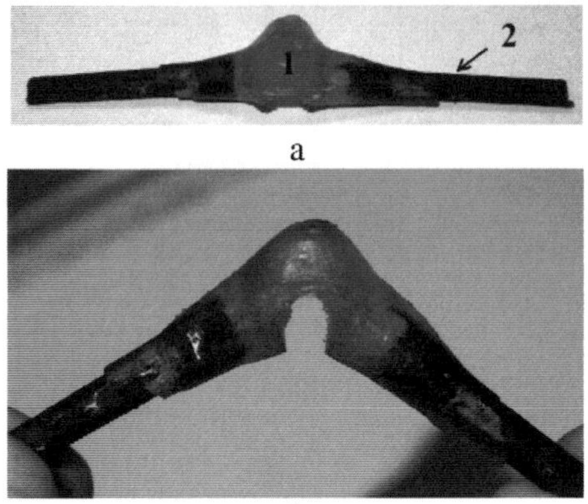

a

b

Figure 4. Interphalangeal joint prosthesis of the finger. a- undeformed state, b - finger flexion

The prosthesis parts are shown in figure 4: 1- polyurethane working part (performing the role of the joint) 2- bone fixators that are made of a carbon-carbon composite material. Carbon-carbon composites have unique compatibility with bone tissue. Their stiffness and strength properties are sufficiently high so that they are ideally suited for use in this type of prosthesis.

The prosthesis is implanted as follows: the interphalangeal joint is opened from the rear side. The damaged ends of the phalanges are removed; the bone anchor is inserted in the advanced medullary canals. The polyurethane working part is located at the position of the natural joint. The living tissue around the joint must remain at its natural location. That is why form of the prosthesis should be very close to the natural form of the joint.

The original version of the prosthesis is shown in Figure 5.

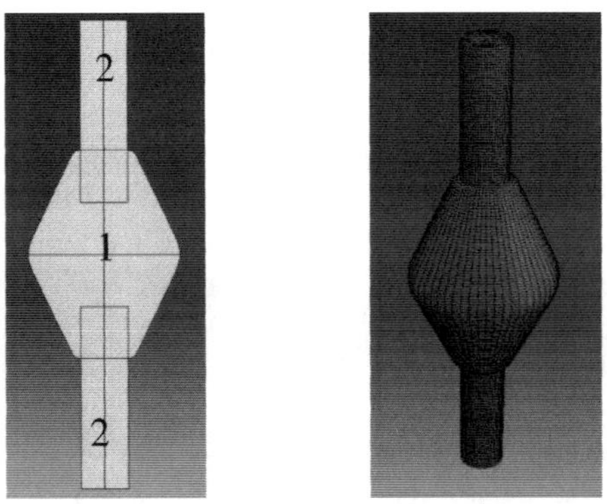

a b

Figure 5. The scheme of the original version of the prosthesis. a- schematic representation (1- working part, 2- bone fixators), b- finite element implementation

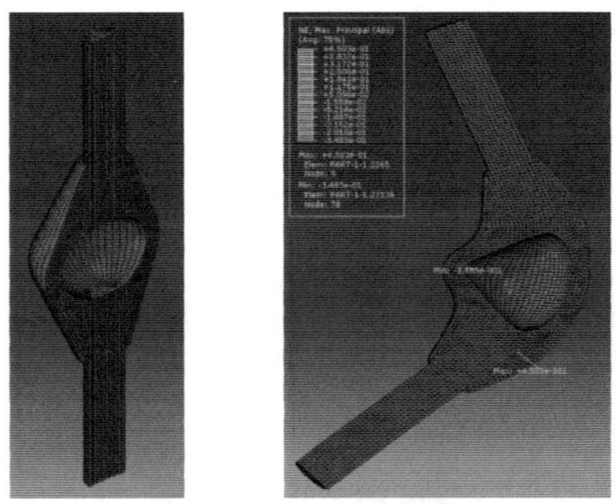

a b

Figure 6. Model of endoprosthesis with a cavity inside the working part. a- finite element model, b-deformed state of bending at an angle of 90° (cut view)

The disadvantage of this geometry is that the working part is too thick and therefore stiff. It was necessary to change the geometry so as to reduce the stiffness of the working part but retain the shape of the prosthesis as close to the physiological version. Additionally, a crease formed at the bending site of the working part when flexed which could compress the living tissue and cause discomfort to the patient during exploitation of the prosthesis. This fold concentrates compressive strain which can negatively affect the state of the carbon layer. Therefore, the revised implant had the shape shown in Figure 6a.

Boundary conditions were set as follows: the end of one bone holder was fixed rigidly and an angular displacement of 90° was applied on the end of the second end. The results of the model calculations show that the presence of the cavity inside the working part doesn`t solve the problem of the appearance of a crease upon flexure. The maximum tensile deformation on the surface of the implant is approximately 42% (Fig. 6b). It was therefore decided to consider a different geometry of the prosthesis. Figure 7a shows the next model of the interphalangeal prosthesis where the cavity is opened on the rear side forming a cut-out section.

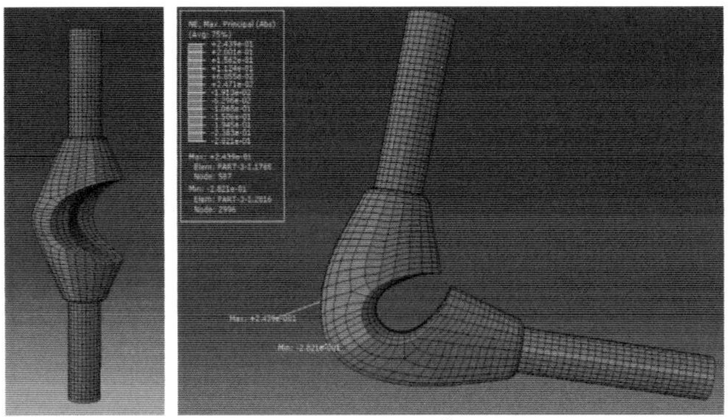

a b

Figure 7. Model of the interphalangeal prosthesis with a cavity cut into the working part to eliminate the crease and reduce the strain on the surface of the prosthesis. a- finite element realization, b- strained state calculation result

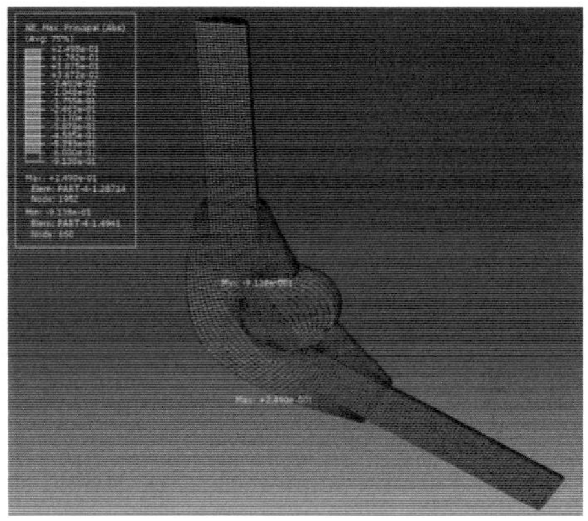

Figure 8. Calculation of strain state of interphalangeal prosthesis with soft, compressible hyperelastic material part

According to the calculation results it can be seen that the maximum tensile stress on the surface of the prosthesis decreased to 24%, and the compressive deformation inside the cut is 28% (which is 6% lower than that observed in the vicinity of creases in the previous model). Furthermore this geometry greatly reduced the stiffness of working part. This geometry eliminates some of the problems associated with high deformations at the sample surface. Reduction of compressive deformation is insignificant but they are not as critical for the normal functioning of the carbon layer as tensile deformations.

However, it should be noted that this cut-out version is not functional. Living tissue surrounding the prosthesis can be clamped inside the cut-out section causing pain to the patient. The problem can be solved by filling the cut-out with soft, hyperelastic and compressible material. In practice, it may be closed-pore polyurethane foam that can also be treated by the plasma to improve the biocompatibility of the prosthesis. Figure 8 shows the calculation of the strain state of this version of the prosthesis. The calculation shows that the maximum tensile strain corresponds to the previous version of the geometry of the prosthesis (24%). The maximum compressive strain in the porous part reaches 91% but this applies to areas under the surface. On the surface strain does not exceed 35% which is not critical to the carbon layer. Thus the introduction of a soft, compressible part solves the pinching problem of the tissue and additionally to reduces strain on the implant surface in comparison with the original version.

4. Conclusion

An investigation of ion-treated polyurethane using the DMA method showed that the treatment had an insignificant effect on its elastic properties. It should be noted that the DMA analysis is performed at strains that do not exceed 10%. Since the strain on the surface of the prosthesis is not much higher than the strain in the experiment this estimate can be considered to be adequate.

To design a different type of prosthesis working for large deformations more research is needed. Also it is necessary to carry out a series of studies to determine how the strength characteristics change. In this phase of the study a form of the implant which is geometrically close to a natural interphalangeal joint was proposed. Some simple design revisions managed to significantly reduce deformation of the surface. This reduction in surface deformation positively affects the functioning of the active carbon layer.

Acknowledgements
This work was supported by the Russian Foundation for basic research (grant 13-01-96009_r_ural_a and grant 14-08-96003_r_ural_a) and the Ministry of Education of Perm Region under agreement (S-26/632).

References
[1] van de Ven P J, Kaatee R, Beutler J J, Beek F J, Woittiez A J, Buskens E, Koomans H A and Mali W P 1999 *Lancet* **353**282
[2] Lambert T L, Dev V, Rechavia E, Forrester J S, Litvack F and Eigler N L 1994 *Circulation* **90**1003
[3] Kondyurin A, Romanova V, Begishev V, Kondyurina I, Guenzel R and Maitz M F 2005 *Journal of Bioactive and Compatible Polymers* **20** 77
[4] Ward R S and Jones R L 2011 *Comprehensive. Biomaterials*, Oxford: Elsevier 431
[5] Kurella A and Dahotre N B 2005 J. *of Bioactive and Compatible Polymers* **20** 5
[6] Lelah M. D., Cooper S.L.1986 *CRC Press* 225
[7] de Groot JH, de Vrijer R, Pennings AJ, Klompmaker J, Veth RPH, Jansen HWB.1996 *Proc. R. Soc. Lond.* **A 17** 163
[8] Handel N, Gutierrez J.2006 *Aesth. Surg.* J.**26**265
[9] Stachelek SJ, Song C, Alferiev I, Defelice S, Cui X, Connolly JM, et al.2000 *Gene. Ther.***11** 15
[10] Ashton JH, Mertz JAM, Harper JL, Slepian MJ, Mills JL, McGrath DV, et al. 2011 *Acta Biomaterialia***7** 287
[11] Thomas V., Kumari T.V., Jayabalan M.2001 *Biomacromolecules.* **2**588
[12] Theron JP, Knoetze JH, Sanderson RD, Hunter R, Mequanint K, Franz T, et al.2010 *Acta Biomaterialia* **6** 2434
[13] Chuang T-W, Masters K.S.2009 *Biomaterials* **30**5341
[14] Kondyurin A.V., Maitz M.F., Romanova V.A., Begishev V.P., Kondyurina I.V., GuenzelR. 2004 *J. Biomater. Sci. Polymer Edn.* **15(2)** 145
[15] Colas A, Curtis J. Ch.2004 Biomaterials Science: An Introduction to Materials in Medicine. 2nd Edition. New York: Elsevier. 80
[16] Colas A, Curtis J. Ch.2004 Biomaterials Science: An Introduction to Materials in Medicine. 2nd Edition. New York: Elsevier.697
[17] Suzuki Y., Kusakabe M., J. S. Lee, M. Kaibara, M. Iwaki and H. Sasabe 1992 *Nucl. Instr. Methods* **B 65** 142
[18] Begishev V., Gavrilov N., Mesyats G., Klyachkin Yu., Kondyurina I., Kondyurin A., Osorgina I.1998 Proc. of 12th Intern. Conf. on High-Power Particle Beams **2** 997
[19] Kondyurin A., Maitz M.F., Surface Modification of ePTFE and Implants using the same, US patent WO 2007/022174 A3, 2007
[20] Bilek M.M.M., Bax D.V., Kondyurin A., Yin Y.B., Nosworthy N.J., Fisher K., et al. 2011Proceed. of the National Academy of Sciences of the United States of America**108**14405
[21] Kondyurina I., Nechitailo G.S., Svistkov A.L., Kondyurin A., BilekM. 2015Nuclear Instruments and Methods in Physics Research B**342**39
[22] Kondyurin A., Bilek M.M.M., Ion beam treatment of Polymers, Elsevier, 2014.

Titania sol-gel coatings with silver on non-porous titanium and titanium alloys

D Horkavcova[1], M Cerny[1] , L Sanda[1] , P Novak[2] , E Jablonska[3] Z Zlamalova-Cflova[3] and A Helebrant[1]

[1] Department of Glass and Ceramics, Faculty of Chemical Technology,
[2]University of Chemistry and Technology Prague, Czech Republic
[3]Department of Metals and Corrosion Engineering, Faculty of Chemical Technology, University of Chemistry and Technology Prague, Czech Republic
[3]Department of Biochemistry and Microbiology, Faculty of Food and Biochemical Technology, University of Chemistry and Technology Prague, Czech Republic

E-mail: diana.horkavcova@vscht.cz

Abstract. The objective of the work was to prepare and characterize titania sol-gel coatings on non-porous titanium and newly developed titanium alloys. Basic titania sol contained two forms of silver. Titania sol without silver was used as a reference sample. Coatings were prepared by dip-coating technique during stirring and fired. Coatings after firing were characterized by scanning electron microscopy. All titania coatings were measured to determine their adhesive and bactericidal properties. Adhesion of the coatings to the substrate was measured by tape test. Gram-negative bacteria *E. coli* was used for the bactericidal test. Coated substrates were immersed into suspension of *E. coli* in physiological solution for 24 hours. The *in vitro* cytotoxicity test was performed after one day. The bactericidal effect without toxicity was confirmed for selected coatings.

1. Introduction

Titania coatings containing various ions prepared by a solo-gel method on metallic, glass or other substrates have been frequently used in photocatalysis, solar cells, sensors and biomaterials [1, 2]. The sol-gel method which is based on a controlled mixing process of organic compounds in alcohols and water with an added catalyst is suitable for formation of thin homogenous coatings using techniques like dip-coating [3], spin-coating [4] or spray-coating [5]. One advantage of dip-coating is the possibility to coat substrates of complicated shapes under constant conditions. By using suitable reagents and appropriate coating conditions it is possible to prepare coatings of various thickness [1] with bioactive [4] and adhesive [6] properties. Many authors have investigated effects of silver in titania sol-gel coatings on antibacterial effects against various microorganisms, most frequently against *Escherichia coli* and *Staphylococcus aureus* while in most cases silver was added in form of silver nitrate [7]. There are also other forms of silver suitable for the purpose, e.g. silver acetate or silver phosphate [8,9]. The effect of silver ions on microorganisms is well known, however, the mechanism of the effect has not been fully explained yet. It is assumed that silver ions interact with three components of the bacterial cell to produce the bactericidal effect: with the peptidoglycan cell wall and plasma membrane, with bacterial (cytoplasmic) DNA and with bacterial proteins [10]. Due to the effect of silver ion, DNA may have lost its replication ability and cellular proteins became inactive [11].

The objective of our work was to prepare thin titania sol-gel coatings containing silver on titanium substrate by dip-coating technique. Antibacterial coatings prepared in this manner on implants might be in the future used for orthopedic applications in order to reduce the risk of post-surgery infection.

2. Experimental

2.1. Materials and methods

Substrates made of pure titanium (Grade 2, ASTM B265) and newly developed titanium alloys TiSi5, TiSi10 sized 30x10x1mm were ground with SiC paper No. 80, 160, 400, 800 and washed with acetone and then in ethanol.

Titania sol was prepared by gradual mixing of tetra-n-butyl-orthotitanate, acetylacetonate, triton X-100, ethanol and 1 $mol.l^{-1}$ HNO_3. The basic titania sol was divided into three parts and various forms of silver were added into two of them. The concentration of silver in the sols samples was always the same ($c_{Ag} = 0.015\ gml^{-1}$). Silver was added into the sols in form of AgNO3 and Ag3PO4 immediately before the coating process. The substrates were dipped once into the individual sols at the rate of 20 $cmmin^{-1}$ and remained in the sols for 30 s. Then the substrates were pulled out from the sols at the rate of 6 $cmmin^{-1}$. Substrates coating was performed by continual mixing on a magnetic stirrer to ensure homogenous distribution of silver in the sols and subsequently in the coatings. All the coated substrates were fired at 400 °C for 2 hours. Cooling was performed in an oven. The substrates with basic titania coating were identified as Ti-T, TiSi5-T, TiSi10-T, substrates with $AgNO_3$ as Ti-TAN, TiSi5-TAN, TiSi10-TAN and substrates with Ag3PO4 as Ti-TAP, TiSi5-TAP, TiSi10-TAP.

Surfaces of individual types of coatings after firing were investigated with the electron microscope (SEM) Hitachi S4700 with an SDD detector.

2.2. Conditions of the bactericidal, adhesion and cytotoxicity test

The test of bactericidal properties was performed with gram-negative bacteria *Escherichia coli (E. coli,* strain DBM 3138). Bacterial culture was incubated in liquid LB medium and then diluted in physiological solution to the bacteria concentration 10^4 $CFUml^{-1}$. The test was performed by dipping of the coated substrate into suspension of *E. coli* in physiological solution. Then the substrate was removed and amount of the suspension was spread on a Petri dish with LB agar. Sole suspension without any substrate served as a reference. The dish was placed into a biological thermostat set up at 36.5°C for 24 hours. After the incubation period the dish was photographed and the quantity of surviving *E. coli* was counted.

Adhesion of the coating to the substrate was measured with a cross-cut tape test under ASTM D 3359-2 [12]. Cuts were made into the coatings arranged into a grid shape and a tape (Permacel 99) was applied on the area with cuts. The tape was peel off and the area with the cuts was evaluated visually by comparison with a standard scale. The percentage of the area removed was determined and the classification grade was assigned.

Test for *in vitro* cytotoxicity was performed according to ISO 10993-5 standard [13]. The samples were submersed into MEM + 5% FBS (fetal bovine serum) with antibiotics at 37 ° C with shaking for 24 hours. L929 cells (mouse fibroblasts, ATCC® CCL-1™) were seeded at a concentration of $1\ 10^5$ cells ml^{-1} into a 96-well plate and cultured in MEM (Minimal Essential Medium) with 10% FBS at standard conditions (37 °C, 5% CO_2). After 24 hours, the medium was removed and the extracts were applied onto the subconfluent cell monolayer. Sole medium served as a control. Cytotoxic effect of the extracts on the cells was measured after 24 hours of incubation using WST-1 (water soluble tetrazolium salt 1) assay. Cells were incubated with 5% WST-1 in MEM (with 10% FBS) for 4 hours and the absorbance of the yellow formazane formed by metabolic reduction of the reagent was measured at 450 nm. Reduced metabolic activity of cells by over 30 % compared to the control was considered as a cytotoxic effect.

3. Results and discussion

Figures 1d-i show surfaces of the titanium substrates with fired titania coatings with silver in two forms. For comparison figure 1a-c shows the fired substrates with the basic titania coatings without silver.

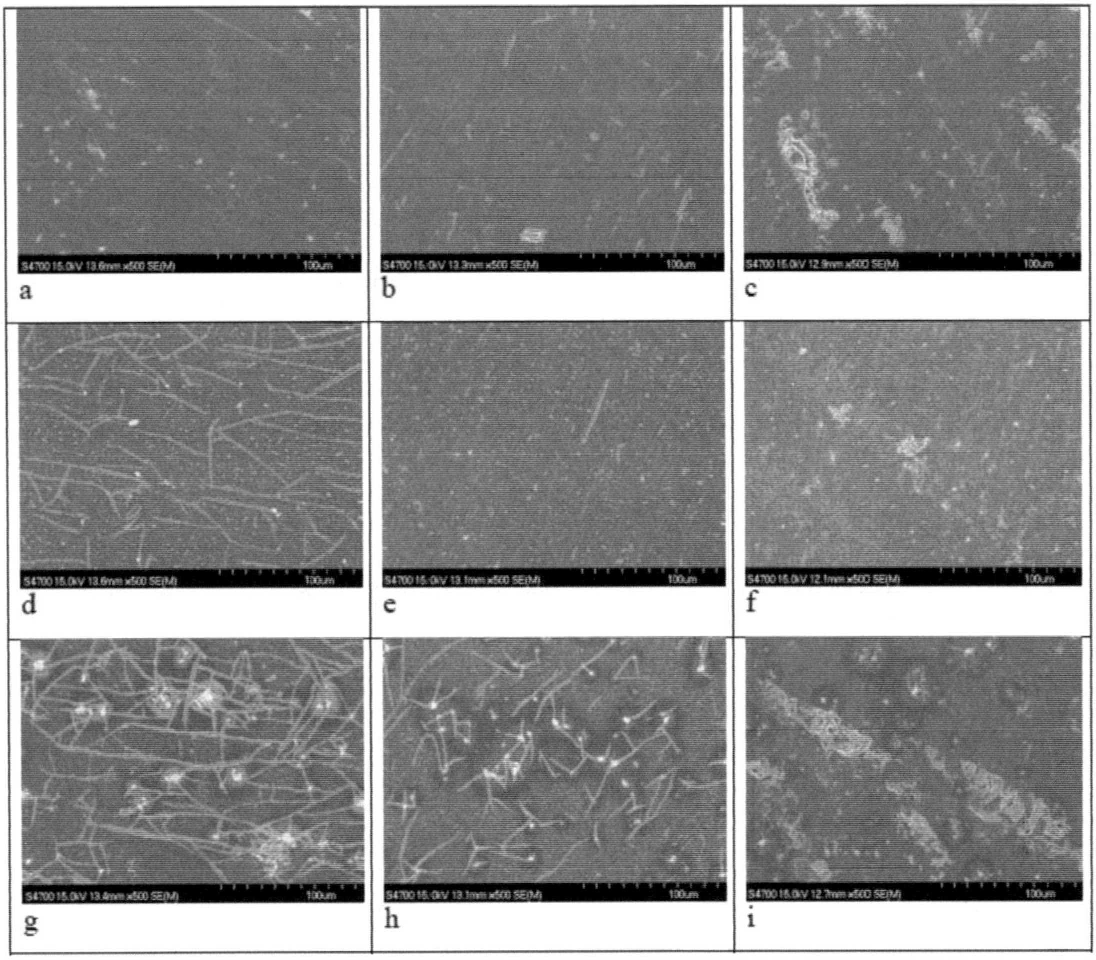

Figure 1. Titania coatings on titanium substrates after firing: a) Ti-T, b) TiSi5-T, c) TiSi10-T, d) Ti-TAN, e) TiSi5-TAN f) TiSi10-TAN, g) Ti-TAP, h) TiSi5-TAP and i) TiSi10-TAP

The figure 1a-c indicates that the basic titania coatings after firing were partly cracked only on substrate TiSi10. Spherical silver particles in all the other types of coatings (Fig. 1d-i) were nearly evenly distributed all over the surface. The size of silver particles added into the sol as $AgNO_3$ (Fig. 1d-f) was in units of nanometers and for particles added as Ag_3PO_4 (Fig. 1g-i) in units of micrometers. The coating TAN on TiSi5 (Fig. 1e) was homogeneous nearly without cracks. In case of the titania coatings with added Ag_3PO_4 (Fig. 1 g-i) the cracks propagated always from particles of the mentioned compound.

Adhesion of the coatings to the substrates measured by tape test was very good, grade 5B as compared to the classification scale. The cracks in the coatings Ti-TAN, Ti-Tap and TiSi5-TAP had no negative effect on the adhesion.

After 24-hours TAN and TAP coatings showed very good bactericidal effect (the number of surviving colonies was 0 - 25) whereas in case of the basic coating T no bactericidal effect was demonstrated (the number of surviving colonies was 800 - 1200 which is comparable with the reference sample). Figure 2 shows the cytotoxicity of the coatings extracts (T, TAN, TAP) on cell line L- 929 (mouse fibroblasts).

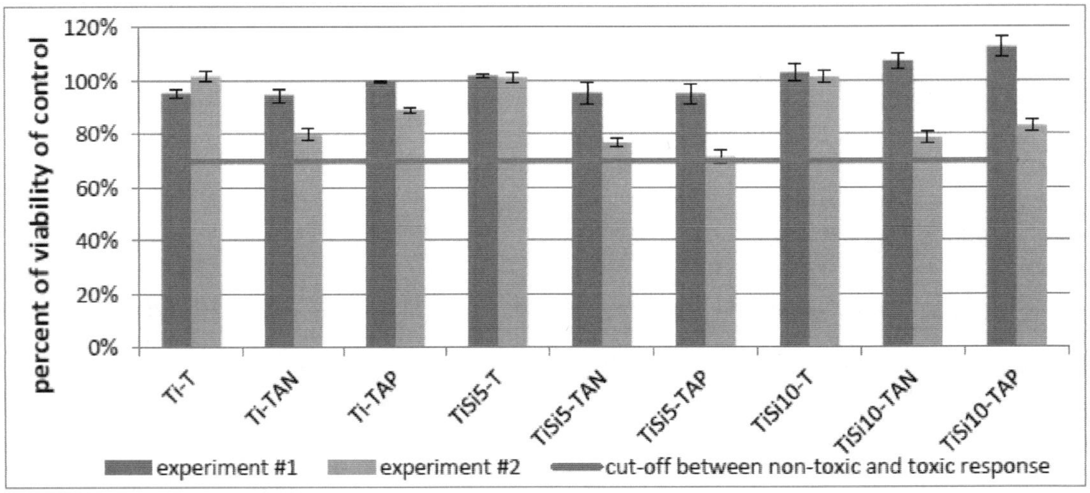

Figure 2. Cytotoxicity of the extracts (T, TAN, TAP) on cell line L-929 (mouse fibroblasts) after 24-hours incubation

After the 24-hours cytotoxicity test, neither of the coatings T, TAN or TAP showed cytotoxic effect, irrespective of the substrate (titanium or titanium alloy TiSi5, TiSi10) used.

4. Conclusion

Titania sol-gel coatings contained silver in form of dissolved AgNO3 and Ag3PO4 particles were successfully prepared by dip-coating technique on three types of titanium substrates. Adhesion of all types titania coatings to the substrates was very good (grade 5B). The 24-hours bactericidal test confirmed significant bactericidal effect of titania coatings containing $AgNO_3$ and Ag_3PO_4 particles. No cytotoxic effect of the coatings was observed.

Acknowledgement

This work has been supported by the Technology Agency for the Czech Republic within the project TE01020390 Center for development of modern metallic biomaterials for medicinal implants.

References

[1] Yun Y J *et al* 2004 *Mater. Let.* **58** 3703
[2] Tang H *et al* 2011 *Surf. Coat. Tech.* **206** 8
[3] Milella E *et al* 2001 *Biomaterials* **22** 1425
[4] Xu W *et al* 2006 *Mater. Let.* **60** 1575
[5] Lopez L *et al* 2010 *Surf Coat. Tech.* **205** 251
[6] Piveteau L-D *et al* 2000 *Biomaterials* **21** 2193
[7] Akhavan O 2009 *J. Coll. Inter. Sci.* **336** 117
[8] Bi Y *et al* 2011 *J. Am. Chem. Soc.* **133** 6490
[9] Buckley J J *et al* 2010 *J. Mater. Chem.* **20** 8056
[10] Chaloupka K *et al* 2010 *Trends Biotech.* **28** 580
[11] Raffi F H M *et al* 2008 *J. Mater. Sci. Technol.* **24** 192
[12] ASTM D 3359-2: Standard Test Methods for Measuring Adhesion by Tape Test, 2006
[13] ISO 10993-5: Biological evaluation of medical devices, Part 5: Tests for in vitro cytotoxicity, 2009

Experimental investigation of plasma-immersion ion implantation treatment for biocompatible polyurethane implants production

R I Iziumov, AY Beliaev[1], I V Kondyurina[2], I N Shardakov[1], A V Kondyurin[3], M M Bilek[3] and D R McKenzie[3]

[1]Institute of Continuous Media Mechanics UB RAS, Perm, Russia
[2]School of Medicine, University of Sydney, Sydney, Australia
[3]School of Physics, University of Sydney, Sydney, Australia

E-mail: izumov@icmm.ru

Abstract. Modification of the surface layer of polyurethane with plasma-immersion ion implantation (PIII) and studying its physical and chemical changes have been discussed in this paper. The goal of the research was to obtain carbonized layer allowing creating biocompatible polyurethane implants. The experiments of PIII treatment in various modes were performed. The investigation of the modified surface characteristics was carried out by observing the kinetics of free surface energy for two weeks after treatment. The regularities between treatment time and the level of free surface energy were detected. The explanation of high energy level was given through the appearance of free radicals in the surface layer of material. The confirmation of the chemical activation of the polyurethane surface after PIII treatment was obtained.

1. Introduction

The necessity of restoring or substitution the damaged human tissues by artificial device produced using different polymers or metal alloys, has determined the development of a separate section of modern medicine – the implantology.

The main purpose of such type of surgery is to restore lost characteristics and properties of tissue or, if possible, its improvement. The most important requirements to the characteristics of the implantable material are biocompatibility, mechanical properties (strength, elasticity or hardness, coefficient of friction, durability), chemical properties (corrosion resistance, the absence of toxic emissions), the weight and the economic aspect [1].Nowadays many research groups are carried out investigations aimed at studying the possibility of applying polymer materials in medicine (food grade polypropylene (Anulon-92) [2], polystyrene [3], polymethylpentene [4, 5], polyethylene [6, 7], polyethylene terephthalate (PET) [8], thermo-plastic polyurethane (Elastollan ® 1180A50) [9], polyhydroxybutyrate (PHB), poly-L-lactic acid (PLLA) [10, 11], polytetrafluoroethylene (PTFE) [12], biodegradable polymer [13]). Polymeric materials are becoming more popular due to the fact that it satisfies almost all specified requirements [14-16]. Its surface may provide immobilization of biologically active molecules and living cells, however, this quality requires significant improvements [17,18]. Rational way to solve this problem is to develop a technique of treatment the outer layer, modifying it so that it becomes a safe, friendly, biocompatible capsule for the body and the inner part of the implant would have the rest of the required parameters.

Among a number of ways to modify the surface (chemical exposure, temperature processing, the corona discharge, ultraviolet irradiation [19, 20]), one of the most promising areas of research is also the use of plasma-immersion ion implantation (PIII). With this technology, it is possible to create on the material surface the layer having the property of biocompatibility; moreover the process itself is clean and leaves no traces on the surface of by-products. Also, comparing with other conventional techniques, PIII advantage consists in the ability to precisely control the surface modification, regardless of its geometry [21].

In numerous papers related to the study the possibilities of PIII technology, it is shown that the structure and surface chemical properties of the treated polymer are dependent on the gas from which the plasma is generated, and the time and intensity of treatment. It is considered the application of the following gases and its mixtures: helium [22], nitrogen [8], oxygen, argon and argon + oxygen mixture [23]. There is often the interest to analyse the influence of the experimental parameters within some range (the pressure of the plasma, gas flow, power, time of treatment). In our experiments we performed the treatment in the nitrogen plasma with ion energy of 20keV. Treatment time was varied from 80 to 800 seconds that corresponds to$5 \cdot 10^{14}$– 10^{16} ions/cm^2.

To analyse the effect of ion-plasma treatment on the surface of the material it is often used a combination of several methods of experimental studies such as optical microscopy [3], atomic force microscopy (AFM) [2, 4], scanning electron microscopy (SEM) [2, 3], x-ray photoelectron spectroscopy (XPS) [3, 4], infrared spectroscopy (FTIR) [2], the study of the surface energy by measuring the contact angle [2, 4], tensile test [2], thermal analysis (Differential Scanning Calorimetry TA) [2], spectrophotometry [2], Rutherford backscattering spectroscopy (RBS) [4], electrokinetic analysis [4], goniometry [4], ablation of PMP surface layers determined by gravimetry [4].For biocompatibility study experiments with mouse embryonic fibroblasts [2, 6], human osteosarcoma cells [3] and human blood (haemolytic potential, thrombogenicity) [2, 8] are performed. In our work, we have applied the analysis of the kinetics of the surface energy, including its polar and dispersive components, as a function of the past time after treatment; in addition we plan to take advantage of XPS and infrared spectroscopy.

The above studies have shown that, depending on the purpose, in one case, ion-plasma treatment can dramatically increase the number of adhered cells or bioactive molecules as well as positively affect the uniform distribution of cells on the surface of the material and their viability. In another case, decrease of platelet adhesion is confirmed that causes the improvement of the blood compatibility.

2. Materials and methods

2.1. PIII treatment of PU samples

For the treatment on PIII device we used the samples of size 1×2 cm^2. Experimental investigations include set of PIII-treatments of polyurethane samples for 40, 80, 400 and 800 s (corresponding to an ion fluency of $5 \cdot 10^{14}$, 10^{15}, $5 \cdot 10^{15}$, 10^{16} ions/cm^2). The source of ions was inductively coupled radio-frequency nitrogen plasma powered at 13.56 MHz. The nitrogen gas used for PIII was 99.99% pure. The plasma power was 100W with reverse power of 12W when matched. The plasma density during treatment was monitored by a Langmuir probe with rf-block from Hiden Analytical Ltd. The base pressure was 10^{-6}Torr (~10^{-4} Pa); the pressure of nitrogen during implantation was $2 \cdot 10^{-3}$Torr (~$4.4 \cdot 10^{-2}$ Pa).

The samples were treated in plasma chamber by ion beam, which was extracted from plasma by the application of high voltage (20kV) at a frequency of 50Hz with duration of pulses of 20μs. The polyurethane samples were placed on a holder with a mesh. The mesh was electrically connected to the holder and held in front of the sample on the distance of 5 cm.

2.2. Contact angles and surface free energy

The contact angles of two liquids (de-ionized water and diiodomethane) on the surface of PU were measured by contact angle analyzer Kruss DS10 based on the sessile drop technique. For calculating the surface free energy and its polar and dispersive components the Rabel model with regression method was used. There was a set of 11 measurements of 5 drops of each liquid per sample. After ion-plasma treatment the kinetics was being monitored during 2 weeks (20, 30, 40 minutes, 1, 2, 5, 7, 23 hours, 2, 4, 11 days after treatment).

3. Results

The first contact angle measurement was carried out immediately after extracting the sample from the PIII device, which process included injection of air into a vacuum chamber and took 20 minutes. The obtained data showed an inverse dependence between the values of contact angles of water and treatment time. The contact angle for the sample treated for 40 seconds was equal to 48±2°, for 80 seconds: 47±1.5°, for 400 seconds: 45±0.9°, for 800 seconds: 38±0.4°.After 2 weeks the contact angle has increased and reached a stable value. For the sample treated for 40 seconds, the contact angle became equal to 63±2.3°, for 80 seconds: 73±1.7°, for 400 seconds: 66±1.3°, for 800 seconds: 64±1.4° (figure 1).

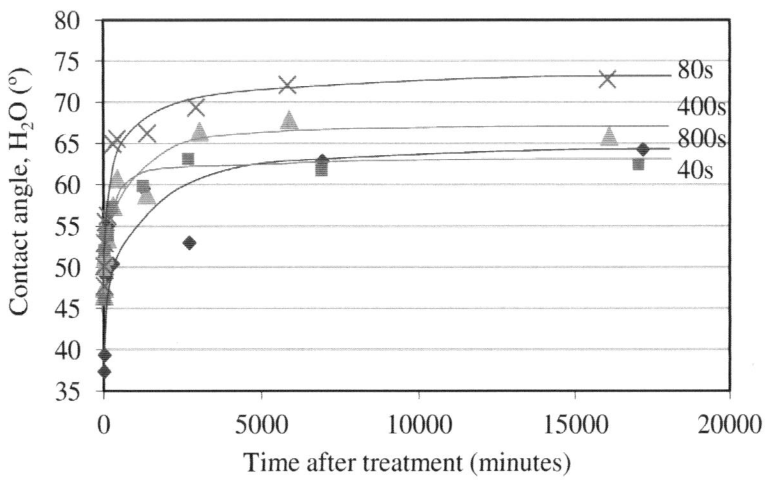

Figure 1. Kinetics of water contact angle after treatment samples in 4 modes of PIII

In this case, the correlation between the treatment time and the contact angle remained only for three treatment modes (80, 400, 800 seconds). The sample treated over the shorter time of 40 seconds, has showed the lowest value of the contact angle, standing out from the overall trend.

Kinetics of diiodomethane contact angle was significantly different from water and it was nonmonotonic. Although as in the case of water the treatment also decreased the contact angle but to a lesser degree. The correlation between the angle and the plasma treatment time is not observed, and after two weeks after treatment the contact angle has reached the level of 44±0.5° for all the samples, regardless of treatment time (figure2).

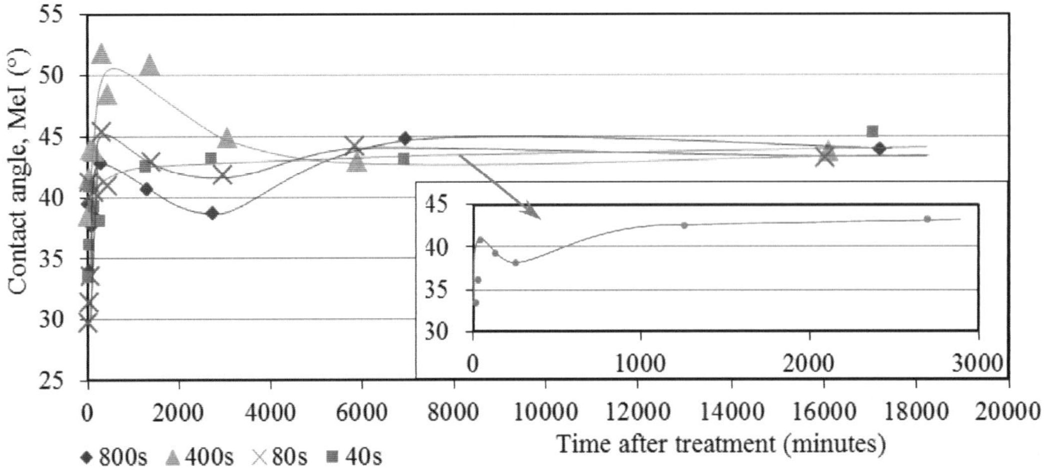

Figure 2. Kinetics of diiodomethane contact angle after PIII treatment during two weeks

The dependence between the total free energy and the treatment time has the correlation similar to the dependence of water contact angle. After 2 weeks for 3 PIII modes there was a direct dependence between the treatment time and energy value: the sample treated for 80 seconds had 45.5 mJ/m^2, for 400 seconds: 48 mJ/m^2, for 800 seconds: 49.5 mJ/m^2. But the treatment during 40 seconds gave the highest value: 50.5 mJ/m^2.

Analysis of the experimental data was carried out by approximation (figure3) using a function (equation (1)) represented as a sum of two exponential functions:

$$f(t) = a_0 + a_1 \exp(-t / k_1) + a_2 \exp(-t / k_2) \quad (1)$$

where f is a total free energy;t is time after PIII treatment; $k_1$and $k_2$are the rate constants of the two characteristic stages; a_0, $a_1$and $a_2$are the constants to be defined.

The effectiveness of such method of approximation was demonstrated in [24], in which the observed phenomenon isqualitatively similar.

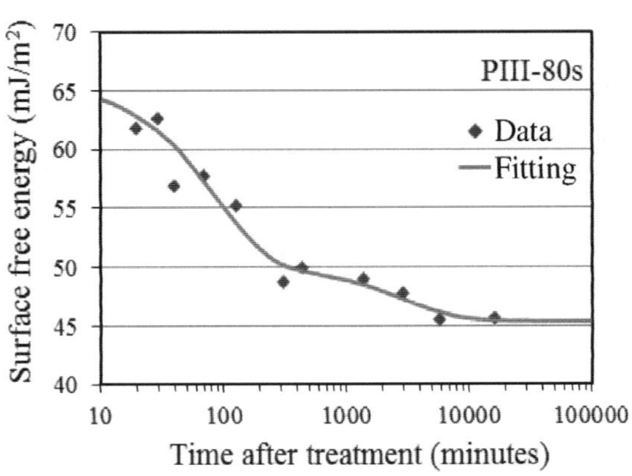

Figure 3. Exampleof data and its fitting

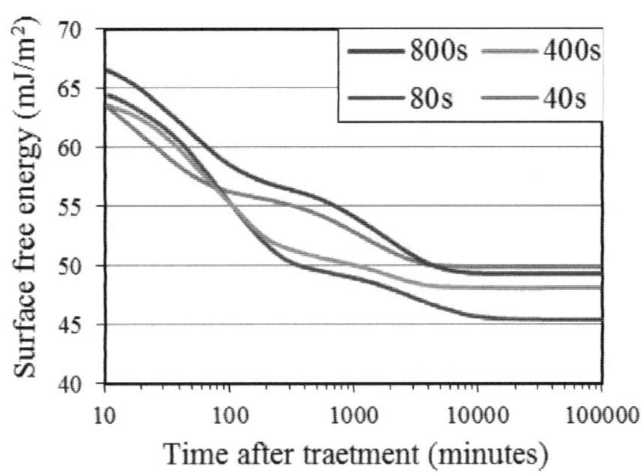

Figure 4. Kinetics of total surface free energy

Analysis of the kinetics of the surface free energy (figure4) revealed that after PIII treatment of PU two characteristic relaxation processappeared, the first of which was predominant over the second one, but faded rapidly,after which only the second process occurred.This is confirmed by the fact that the characteristic time for the first process (k_1) although has a large spread (from 24.3 to 45 min) but comparing to the second one ($k_2$variesfrom 1258.5 to 3177.8 min) has significantly (by two orders of magnitude) smaller value (Table 1).

Table 1. The constants of fitting function and its values in limiting cases fortotal surface free energy

Time of treatment, s	$f(0)$, mJ/m^2	$f(\infty)$, mJ/m^2	k_1, min	k_2, min
40	67±1.1	50±0.7	24.3±5.3	1258.5±438.3
80	65.5±0.9	45.5±0.5	89.9±14.3	3177.8±1805.9
400	65±0.5	48±0.2	80.5±9.6	1638.6±841.7
800	66±1.2	49.5±0.5	45±10.8	1789.9±815.2

Extrapolation of the data allows determining the value of free energy at the time immediately after treatment, f (0) (Table 1). This value is essentially independent of the treatment time and equals 66±1mJ/m^2. Similar way we determined the level, which energy reached at the expiration of arbitrarily large period of time, $f(\infty)$. It is equal to 50.5, 45.5, 48, 49.5 mJ/m^2, respectively for 40, 80, 400, 800 seconds oftreatment.

The qualitative behavior of the total energy is caused by polar component, because a dispersivecomponent remains during two weeks practically constant.Its value was inside the range of 40±5 mJ/m^2 and had no dependence on the treatment time.

Polar component after 2 weeks decreased significantly (by 35-55%).Moreover, there was a direct correlation between the reduction of energy and the PIII mode: for the sample treated for 40 seconds, reduction was 35% (from 20 to 13 mJ/m^2) for 80 and 400 seconds: 53% (8 to 17 mJ/m^2 and from 21 to 10 mJ/m^2, respectively) for 800 seconds: 55% (from 26 to 12 mJ/m^2).

After 2 weeks remaining values of polar components were 3, 8, 10, 12 mJ/m^2, respectively for 40, 80, 400, 800 sec. Thus, we can trace the situation similar to total free energy, where the 40 seconds treatment is out from theoverall trend (Table 2).

Table 2. The constants of fitting function and its values in limiting cases for polar component of the surface free energy

Time of treatment, s	$f^P(0)$, mJ/m^2	$f^P(\infty)$, mJ/m^2	k_1^P, min	k_2^P, min
40	24±0.4	13±1.1	20±4.8	767±388.1
80	20.5±0.3	7.5±0.5	107±21.2	3121±1265.3
400	23±0.8	10±0.7	42.9±16.8	2224±574.5
800	28±0.5	12±1.3	43.8±16.9	5494.2±4559.4

4. Conclusion

In this work the investigation of polyurethane modification to create chemically activesurface layer in material was performed. By monitoring the kinetics of surface free energy and its polar and the dispersivecomponents itwas established that treatment of the material leads to significant and irreversible changes in the surface layer.It results in a sharp increase of the surface energy after PIII treatment and preserving its sufficiently high, in comparison with untreated PU, level thatreflectssignificant chemical changes on the surface of polymers.At the same time, such a considerable increase of energy (up to a level of about 67 mJ/m^2)and behavior of its polar and dispersive components can be attributed to only the appearance of free radicals in the material, which in turn makes it possibleto add, using existing techniques, to polyurethane the property of biocompatibility.Using a level $f(\infty)$, which is reachedby energy after a sufficiently long period of time, as a criterion of selecting a treatment mode, we can make the assumption thatamong all variants of treatment the briefest one (40s) may be the most preferable for us by the fact that stored energy and, therefore, the concentration of free radicals onto the surface has the highest level among all the considered variants.

At this stage of research it can be claimedthat the application of PIII technology has enabled to activatesurface layer of polyurethanethat offers the prospect of further modification of the surface to the level of biocompatibility.

5. Acknowledgements

The work was supported by the Russian Foundation for Basic Research (grant 13-01-96009_r_ural_a and grant 14-08-96003_r_ural_a)and the Ministry of Education of Perm Region under agreement (S-26/632).

6. References

[1] Gomathi N, Sureshkumar A and Neogi S 2008 *Curr. Sci.* **94** 1478–1486.
[2] Gomathi N, Rajasekar R, Babu R, Mishra D and Neogi S 2012 *Mat. Sci. and Eng.* C **32** (7) 1767
[3] Recek N, Mozetič M, Jaganjac M, Milkovič L, Žarkovic N and Vesel A 2013 *Vacuum* **98** 116
[4] Slepička P, Trostová S, Slepičková Kasálková N, Kolská Z, Malinský P, Macková A, Bačáková L and Švorčík V 2012 *Polymer Degradation and Stability* **97** 1075
[5] Michaljaničová I, Slepička P, Slepičková Kasálková N, Sajdl P and Švorčík V 2014 *Vacuum* **107** 184
[6] Rimpelová S, Slepičková Kasálková N, Slepička P, Lemerová H, Švorčík V and Ruml T 2013 *Mat. Sci. and Eng.* C **33** 1116
[7] Kondyurin A, Naseri P, Fisher K, McKenzie D R and Bilek M M 2009 *Polymer Degradation and Stability* **94** 638
[8] Junkar I 2013 *Vacuum* **98** 111
[9] Alves P, Cardoso R, Correia T R, Antunes B P, Correia I J an Ferreira P 2014 *Colloids and Surfaces* B **113** 25
[10] Slepičková KasálkováN, Slepička P, Bačáková L, Sajdl P and Švorčík V 2013 *Nuclear Instruments and Methods in Physics Research Section B* **307** 642
[11] Slepicka P, Trostová S, Slepicková Kasálková N, Kolská Z, Sajdl P and Švorcík V 2012 *Plasma Processes Polym.* **9** 197
[12] Reznícková A, Kolská Z, Hnatowicz V, Stopka P and Švorcík V 2011 *Nucl. Instr. Meth.* B **269** 83
[13] Morent R, De Geyer N, Desmet T, Dubruel P and Leys C 2011 *Plasma Processes Polym.* **8** 171
[14] Silva E A and Mooney D J 2004 *Current Topics in Developmental Biology* **64** 181
[15] Shin H, Jo S and MikosA G 2003 *Biomaterials* **24** 4353
[16] Ma Z, Kotaki M, Yong T, He W and Ramakrishna S 2005 *Biomaterials* **26** 2527
[17] Küttel O M, Martinu L, Poitras D, Klemberg-Sapieha J E and Wertheimer M R 1992 *Materials Science and Engineering* B – Advance 321
[18] Aflori M, Drobota M, Timpu D and Barboiu V 2008 *Journal of Optoelectronics and Advanced Materials* **2** 291
[19] Behnisch J, Hollander A and Zimmermann H 1993 *Surf. Coat. Technol.* **59** 356
[20] Mirabedini S M,, Rahimi H, Hamedifar S and Mohseni M 2004 *Int. J. Adhes.* **24** 163
[21] Boenig H V 1882 Plasma Science and Technology, *Cornell University Press,* Ithaca, New York
[22] Aflori M, Drobota M, Dimitriu D, Stoica I, Simionescu B and Harabagiu V 2013 *Mat. Sci. and Eng.* B **178** 1303
[23] Chen Y, Gao Q, Wan H, Yi J, Wei Y and Liu P 2013 *Applied Surface Science* **265** 452
[24] Kosobrodova E, Kondyurin A, McKenzie D R and Bilek M M M 2013 *Nuclear Instruments and Methods in Physics Research* B **304** 57

Electrochemical synthesis of cellulose mesylate

Sh Sh Khidirov[1], M A Akhmedov[1] and H S Khibiev[1]

[1]Dagestane State University, Chemical, Physical and Organic Chemistry Department, Republic of Dagestane, Makhachkala

E-mail: muhamadahmedov@mail.ru

Abstract. The article deal with the possibility anode modification of cellulose to form its ester – mesylate by voltametric measurement method and preparative electrosynthesis on a platinum electrode in the system cellulose – dimethyl sulfoxide - methanesulfonic

1. Introduction

Oxidation of cellulose in which there is no the polymer degradation leads to a modifying to form various functional groups [1]. The oxidation products of cellulose are used for different application in medicine and pharmacy [2-4]. Mesylates are the esters, which are formed at interaction of the functional hydroxyl groups of cellulose with methylsulfogroups. It is also known that the mesylates by used in organic synthesis for the protection of hydroxyl groups [5]. Previously, the authors [6,7] proposed the electrochemical modification of cellulose in copper - ammonia solution (reagent of Schweitzer - $[Cu(NH_3)_4](OH)_2$). The paper [8] in a wide range of positive potentials studied electrochemical modification of another representative of sugar-starch polysaccharides. Cellulose is difficult to dissolve in organic solvents. Selection of the solvent and supporting electrolyte aqueous solution is a difficult task. Found that aqueous solutions of dimethyl sulfoxide (DMSO) is well dissolved cellulose [9].The authors of [10-11] it was stated that DMSO is oxidized to methanesulfonic in this field potentials.

In this paper, the possibility of formation of mesylates by means of anode modification of cellulose by dissolving in dimethyl sulfoxide (DMSO) is shown.

2. Formulation of the problem and method of solution

2.1. Subject of investigation

Anode modifying processes of cellulose pulp in the system –dimethyl sulfoxide- methanesulfonic acid are studied using a potentiostatic methods and preparative electrolysis. Potentiostatic measurements were carried out using a potentiostat IPC – Pro MF.

2.2. Materials and reagents

All measurements were performed on a platinum electrode. We used the following reagents: DMSO (chemically pure), 75% solution of methanesulfonic acid (BASF), a cellulose - ashless filter of the brand FILTRAK and bi distilled water.

2.3. Preparative methods of electrosynthesis

Preparative electrolysis was carried out in a diaphragm electrolyze with anode and cathode separation cationiteof membrane brand MK-40. All measurements were performed at room temperature. As a source of direct current with adjustable voltage by used TYPE: TR-9252. Anodic current density was set at a preparative electrosynthesis according to the oxidation potentials of cellulose in stationary conditions.

The anode compartment of the electrolytic cell was filled with 50 ml of DMSO solution containing CH_3SO_3H (0.4 M) and dissolved of cellulose (0.75 g/ml), the cathode compartment was billed solution water solution. Electrolysis was carried out at a temperature 40- 60□C in current of 4 hours at a current density of 5.15 mA/cm^2. After electrolysis, the anolyte became viscous without discoloring.

2.4. Methods of chemical analysis

Purification of the final product - a modified of cellulose, prepared by electrochemically, was carried out as follows: an electrolyte - methanesulfonic acid was is neutralized with aqueous ammonia solution, water and DMSO were distilled off the solvent by vacuum distillation, samples were thoroughly washed with water bidistilled water and supported on the slide glass and dried in an oven at 75°C for 1 hour.

Products of electrooxidation were analyzed by infrared (IR) and quantitative X-ray spectral fluorescence spectrometry (XRF) spectrometer for IR «Nicolet 6700» the company «Thermo Nicolet» USA and «Shimadzu EDX-800 HS».

The qualitative and quantitative elemental analysis by X-ray fluorescence spectrometry «ShimadzuEDX-800 HS» was carried out in the following conditions: voltage 50 kV for the range from the titanium to uranium, 15 кВ for a range from the sodium to scandium time set 100 seconds, calculation of concentrations - by the method of fundamental parameters of the software to the spectrometer using a standard database and processing of the X-ray fluorescence spectra of the program DXP-700E (Version 1.00 Rel.017).

3. Results and Discussion

3.1. The results of IR analysis

The presence of methylsulfogroups in the product electrochemically modified of cellulose is found by IR spectroscopy. Terms of: The number of scans 32, resolution - 4 reciprocal centimeters (identification of the material). Absorption spectra taken with the prefix of frustrated total internal reflection (FTIR) in the range of wave numbers 4000 – 640 cm^{-1} and in the transmission mode in the wavenumber range 4000 - 400 cm $^{-1}$ was recorded. The obtained spectra are interpreted in terms of a separate absorption maxima, and was directly compared with the spectra library standards using the program «Omnic».

The resulting product was identified as a modified cellulose mesylate (methanesulfonic acid ester of cellulose) having a gross formula $[C_6H_7O_2(OH)_{3-X}(SO_3CH_3)_3]_n$ of the following structure (Fig. 1).

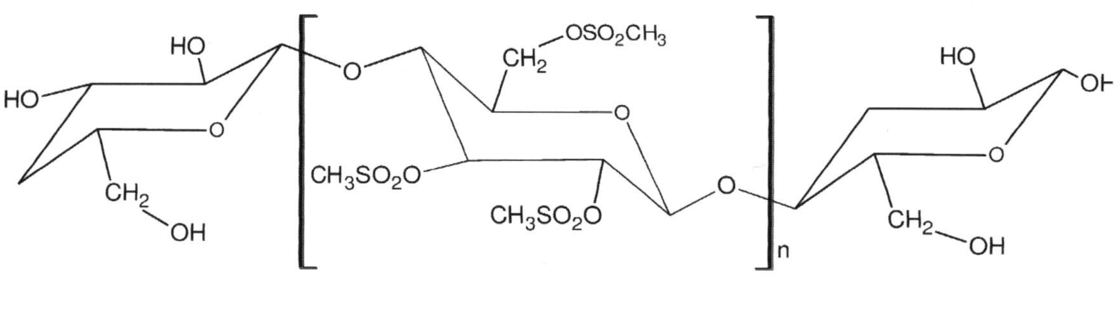

Figure 1. The structural formula of cellulose mesylate

The values of characteristic absorption bands of the functional groups are shown in table1.

Table 1. Values of the characteristic absorption bands of functional groups of the product electrochemically modified of cellulose

The region of absorption, cm^{-1}	Characteristic groups
1480-1450	CH_2–
1280-1230	R–SO_2–R^1
1100-1050	S=O; R–S(O) –OH
850-800	S–O
600-580	C–S

3.2. The results of EDX -analysis

The presence of sulfur atoms in the methylsulfogroups in the resulting product of the modified pulp was found by energy dispersive analysis (EDX).

EDX-analysis of data mesylate of indicate shows the presence of of sulfur atoms - 84.1% oxygen - and 13.1% carbon - 2.8%. From the results of obtained by X-ray fluorescence analysis, it is seen that the ratio of sulfur to carbon in the resulting product is 40: 1, which are close to the theoretical values calculated for the molecular level mesylate of cellulose. Such coincidence is observed when complete substitution residues methylsulfogroups three hydroxyl groups of of cellulose monomer.

Figures 2-3 are images structural changes of cellulose (before and after electrolysis) obtained by the scanning electron microscope ACPEX PSEM Express.

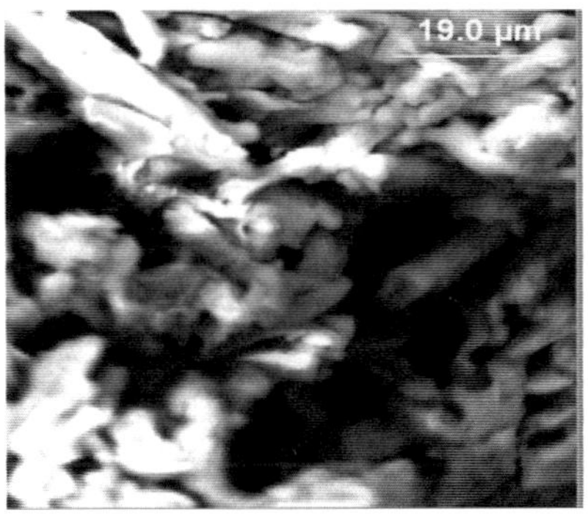

Figure 2. Image of scanning electron microscopy the surface of cellulose before electrolysis

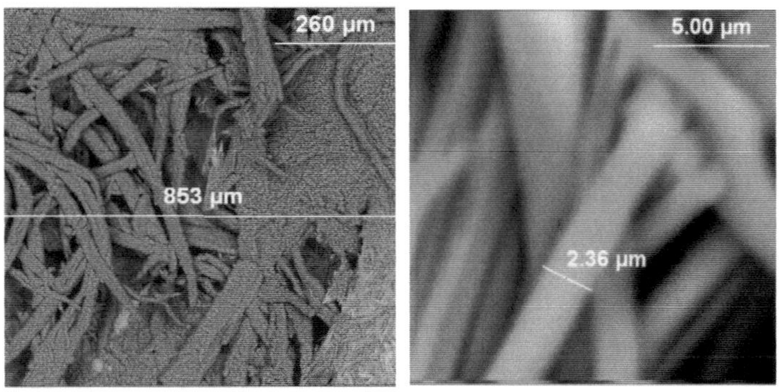

Figure 3. Image of scanning electron microscopy cellulose surface after electrolysis.

4. Conclusion

His found that by electrochemical the oxidation cellulose dissolved in dimethyl sulfoxide on the background of the electrolyte methanesulphonic acid is to modification to form esters - mesylate.

Acknowledgements

The work was supported by the Russian Foundation for Assistance to small innovative enterprises in the scientific and technical field (FASIE) grant № 3326GU1/2014.

References

[1] http://www.alhimikov.net/organikbook/celluloza
[2] Ovchinnikov Yu.A. 1987 *Bioorganic chemistry* M: Enlightenment (Russian addition) 457
[3] Budaeva V.V.; Mitrofanov R.Y.; and et. 2010 J. *Polzunovsky Herald* **4** (1) 158

[4] Chubenko V.A. 2012 *J. Practical Oncology* **13** (1) 47
[5] Butrim S.M. and Bildyukevich T.D. 2010 *J. Herald BGU* **2** (3) 18
[6] Akhmedov Sh. V.; Khidirov Sh. Sh; Khibiev H.S. 2011 *Proc. Int. Conf. "Inovatika*-2011" **2** 156
[7] Rabadanov M.H.; Khidirov Sh.Sh.; and et. 2012 Patent RU 2443631
[8] Khidirov Sh.Sh.; Hibiev H.S.; and Akhmedov Sh.V. 2014 *J. Herald DSU* **1,** 159
[9] Berezin A.S.; Tuzhikov O.I. 2010 J. *Izvestiya VSTU* **2** (7), 5-23
[10] Khidirov Sh.Sh.; Omarova K.O.; and Hibiev H.S. 2012 *J. Herald DSU*, **1** 235
[11] Omarova K.O.; Khidirov Sh.Sh.; and Hibiev H.S. 2013 *J. Herald DSU*, **1** 194

Researches of mechanical behavior of bone tissues for development and selection of individual ceramic implants

T V Kolmakova, S P Buyakova and S N Kulkov

Institute of Strength Physics and Materials Science (ISPMS) SB RAS, Tomsk, Russia
Tomsk State University, Tomsk, Russia

E-mail: kolmakova@ftf.tsu.ru

Abstract. The researches of mechanical behavior were conducted and the effective mechanical properties of model compact bone micro volumes under uniaxial compression were obtained taking into account the structural characteristics and the mineral content; experimental analysis of the mechanical behavior and the effective mechanical parameters of obtained porous zirconia ceramics were conducted. The comparison of obtained in paper calculated and experimental mechanical properties of bone and ceramics was carry out and the recommendations on the use the ceramics with certain porosity to replace the compact bone fragment with a certain structure and mineral content were suggested.

1. Introduction

Many years of experience in the use of artificial materials to restore elements of the bone through the endoprosthesis replacement indicates that the biochemical compatibility of implant with the organism is necessary but not sufficient condition for the functioning of the implant - bone system as a whole. Important condition is the biomechanical compatibility of implant and bone. It is known that a significant difference in the elasticity of bone and the implant can result in the loss of the latter due to the subsequent resorption of a bone tissue in contact with implant. For Creation of artificial materials that meet the conditions of biomechanical compatibility with bone, requires as much as possible a better understanding of the mechanical behavior of the replaced bone fragment, that defined by its hierarchically organized structure [1-5].

The mechanical behavior and mechanical parameters of model fragments of compact bone under uniaxial compression with consideration structural features and mineral content were studied; the experimental studies of the mechanical behavior and mechanical parameters of porous zirconia ceramics also have been executed.

2. Materials and methods

Geometrical model of compact bone tissue is constructed on basis of a real image of the microstructure of compact bone [6] (Figure 1, Figure 2).

The model microvolume of compact bone contain the osteons and Haversian canals, the matrix and cement line that separates osteons from the matrix (Figure 3).

The total porosity of the compact bone fragment P determined by Haversian canals which are modeled explicitly (Figure 4), and by Volkmann's canals, which taken into account implicitly in the calculation of the effective mechanical characteristic of the structural components of the model microvolume.

IOP Conf. Series: Materials Science and Engineering **123** (2016) 012005 doi:10.1088/1757-899X/123/1/012005

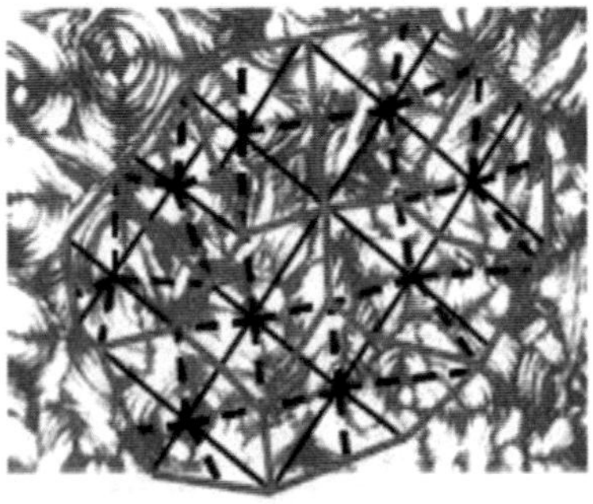

Figure 1. Division of picture of compact bone microstructure on the structural elements

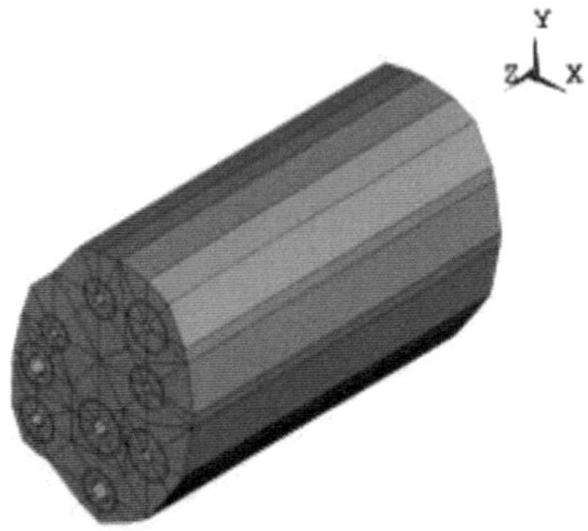

Figure 2. Geometrical model of compact bone tissue

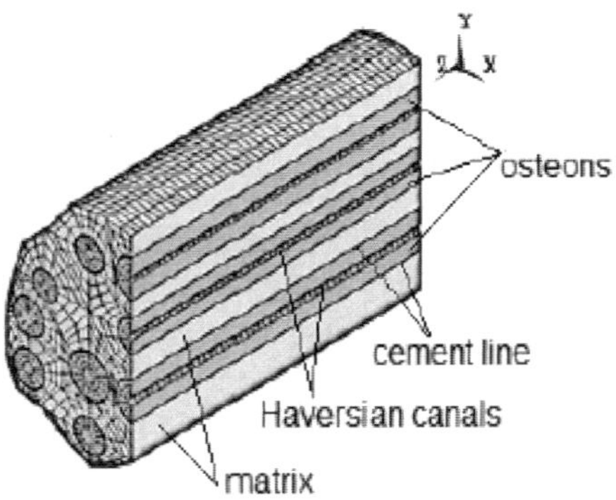

Figure 3. Finite element model of compact bone tissue

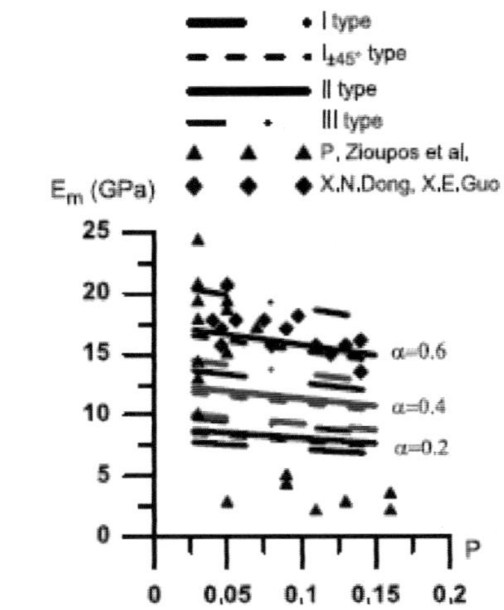

Figure 4. The dependence of the longitudinal modulus of elasticity of compact bone fragments of different types with different mineral content α from porosity

The material of structural components of the model microvolume compact bone was considered transversely isotropic. Effective mechanical properties of the matrix material and the reinforcing elements, osteons, are calculated taking into account the volume fraction of collagen and mineral component, the orientation of collagen and mineral fibers, the porosity due to the presence Volkmann's canals.

Calculations of the stress-strain state of the model compact bone fragments were performed in the software package ANSYS.

Microvolume was subjected to axial compression along the axis Z. In Figure 2, the plane of loading is shown in the foreground, plane of fixing in the background. The problem was solved in the framework of the linear theory of elasticity.

Samples of ceramic were prepared from the finely divided powder of zirconia partially stabilized with yttrium oxide, with a porosity of 10 to 70% obtained using pore-forming material. Uniaxial compression tests were carried out according to standard procedures on a universal testing machine "Instron-1185" with a constant loading rate 3·10-4 s-1.

3. Results and Discussion

Figure 5 shows the results of calculations of longitudinal modulus of elasticity of compact bone fragments with different directions collagen-mineral fibers (I type - transverse fibers, I ± 45° type - fibers are arranged at angles of ± 45° in adjacent lamellas of osteon, II type - fibers orientation change with a perpendicular to the parallel to the bone axis, III type - longitudinal fibers) with different mass fractions α of minerals and porosity P at axial compression in comparison with the experimental results P. Zioupos et al. [7], XN Dong, XER Guo [5].

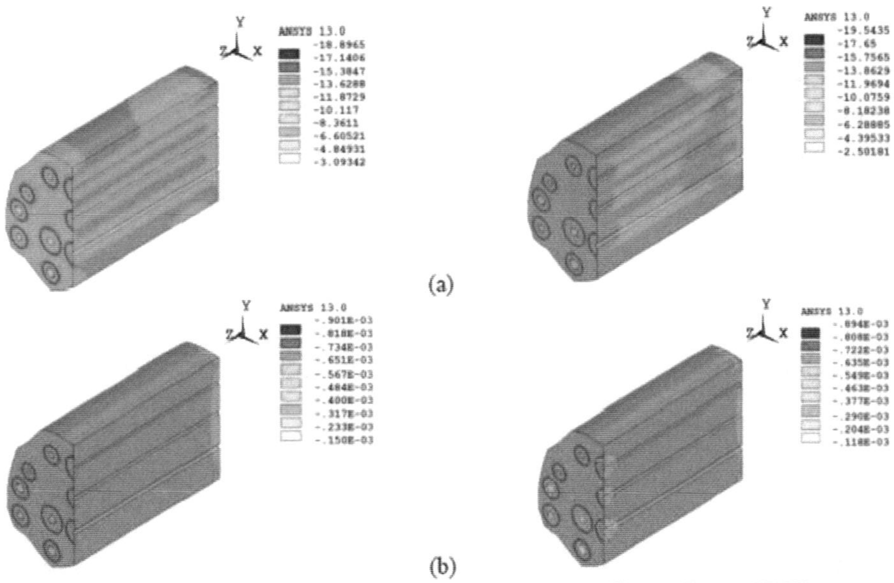

I±45° type, Em $_{I\pm45°}$ = 16.5 GPa   II type, Em$_{II}$ = 17 GPa

Figure 5. Distribution of the stress a) σz (MPa), and strains b) εz in model microvolumes of compact bone tissue of different types with P = 0.027, α = 0.6, which were clipped by YZ plane, at compressive stress σm = -6.5 MPa

The cortical bone microvolumes of I± 45 type and type II regardless from their porosity P have the same longitudinal modulus of elasticity (Em), but differ in the nature of the distribution of stresses a) and strains b), that is determined by various non-uniform deformation response of microvolumes in mutually perpendicular directions relative to the direction of load action, i.e., in the directions of axes X and Y.

This result demonstrates the need of accounting of effective parameters responsible for the nature of the distribution of stresses and strains of bone tissue.

As a result of the tests on uniaxial compression of zirconia ceramic ZrO2 (Y2O3) were obtained dependence of longitudinal modulus of elasticity and ultimate compressive strength of the ceramic samples from the specific volume of pore θ.

Comparison of the calculated modulus of elasticity and the ultimate compressive strength of bone microvolumes, calculated using the Hernandez formula [8], and the experimentally obtained mechanical characteristics of ceramic (Table 1) showed that the ceramics with a pore volume from 50 to 60% has the same modulus of elasticity as a compact bone with mass concentration of minerals from 40 to 55% and a porosity from 3 to 15%.

Table 1. Calculated and experimental effective mechanical properties of compact bone and ceramics

Ceramics $ZrO_2(Y_2O_3)$ (experiment)			Compact bone (calculation)				
θ	E, GPa	σ_c, MPa	α	P	Type	Em, GPa	σ_c, MPa
0.50	16.87	110.0	0.55	0.15	III	16.4	109.7
0.52	15.50	98.0	0.50	0.10	III	15.8	93.8
			0.45	0.027	III	15.5	81.2
0.54	14.17	87.3	0.45	0.15	III	13.7	62.7
			0.40	0.027	III	14.2	58.5
0.56	12.98	77.7	0.40	0.10	III	13.2	50.3
			0.40	0.15	III	12.6	45.1
0.58	11.89	69.2	0.45	0.15	II	11.6	62.7
			0.40	0.027	I±45°	11.9	58.5
0.60	10.89	61.6	0.40	0.15	II	10.7	45.1
			0.40	0.15	I±45°	10.5	45.1

4. Conclusion

The ceramics with porosity 50-56% may be candidate for replacement of compact bone fragment with the location of the collagen- mineral fibers parallel to the axis of the bone (III type), and the ceramics with a porosity of 58 to 60% may be candidate for replacement of the compact bone fragment with the direction of collagen-mineral fibers, which change the orientation of the transition from one lamella to another (II type), or placed at angles of ± 45 degree(I± 45 type).

Acknowledgements

The researches were executed in the framework of the project RFMEFI60714X0069 with the financial support of the Russian Ministry of Education and Science.

References

[1] Knets, I. 1999 *Acta of Bioengineering and Biomechanics.* **1** 55
[2] Reilly, D.T.; Burstein, A. H. 1974 *J. Bone Joint Surg.* **56a** 1001
[3] Zioupos, P.; Kaffy, C.; Currey, J. D. 2006 *Int. J. Fracture.* **139** 407
[4] Sevostianov, I.; Kachanov, M. 2000 *J. Biomech.* **33** 881
[5] Dong, X.N.; Guo, X.E. 2004 *J. Biomech. Eng.* **126** 387
[6] Reid, S. 1986 *Anat. Embryol.* **174** 329
[7] Zioupos, P.; Cook, R.B.; Hutchinson, J.R. 2008 *J. Biomech.* **41** 1961
[8] Hernandez, C.J. ; Keaveny, T.M. 2006 *Bone.* **39** 1173

IOP Conf. Series: Materials Science and Engineering **123** (2016) 012006 doi:10.1088/1757-899X/123/1/012006

Electrochemical behavior of 45S5 bioactive ceramic coating on Ti6Al4V alloy for dental applications

M M Machado López[1], M I Espitia Cabrera[2], J Faure[3], M E Contreras García[1]

[1]Instituto de Investigaciones Metalúrgicas, UMSNH, C.U. Edificio "U", Morelia, Michoacán, México.
[2]Facultad de Ingeniería Química, UMSNH, C.U. Edificio "E", Morelia, Michoacán, México.,
[3]INSERM UMRS-926, Laboratoire de Microscopie Electronique Analytique 21 rue Clement ADER 51685 REIMS cedex, France.

E-mail: eucontre@gmail.com

Abstract. Titanium and its alloys are widely used as implant materials because of their mechanical properties and non-toxic behavior. Unfortunately, they are not bioinert, which means that they can release ions and can only fix the bone by mechanical anchorage, this can lead to the encapsulation of dense fibrous tissue in the body. The bone fixation is required in clinical conditions treated by orthopedic and dental medicine. The proposal is to coat metallic implants with bioactive materials to establish good interfacial bonds between the metal substrate and bone by increasing bioactivity. Bioactive glasses, ceramics specifically 45S5 Bioglass, have drawn attention as a serious functional biomaterial because osseointegration capacity. The EPD method of bioglass gel precursor was proposed in the present work as a new method to obtain 45S5/Ti6Al4V for dental applications. The coatings, were thermally treated at 700 and 800°C and presented the 45S5 bioglass characteristic phases showing morphology and uniformity with no defects, quantification percentages by EDS of Si, Ca, Na, P and O elements in the coating scratched powders, showed a good proportional relationship demonstrating the obtention of the45S5 bioglass. The corrosion tests were carried out in Hank's solution. By Tafel extrapolation, Ti6Al4V alloy showed good corrosion resistance in Hank's solution media, by the formation of a passivation layer on the metal surface, however, in the system 45S5/Ti6Al4V there was an increase in the corrosion resistance; i_{con}-, E_{corr} and corrosion rate decreased, the mass loss and the rate of release of ions, were lower in this system than in the titanium alloy without coating.

1. Introduction

Titanium and Titanium alloys are extensively used as biomaterials for implants applications, due to their excellent mechanical properties and non-toxic behavior. Unfortunately, in general, metallic implants are not bioinert at all, which means that they can release ions and they can only have mechanical anchorage with bone, leading to encapsulation with fibrous tissue inside the body. In most of the cases the osseointegration is required for orthopedic and dental applications [1, 2], so it is necessary to coat the metal in order to obtain good interfacial union between metal substrate and bone which increases the bioactivity [3]. Recently, electrochemical deposition methods (EPD) have gain increased interest due to the simplicity of their application and the possibility to obtain uniform coatings inclusive in irregular form substrates at room temperature with low cost equipment [4].

Bioactive glasses, mainly the ceramic formulation Bioglass 45S5 (45% SiO_2, 24,5%, Na_2O 24,5% CaO y 6% P_2O_5 weight %), have been of interest as biomaterials because their osseointegration capacity [5, 6, 7, 8]. It has been reported that bioglass 45S5 can be inserted at high scale in bone hurted zones to help the restoration and it also brings structural support establishing hard bonding to bone [9, 10]. Elsewhere bioglass has been used for other applications as implants, to release the conductive auditive loss, for the maintenement of endosseus prosthesis and as particulate material to enhance the natural restoration process in patients with periodont al disease [10, 12]. Nevertheless, those processes imply the formation of hydroxiapatite calcium phosphate layer on the implanted glass material [13]. In this work, electrophoretic deposition method (EPD) was used to obtain 45S5/Ti6Al4V system, starting with a colloidal precursor suspension obtained by sol-gel synthesis. The obtained films were structurally characterized and the corrosion resistance in Hank's solution at 38°C was determined by potentiometric tests.

2. Experimental procedure

2.1. Ti6Al4V substrates preparation

1 x 1 cm Ti6Al4V plates were used as substrates because this alloy is the most used for implants and dental membranes, excellent mechanical properties of this alloy have been reported by different authors, fatigue strength of 707 MPa in 10^7 cycles, Young's modulus of 110 GPa, creep resistance of 860 MPa and tensile strength of 930 MPa [14, 15, 16, 17, 18, 19]. Alagic et al. [20], studied the wear resistance of Ti13Nb13Zr and Ti6Al4V alloys and reported a wear volume loss of 5.19 y 0.0751 mm^3, and hardness values of 231 and 575 HV, respectively, so the Ti6Al4V presented the best mechanical behavior. Substrate plates were previously polished with SiC from 240, 320, 400 to 600 mesh then, they were ultrasonic washed in acetone bath, rinsed in distilled water and wetted in alcohol.

2.2. 45S5 bioglass coating

45S5 bioglass was sinthetized by sol-gel, starting with a HNO_3 1M solution in which, the precursors were hydrolized tetraetilortosilicate (($C_8H_{20}O_4Si$), 99.9%, Sigma Aldrich, US), 6% del trietil fosfate ((C_6Hi_5O4P), 99.8%, Sigma Aldrich, US), then 24.5% sodium nitrate ($NaNO_3$, 99%, Sigma Aldrich, US) and 24.5% calcium nitrate (($Ca(NO_3)_2.4H_2O$, 99%, Sigma Aldrich, US) were added. After hydrolysis and polycondensation take place, the suspension was taken to the electrophoretic cell to carried out the deposition on the Ti6Al4V substrates. The hydrolized deposits were thermally treated at four different temperatures 500, 600, 700 and 800°C to obtain the 45S5 bioglass coatings. The four samples were labeled as follows: BG500, BG600, BG700 and BG800. The heating rate (5°C/min), time at calcination temperature (120 min) and cooling rate (5°C/min) until room temperature, were the same and constant for all the samples. The obtained phases in the 45S5 bioglass coatings were characterized by X Ray Diffraction (XRD), using a SIEMENS D-5000 model difractometer with CuKa radiation. The nanostructural morphology characterization was followed by scanning electron microscopy (SEM) using a field emission scanning electron microscope JEOL JSM.7600F, The obtained coatings elemental composition was determined by EDS.

2.3. Electrochemical test

Ti6Al4V alloy and 45S5/Ti6Al4V system were evaluated by electrochemical techniques in a GAMRY potenciostate in order to know the corrosion behavior in Hank's solution at 37°C simulating the human body ambient. Tafel extrapolation technique was used applying a potential scanning from -250mV to 1500 mV with a scanning rate of 1 mV/s, over the open circuit potential (OCP). Exposition area of 0.31669 cm^2 was used for all samples, equivalent weight of 11.44 g. was used for the titanium alloy, according with ASTM G102-89 (ASTM 1999) norm [21]. The electrochemical cell was conformed to a saturated AgCl electrode as reference electrode, high purity graphite electrode as counter electrode and Ti6Al4V alloy and 45S5/Ti6Al4V samples as working electrodes. Deaereated Hank's solution at pH=7.4, and at 37 °C, was the media used, the composition from SIGMA is shown in Table 1, electrochemical tests were carried out after 1 hour of inmersion in Hank's solution in order to stabilize the open circuit potential and temperature.

Table 1. Hank's solution chemical composition (g/l) (ZIGMA)

NaCl	KCl	Na_2HPO_4	KH_2HPO_4	$NaHCO_3$	D-Glucose
8.0	0.4	0.047	0.06	0.35	1.0

3. Results

3.1. Structural characterization

In order to study phases evolution during thermal treatment of the sol-gel electrodeposited bioglass precursors, structural characterization was carried out by X Ray Diffraction of the 45S5 bioglass xerogels thermally treated at 700, 800 y 1000°C and are shown in Figure 1. Crystalline phases peaks of $Na_2Ca_3Si_6Oi_8$ (JCPS #77-2189), began to appear in the xerogel treated at 700°C, in the difractograms of the xerogels treated at 800 and at 1000°C, those peaks appear with higher intensity and also another peak, corresponding to the crystalline phase $Na_2Ca_4(PO)_42SiO_4$ (JCPS #32-1053) is present. Those phases coincide with the ones reported by other authors who synthetized bioactive glasses with the 45S5 bioglass composition [4, 5, 6, 7, 10, 22, 23].

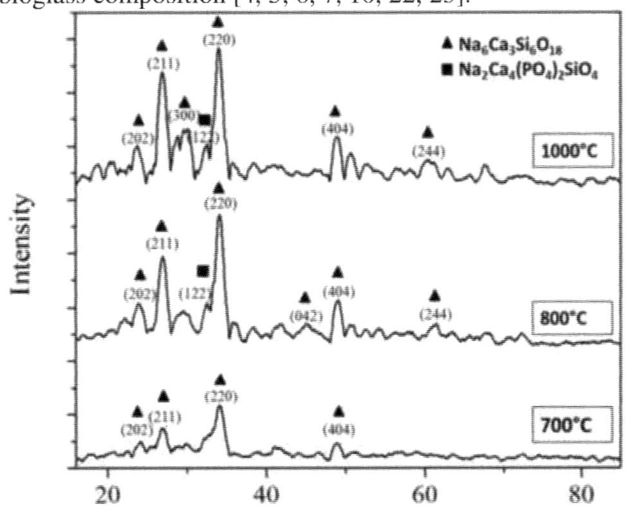

Figure 1. X Ray diffraction patterns of the xerogels treated at 700, 800 and 1000°C

Morphologies obtained by Field emission SEM of the 45S5 Bioglass xerogels, thermally treated at 700, 800 and 1000°C are presented in Figures 2a-d. Xerogel treated at 700°C presents a fiber like structure, as it was treated at higher temperature than T_{g1} and T_s. the phases detected in the corresponding X Ray difractogram, which are reach in phosphorus and silicon, are present [10]. In the micrographs of Figures 2b and 2c, corresponding to the xerogels treated at 800 and 1000°C, a growing of the particles and an evident change in the morphology is observed. This change may be due to the development of an amorphous matrix in which there are immersed crystalline phases and a great densification due to the higher temperature treatment over the T_{g1}, T_s and T_{g2} at which the phases rich in phosphorous and silicon are obtained. Figure 2d show the porous morphology of the xerogel treated at 1000°C, at lower magnification.

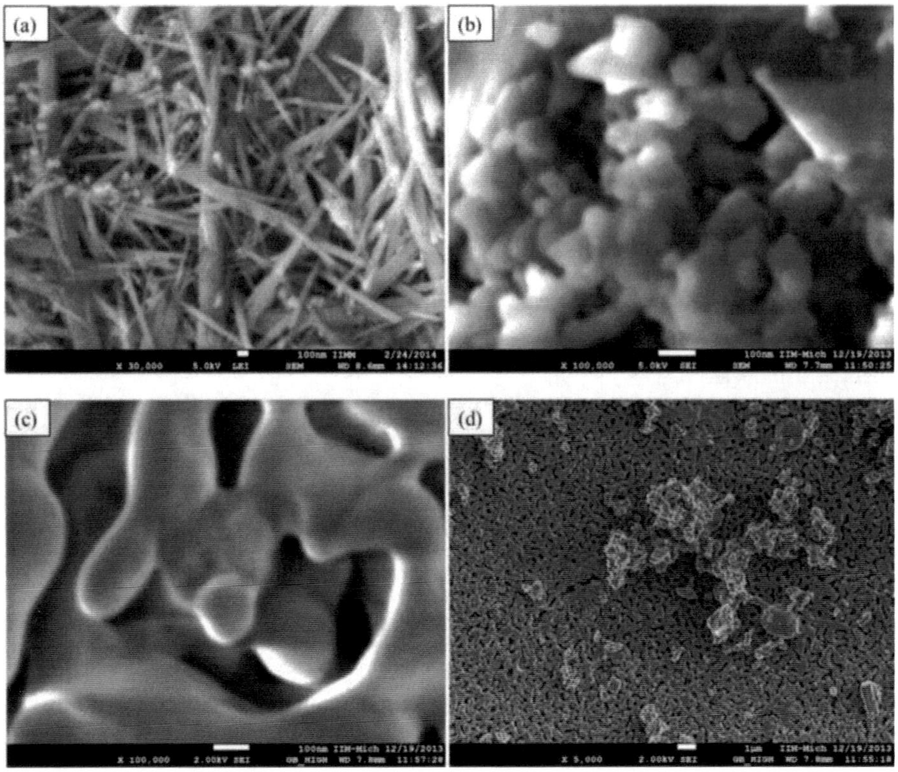

Figure 2. Field emission SEM micrographs of the 45S5 xerogels thermally treated at (a) 700°C (30000X), (b) 800°C (10000X), (c) 1000°C(100000X) and (d) 1000°C (5000X).

The microanalysis of the obtained 45S5 xerogels treated at 700, 800 and 1000°C, presented the signals of O, Si, Ca, Na y P, all of them are characteristic elements 45S5 bioglass. In xerogel treated 800°C, the same elements of the bioglass were detected, observing an increase in Calcium and a decrease in Sodium, because this xerogel was treated at higher temperature than T_{g2} at this temperature, it is possible to have diffusion and this occurs in the transition between the second and the third step of the 45S5 bioglass sinterization process. The microanalysis of the xerogel treated at 1000°C, showed evidence of more diffusion processes, enhancing the calcium content and decreasing the sodium content. Obtained 45S5 bioglass coatings morphologies were also observed by FESEM, at 5000X and are shown in Figure 3. Figure 3a, shows the coating treated at 500°C which presents inhomogeneities and defects due to the low thermal treatment temperature bellow the glass transition temperature T_g. Figure 3b, shows a more homogeneus coating treated at 600°C, which is distributed on all the substrate surface. Nevertheless it is specter that this coating is not well densifiqued because it was only treated at temperature 20°C up to the glass transition temperature T_s, so it is only partially dense. Figure 3c, shows the coating treated at 700°C, the change in the structure is evident appearing small and well distributed crystals, covering uniformly all the substrate surface, no defects were observed in this coating. Berbecaru et al. [23] obtained a 45S5 bioglass coating on Ti6Al4V alloy by magnetron sputtering technique and stablished that thermal treatment at 700°C by 2 hours, was adequate to obtain a glass crystallized coating according with their XRD and FTIR results which is in agreement with the present results, they also found that immersion in body simulated fluid (SBF) generates the band corresponding to Si-OH bonding. Figure 3d, shows the coating treated at 800°C, which presents a growing structure on the substrate formed by bigger aggregates originated at temperatures over the glass transition temperature T_{g1}, where phases richer in phosphorus and silicon like $Na_2Ca_3Si_6Oi_8$ and $Na_2Ca4(PO)_42SiO4$ crystallize, originating a substantial chance in the morphology, which generates separation on that surface coating. The micrograph at higher magnification (Figure 3e), allows to see the acicular morphology of these crystals. Plewinski et al. [5] established that bioglass 45S5 crystallization allows better formation of the apatite layer in comparation with the amorphous bioglass and concluded that this vitro-ceramic should be consider as a material for coating dental implants.

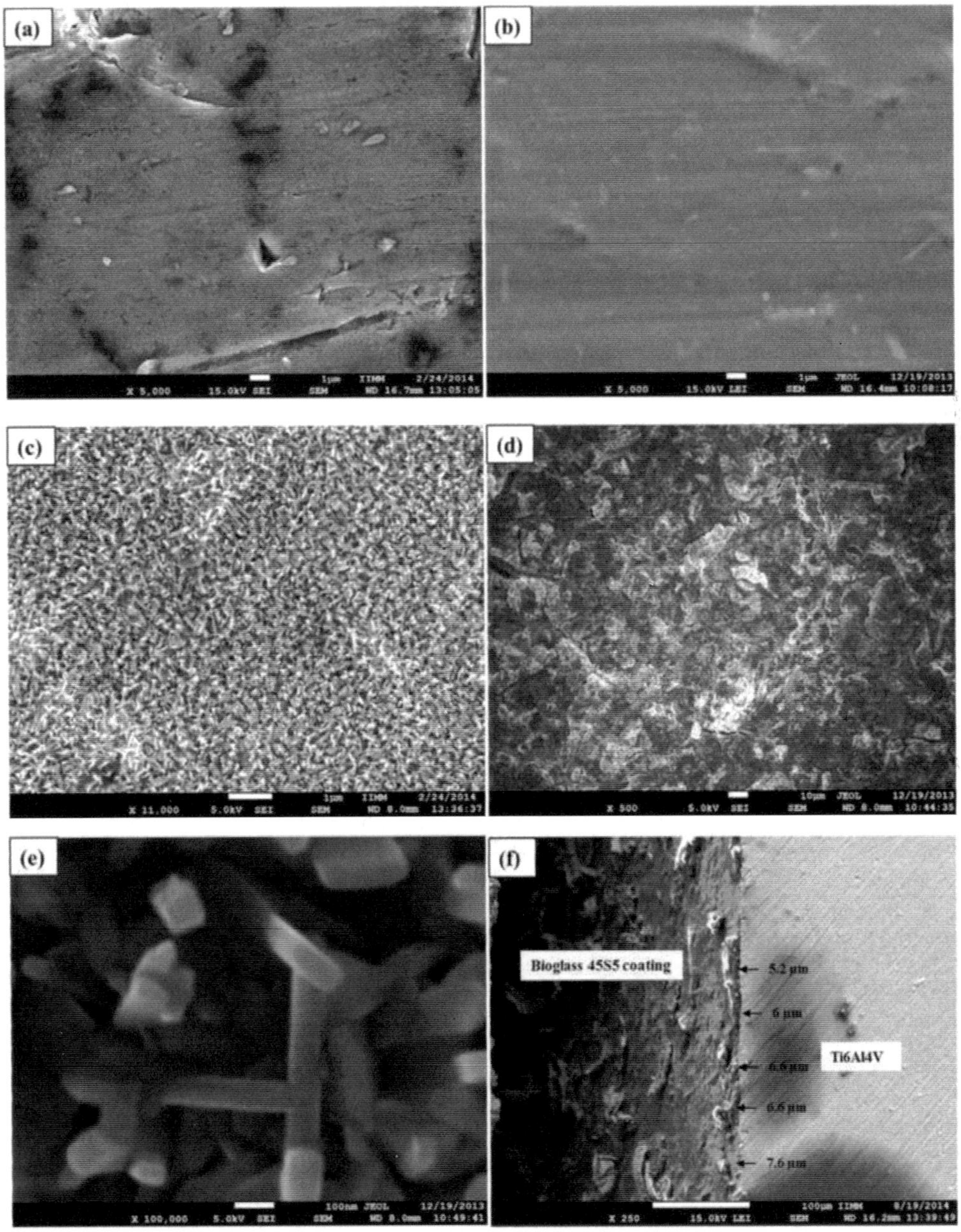

Figure 3. FESEM Micrographs of 45S5 bioglass coatings thermally treated at: a) 500°C (5000X), b) 600°C (5000X), c) 700°C (11000X), d) 800°C (500X), and e) 800°C (100 000X), f) coating thickness

3.2. Tafel extrapolation

The behavior of the potentiodynamic polarization curves at open circuit potentials, as a result of the inmersión in Hank's solution media of the Ti6Al4V alloy and the 45S5 bioglass/Ti6Al4V system are discussed here. At the beginning, the simple BG700, presents better corrosion resistance than the Ti6Al4V alloy, because it had a current intensity (i_{corr}) value of 0.001 A/cm^2 Alves et al. [25] evaluated this alloy polished with SiC 600 mesh in Hank's solution at 25 and 37°C finding i_{corr} values of 0.0877 y 0.0309 A/cm^2, respectively, similar value to the one obtained in this work for the Ti6Al4V alloy which was a i_{con}-of 0.013 A/cm^2 According to this result, the bioglass obtained coating inhibits the electrochemical attach and avoid the Al_2O_3 and V_3O_5 oxides liberation, which origins Al+ and V+ liberation to the blood current and to the osseous tissue inside the body [26,27]. This effect is evident by the displacement to the left of the anodic and cathodic curves obtained for the BG700 sample. The curve displacement indicates the diminishing of the corrosion current intensity (i_{co}ir), for the BG700 coating, in comparation with the obtained values for the Ti6Al4V alloy, which means that there are less electrochemical reactions generating the bioglass degradation than the ones for the Ti6Al4V alloy. There was also observed an important change in the corrosion potential (E_{corr}) taking more noble values, diminishing the activity of the 45S5/Ti6Al4V system treated at 700°C, also this system presented the lowest values of the anodic (B_a) and cathodic (B_c) slopes, which indicates the highest protection barrier in Hank's solution (pH=7.4) at temperature of 37 - 38°C, in comparation with the other coatings and the Ti6Al4V substrate. The less noble corrosion potential (E_{corr}) was obtained by the Ti6Al4V alloy with a negative value of -292.9 mV, presenting the higher activity, which means that an activation process is carried out allowing the formation of a passivation layer The bioglass coating with the highest current corrosion intensity (i_{co}ir) was BG800 due to the higher crystalline growing detected in the morphology observation leading to the formation of active sites. As it was observed in Figure 3e. Nevertheless the increase in i_{coiT} value was not really important as it could be expected, since it did not presented an important activity. BG700 sample had the more noble value for corrosion potential (E_{corr}) of -50.32 mV, which means less activity in the system. BG500 and BG600 samples showed more negative values of corrosion potentials (E_{corr}), than BG700 and BG800 samples, indicating that in BG500 and BG600, first there is a higher activity and then there is stabilization. In the case of the sample BG500 there was a thermal oxidation which generated growing in the passivation layer [28], so its behavior was very similar to the one of the Ti6Al4V substrate. BG700 sample presented the lowest corrosion rate, in millimeters per year (mpy), having 1.317x10^{-5} mm/year and a weight loss of 0.028x10^{-9} g/cm^2, compared with Ti6Al4V alloy that had 1.161x10^{-4} mm/year and 140.9x10^{-9} g/cm^2, respectively. Ions liberation for sample BG700 was estimated of 1.15x10^{-9} g/cm^2, compared with Ti6Al4V alloy that had 11.78x10^{-9} g/cm^2 Balamurugan et al. [3] reported a bioactive glass coating obtained by electrophoretic deposition method with increased corrosion resistance in physiologic media. Krzakala et al. [19] anodized the Ti6Al4V alloy in $Ca(H_2PO_2)_2$ and NaOH baths obtaining bioactive films with low values of i_{co}ir and the activity reduction with possitive values of E_{corr} in comparation with the alloy, and also reported high polarization resistance.

4. Conclusions

- 45S5 bioglass thin and homogeneous coatings on Ti6Al4V alloy were obtained by sol-gel synthesis and electrophoretic deposition methods.
- The coatings were nanostructured and had mean thickness of 1 -5 microns.
- The obtained bioglass coatings presented good protection against ionic attach of Hank's Solution specially the one thermally treated at 700°C (BG700).
- According with the extrapolation Tafel results, there was a reduction of one order of magnitude in the current density of corrosion (i_{co}ir), in the corrosion rate and in the weight loss of the obtained 45S5 Bioglass coating thermally treated at 700°C compared with the Ti6Al4V alloy and the other obtained coatings.
- The more noble behavior was presented by the BG700 sample, which change the corrosion potential to more positive values, reducing the 45S5/Ti6A4V system activity.

- Relating the obtained extrapolation Tafel, the lowest value for ions liberation was also presented by the bioglass coating treated at 700°C (BG700).

References

[1] A. Balamurugan, G. Sockalingum, J. Michel, J. Fauré, V. Banchet c, L. Wortham, S. Bouthors, D. Laurent- Maquin, G. Balossier, 2009 *Materials Letters* **60** 3752
[2] L. Ding, Y. Fuqian, J. Nychka, 2008 *Engineering Fracture Mechanics* **75** 4898
[3] A. Balamurugan, G. Balossier, J. Michel, J.M.F. Ferreira, 2009 *Electrochimica Acta* **54** 1192
[4] F. Pishbin, A. Simchi, M.P. Ryan, A.R. Boccaccini, 2011 *Surface & Coatings Technology* **205** 5260
[5] M. Plewinski, K. Schickle, M. Lindner, A. Kirsten, M. Weber, H. Fischer, 2013 *Dental Material* **29** 1256
[6] O. Bretcanu, X. Chatzistavrou, K. Paraskevopoulos, R. Conradt, I. Thompson, Aldo R. Boccaccini, 2009 *Journal of the European Ceramic Society* **29** 3299
[7] H. Li, D. Wang, J. Hu, C. Chen, 2013 *J. Sol-Gel Sci. Technol.* **67** 56
[8] Imran Farooq, Z. Imran, U. Farooq, A. Leghari, H. Ali, 2012 *World Journal of Dentistry* **3** 199
[9] Q. Chen, G. Thouas, 2011 *Acta Biomaterialia* **7** 3616
[10] L. Lefebvre, L. Gremillard, J. Chevalier, R. Zenati, D. Bernache-Assolant, 2008 *Acta Biomater.* **4** 1894
[11] M. S. Bahniuk, H. Pirayesh, H. D. Singh, J. A. Nychka, L. D. Unsworth, 2012 *Biointerphases* **7** 1
[12] H. Sheng, J. Chang, L. Mingqiu, N. Congqin, 2009 *J. Mater. Med: Materials Medicals* **20** 281
[13] T. Kokubo, H. Takadama, 2006 *Biomaterials* **27** 2907
[14] S. Bauer, P. Schmuki, K. Von der Mark, J. Park, 2013 Progress in *Materials Science* **58** 270
[15] S. R. Paital, N. B. Dahotre, 2009 *Materials Science and Engineering* R **66** 7
[16] L. Guéhennec, A. Souidan, P. Layrolle, Y. Amouriq, 2007 *Dental Materials* **23** 844
[17] B. D. Ratner, A. S. Hoffman, F. J. Schoen, J. E. Lemons, *Biomaterials Science: An Introduction to Materials in Medicine*, Academic Press, 1996, 3-17.
[18] C. E.B. Marino, S. R. Biaggio, R. C. Rocha, N. Bocchi, 2006 *Electrochimica Acta* **51** 6580
[19] A. Krza, K. Sluz, G. Dercz, A. Maciej, A. Kazek, J. Szade, A. Winiarski, M. Dudek, J. Michalska, G. Tylko, A. Osyczka, W. Simka, 2013 *Electrochimica Acta* **104** 425
[20] C. Alagic, Z. Cvijovic, S. Mitrovic, V. Panic, M. Rakin, 2011 *Corrosion Science* **53** 796
[21] ASTM-G102-89. Standard Practice for Calculation of Corrosion Rates and Related Information from Electrochemical Measurements.
[22] D. Krause, B. Thomas, C. Leinenbach, D. Eifler, E. Minay, A. Boccaccini, 2006 *Surface & Coatings Technology* **200** 4835
[23] C. Berbecarua, H. Alexandru, G. Stan, D. Marcov, I. Pasuk, A. Ianculescu, 2010 *Materials Science and Engineering* B **169** 101
[24] 0. Peitl, E. Dutra, L. Rench, 2001 *Journal of Non-Crystalline Solids* **292** 115
[25] V. Alves, R. Reis, I. Santos, D. Souza, T. Goncalves, M. Pereira-da-Silva, A. Rossi, L. da Silva, 2009 *Corrosion Science* **51** 2473
[26] D. Estévez, E. Farfán, L. Forero, F. Rincón, 2007 *Scientia et Technica* Año XIII **36** 157
[27] N. Mobarak, A. Swayih, F. Rashoud, 2011 *International Journal Electrochemical Science* **6** 2031
[28] Prateek Jain, Tapendu Mandal, Prem Prakash, Ashish Garg, Kantesh Balani, 2013 *J. Coat. Technol. Res.* **10** 263

IOP Conf. Series: Materials Science and Engineering **123** (2016) 012007 doi:10.1088/1757-899X/123/1/012007

Artificial food lump from porous neoprene and the method of its use for the evaluation of adaptation patients to the dental constructions

A Reshetnikov[2], A Urakov[2], A Kasatkin[2], M G Soiher[1], M Kopylov[1]

[1]Institute of Biotechnology and Interdisciplinary Dentistry, Moscow, Russia
Izhevsk State Medical Academy, Izhevsk, Russia

E-mail: areshetnikov@list.ru

Abstract. New dental product called artificial food lump is offered for dental practices. In its size and shape it is similar to the natural food bolus, which is formed in adult's mouth when chewing white bread. This innovative product resembles an inedible and non-swallowable chewing gum. Artificial lump is made of porous neoprene; it is elastic and has food flavor. It is not destroyed by chewing and has stable elasticity during chewing. Besides, artificial lump is manufactured in a way that it can be attached to the patient's clothes with a braid line. New medical device is intended to create the masticatory loading in patients' mouth in order to evaluate the quality of mounted dental restorations as well as patient's adaptation to it during the chewing process.

1. Introduction

While providing dental care to a patient with pathology of oral hard and soft tissues, we mount various types of dental restorations with different strength and irritation characteristics [1]. Using temporary or permanent restorations made of advanced materials and ensuring high biological compatibility with human tissues, allows us to minimize their irritating effect for the patients. However, on the one hand there are some cases of partial and/or complete dental restorations damage, and oral cavity tissues damage on the other hand. All this leads to decreased efficiency and safety, and increases dental treatment duration and costs [2-5].

Intense mechanical interaction between tissues and dental restorations while chewing could be one of the reasons for these complications development. Moreover, the food damaging effect is related to its hardness and temperature. However, current standards for assessing the quality of dental restorations materials do not include the assessment of their suitability for chewing with the use of reference loading [6-11]. However, nowadays the only way to solve this problem is to use the reference chewable product. It is clear that in order to create a reference masticatory loading we should use a medical device in the form of natural bolus, but having "reference" characteristics. Here we mean constant elasticity during chewing and the product hydrophobicity.

In our opinion, infrared thermography could be used as a possible method for assessing patient's adaptation to the dental restorations [12]. This method allows us to detect artificial constructions defects as well as soft tissues local inflammation in its normal and irritated states and with different pathologies [13-15]. In particular, infrared monitoring of skin local temperature dynamics at the injection site allows us to define the local inflammation that occurs in the chemical, physical and biological irritation [16-18].

Thus, the infrared monitoring of the patient's oral cavity which is performed after dental restorations mounting and followed by chewing the artificial lump, can provide the information on the mounted dental restorations strength when masticating food as well as the patient's oral tissues durability when he has his dental restorations mounted and is masticating food with it. All this provides a new way for evaluating the quality of dental treatment.

The aim of our research is to study the possibility of preventing the complications in the chewing process after mounting the dental restorations by means of well-timed chewing of artificial lump with constant and safe elastic properties, and defining the local temperature dynamics in oral cavity tissues.

2. Materials and Methods

At first we defined the size of natural food bolus which was formed in adult's mouth when chewing natural food. For this purpose we asked healthy adult volunteers (n = 80) to chew pieces of wetted solid food for 30 seconds, after which the pieces were taken out for examining their mechanical and physical and chemical properties. In particular, the lumps shape, size and volume were defined with the use of proper measuring instruments at ambient temperature of +25°C.

Infrared thermography of oral tissues for 80 healthy adult volunteers and 100 adult patients was performed at «ReSto» dental clinic (Izhevsk, Russia) using a thermal ThermoTracer TH9100XX (NEC, USA) in the temperature range of +25 - +36° C and at temperature of air +25° C. We studied temperature and infrared spectrum before and after chewing the sample. The data were processed using Thermography Explorer and Image Processor computer software programs. Dental restorations in accordance with International Organization for Standardization (ISO) were mounted as indicated in agreement with international recommendations and standards.

The study was approved by the Ethics committee at Izhevsk State Medical Academy and complied with the Declaration of Helsinki. All patients gave informed consent.

3. Results

Our results showed that the natural bolus had a cylindrical shape with a diameter of 1 cm and a length of 4 cm and a hemisphere shape at its ends. Such shape and dimensions allow easy lump placement inside the mouth of an adult person. The results analysis showed that the estimated bolus size and shape were optimal for solving the problem, because the smaller size doesn't not provide a proper simulation for the physiological process of chewing semi-coarse and coarse food, especially bread products, and does not insure the mechanical impact on the entire premolars and molars chewing area. The bigger sizes, in its turn, exceed safe and comfortable bolus size.

While studying the elastic properties of synthetic materials we found that porous neoprene with a porosity of not more than 30% could be chosen as an elastic base for the artificial lump. Neoprene is the most inert synthetic polymer, and it does not react with water, alcohol and other organic solvents. The porosity of neoprene is not more than 30%, and pore air filling provides safe physical and chemical conditions and elasticity level.

We conducted a clinical trial of artificial food lump made of neoprene and with size and shape similar to natural bolus. The results confirmed our assumptions. In particular, all the healthy volunteers and patients with mounted dental restorations admitted that they could easily move the artificial food lump in the mouth and chew it without feeling of discomfort. Clinical studies showed that during chewing, artificial lump should be provided with a thread attached to it with one end, and with another end fixed with a clip to the patient's clothing. This prevents accidental ingestion of artificial food lump during evaluating patient's adaptation to the dental restorations.

The study of the local temperature dynamics of patients' oral tissues immediately after mounting new dental restorations proved infrared thermography method to be highly informative for assessing the patient's adaptation to dental restorations elements. As a result, we developed a special method for instant assessing of dental restorations resistance to the patient chewing. The method is based on the detection of heat radiation of oral cavity tissues, carried out by means of infrared thermography of oral cavity tissues after chewing the "sample for chewing" [19].

Instant diagnosis of patient adaptation to the dental restorations is based on using two artificial food lumps preheated to +37°C. The first lump is used before mounting dental restorations, and the second one is used after it. In both cases the patient is asked to put a lump into his mouth and chew it for 30 seconds, consistently moving the lump along the tooth alignment. When chewing is finished each lump is taken out and examined with the help of thermal camera in order to analyze heat radiation dynamics of oral tissues. In this case, if local hyperthermia or short uniform and symmetrical temperature rise in the oral tissues are not detected after chewing the first lump, the patient's adaptation to chewing is evaluated as excellent, predicting high patient's durability regarding dental restorations, and in this case the decision about its safe mounting could be positive. Dental restorations are mounted together with controlling the dynamics of tissues thermal radiation after chewing the second identical artificial lump for 30 seconds. After that, we compare these data on thermal radiation dynamics with the original, and if hyperthermia values and its duration are higher, then we evaluate patient's adaptation to dental restorations as poor and predict tissue damage.

4. Conclusions

In summary, the artificial food lump and its application method for instant assessment of patient's adaptation to the dental restorations expand the evaluation scope and increase its urgency, informative value and reliability, as well as prevent complications in the chewing process after mounting dental restorations due to chewing artificial food lump with constant safe and elastic properties at proper time and by defining the temperature dynamics in oral tissues during dental care.

References

[1] Zafar N and Ghani F 2014 *Pak J Med Sci* **30(3)** 619-25

[2] Anusavice K J 2012 *Dent Mater* **28(1)** 102-11

[3] Urakov A, Reshetnikov A, Reshetnikova A, Reshetnikova N 2011 Method of splinting brackets Patent RU 2437632 *Bull.* **36**

[4] Rammelsberg P, Schwarz S, Schroeder C, Bermejo JL, Gabbert O 2013 *Clin Oral Implants Res.* **24(7)** 758-62

[5] Reshetnikov A, Soiher E M 2013 Physical and chemical factors of aggression in dentistry Applied and Fundamental Studies Proceedings od the 3rd International Academic Conference. *(August 30-31, 2013, St. Louis, USA).* St. Louis: Publishing House "Science & Innovation Center" 169 – 172.

[6] ISO/TC 106/SC 4 Dental instruments

[7] ISO/TC 106/SC 7 Oral care products

[8] ISO/TC 106/SC 8 Dental implants

[9] ISO 6892 Metallic materials – Tensile testing at ambient temperature

[10] Urakov A, Urakova N, Reshetnikov A, Nazarova Eh, Lipanov A, Dementyev V 2012 Method for making and inserting dental structures Patent RU 2469640 *Bull.* **35**

[11] Reshetnikov A, Urakov A, Strelkov N 2009 Way of tooth restoration Patent RU 2360640 *Bull.* **19**

[12] Ammer K 2008 *Thermology international* **18(4)** 125-129

[13] Urakova N, Urakov A 2014 *Biomedical Engineering* **48(3)** 111-115

[14] Urakov A, Urakova N, Dementyev V 2013 Resuscitation **84S** S73

[15] Kasatkin A, Reshetnikov A 2014 2014 *Thermology international* **2** 64-65

[16] Urakov A, Kasatkin A, Urakova N, Ammer K 2014 *Thermology international* **24(1)** 5-10

[17] Soiher M G, Soiher M I, Urakov A, Reshetnikov A, Kopylov M, Fisher E 2014 *Thermology international* **2** 60

[18] Urakov A, Urakova N, Kasatkin A 2013 *Thrombosis Research* **131 S1** 79

[19] Urakov A *et al* 2014 Artificial food bolus and method forinstant assessment of dento-facial health with using artificial food bolus Patent RU 2533840 *Bull.* **32**

Development of new materials and structures based on managed physical-chemical factors of local interaction

A L Urakov

Department of General and Clinical Pharmacology, Izhevsk State Medical Academy, Izhevsk, Russia

E-mail: urakoval@live.ru

Abstract. The paper states that assigning certain physical and chemical characteristics to pills and medical drugs solutions can substitute for the development of new drugs (which is essentially equivalent to the creation of new medicines). It is established that the purposeful change of physical and chemical characteristics of the standard ("old") materials (in other words, the known substances) is fundamental for the production of solid and liquid medicines, which allows us to get "new" structures and materials. The paper shows that assigning new physical and chemical properties to "old" materials and their further usage for the production of tablets and solutions from the "old" and well-known medicines can turn even very "old" medicine into very "novel" (moreover, even very fashionable) one with unprecedented (fantastic) pharmacological activity and new mechanisms of action.

1. Introduction

As an example let us consider the evolution of materials and structures of drugs and medical devices introduced into the body. Today they don't usually analyze the mechanism of drugs action taking into consideration the physical and chemical properties of the powders, tablets, solutions, gases or other objects and forms in which medicines are administered to patients [6,23,24]. Besides, they ignore the change in a medicine physical and chemical properties caused by the fact that it's not made from the medicines but from excipients [17]. Nevertheless, most of the ready products today often consist of excipients rather than chemical reagents [9].

It is no secret that final pharmaceutical product is completely different from "pure" chemical reagent, which was historically used for examining its toxicity in the beginning [10]. The matter is that "pure" chemical reagent is deliberately made very much "pollute" in the process of tablets, solutions and sprays manufacturing [9,10]. It is deliberately mixed with large amount of excipients and other substances. As a result, the medicine itself often has less than 1% of volume in the resulting mixture or solution which is used to prepare the final pharmaceutical product [10,12,14].

Besides, medicines and medical devices are manufactured by different pharmaceutical companies using different technologies and different starting materials. Therefore, products made by different plants cannot have same physical and chemical properties [10,14,17].

Our results show that these features can be used to distinguish the drugs and medical devices (syringes, probes, catheters, drainage tubes and other devices) between each other in quality and local irritant action on patients' tissue when administered [20,23]. Moreover, ignoring the role of these factors reduces not only the efficiency and accuracy of diseases prevention, diagnosis and treatment method, but also their safety [5]. Besides, it turned out that it was the local physical and chemical drugs and medical devices aggressiveness that was the base of their local irritating and damaging effect on tissue [9,10,16,18]. It is also responsible for the development of medical iatrogenic diseases [12,19,23,24].

However, solving the local physical and chemical interaction problems for drugs and medical devices when they are administered to the patient is not only based on their variety and the fact that they are made of different materials with very unusual physical and chemical properties. The fact that all human body tissues have different physical and chemical properties, which, in addition, can change significantly at the period of diseases is also very important. Therefore, the same medicine or medical device with certain physical and chemical properties (quality indicators) may correspond to physical and chemical properties of any body tissue in its normal state, and at the same time may not correspond to it in the health problem situation [13,21]. It may be a friend for one body area and at the same day an enemy for the others [3,25].

The matter is that the main cause of local irritant actions of drugs and medical devices is the difference in their physical and chemical properties [12,24].

Which physical and chemical local interactions factors are the most important for the development of new materials and devices for medical use?

2. Materials and methods

We studied physical and chemical properties of 200 venous blood samples, 300 limbs veins, 300 intravascular catheters which were installed in them and devices used for infusions for 200 patients who were being treated in Izhevsk Hospitals within the period from March 2008 to October 2014. In addition, we studied the blood conditions in major vessels and subcutaneous veins in the injection site and in catheterization area of the wrist veins, forearm and shoulder of 10 corpses of adult patients who had died in the hospitals. Blood viscosity measurements were performed using a cone-plate viscometer Brookfield DV-II, coagulation activity of the blood is studied by Lee-White, status of the lumen and the walls of the veins are studied in vena section, the state of catheter lumen in situ is studied through visual observations, and with the help of the ultrasonic device Logik Book XP, equipped with linear transducer 8L [20].

The dynamics of temperature in different parts of the body in 200 patients and in 100 healthy volunteers was studied using advanced infrared radiation with the help of thermal imager TH91XX (NEC, USA). Further data processing was performed with Thermography Explorer and Image Processor software programs. Ambient temperature in the examination room was 24 ± 25°C, the temperature range in the thermal camera was set as 25 - 36°C. At the examined area we determined the mean temperature, standard deviation and distribution of isotherms at the thresholds of 34, 32, 30, 29 and 25°C [21].

We studied pus rheology in 100 patients with purulent peritonitis, purulent conjunctivitis and tuberculosis of the lung and pleura. Pus rheology was studied through visual observations in vivo and in vitro before and after the introduction of medicines solutions in pus with regard to the interacting masses volume, their interaction duration, gravity, specific gravity (density), temperature, pH and osmotic characteristics. Pus was taken for analysis before and 15 minutes after the interaction with plasma-substituting solutions and antiseptic funds began. Smears were prepared and stained similarly to blood smears using standard laboratory methods of dyeing with 0.5% Mine-Gryunval paint prepared with 96° ethyl alcohol, and Romanovsky-Giemsa staining [14].

We have studied the peculiarities of tablets local interaction in the oral cavity, esophagus and stomach in 10 adult volunteers and 10 conscious piglets. The movement of the tablets in the stomach cavity was studied using ultrasound. Tablets were administered before and after the introduction of water and food lumps (white bread) through the mouth and/or stomach probe and/or intestinal probe. The body position in space was taken into account. Peculiarities of tablets local interaction inside the stomach were observed in 10 anesthetized cats after cutting the stomach through surgical operations.

For this purpose we have used 0.1 g of ascorbic acid tablets with glucose and/or 0.5 g of calcium gluconate tablets.

We have identified the most important physical and chemical characteristics of 200 quality medicines produced by pharmaceutical companies in the form of tablets and solution for injection. The medical drugs which were examined belonged to various pharmacological groups. The medicine quality was studied by taking into account the manufacturers and serial number of medicine. To analyze medicine quality we compared quality characteristics values which had been specified by the manufacturer in the medicine passport with the observed ones. Additionally, for the tablets we identified specific deforming hardness in Rockwell units HB (in a Brunel scale).

When evaluating the quality of drugs solutions we were taking into account their osmotic activity and the degree of gas saturation. Osmotic activity of aqueous solutions was defined with cryoscope using vapor-pressure OSMOMAT-030 RS brand osmometer of ANSELMA Industries (Austria) company [10]. The visualization of gas bubbles in carbonated solutions was held in visible spectrum of radiation through visual observation and ultrasound method using "ALOKA SSD-ALPHA 10" ultrasound device and convection sensor with a frequency of 3 to 7 MHz [3,20].

Statistical analysis of the results was performed using the BIOSTAT software program using standard methods.

3. Results

A very important factor is the local temperature of human body tissues. The matter is human body temperature equals 36.6° C only in theory. Our results show that in real life there are no conditions under which the temperature of all body parts would be equal to 36.6° C. The reality is that the temperature of various body parts of different patients can vary ranging from 0 to +42° C. At the same time Arrhenius and Van't Hoff proved that 10° C temperature change can change the chemical reactions flow rate from 2 to 4 times.

Another important factor is the flow and transferability characteristics of the body tissues. The reason is that the human body is constantly changing its position in space, but almost all of its body tissues move insignificantly relatively to each other. Therefore, we assume that a man and his body are relatively mobile, and most of his inside organs and tissues are relatively stationary (stay in one place). At the same time, many of them have moderate and high elasticity. Due to this human body maintains its shape independently of body orientation in space. This is not true for blood, because blood doesn't have permanent form and doesn't stay in one place even after person's death and efflux from his body. Therefore, when injected into the skin, subcutaneous fat, skeletal muscle, bone, cerebrospinal fluid, pus and other organs and tissues, these products may remain at the same place for a long time. At the same time, after being injected into blood the products begin immediately to flow with blood, and sometimes are carried far away from the injection point.

Among other important and rather uncertain local interaction physical and chemical factors are its duration, the specific pressure, reduced specific gravity, pH (acidity or alkalinity), concentration, osmotic activity and gas saturation.

In experiments in vitro and clinic observations in vivo the dynamics of the lumen state in subcutaneous veins was studied, as well as rheological properties of blood and pus dense in patients after their interaction with vascular catheters, pleural catheter and plasma-substituting solutions and antiseptics with respect to physical and chemical factors of the local interaction.

It is shown that solutions of medicines are introduced in a body at room temperature (as a rule, at a temperature of +24 - +26°C). However, some solutions may be of lower or higher temperature, as some medicines are deliberately cooled, because they are stored in the refrigerator at a temperature of +4 - +8°C, and some medicines are heated due to ignorance, as there is no temperature control even in the presence of heat sources (solar rays, infrared radiation from incandescent lamps and medical devices). As a result, the intravenous introduction of solutions for injections, executed through the conventional technique often changes the blood temperature inside the veins, which, in turn, changes the temperature of the venous wall and the tissues surrounding it, including the skin in the area of subcutaneous veins projection. We called this phenomenon as "thermo – contrast tissues", and it allows us to diagnose the subcutaneous veins localization without x - ray contrast media and without physical contact with the body of the patient due to the capabilities of infrared skin thermography.

It turned out that 10% of the patients' skin above the location of catheter free end in vein is heated and in this place a hyperthermia hotbed is formed, which continues, as a rule, not less than 2 days. In addition, in the vein located under the local hyperthermia hotbed we found the wall surface clot. Moreover, the blood clot is firmly attached to the part of the venous wall, which turns out to be the opposite and close to the open «nozzle» of vascular catheter and which is the least protected by the blood from «pure» drugs (drugs not diluted in blood) irrigation. However, the most amazing is the fact that this part of the venous wall is under the strong influence of heparin, expiring inside of vein from the catheter!

It appeared that 5 - 8 minutes before the local hyperthermia hotbed 0.1 ml of Dopamine solution (2-(3, 4dihydroxyphenyl)-athylamin4-2(2-aminoethyl)benzen-1,2-diol) or to 2 ml of 2.5% of Chlorpromazine hydrochloride (2-chloro-10[3-(diethilaminopropil)] phenothiazines hydrochloride) was introduced for each patient of this group. At the same time, the experiments revealed some patients in whom the intravenous administration of these drugs did not cause the local hyperthermia hearth appearance. It turned out that at the same time these patients were introduced 200 ml of 4% sodium bicarbonate in vein. In this regard, we measured the acidity of these drugs, and found out that the standard dopamine and 2.5% chlorpromazine hydrochloride solutions are acidic, and the pH is in the range of 4.2 to 4.7. However, after diluting these medicines solutions with 4% sodium bicarbonate solution in a ratio of 1:2 or more, drugs become alkaline and acquire a pH in the range of 7.0 to 8.0.

On the basis of the obtained data the assumption is made that the local aggressiveness sour medications can reduce by combining them with a solution of 4% sodium bicarbonate. To check this assumption we held infrared thermography of 20 patients hands with multiple intravenous introduction of standard solutions dopamine or chlorpromazine hydrochloride accompanied by the introduction of the solution of 4% sodium bicarbonate. It turned out that the immediate intravascular catheters and introduction in veins of the 200 ml 4% sodium bicarbonate solution at room temperature (+24 - +26°C), produced after each dopamine or chlorpromazine hydrochloride injection, allowed to exclude the appearance of local hyperthermia hearth «in vascular catheter course», a clot (thrombus) and clogging of intravascular catheters for 5 days in all 20 patients.

Following this, we conducted similar experiments in vitro with fresh venous blood, fresh blood clots, with cadaveric blood (devoid of the ability to natural coagulation) and with dried blood of patients. The results showed that in vitro fresh blood clots at a temperature of +24, +37 +42° C over 15, 7 and 5 minutes (respectively). In addition, while injecting 1 ml of similar fresh venous or cadaveric blood into 0.1 ml of quality fluids drugs such as dopamine or chlorpromazine hydrochloride, in this blood clots appear immediately. This happens, most likely due to acid coagulation (denaturation) of the plasma proteins, as administered solutions are very acidic. At the same time, the preliminary introduction in fresh blue blood solution of 4% sodium bicarbonate in a ratio of 1:1 or more retains fluid properties of blood at a temperature of +24, +37 +42°C for 30 minutes of observation. The subsequent introduction in 1 ml of this blood 0.1 ml dopamine or chlorpromazine hydrochloride almost completely prevents the coagulation effects on blood.

In addition, it is shown that irrigation fresh blood clots and dried blood spots solution of 4% sodium bicarbonate in comparable volumes dissolves blood clots and softens the dried blood within 1-4 minutes in all studied temperatures. Moreover, it turned out that the ability of a solution of 4% sodium bicarbonate to dissolve blood clots depends on temperature as follows: the higher the temperature of interaction, the higher the efficacy.

However, the results of the experiments have shown that a solution of 4% sodium bicarbonate is not able to completely dissolve all the dried blood, to destroy 100% of residuals spot of blood for 1-4 minutes of interaction even at a temperature of 42°C. In this regard, for the potentiation of hemolytic activity of this solution we decided to add a 3% hydrogen peroxide. The results of experiments show that irrigation residues blood stains solution of 4% sodium bicarbonate and 3% hydrogen peroxide provides practically instant interstitial application only cold «boiling» in the rests of blood stains, leading in 1-2 seconds to their complete destruction, coupled with the initial brief clarification (change of dark red color of blood on a light red color), quickly changing on the complete and final discoloration of biological mass. These data prove that the solution of 4% sodium bicarbonate is an effective and safe «chemical» anticoagulant and fibrinolysis drug, which can prevent blood clots and blockage of veins and intravascular catheters blood clots when repeated intravenous injections «neutral» and «sour» drugs. To do this, the data of a solution fill needle catheter before introducing them to vein, and then re-filled catheter immediately after removal from the needles and after each entered medication [8].

Our results showed that after injection into a vein of acidic drugs, blood clot can be formed in vein and appearance of the skin of limbs hearth of local hyperthermia above the end of the vascular catheter. However, a quick wash of the area of vein for 3 minutes with a warm solution of 4% sodium bicarbonate eliminates local hyperthermia in the place of location of catheter working end and blockage of veins and catheter blood clots.

It is shown that the rheological properties of thick pus and solutions of antiseptics and substitutes plasma at their local interaction depends mostly on conformity of volume and concentration, density, temperature, alkaline, osmotic and turbulent activity of medicines. The effect of the following factors was studied: gravity, specific gravity, temperature, turbulence, strength, internal pressure, carbonation, pH, osmotic activity, the total concentration of ingredients, the surface activity and the amount of medication.

After placing a thick pus and known plasma substitutes and hygienic liquids in laboratory test tubes in equal amount of samplings at a temperature of 24 or 37 °C pus was always located at the bottom of the tubes and liquids is always located above the pus. In particular, boiled water, tap water, water for injection, solution 0.9% sodium chloride solution, a solution 5% glucose, solution 20% glucose, solution 40% glucose, solution 10% sodium chloride, solution 10% sodium sulfatsil, solution 20% sodium sulfatsil, solution 0.02% furatsilin, solution 0.5% chlorhexidine, solution 70% ethyl alcohol and solution 1% sodium hydrocarbonate were always on top and not mixed with pus during 15 minutes of observation. Therefore rheological properties and other physical and chemical characteristics of pus were not changed. Slight decrease in dense purulent mass viscosity and pus microstructure change during 15 minutes of interaction with these solutions happened only when heated to 42°C.

Then we have defined the indicators for the acidity of water and solutions of antiseptics and plasma substitutes. The obtained results showed that all of them except 4% sodium bicarbonate solution and solutions of 10% and 20% sodium sulfatsil have a pH below 7.0, and so are acidic. In particular, indicators of acidity of tap water, boiled water, water for injection, solution furatsiline 1:500, solution of 3% hydrogen peroxide, solutions 0.9% and 10% of sodium chloride are in the pH range of 5.0-6.0, and the performance of solution acidity of 0.25% novokaine and solutions 5%, 20% and 40% glucose are in the pH range of 2.7 to 4.0 (Table 1).

In the next series of experiments in vitro at a temperature of 24 °C we studied the change of rheology and microstructure of thick pus during 15 minutes after administration into pus water or water solutions of medicines after preliminary artificial amplification of acidification to pH 2.0 or alkalinity to pH 12.0 by introducing them accordingly with hydrochloric acid or sodium hydroxide. Our results showed that only alkalization decreases viscosity and microstructure purulent masses dense.

Along with this, we studied the features of thinning and removal of thick and sticky pus under the influence of water and water solutions of medicines in the conditions of their increased turbulence, which was reached by heating and mechanical wobbled from side to side model cavity (test tube), filled commensurate amounts of pus and one of the investigated solutions. It is shown that the continuous rocking from side to side tubes with interacting environments, namely, with pus and high alkaline liquids (water for injection, solutions 0.9% and 10% sodium chloride, 10% and 20% sulfatsil the sodium, 0.02% furatsilin at pH 12.0), causes continuous reciprocal progressively offset medicated fluid, which is the top layer, with respect to the stationary mass of thick pus, which is the lower layer), and accelerates the process of pus liquefying on the media separating boundary. Herewith, the shorter the time interval between the beginning of the interaction and the onset time of removal of the main pus mass at a temperature of 24°C and 42°C in 4 and 5 times, respectively, compared with the control (in terms of their physical immobility).

Table 1. Values of pH of the water and of water solutions of modern antiseptic funds and plasma substitutes

N	Liquid medication	Manufacture, Series №.	pH
1	Tap water		5.95 ± 0.11*
2	Drinking water from the kettle		6.05 ± 0.07*
3	Water for injection in vials of 2 ml	SE "Lvivdialect" Series № 1151205	5.02 ± 0.08*
4	Solution 0.9% sodium chloride for injection, 200 ml	JSC "Galichpharm" Series № 300604	5.34 ± 0.04*
5	Solution 10% sodium chloride for external use 200 ml	Pharmacy № 131 (Izhevsk)	6.05 ± 0.09*
6	Solution furacillin 1:5000	Pharmacy № 131 (Izhevsk)	5.70 ± 0.10*
7	Hydrogen peroxide 3% 40 ml	LLC "Batхэм-pharmacy" Series № 841101	5.45 ± 0.25*
8	Solution 5% glucose for injection 500 ml	JSC "Biosynthesis" Series № 48102000	4.05 ± 0.09*
9	Solution 20% glucose for injection 500 ml	"Novosibirskpharm" Series № 30703	3.50 ± 0.15*
10	Solution 40% glucose for injection 500 ml	JSC "Dalhimpharm" Series № 210101	2.70 ± 0.07*
11	Solution 0,25% novokaine for injection 500 ml	Pharmacy № 131 (Izhevsk)	4.05 ± 0.03*
12	Solution 10% sodium sulfatsil 10 ml	FSUE "Moscow Endocrine Factory" Series № PN 001084/01	7.90 ± 0.10*
13	Solution 20% sodium sulfatsil 10 ml	FSUE "Moscow Endocrine Factory" Series № PN 001084/05	8.30 ± 0.15*
14	Solution 4% sodium bicarbonate 200 ml	Pharmacy № 131 (Izhevsk)	8.20 ± 0.09*

Note: * - $P < 0{,}05$, $n = 5$.

However, it appeared that artificially supported for 30 min increased turbulence of one of the above solutions doesn't lead to absolutely complete dissolution of the entire mass of pus.

To check the value of high alkalinity activity of the solutions we have studied rheology thick pus under the influence of such famous alkaline drugs as solutions of 4% and 10% sodium bicarbonate (pH 8.0 and 8.0) and solutions of 2.4% and 24% aminofillin (pH 9.0 and 12.0, respectively). Our results showed that after 15 min of interaction of each of these solutions with thick pus, pus completely lost its viscosity and became liquid and very fluid. Solution of 10% sodium bicarbonate has the highest ability to dissolve thick pus.

To check the value of hyperturbulence and hyperthermal activity of the solutions we have studied rheology thick pus under the influence of solution of 4% sodium bicarbonate in conditions of continuous swinging tube containing equal amounts of interacting environments, 24, 37 and 42°C. Found that full pus liquefaction occurred at a temperature of 24 °C for 15 min, at a temperature of +37°C, through of 12.5 min, and at a temperature of 42 °C, 12 min of interaction.

Then we explored a range of specific weight, osmotic and acid activity of purulent masses obtained in patients with purulent peritonitis, purulent pleuritis, purulent conjunctivitis, purulent rhinitis and purulent abscesses. The obtained results showed that all the festering mass is a relatively heavy, isotonic and acidic biomass with a specific gravity within 1.030-1.040 g/cm3, with osmotic activity within 280-300 mOsmol/l of water and acid activity within the pH 5.8-6.2. Following this we defined specific weight of the liquids we used. They all have specific gravity of less than 1.30 g/cm3, except for the solution of 10% sodium chloride and solution 10% sodium bicarbonate. The extensive research has shown that the solutions of sodium bicarbonate in concentrations above 4%, have specific weight, exceeding 1.040 g/cm3 (meaning that they are "heavier" than the "heaviest" pus), pH 8.0 (that is, are alkaline) and osmotic activity above 450 mOsmol/l of water (i.e., are weak hypertonic solutions).

These data allowed to explain why purulent mass sink in water and in solutions with total concentration of ingredients of less than 3%, as well as to suggest that the festering mass will float up. Our studies have confirmed this assumption. Solution 4% of sodium bicarbonate is able to sink in purulent masses under the gravity force implemented in thickness of purulent masses.

Consequently, the high dense, high alkalinity and high osmotic activity, that is inherent in a solution of 4% sodium bicarbonate, ensures a high ability to dilute thick pus. In addition, as it is shown by our results, ability to dissolve thick pus of this solution can be enhanced by high temperature and high turbulence.

It is clear that the maximum permissible hyperthermia can be given a solution to its simple heating up to 42°C, but the maximum high turbulence needed to accelerate the process of liquefaction and dispersion of pus, cannot be achieved by «manual» jiggle the capacity of interacting environments from side to side. Therefore to give a solution the maximum possible turbulence we decided to increase the concentration and the pressure of gases, in particular, due to carbon dioxide similarly carbonated mineral water and due to the hydrogen peroxide. We hypothesized that the high saturation by gas of the solution 4% sodium bicarbonate will have a powerful aggressive action on pus, because warm, heavy, alkaline and high osmotic action of solution is to be able to intensively penetrate into a pus and high saturation solution by gas and rapid formation of bubbles of carbon dioxide within a pus is to be able to blasting it from the inside.

To verify this assumption, we initially added in a solution of 4% sodium bicarbonate carbon dioxide under excessive pressure of 0.2 ATM. Infusion of the solution at a temperature of 42 °C in a test tube with a festering mass withdrawn from the pleural cavity of a patient suffering purulent pleural empyema, led to the rapid formation of purulent foam and for the expulsion of tubes almost all purulent mass. Also intensively formed purulent foam was thrown out of the tube like a geyser. We found that high turbulence and high saturation by gas of warm solution of 4% sodium bicarbonate gave it an ability to effectively and safely dispose of pus out of purulent fistula when pancreatic necrosis.

It also found that high saturation by carbon dioxide of solution 0.9% sodium chloride provides visualization using ultrasound vector direction and speed of movement of streams of a solution into the abdominal cavity is closed when it is flushed with the conditions of purulent peritonitis by identifying and monitoring of the movement of bubbles of carbon dioxide. In addition, it is shown that the visualization of ultrasound move process of gas bubbles in a moving solution of 0.9% sodium chloride in the abdominal cavity allows you to monitor and change the flow of a fluid due to changes in the location of the patient's torso in space together with a cavity. The point is that changing the location of the torso and abdomen in space allows you to change the direction of fluid in the abdominal cavity, necessary to wash better chosen site. The ultrasound provides visualization move the liquid on the change of movement of gas bubbles.

Then instead of the carbon dioxide we added to a solution of 4% sodium bicarbonate 3% hydrogen peroxide. After this activity of dilute pus of this solution has been studied when heated to a temperature of 42°C and introduction in the tube with pus. It turned out that infusion into a test tube with a thick pus equal amount of warm solution of 4% sodium bicarbonate and 3% hydrogen peroxide at a temperature of 42°C leads to 5 min to complete transformation of two interacting environments in one turbid liquid with fluid properties.

In parallel we investigated changes in the state of the blood through visual observation and ultrasound after a single intravenous injection of aqueous isotonic isoalcalaine solutions of sodium chloride and sodium bicarbonate, which additionally contains hydrogen peroxide in various concentrations. We determined the ability of hydrogen peroxide to saturate venous blood with oxygen. The formula has been developed and we patented a new quickening tool that represented a hydrogen peroxide solution with osmotic activity 280 mosmol/l of water and pH 7.4. This tool is intended to be injected into the portion of the venous blood, in order to saturate it with oxygen immediately before intravenous injection into the blood stream of the patient and increase the effectiveness of preventing hypoxic damage to the cortex of his brain.

After that we explored the dynamics of color cotton-gauze swabs and skin, soaked venous blood in norm and after injection of solutions of drugs. We investigated decolorizing activity of aqueous solutions of various chemical and medicinal substances with their concentrations in the range of 0.05 - 0,008%, temperatures in the range +20 to +40°C and pH (alkalinity) in the pH range of 6.0 to 8.5. It is revealed that the increase in pH above 7.4 for, or heated above 37°C increases, and the decrease in pH below 7,4, or cooling below 37°C on the contrary reduces the ability of the solutions to discolor bloody tissue. We found that at concentrations less than 0.05%, hydrogen peroxide has the strongest decolorizing effect. We determined the optimal composition of ingredients bleach solution, which formed the basis of medicines called "Bleach bruises".

On the other hand, the physical and chemical characteristics of the tablets also affect the action of drugs when taking them inside. It turned out that the physical and chemical properties of modern tablets significantly differ in their physical and chemical properties of the tissues of the gastrointestinal tract. This difference is the reason that when the local interaction of the tablets have a local physical and chemical effect on the tissues. We found that all the tablets are very heavy (sink in water and gastric juice), many tablets are very solid, slowly decaying, very acidic and/or very salty. However, producers and consumers of drugs are not aware of this!

So today due to the ignorance of doctors and patients many pills damage the enamel of the teeth, established dental designs, localauth and burn the mucous membranes of the oral cavity, esophagus and stomach. A lot of iatrogenic pills cause such damage as medical caries, medication, stomatitis, esophagitis, gastritis and ulcer pyloric stomach.

At the same time, it is shown that the "correct" changes in the physical-chemical properties of the tablets can make them safe. In particular, we invented the "easy" (floating) tablet, which is a thick foam. This pill does not cause an ulcer in the pyloric stomach.

Our data show that the values of specific deforming pressure pills now occupy the range from 0.03 ± 0.0001 Newton/mm2 (tablets Xefocam (lornoksikama 4 mg, Nycomed), 160 ± 0.3 Newton/mm2 (tablets Ketorol, Dr.Reddy's, India). In other words, all tablets have different hardness. Moreover, the value of specific deforming pressure tablets can distinguish them from one another by 5000 times! That's why when being chewed, some tablets hurt the gums, lips, tongue, teeth, fillings, crowns, dentures, dental implants and braces. In this regard, we proposed to produce only soft tablets, and the value of specific deforming the hardness of the tablets should be included in a legitimate list of monitorable indicators for the quality of medicines.

It was found that rheology of liquid, viscous and dense biological tissues may improve medicines having the following physicochemical characteristics: hypertermia, high alkaline, high turbulence and high saturation by gas. We found that the leaders of improving rheology medicines and biological tissues are sodium bicarbonate, hydrogen peroxide and carbon dioxide, introduced in medicines similar carbonated beverages. The data allowed to develop a new hygienic medicines designed to liquefy thick purulent masses in patients with pleural empyema, peritonitis, rhinitis, sinusitis, conjunctivitis, tearful stones, osteomyelitis and sulfur tubes. New sanitary preparations are heated to 42°C aqueous solutions 0.5-10% sodium bicarbonate, 0,5%- 3% peroxide of hydrogen and carbon dioxide, which is entered into the solution at a pressure of 0.2 ATM.

A new page in the pharmacology of drugs associated with the presence of gases is opened: it is shown that the tablet is made in the form of a solid foam can float on the surface of the gastric juice, the solution representing the liquid foam can serve as an ultrasound contrast agent and the alkaline solution diluted hydrogen peroxide can whiten bruises in the skin.

4. Conclusion

Thus, the physical and chemical local interaction factors for drugs and medical devices are crucial to many diagnostic and therapeutic procedures outcome. High compliance of physical and chemical properties of pharmaceutical and medical products with the properties of human body tissues, which they are injected into, is the basis of safe medical technologies. Considerable incompliance of physical and chemical properties of the products with the properties of the tissues in contact is the reason for local irritant, inflammation and even subsequent necrosis. The severity of local damages and their development probability rise together with the increasing difference in factor criteria, interaction period and the local temperature.

It is shown that the physical and chemical characteristics of tablets and solutions of drugs can have a significant impact on the state of the tissues of the human body when the local interaction with them. In particular, sour and hot tablets and solutions can seal, cauterize and "weld" blood proteins, causing thereby denaturing (i.e. chemical) blood clotting, blood clots and blockage of the lumen of the vein and catheter. In turn, alkaline and warm medicines may soften, dilute and dissolve thick pus, blood clots, and dried blood spots. This improves the fluidity of pus and blood and retains for a long period of patency of the blood vessels, vascular catheters, fistulas and probes. On the other hand, cooling not acidic aqueous solutions of drugs can slow down the natural processes of inflammation and blood coagulation. This can cause local bleeding and infection of the injection site. And, finally, the enrichment of tablets and solutions gas up to the formation of these foam (respectively solid or liquid), heating and alkalization gives them their destructive capacity at the local interactions: medications can cause cold interstitial "boiling", "explode" thick and solid clots. The most amazing is that the saturation of medicines with various gases gives them such an incredible pharmacological properties, such as ability to saturate the blood with oxygen and bleach bruises!

References

[1] Chereshnev V A et al. 2008 Hyper-gazation and hyper-osmotic antiseptic. Invention RU Patent 2331441 C2

[2] Chereshnev V A et al. 2008 Method of peritoneal dialysis from fizzy solution. Invention RU Patent 2336833 C2

[3] Radzinsky V E, Urakov A L, Urakova N A 2014 The way obstetric aid efforts. Invention RU Patent 2502485 C2

[4] Reshetnikov A P et al. 2009 The method to express remove blood stains from clothes. Invention RU Patent 2371532 C2

[5] Strelkov N S et al. 2002 The method of treatment of long-term psychological wounds. Invention RU Patent 2187287 C2

[6] Urakov A L 1989 *Science and Life* ***9*** 38-42

[7] Urakov A L et al. 2009 Tool to thin thick and sticky pus. Invention RU Patent 2360685 C2

[8] Urakov A L et al. 2010 The way the cubital vein catheterization and multiple intravenous injection of drugs. Invention RU Patent 2387465 C2

[9] Urakov A L, Urakova N A, Yushkov B G, Zabokrickiy N A, Hausknecht M Y 2011 *Bullitin of Ural Academy Medical Sciences* **1 33** 84 - 87

[10] Urakov A L, Urakova N A, Kozlova T S 2011 *Bullitin of Ural Academy Medical Sciences* ***1 33*** 105 – 108

[11] Urakov A L et al. 2012 The way and means to remove sulphuric tube. Invention RU Patent 2468776 C2

[12] Urakov A L, Urakova N A 2012 After injection bruising, infiltrates, necrosis and abscesses can cause medicines because of the lack of control of their physical-chemical aggressiveness. *Modern Problems of Science and Education* ***5*** 5 - 7; URL: www.science-education.ru/105-6812

[13] Urakov A L, Urakova N A 2013 Original hygiene for prevention of postoperative adhesions, efficient liquefaction of thick purulent masses, sulfuric tubes and tear stones. *Modern Problems of Science and Education* ***1*** URL: http://www.science-education.ru/107-7607 (Date of access: 04.12.2013)

[14] Urakov A, Urakova N, Chernova L 2013 *Journal of Materials Science and Engineering* B ***3 11*** 714 – 720

[15] Urakov A L et al. 2014 Hyperoxygenation tool E. M. Soicher for saturation of venous blood with oxygen. Invention RU Patent 2538662 C2

[16] Urakov A L, Urakova N A 2014 *Thermology International* ***2*** 63 – 64.

[17] Urakov A, Urakova N, Kasatkin A, Chernova L 2014. *Journal of Chemistry and Chemical Engineering* ***8 01*** 61 – 65

[18] Urakov A L, Reshetnikov A P 2014 *Success of Modern Natural Science* ***9 2*** 33 - 37

[19] Urakov A L, Nikityuk D B, Urakova N A, Soicher M I, Soicher M G, Reshetnikov A P 2014 *Doctor* ***7*** 56 - 60

[20] Urakov A L, Kasatkin A A, Urakova T V *Ultrasound and Functional Diagnostics* ***4*** 61 – 65

[21] Urakov A L, Kasatkin A A, Urakova N A, Ammer K 2014 *Thermology International* ***24 1*** 5 – 10

[22] Urakov A L, Urakova N A, Chernova L V, Fischer E L 2014 Bleach bruises. Invention RU Patent 2539380 C2

[23] Urakova N A, Urakov A L 2013 *Success of Modern Natural Science* ***1*** 26 – 30

[24] Urakova N A, Urakov A L 2013 Injecting a skin disease. *Modern Problems of Science and Education* ***1*** 19 - 23; URL: http://www.science-education.ru/107-8171 (Date of access: 22.01.2013)

[25] Urakova N A, Urakov A L 2014 *Biomedical Engineering* ***48 3*** 111 – 115

The particle size distribution function in the composite films and microwave magnetic properties

L Kotov[1], V Ustyugov[1], V Vlasov[1], V Turkov[1], M Lasek[1], Yu Kalinin[2], A Sitnikov[2] and E Golubev[3]

[1]Syktyvkar State University, Syktyvkar, Russia
[2]Voronezh State Technical University, Voronezh, Russia,
[3]Komi Scientific Center of UB RAS, Syktyvkar, Russia

E-mail: kotovln@mail.ru

Abstract. Nano- and microstructure of $\{(Co_4\text{-}Fe_4\text{-}Zr_{0.7})_x+(Al_2O_3)_{1-x}\}$ (A1 series) and $\{(Co_1\text{-}Nb_{0.2}\text{-}Ta_{0.05})_x+(SiO_2)_{1-x}\}$ (A2 series) films was investigated by atomic force microscopy. The distributions of the metallic granules effective size by various concentrations of the metal phase were obtained. Microwave magnetic characteristics of composite films in respect of obtained distributions were calculated.

1. Introduction

The microwave magnetic and magneto-optical characteristics of magnetic composite films are currently intensively investigated. Such films have unusual properties, such as giant magneto resistance, the anomalous Hall effect, and others [1, 2]. The films have a highly heterogeneous structure defined by the ratio of the concentration of a metal and a dielectric in a film. This fact complicates the explanation of the dependency of the microwave magnetic characteristics of the films from the external magnetic field and structure. Determination of structural and magnetic properties of the film granules makes it possible to construct theoretical models describing its magnetic properties.

2. Preparing of films and investigation methods

The investigated films were prepared by ion beam sputtering on the polycrystalline glass substrate. For film of A1 series used a composite CoFeZr target with size 280x80x10 mm. On the target surface had fixed 12 batched of Al_2O_3 with size 80x10x1 mm. Films of A2 series were obtained with using a composite CoNbTa target and dielectric Zr_2O_3. The uneven arrangement of the polycrystalline glass plates of silicon oxideon the target surface allow to obtain the continuous concentration change of the metallic and insulating phases of the composite. Deposition time was 120 minutes. Time of cleaning the surface by ion beam was 30 minutes. The total pressure in the chamber before the depositon and after purification of polycrystalline glass substrate was $P = 2.4 \times 10^{-4}$ Torr. Nitrogen pressure in chamber was $P = 7.8 \times 10^{-5}$ Torr.

The chemical composition and thickness of the films were determined using a scanning electron microscope JSM-6400. Used metal alloys have a eutectic composition, and one could expect that the structure of the metallic phases on the film highly likely to be amorphous.

Films were annealed at a temperature of 823 K in a normal atmosphere. Determination of relief and phase contrast surface films before and after annealing was performed on an atomic force microscope (AFM) ARIS-3500.

3. Size distributions of particles

Consider the features of images obtained using an AFM. Topology and phase contrast surface are shown in Fig. 2a-b for A1-series of films with a metal concentration of x = 0.54 (Fig. 1a, b) and A2-series with x = 0.51 (Fig. 1c, d). It can be seen that the films have a non-uniform surface.

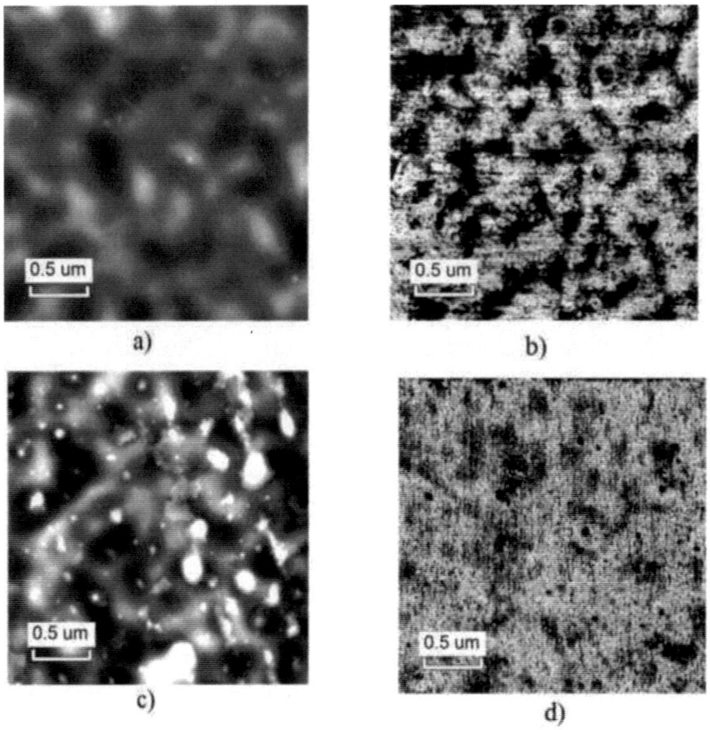

Figure 1. AFM Images of topology and phase contrast of A1 and A2 series composite films

Using the software Gwyddion were built distribution function of metal pellets on the effective size (granules approximated by an ellipsoid of rotation) at different concentrations of the metal phase. Distribution curves of the effective dimensions (radius of ellipsoid on the film plane) of the granules are shown in Fig. 2. On the abscissa in Figure 2 shows effective average size of the granules obtained from the AFM topographic images, and on the ordinate their relative concentrations. It can be seen that increasing of the concentration of the metal alloy the relative amount of small particlesize decreases and the number of large particles increases. With increasing concentration of the metal phase x the size and shape of a magnetic granules increases. Changing the shape of the granules leads to the change of the granules demagnetizing fields and respectively, the magnetic characteristics of the films changes.

Annealing of the composite films reduces the heterogeneity of the surface film due to fusion of small areas of metal and dielectric to a larger structures form. This decreases the relative fraction of particles with a small effective size (Figure 3), the particles are formed of a large amount exceeding the maximum particle size observed in the films, not annealed.

The knowledge of the film thickness and the effective radiuses of metal magnetic areas can define their distribution function of field demagnetization value.

The films were cut along the concentration gradient of the metal phase into samples sized 2.5x5 mm^2. Magnetic resonance characteristics of the films at room temperature were determined by EPR-microwave spectrometer at a frequency of 9.36 GHz of rf field.

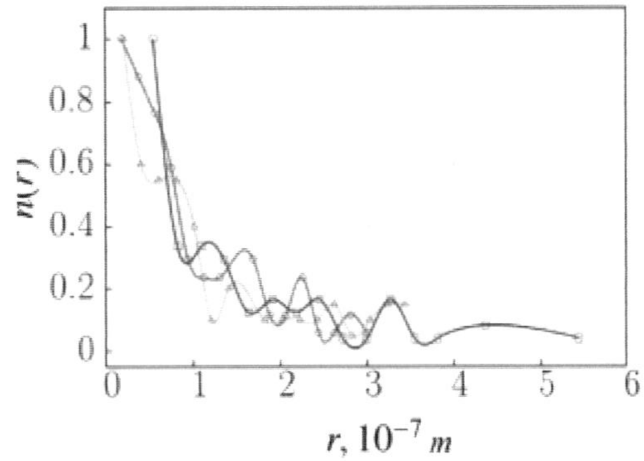

Figure 2. Particle size distribution of A1 film for concentrations of magnetic phase x = 0.45 (red line), 0.51 (green line), 0.57 (blue line)

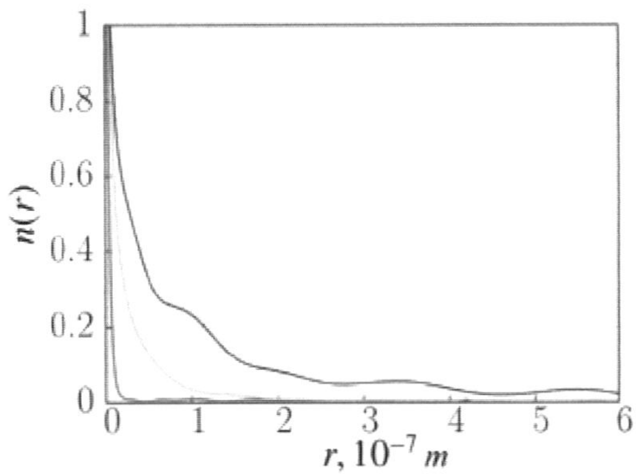

Figure 3. Particle size distribution in A1 film without allealing (red line) and after annealing by T=700 K (green line), T=843 (blue line)

Dependence of the resonance absorption line width on dc magnetic field is shown in Figure 4 (circles). Fig.4 shows that increasing the concentration of the metal magnetic phase x = 0.3/0.34 results in first increasing of the resonance line width and then decreasing after reaching the maximum. The increasing of the line width with x increasing is associated with a random distribution of granules in volume of the film and granules size difference, leading to the presence of various small local demagnetizing fields and weak overlap of neighboring granules (the presence of superparamagnetic state). In small x area one of the contributions can be estimated according to the equation:

$$\Delta H_{por} = 1.5(4\pi M_s)\left(\frac{\lambda}{1+\lambda}\right), \tag{1}$$

where M_S - the average magnetization of the film saturation, λ- the material porosity. This contribution is shown in Figure 4 by the curve at low concentrations of magnetic phase. You can see that this dependence is qualitatively consistent with the experimental results.

Reduction in the resonance line width at concentrations of the metal phase exceeding the percolation threshold can be explained by "exchange narrowing". Kurtosis coefficient was chosen as a measure of the resonance line width. For a polycrystalline sample of the spherical shape the degree of resonance peak sharpness is proportional to the square of the magnetization:

$$E_x \propto 0.12\left(\frac{4\pi M_s}{H_{an}}\right)^2 - 3, \tag{2}$$

where M_S - the magnetization of the sample, H_{an} - induced anisotropy field.

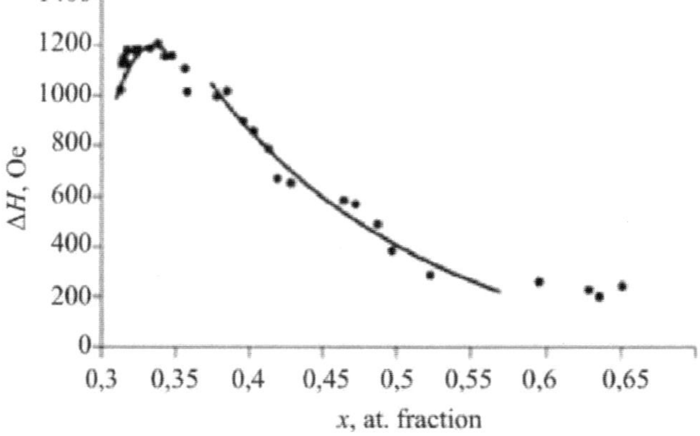

Figure 4. Resonance absorption line width of A1 films

You can see that under the condition of magnetic phase high concentrations the line narrows. To take into account the exchange interaction we need add to the right side of equation (2) the component proportional to

$$\frac{1}{r_0}\left(\frac{\omega_{ex}}{\gamma H_{an}}\right)^2, \tag{3}$$

where r_0 - the average grain size, $\omega_{ex} = 2E_{ex}/\hbar$ - the exchange frequency, E_{ex} - the exchange interaction energy.

To calculate the width of the line, you can use the experimental dependence of the FMR linewidth of the resistivity and magnetization of the film:

$$\Delta H \propto \frac{R_c}{M_s}, \tag{4}$$

where R_c — the resistivity of the film. Concentration dependence of the FMR linewidth calculated by the formula (4) is shown in Figure 4.

4. Theoretical calculations

For the theoretical calculation of the resonance fields we use the Dubowik model. Let the plane of the composite film coincides with the plane XY, magnetic bias field is directed along the axis Z. The particles included in the dielectric matrix is approximated by ellipsoids of rotation with the demagnetization factor $\frac{N_x}{4\pi} = \frac{N_y}{4\pi} = N_\parallel$ and $\frac{N_z}{4\pi} = N_\perp$. The values of the demagnetizing factors are determined by the known formulas Osborn based on the data obtained from the AFM images of films.

We write the free energy of an ensemble of particles in the following form:

$$\mathrm{w} = -\mathrm{M}\cdot\mathrm{H} + \frac{1}{2}\mathrm{f}^2\mathrm{M}\mathrm{N}_{\mathrm{film}}\mathrm{M} + \frac{1}{2}\mathrm{f}(1-\mathrm{f})\mathrm{M}\mathrm{N}_{\mathrm{part}}\mathrm{M} \tag{5}$$

where f – the volume concentration of the metal phase in the composite, N_{film} – demagnetizing factor of film, N_{part} - demagnetizing factor of one particle. To find the ferromagnetic resonance equation for described geometry of the problem the polar axis OX was chosen. In this case, the free energy of an ensemble of particles can be written as:

$$w = -H_0 M_z f \sin\varphi \cdot \sin\theta + \frac{1}{2} M_S^2 f^2 \sin^2\varphi \cdot \sin^2\vartheta + \frac{1}{2} M_S^2 f(1-f)(N_p(\cos^2\varphi \cdot \sin^2\vartheta + \cos^2\vartheta) + N_\perp \sin^2\varphi \cdot \sin^2\varphi$$

Q=(f-(1-f)($N_\parallel - N_\perp$)), then the equilibrium conditions of the magnetization vector equations can be written:

$$\varphi_0 = \frac{\pi}{2}, \qquad \sin\vartheta_0 = \frac{H_0}{4\pi M_S Q}$$

when $H_0 < 4\pi M_S Q$ and

$$\varphi_0 = \frac{\pi}{2}, \vartheta_0 = 0$$

when $H_0 \geq 4\pi M_S Q$.

Using Smith-Beljers equation we can obtain:

$$\left(\frac{\omega_{res}}{\gamma}\right)^2 = H_0^2(2f^2 + 3f + 1) - (4\pi M_S)^2 Q^2 (f^2 + f)$$

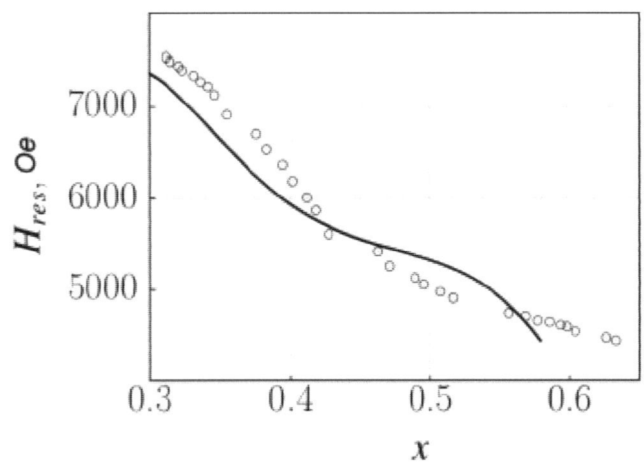

Figure 5. Resonance field of A1 films

Let compare values of the resonance fields obtained from the formula (9) with the experimental data. From Fig. 5 can be seen that the chosen model gives a satisfactory result in the concentration range f = 0.3 ^ 0.6 . The discrepancy between theory and experiments in the area of low concentrations of magnetic metal due to the limitation of the scope of the selected model, which consists in the fact that the relation (5) adequately describes the ensemble of identical and regularly spaced particles. In real films at low concentrations of metals the particle size and shape widely changes, as well as a random location in the volume of the film. At high concentrations of metal f>0.6, the formation of adhesions complex shape of the particles also remains outside the scope of applicability of the model.

5. Acknowledgements
This work was supported by RFBR (grant 13-02-01401-a).

6. References

[1] Kotov, L. 2008 *Advanced Materials Research* **47** 706
[2] Kotov, L. et al 2012 *Journal of Nanoscience and Nanotechnology* **12** 1696

IOP Conf. Series: Materials Science and Engineering **123** (2016) 012010 doi:10.1088/1757-899X/123/1/012010

Explosive Surface Hardening of Austenitic Stainless Steel

T Kovacs-Coskun

Obuda University, Banki Donat Faculty of Mechanical and Safety Engineering - Material Science Department - 1081 Budapest - Hungary

E-mail: kovacs.tunde@bgk.uni-obuda.hu

Abstract. In this study, the effects of explosion hardening on the microstructure and the hardness of austenitic stainless steel have been studied. The optimum explosion hardening technology of austenitic stainless steel was researched. In case of the explosive hardening used new idea mean indirect hardening setup. Austenitic stainless steels have high plasticity and can be easily cold formed. However, during cold processing the hardening phenomena always occurs. Upon the explosion impact, the deformation mechanism indicates a plastic deformation and this deformation induces a phase transformation (martensite). The explosion hardening enhances the mechanical properties of the material, includes the wear resistance and hardness. In case of indirect hardening as function of the setup parameters specifically the flayer plate position the hardening increased differently. It was find a relationship between the explosion hardening setup and the hardening level.

1. Introduction

Shock hardening is a very useful and common technology. Explosive hardening of railway frogs from Hadfield steel (Mn steel) is a common technology in the world, which allows to increases a surface and subsurface hardness of frog [2]. This hardening technology is also able to increase the hardening and wear resistance of the austenitic stainless steel. This steel has a great ductility, low hardness and very good corrosion resistance. It can't increase the hardness by the way of simple heat treating. Cold working and aging heat treatment involve hardening in case of this steel. That static strain aging is well-known phenomenon frequently observed in bcc metals and alloys [3].

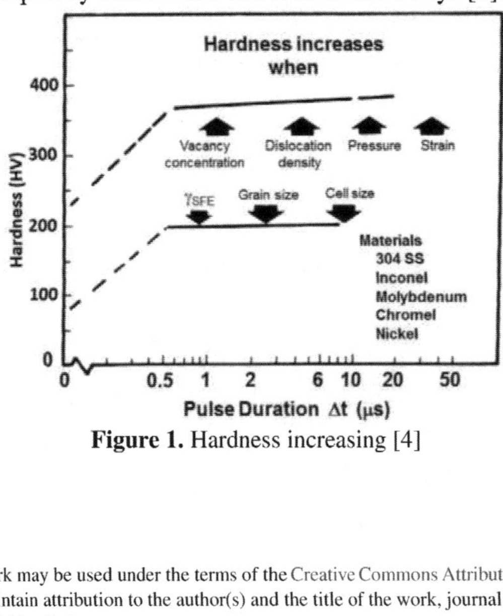

Figure 1. Hardness increasing [4]

It knows that the explosion caused shock also occurs a hardness rising but the parameter of this process not well understood yet. The effect of strain rate on the Y-a' transformation in stainless steels has been of interest for a number of years. The early work simply noted that an increase in rate decreased the amount of martensite [4]. The aim of this study is to report the results of the hardness improving of an austenitic stainless steel treated using explosive treatment. Among the treatments intended to improve the surface properties of materials, shocks are known to induce an important hardening, either by flyer plate impact [5, 6].

2. Materials and explosive hardening setup

2.1. Used materials

In case of our tests we used austenitic stainless steel (X5CrNi1810, EN 1.4301, AISI 304) and like flayer plate an unalloyed low carbon steel the chemical composition of them is given in Table 1. The flayer plate material was unalloyed low carbon steel (S235JR, EN 1.0037). The Table 1. shows the basic parameters of the used metals.

Table 1. Chemical composition of the used steels (at.%)

DIN;	C	Si	Mn	Cr	Ni	N	P	S	Al	Cu
X5CrNi1810	G,G7	1	2	1S,25	9,25	G,11	G,G45	G,G15		
S235J2	<0,2	<0,6	<1,4	<0,3	<0,4		<0,045	<0,045	<0,1	<0,3

Table 2. Basic performance of the steels

	Sign by DIN	Thickness	Size	HV30	Yield stress (MPa)
Flayer plate	S235J2	1,5 mm	SGxSG mm	3GG HV	235
Base plate	X5CrNiISIG	5 mm	DSGx4G mm	215 HV	220

2.2. Explosive hardening parameters

The used explosive hardening technology was a new setup. The base plate and the flayer plate are parallel and the explosive find directly on the surface of the flayer plate without buffer see Figure 1. [5]. In case of this setup the flayer plate worked like hammer. The surface of the austenitic steel had thin plastic coating, to prevent the joining of the flayer plate and the base plate.

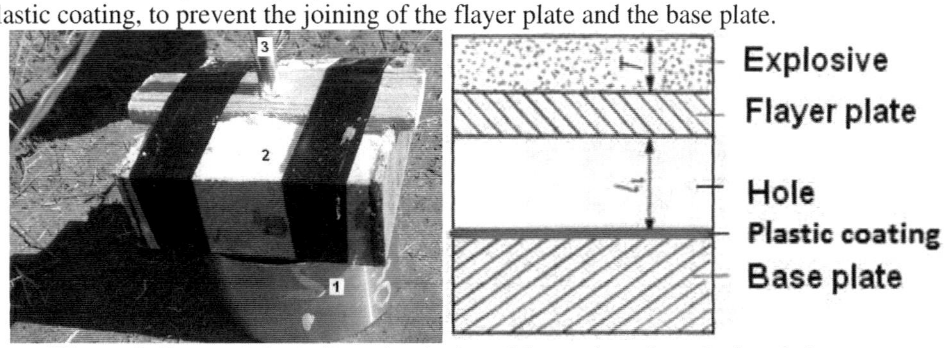

Figure 2. Setup of the indirect hardening (1 base plate, 2 explosive, 3 detonator)

Table 3. Parameters of used explosive

Explosive PERMON 10T (powder)			
Volume of the gas	92S dm^3/kg	Distance between the	1,5 mm
Detonation rate	32GG m/s	Thickness of the explosive	3G mm
Density	S5G kg/m^3	Weight	319 g

The pressure of the nascent gases, calculate by the equation (1) [7]:

$$p = v_d^2 \rho_0 \frac{(\rho - \rho_0)}{\rho} \tag{1}$$

Where:

v_d: detonation velocity of explosive [m/s];
ρ_0: density of explosive [kg/m^3];
ρ: density of the nascent gases [kg/dm^3];
p about 10^9 [Pa].

The explosion impact force on the surface (2) [7]:

$$J = \int p\,dt \tag{2}$$

Where:

J: impulse on the surface unit [N/m^2];
p: nascent gases pressure from the equation (1) [Pa].

The p pressure quantity depends on the parameters of explosive material and the effect time depends on the amount (thickness) of the explosive material. The velocity of the collision (v_c) must be lower than the speed of the sound (v_x) (1), that mean it need to use for this technology a low speed explosive. The interfacial pressure at the collision front also must exceed of the materials yield strength to occur a plastic deformation. This is the surface hardening under extreme pressure [6,7].

$$\frac{v_c}{v_s} < 1 \tag{3}$$

In case of the setup parameters optimization it was used some empirical parameters with the density of explosive, base plate and flayer plate (setup see in Figure 2). The thickness of the explosive powder was optimized on base of practice. It knows that it needs a minimal amount of explosive, that about 0,017 (g/mm^2) Permont 10T [7].

The used parameters in case of the setup posed by (2) when l_b is the thickness of the flayer plate and ;! is the distance (hole) between the base and flayer plate. The collision velocity depends on this distance [4]. The hardness increasing depends on the pressure of the nascent gases (see Figure 3). The explosion kinetic shows the Figure 4. It can see that the nascent pressure in the first and second period increase us function the time and in the third period the pressure is constant.

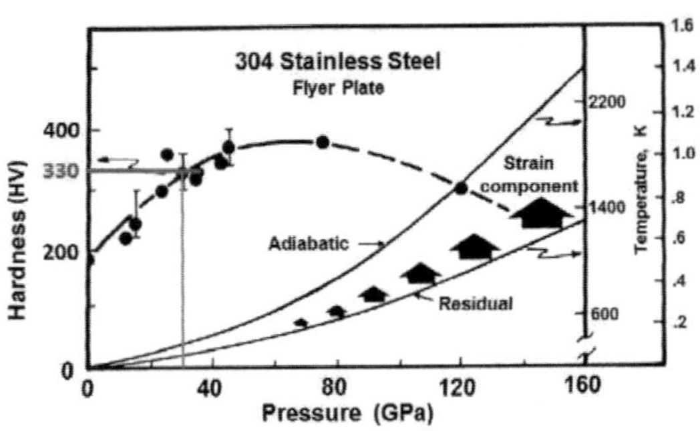

Figure 3. The hardness as function the pressure of nascent gases [6]

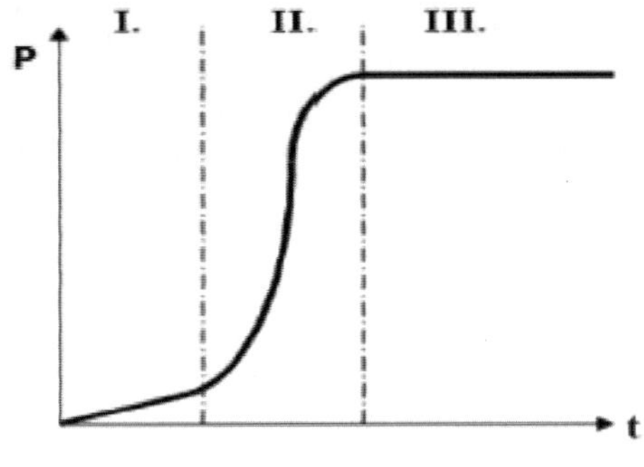

Figure 4. The kinetic of the detonation (I.: burning period, II.: explosion period, III. detonation period) [7]

3. Results and Discussion

3.1. Hardness testing

The used tester was in case of experiments Vickers Hardness tester (30 kg). Results are shows Table 4. Cause of the plastic deformation the hardness increased.

Table 4. Hardness after explosive hardening

Hole size	Distance from detonator (mm)				
	0mm	10mm	20mm	30mm	40mm
	HV30				
2mm	222	217,5	234	233	322,5
3mm	217	222	254,5	254,5	330
4mm	243	227,5	245	251	334,5

The Figure 5 and Figure 6 show the hardness in case of different setup parameters and distance from detonator.

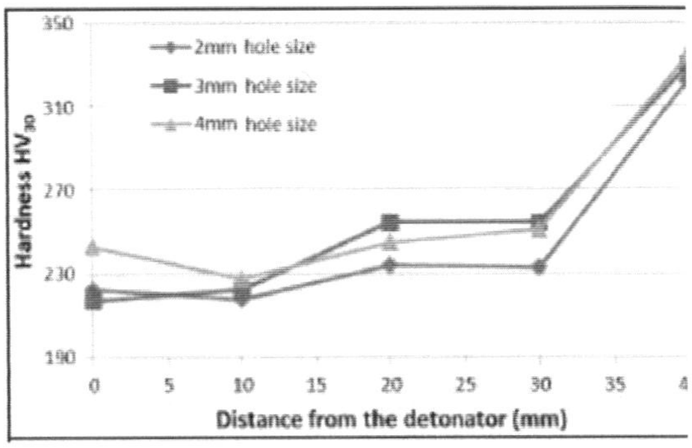

Figure 5. Hardness as function the pressure of nascent gases

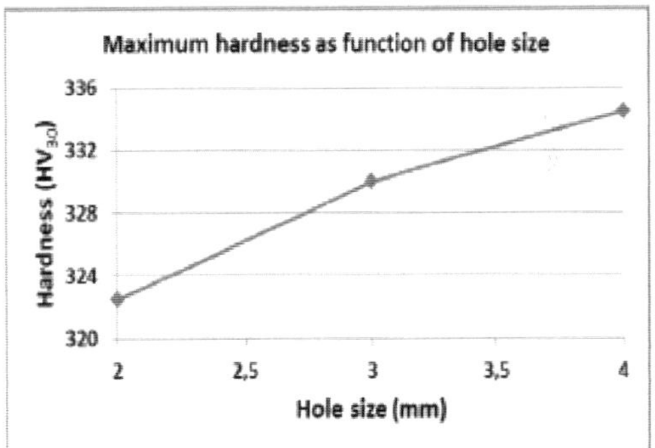

Figure 6. Hardness as function the setup (hole size)

3.2. Wear test

The wear resistance of the hardened samples were tested by a ball cratering tribometer in case of standard parameters (load, rotation per minutes, investigation time, etc.). Wear coefficient calculated by the fallowing Archard equation (4) [8] :

$$K = \frac{C_K^2}{2 \cdot N_C \cdot f} \quad [m^2/N] \tag{4}$$

Table 5. Wear test results

Setup Hole size	Crater depth h (^m)	Wear lost volume V (mm^3)	Wear coefficient K (m^3/Nm)	Wear resistance (1/K)	Hardness (HV30)
2 mm	0,22	$8{,}13*10^{-9}$	$8{,}93*10^{-20}$	$1{,}11*10+^{19}$	322,5
3 mm	0,21	$7{,}54*10^{-9}$	$8{,}29*10^{-20}$	$1{,}20*10+^{19}$	330
4 mm	0,16	$5{,}76*10^{-9}$	$6{,}33*10^{-20}$	$1{,}58*10+^{19}$	334,5

3.3. Discussion

The indirect explosive hardening technology improves the hardness of the used austenitic stainless steel. Hardness increasing depends on the setup of hardening technology and also depends on the distance from the detonator (see Figure 5-6). The wear resistance shows good correlation with hardness and also with the holes sizes of the setup. It finds relationship between the setup, hardness and wear resistance.

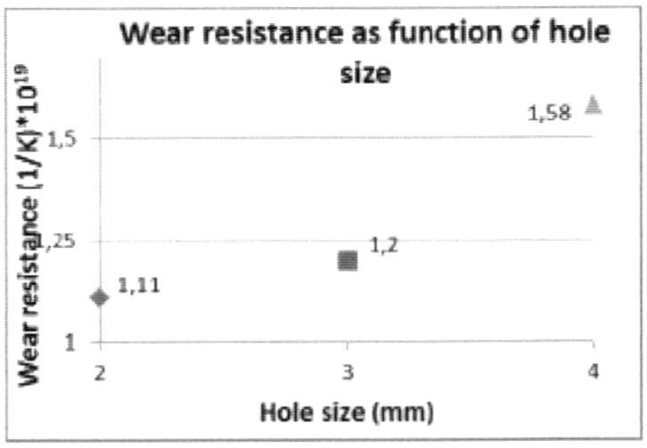

Figure 7. Wear resistance

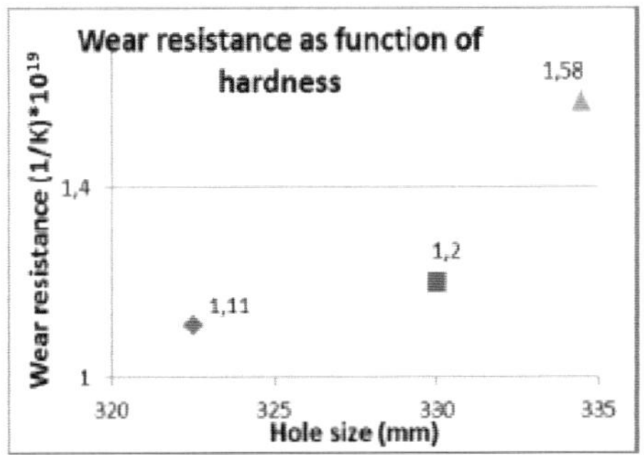

Figure 8. Hardness

4. Conclusion

The used explosion hardening technology results give new information even that the hardness shows good correlation with wear resistance is knows.

I. The hardness increase us function of the setup holes sizes because the result hardness depends on the plastic deformation rate and the plastic deformation depend on the collision energy of the flayer plate.
II. The result hardness also depends on the distance from the detonator, because during the explosion the nascent gases pressure increase by the time (in the first and second period).
III. In case of explosive hardening the used parameters base some empirical equation what are usually secret. The literature of this process is also poor about the determination of the parameters. On the base of the results in case of indirect hardening setup suggestible the biggest holes setup.

Acknowledgements

The author would like to acknowledge all members of this project and especially for Andras Szalay S-Metalltech 98 Kft., Tamas Rigo OE and Laszlo Lukacs NKE.

References

[1] J.R. Davis: 2002 *Surface Hardening of Steels,* ASM International pp 1-16
[2] Petr Havlícek, Katerina Busóvá: 2012 *Experience with explosive hardening of railway frogs from* hadfield steel, *Metal 2012. Brno, Czech Republic*
[3] Sang Hun Leel, Jeom Yong Choi, Won Jong Nam: 2009 *Materials Transactions* **50** (4) The Japan Institute of Metals pp 926-929
[4] M. A. Meyers, L. E. Murr: 1980 *Shock Waves and High-Strain-Rate Phenomena in Metals,* International *Conference on Metallurgical Effects of High-Strain-Rate Deformation and Fabrication, Albuquerque, N.M., pp 91-111*
[5] L. Fouilland-Paill, M. Gerland, P. Violan: 1995 *Materials Science and Engineering A* **201** 32
[6] K.P. Staudhammer, C.E. Frantz and S.S. Hecker, in M.A. Meyers and L.E. Murr (eds.):1981 *Shock Waves* and High Strain Rate Phenomena in Metals, *Plenum, New York, pp 91-112*
[7] Dr. Göbl Nándor - Horváth Dániel - Dr. Kovács-Coskun Tünde - Prof. Dr. Lukács László - Dr. Rácz Pál Szalay András - Dr. Zádor István: 2013 Large eneregy metal processing, Budapest (in Hungarian)
[8] T. Kovács 2007 The effect of microstructure on the local wear behaviour of steels, *Ph.D. Thesis,* Budapest

Characterization of ball-milled carbon nanotube dispersed aluminum mixed powders

M A Maleque, U Abdullah, I Yaacob and Y Ali

Department of Manufacturing and Materials Engineering-Kulliyyah of Engineering, International Islamic University Malaysia, Kuala Lumpur- Malaysia

E-mail: maleque@iium.edu.my

Abstract. Currently, carbon nanotube (CNT) is attracting much interest as fibrous materials for reinforcing aluminum matrix composites due to unique properties, such as high strength, elastic modulus, flexibility and high aspect ratios. However, the quality of the dispersion is the major concerning factor which determines the homogeneity of the enhanced mechanical and tribological properties of the composite. This work study and characterized carbon nanotube dispersion in ballmilled CNT-aluminum mixed powders with four different formulations such as 1, 1.5, 2 and 2.5 wt% CNT under high energy planetary ball milling operations. The ball milling was performed for two hours at constant milling speed of 250 rpm under controlled atmosphere. The characterization is performed using FESEM and EDX analyzer for mapping, elemental and line analysis. The experimental results showed homogeneous dispersion of CNTs in aluminum matrix. The composite mixture showed similar pattern from mapping, elemental and line analysis. Identification of only two peaks proved that controlled atmosphere during milling prevented the formation of inter metallic compounds such as aluminum carbide in the composite mixture. Therefore, this CNT-Al composite powder mixture can be used for new nano-composite development without any agglomeration problem.

1. Introduction

Al matrix composites have wide prospects of application in aviation, spaceflight and automobile industries because of lower density (a requirement necessary for the weight reduction for many components thereby saving fuels and hence energy). Research in the field of carbon was revolutionized by the discovery of carbon nanotubes (CNTs) by Iijima in 1991 [1]. Although CNTs might have been synthesised in 1960 by Bacon, it took the genius of Iijima to realise that they are tubes made by rolling a graphene sheet onto itself. A multiwalled carbon nanotube (MWCNT) is made up of many single walled carbon nanotubes (SWCNT) arranged in a concentric manner. Experiments and simulations showed that CNTs have extraordinary mechanical properties over carbon fibers, e.g. stiffness up to 1000 GPa, strength of the order of 100 GPa [2], and thermal conductivity of up to 6000 W mK. Carbon nanotube reinforced metal matrix composites (CNT-MMC) are prepared using different type of processing techniques. Powder metallurgy (PM) is the most popular and widely used route by the researchers to synthesize CNT-MMC materials. Carbon nanotube reinforced aluminium (CNT-Al) composites are mainly produced using powder metallurgy method. Ball milling is the easier and cheaper means of synthesization of composite materials using powder metallurgy route. Nano composite materials are multiphase materials obtained through the artificial combination of different materials either ceramic, polymeric and metallic matrices with carbon nanotubes to attain properties that the individual components by themselves cannot attain. It is believed that graphitic type of carbon fibre enhances the thermal conductivity, mechanical and tribological property of the material significantly.

However, successful dispersion of CNTs in the metallic matrices is of paramount important before achieving any sort of significant benefits in terms of the mentioned property of the composite [3]. Some work covered ball milling of a mixture of CNT and aluminum powder to break down the CNT clusters and the authors limited the milling speed to 200 rpm and the milling time to 5 min. Morsi and Esawi [4, 5], investigated the effect of CNT wt% (2 and 5 wt%) and the influence of milling time and process control agents on the size and morphology of the mechanically alloyed powder. The results show that particle size and morphology vary with milling time and CNT content and also the addition of methanol as process control agents can aid in controlling the powder characteristics. This paper studied and analyzed carbon nanotube reinforced aluminum nano-composite prepared using powder metallurgy route under high energy planetary ball milling operations. The CNTs dispersion in aluminum matrix were analyzed by energy dispersion spectroscopy (EDX) using energy dispersive X-ray analyzer.

2. Materials and method

Pure Al (99.7%), with particle size of 78 μm which has nearly spherical shape with some satellite subparticles was used as a matrix material. The image of the aluminium powder obtained using scanning electron microscopy (SEM) is showed in Fig.1a. The multi walled carbon nano tubes (MWCNTs) with a nominal diameter of 10 nm, length of 5-15 μm, and surface area of 40-300 m2g-1 was used as a reinforcement. The image of the MWCNTs obtained through field emission scanning electron microscopy (FESEM) is shown in Fig. 1b. Three compositions such as 1.5, 2 and 2.5 wt% CNT with the balance in each case being Al were studied in this investigation. Each composition was place in a tube together with a stainless steel ball of 10 mm diameter to make the ratio of ball-to-powder ratio of 5:1 for preliminary mixing via manual shaking for about 10-15 minutes. The preliminary mixture was then placed in 250 mL stainless steel mixing jars containing stainless steel milling balls of 10 mm diameter, the initial ball-to-powder weight ratio (BPR) is 10:1. The jar was filled with argon gas and the ball mill operations started using planetary mill (FRITSCH pulverisette 5 05.5000/00409) at constant speed of 250 rpm for tree different milling time of 1, 2 and 3hours. The CNTs dispersion were analysed by mapping, elemental and line analysis using EDX analyzer.

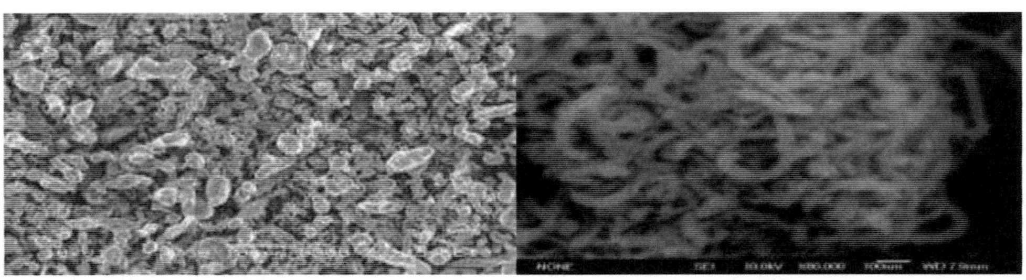

a) b)

Figure 1. Image of the as-received raw materials: (a) aluminum powder and (b) CNT

3. Result and Discussion

The investigation on the uniform and homogeneous dispersion of CNT into the Al matrix was conducted by elemental mapping using FESEM-EDX analyser. Fig. 2 presented mapping results of 1 and 1.5 wt% CNT and Fig. 3 for 2 and 2.5 wt% CNT in CNT-Al mixed powders. The data obtained from mapping result followed the same trend for all the CNT compositions and showed the presence of CNT over the entire surface of Al particles. The scanned image was mapped to study the possible elemental composition of the scanned portion.

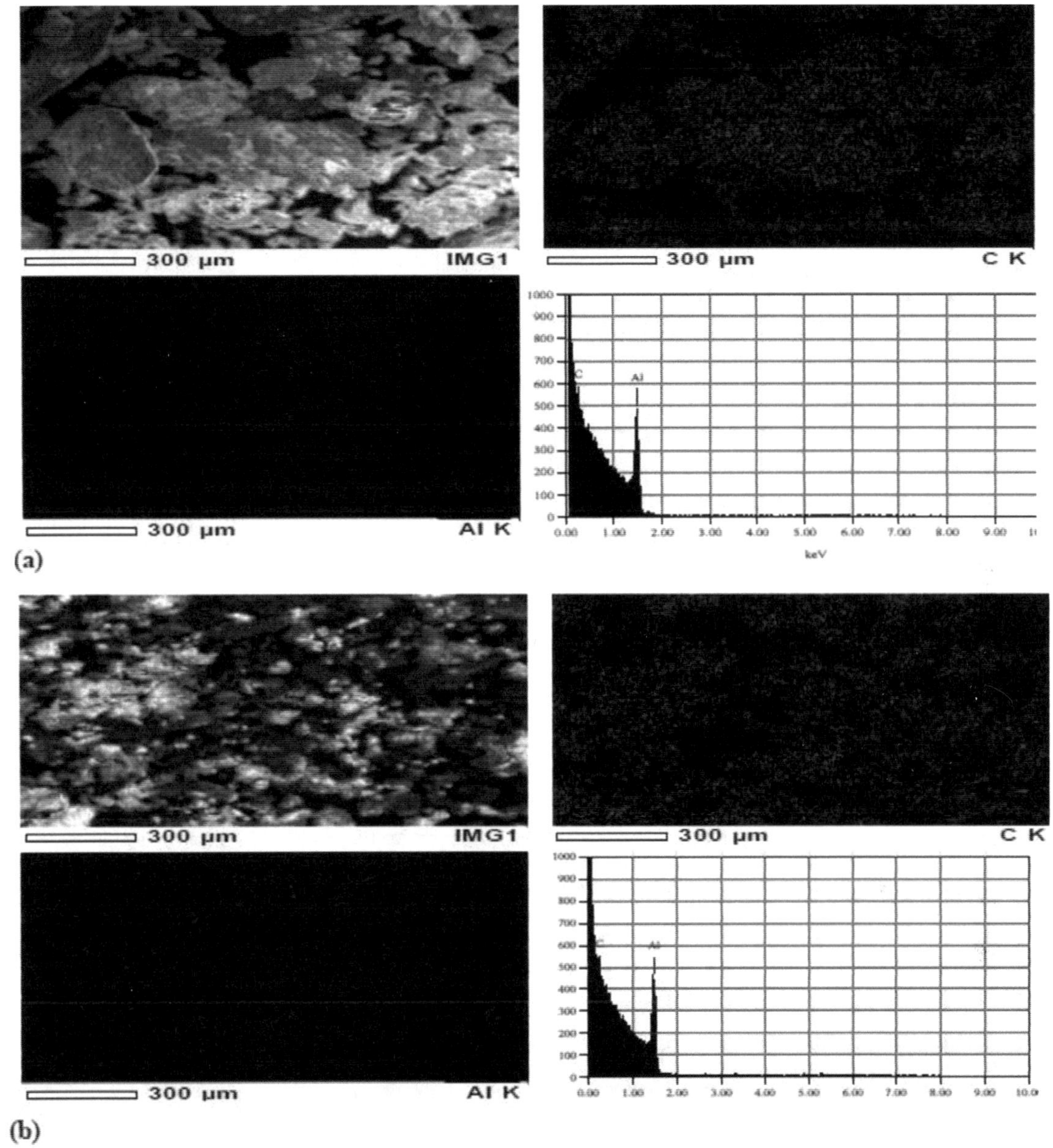

Figure 2. Mapping and elemental analysis of: (a) 1 wt% and (b) 1.5 wt% CNT-Al mixed powders. Ball milling parameters: 250 rpm speed and 2 hrs milling time

From the mapping results, two elements were observed such as aluminium (Al) and CNT but CNT is represented by the symbol (C) and the map proved that there is no formation of compound such as carbides, due to the fact that only two elements were observed and the reinforcement material CNT were well dispersed in the aluminium matrix. Two different maps were shown with two different colours for easy identification of the found materials which was indicated at the extreme right of each map. The analysis graph was within the energy range of 0-20 keV.

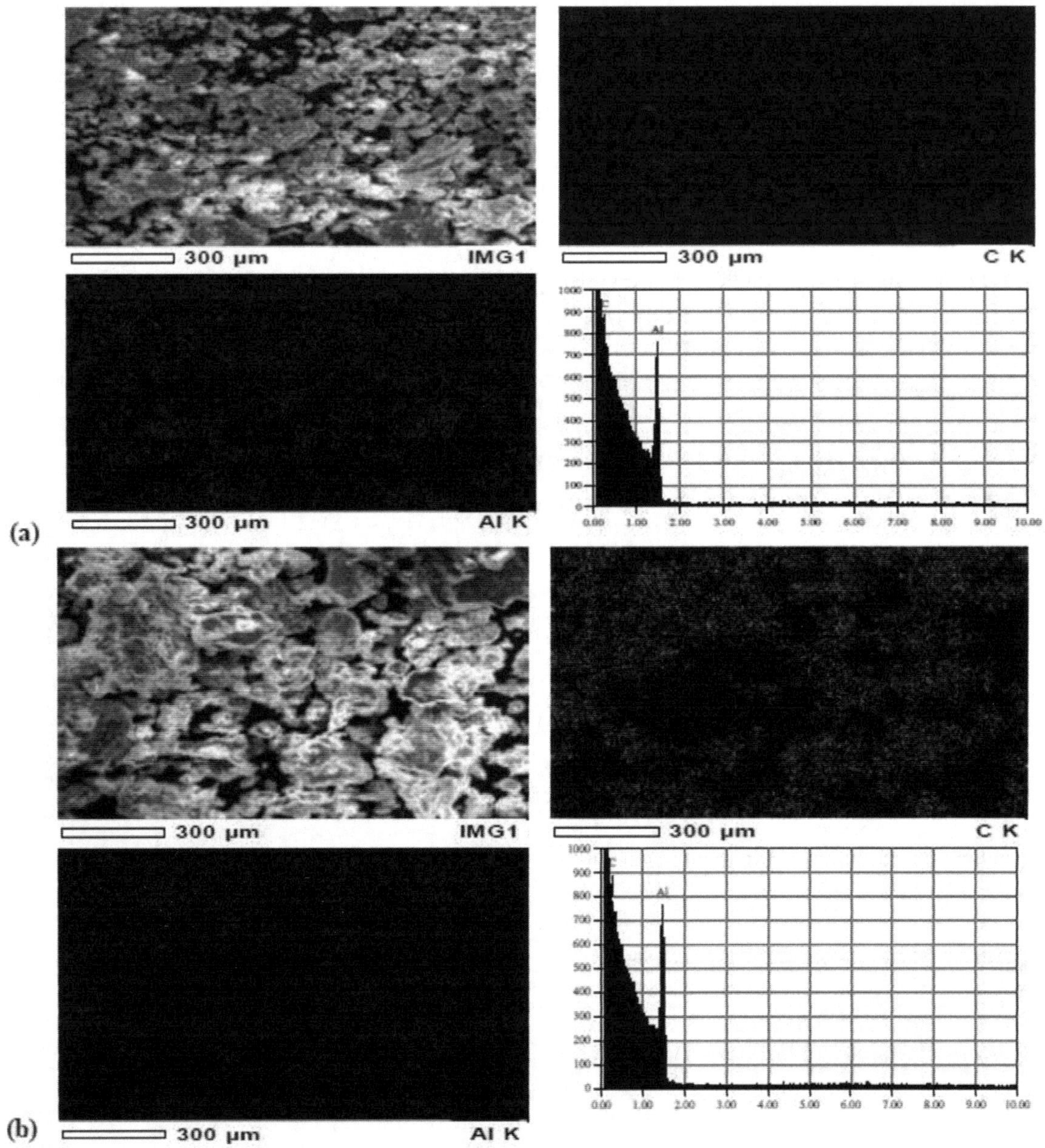

Figure 3. Mapping and elemental analysis of: (a) 2 wt% and (b) 2.5 wt% CNT-Al mixed powders. Ball milling parameters: 250 rpm speed and 2 hrs milling time

Line analysis of the CNT-Al mixed powders is presented in Fig. 4 and showed the effective dispersion of CNTs into the Al matrix during milling operation of mixed and blended CNT-Al powder at a speed of 250 rpm for 2 hrs of milling. The line analysis results were observed at the line intensity of between 0-45 and the distance within the range of 0-1.33 mm.

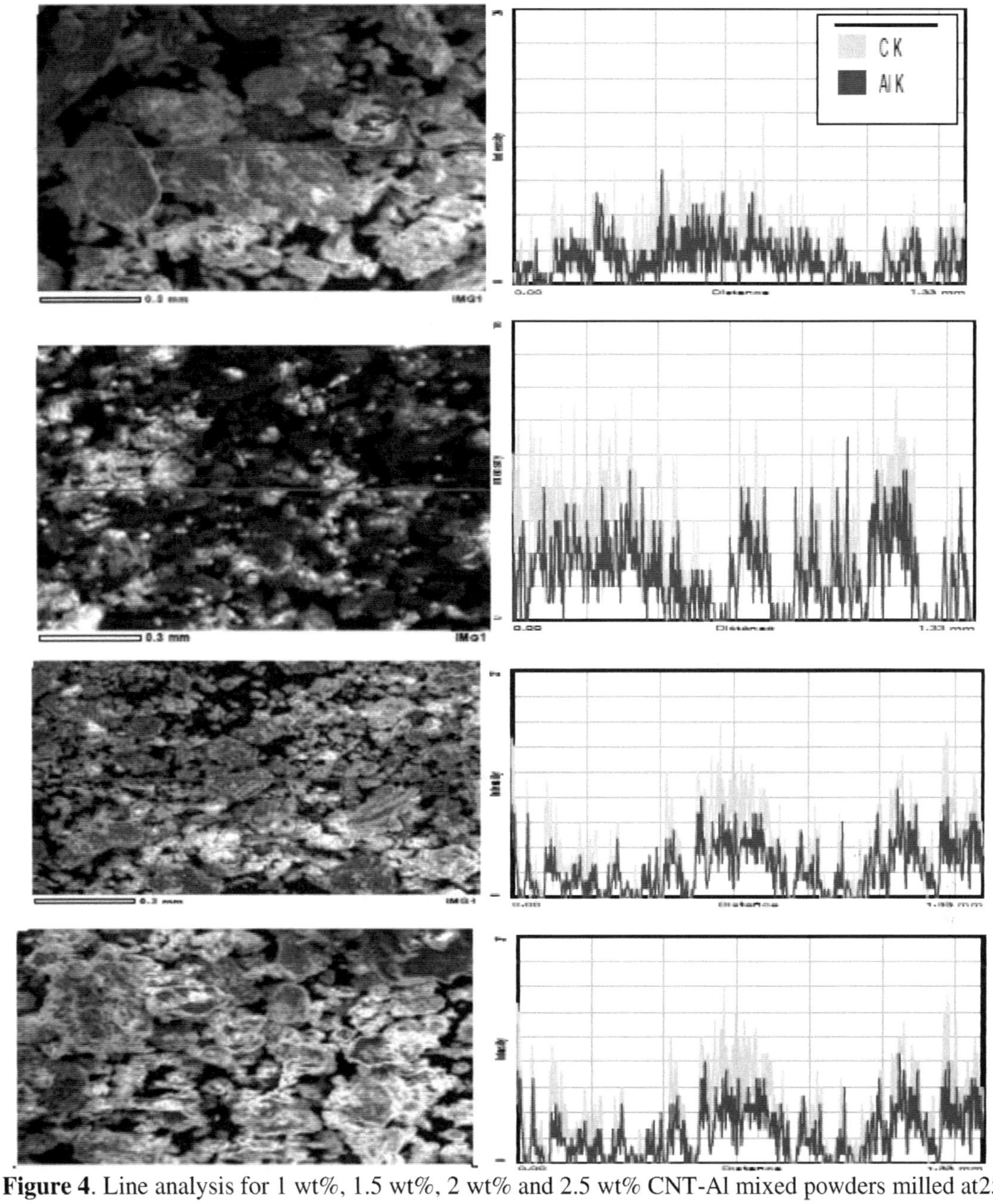

Figure 4. Line analysis for 1 wt%, 1.5 wt%, 2 wt% and 2.5 wt% CNT-Al mixed powders milled at250 rpm speed for 2 hr duration

For all CNT compositions, the result showed similar trend as was obtained from mapping and line analysis also indicated almost similar quality of dispersion of CNT in the aluminum matrix. The line analysis of the mixed CNT and Al powders showed that 2 hrs milling time was sufficient to disperse CNT into the Al matrix at 250 rpm milling speed. This indicates the mutual coexistence of CNT and Al which has been shown in two different colours such as blue and green for aluminum and carbon respectively (Fig. 4). The lines indicated that Al and CNT go along and followed exactly the same trends which indicated the uniform distribution of CNT in the aluminum matrix for all the composition. It can be said that the data obtained from mapping, elemental and line analysis plots for all the CNT compositions indicated that CNT is uniformly and homogeneously dispersed in aluminum matrix without any formation of the significant carbides due to controlled atmosphere of milling. However, the involvement of argon gas during milling operation have eliminated unwanted reactions that are likely to occur during the process, and the use of suitable milling speed and time also help in preventing the contamination. The use of argon gas during milling operation was very vital as it control surface and interface contamination which are likely to occur during the synthesization of any composite mixture via high energy ball milling Esawi et al. [6].

4. Conclusion

The data obtained from mapping, elemental and line analysis plots for all CNT compositions indicated that CNT is uniformly and homogeneously dispersed in aluminium matrix without any formation of carbides due to controlled atmosphere of milling without formation of significant carbide. Therefore, this CNT-Al composite powder mixture can be used for new nano-composite development without any agglomeration problem.

Acknowledgements

The authors would like to thank the Ministry of Higher Education (MOHE), Malaysia for financial support under ERGS research grant scheme No.: ERGS 12-021-0021. Authors are grateful to the International Islamic University (IIUM) which made this study possible.

References

[1] Iijima, S., 1991 *Nature* **354** 56.
[2] Bakshi, S.R.; Lahiri, D.; Agarwal, A. 2010 *Int. Mat. Rev.* **55** 41
[3] Umma, A.; Maleque, M.A.; Iskandar, I.; Mohammed, Y.A. 2012, *Aust. J. of Basic and Apl. Sci.* **6** 69.
[4] Morsi, K.; Esawi, A. 2007 *J. of Mat. Sci.* **42** 4954
[5] Morsi, K., A.; Esawi, P.; Borah, S.; Lanka, A.; Taher, M. 2010 *Mat. Sci. and Eng*: A **527** 5686
[6] Esawi, A. M.; Morsi, K.; Sayed, A.; Gawad, A. A.; Borah, P. 2009 *Mat. Sci. and Eng*: A **508** 167

IOP Conf. Series: Materials Science and Engineering 123 (2016) 012012 doi:10.1088/1757-899X/123/1/012012

Experimental and numerical investigations on adhesively bonded joints

R Negru[1], L Marsavina[1] and M Hluscu[1]

[1]Politehnica University Timisoara, Timisoara, Romania

Email: radu.negru@upt.ro

Abstract. Two types of adhesively bonded joints were experimental and numerical investigated. Firstly, the adhesives were characterized through a set of tests and the main elastic and mechanical properties were obtained. After that, the stress distributions at interface and middle of adhesive layer were determined using a linear elastic FEA. The numerical data were fitted by a power law in order to determine the critical values of intensity of stress singularity.

1. Introduction

Adhesively bonded joints are used in many engineering applications (automotive and aerospace industries) as an alternative to the conventional riveted and bolted joints. Several advantages characterize the adhesive bonding including: good stiffness and strength, reduced weight and cost, capacity to join dissimilar materials. Delicate joining technology (surface preparation, curing procedure etc.) and environmental sensitivity represent real disadvantages [1]. As a new trend, the hybrid joints (bonded/riveted [2]) are of great interest. However, the joint's strength evaluation represents a major challenge.

The geometries of the single- and double-lap tested joints were established according to the ASTM D1002-10 and ASTM D3528-96 recommendations (figure 1, width of 25.4 mm). Aluminium alloy 7075 T651 adherents and two types of adhesives (araldite AV138M-1 with hardener HV 998 and araldite AW106 with hardener HV 953U) were used.

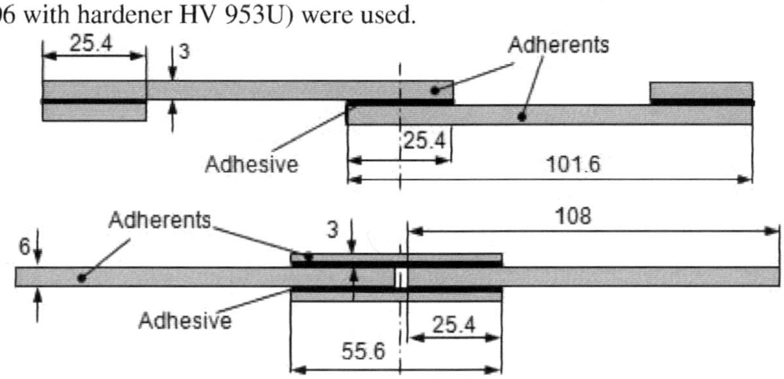

Figure 1. Geometries of adhesive-bonded joints, single-lap (up) and double-lap (down)

2. Experimental investigations

After the mechanical treatment of the surfaces, the adherents were cleaned in an acetone bath. The adhesives were cured at room temperature for 24 hours, under constant pressure applied through weights. The thickness of adhesive layer was carefully controlled at 0.25 (mm), but no attempts were made to control the fillet geometry.

Table 1. The elastic and strength properties of bonded materials

Materials	E (MPa)[a]	ν (-)[a]	σ_u (MPa)[b]
Al 7075 T651	70198	0.330	545.06±8.82
Araldite AV138M-1	3384	0.352	18.27±0.77
Araldite AW106	2585	0.406	37.12±5.08

[a] Determined by impulse excitation technique, ASTM E1876-09
[b] Obtained through tensile tests

Table 2. The apparent shear strength of adhesively bonded joints

Joint type	Adhesive	t (mm)	Fmax (kN)	τ (MPa)
Double-lap	AW106	3; 6	12.956±1.270	10.04±0.98
Single-lap	AW106	3	4.522±0.570	7.01±0.88
Single-lap	AV138M-1	3	3.301±0.216	5.12±0.33

The tests were carried out on an universal testing machine under displacement control at 1.3 (mm/min) speed. The stress-strain curves (figure 2) and the tensile strength σu for adherent and adhesives were experimentally obtained (table 1). Also, the elastic properties were determined using a Resonant Frequency and Damping Analyzer device based on impulse excitation technique, ASTM E1876-09 (table 2). A brittle behaviour was observed for AV138M-1 adhesive and a nonlinear one for AW106 adhesive.

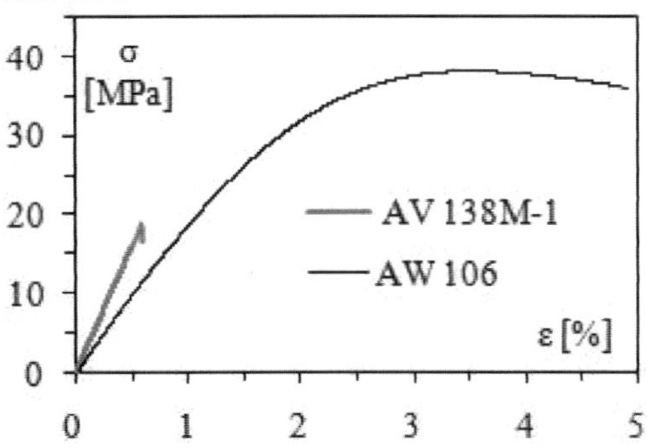

Figure 2. Representative tensile stress-strain curves for adhesives

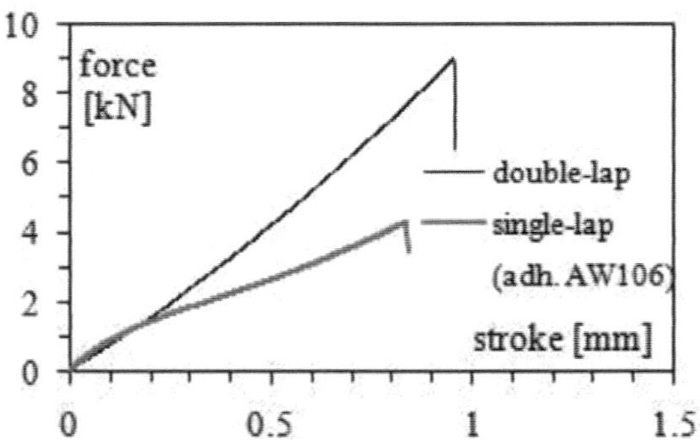

Figure 3. Load-displacement curves for adhesively-bonded joints

3. Analytical solution for stress singularity

As a consequence of changes in geometry and in elastic properties, singular stress fields occur at the edge of interface between the adhesive and the adherent. An investigation in determining how the order of the stress field singularity depends on the material elastic constants and corner angles was conducted by Bogy and Wang [3]. For the free edge geometry presented in figure 4, the stress singularity could be expressed in the form:

$$\sigma_{ik,}=Hr^{\lambda^{-1}}f_{ik,j}(\alpha,\beta,\lambda,\theta) \quad (1)$$

$$u_{ij}=\frac{Hr^{\lambda}}{2G_j}g_{ij}(\alpha,\beta,\lambda,\theta) \quad (2)$$

where: *H* represents the stress intensity factor at joint; *i*, *k* are the polar coordinates (r, θ); *j* is the material number; G_j represents the shear modulus of material *j*; λ-*1* is the stress singularity; f_{ik}, *j* and g_{ij} are non-dimensional functions of λ, θ and Dundurs parameter α and β.

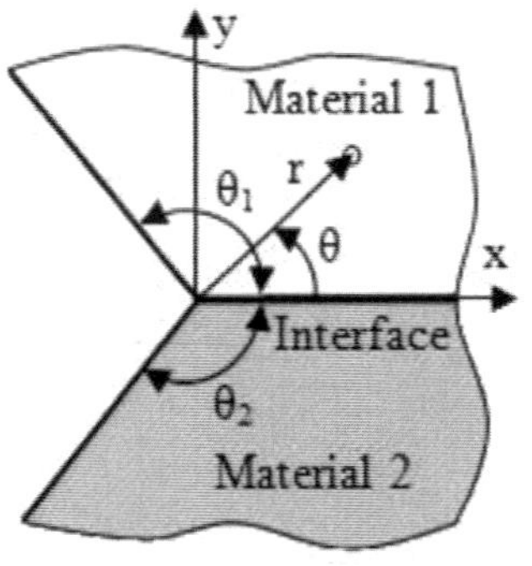

Figure 4. Bonded dissimilar materials

Expressing the stresses and displacements through two complex functions $\Phi(z)$ and $\Omega(z)$ and imposing the appropriate boundary conditions (i.e. the continuity for stresses and displacements on the interface, the values of stresses on the free edges) the problem is reduced to a linear homogeneous system (Sinescu et al. [4]). Because a non-trivial solution is searched, the stress singularity was found by imposing to have null determinant of the coefficients matrix. For computing the stress singularity a Mathcad routine was developed and the results for bonded joints were obtained (table 3).

Table 3. The stress singularity values for investigated adhesive bonded joints

Joints	θ1 (rad)	θ2 (rad)	λ-1 (-)
Al 7075 T651 / AV138M-1	π/2	π/2	-0.2714
	π	π/2	-0.1951
Al 7075 T651 / AW106	π/2	π/2	-0.3127
	π	π/2	-0.2084

4. Numerical analysis of bonded joints

A 2D linear elastic FEA was conducted using commercial Abaqus package. The bonded joints were meshed using 8 nodes isoparametric plane strain quadrilateral elements. A progressive mesh towards the overlap ends was adopted, in order to take into account the stress gradient and the singularities. The minimum size of the element at the overlap ends is 0.008 (mm) with 8 elements in the adhesive thickness and the maximum aspect ratio reaching the maximum value 10:1. The model was clamped at one adherent end and allowed to move only in the longitudinal direction at the other end. The applied load represents the average load obtained experimentally for joints failure.

The results plotted in figure 5 in terms of maximum principal stress $\sigma 1$ (log axes used for distance, in order to highlights the stress distributions at overlap ends) show that the stresses are almost constant across the overlap length excepting the ends, due to the joint singularity. The stress peaks are higher in the case of the single-lap bonded joint. As expected, for the stress distributions through the middle of the adhesive layer the singularity effect is not present, but it is obvious at adherent-adhesive interface.

A method for debonding strength evaluation using two stress singularity parameters *H* and λ-*1* was proposed by Hattori et al. [5,6]. Expressing the maximum principal stress $\sigma 1$ distribution near the bonding edge, in the dominant stress singularity zone, as follows (Sawa [7]):

$$\sigma_1(r)=H_C\cdot r^{\lambda-1} \tag{3}$$

the critical values of intensity of stress singularity *HC* were determined by fitting equation (3) to the stress distributions obtained through FEA, near the bonded edge.

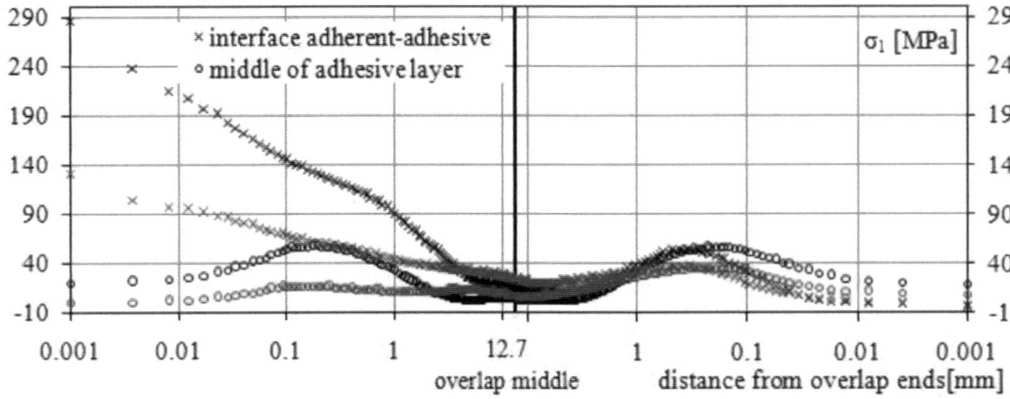

Figure 5. Maximum principal stress σ1 vs. distance from ends to middle of overlap (adhesive AW106, double-lap in blue and single-lap in black)

Considering the single-lap bonded joints, the critical values of *HC* were calculated for these two different used adhesives (in this case $\theta 1=\pi$ and $\theta 2=\pi 2/$) by fitting the numerical data through a power law in the singularity predominant area, resulting the values presented in figure 6.

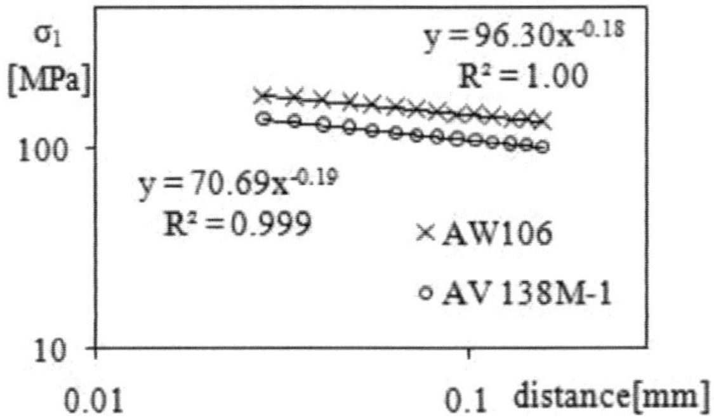

Figure 6. Numerical data fitted by the power law in the case of single-lap bonded joints

5. Conclusions

This paper presents the mechanical characterization of two types of adhesives through tensile and tensile-shear tests (with single- and double-lap bonded joints). The determined properties were applied to the linear-elastic FEA of bonded joints and the stress distributions at the adhesive-adherent interface were studied. Numerical data were fitted by a power law and the critical values *HC* of intensity of stress singularity were determined for used adhesives.

Acknowledgment

This work was supported by a grant of the Romanian National Authority for Scientific Research, CNCS-UEFISCDI, project PN-II-PCCA-2011-3.2-0068, contract number 206/2012.

References

[1] Pocius A V 2012 *Adhesion and Adhesives Technology* 3rd edition (Munich: Hansen) pp 1-16.
[2] Sadowski T, Kneć M, Golewski P 2010 *Int J Adhes Adhes* **30** 338.
[3] Bogy D B and Wang K C 1971 *Int J Solids Struct* **7** 993.
[4] Sinescu C, Marsavina L, Cernescu A 2012 *Comp Mat Sci* **59** 57.
[5] Hattori T, Sakata S and Murakami G 1989 *J. Elect. Pack.* Trans. ASME **1-111** 243.
[6] Hattori T, Nishimura N and Yamashita M 2008 *Eng. Fract. Mech.* **75** 1725.
[7] Sawa T 2008 *Analytical Models with Stress Functions* (Modeling of Adhesively Bonded Joints) ed L F M da Silva and A Öchsner (Berlin: Springer) pp 53-92.

Influence of the NaCl/LiF additives on structure, phasetransitions and dielectric properties of BSPT ceramics

N V Golubko[1], G M Kaleva[1], A V Mosunov[1], E D Politova[1], A H Segalla[2]

[1]Karpov Institute of Physical Chemistry, Moscow,Russia
[2]ELPA Company, Zelenograd, Moscow, Russia

E-mail: politova@cc.nifhi.ac.ru

Abstract. Influence of the NaCl/LiF additives (with ratio 60:40) on structure parameters, phase transitions and dielectric properties of solid solutions close to the Morphotropic Phase Boundary in the system (1-x)BiScO3 - xPbTiO3 (x=0.635, 0.645) has been studied. Using as initial the BSPT powders prepared from nitrate solutions, dense single phase ceramic samples doped by5, 10 and 15 w. % of NaCl /LiF have been prepared at 1313 - 1323 K (2 h). Shift of the BSPT compositions phase content from initial mixtures of Rhombohedral and Tetragonal phases to the region of Tetragonal phase was stimulated by addition of the NaCl/LiF additives. This effect is accompanied by decrease of the unit sell volume, increase in the c/a ratio and increase in the Curie temperature value from 668 to 730 K for x=0.635 and from 672 to 724 K for x=0.645 in compositions doped by 10 w. % of the NaCl/LiF additive.

1. Introduction

Application of widely used ceramic materials based on the lead zirconate-titanate solid solutions is limited by temperatures determined by their depolarization temperature ~ 0.5TC (TC ~ 600 K). This stimulated intensive studied of bismuth containing oxides based on the perovskite structure iMe^{3+}O$_3$oxides with Curie temperatures TC > 700 K which have prospects for the development of materials for high temperature applications [1 - 4]. Regulation of functional properties is possible by substitutions of cations in the A- and B-sublattices, by modification of compositions using over stoichiometric additives, and by optimization of preparation conditions [4 -13]. Ceramics based on solid solutions from the Morphotropic Phase Boundary (MPB) in the system (1-x) $BiScO_3$ - $x PbTiO_3$ (BSPT) may be considered among the most studied as they revealed high TC values > 700 K and high piezoelectric coefficients. It is worse mentioning that investigation of compositions close to the MPB is widely used for the development of new effective piezoelectric materials. Usually, cation substitutions in the A and B-sublattices of perovskite structure are used to regulate functional characteristics. As for the BSPT- based compositions, in order to prevent Bi- and Pb-cation loss, the decrease of sintering temperature comprises one of the most important tasks.

Earlier, positive effect of over stoichiometric additives on the phase content, microstructure, and functional properties of the BSPT ceramics with x = 0.60 - 0.66 was proved [8, 11]. It was shown that addition of KCl with low melting temperature favored to increasing density of ceramics, allowed to regulate the unit cell parameters, microstructure and defined the improvement of functional properties of ceramics with x = 0.64 [11].

In this work, the BSPT ceramics with compositions from MPB with x = 0.635 (BSPT1)and x=0.645 (BSPT2) modified by complex additive NaCl/LiF (with molar ratio 60:40) have been studied.

IOP Conf. Series: Materials Science and Engineering **123** (2016) 012013 doi:10.1088/1757-899X/123/1/012013

This NaCl/LiF composition has low melting temperature ~ 950 K, so decrease of sintering temperature of ceramics may be expected. Moreover, this is important for the prevention of Bi and/or Pb oxides loss during the high temperature sintering, and favors to the stoichiometry preservation and functional properties optimization.

2. Experimental

Initial powders of the BSPT compositions with x =0.635 (BSPT1) and 0.645 (BSPT2) were prepared by the co-precipitation of nitrate solutions following the conventional procedure, and then calcined at973 K (7 h) [9]. The NaCl /LiF additives, with molar ratios 60:40, were added in amounts 5, 10 and 15w. % to the BSPT powders. The milled and pressed into tablets powders were sintered in air at 1323 K(2 h) and then slowly cooled to the room temperature.

Phase content and crystal structure parameters of the samples were studied at the room temperature using the X-ray diffraction method (DRON-3M, CuKa-beam). Spontaneous polarization of ceramics was estimated using the Second Harmonic Generation method (SHG, Nd:YAG laser, l=1.064 μm).Phase transitions and dielectric properties were studied using the dielectric spectroscopy method(Agilent 4284 A, 1 V) in the temperature range of 300 – 1000 K at frequencies 100 Hz – 1 MHz.

3. Results and discussion

According to the X-ray diffraction data, pure perovskite structure samples were prepared at sintering temperature 1323 K (Figure 1). Figure 2 shows parts of the diffraction patterns displaying peaks with*h*2+*k*2+*l*2=4. Increase in the content of tetragonal phase was observed with increasing x value of initial samples. Displacement of the X-Ray diffraction peaks to higher 2q values indicates the unit cell volume decreasing with increasing amount of the NaCl/LiF additive. This picture demonstrates transition to the tetragonal structure in the samples containing ³ 5 w. % of the NaCl/LiF additive as well. So, addition of 5 w. % of the NaCl/LiF additive to the initial compositions with x=0.635(BSPT1) and with x=0.645 (BSPT2) leads to the transformation of the R phase (BSPT1) or mixture of T and R phases (BSPT2) to the tetragonal phases. This means that the NaCl/LiF additive stimulates displacement of the compositions to the region with tetragonal structure.

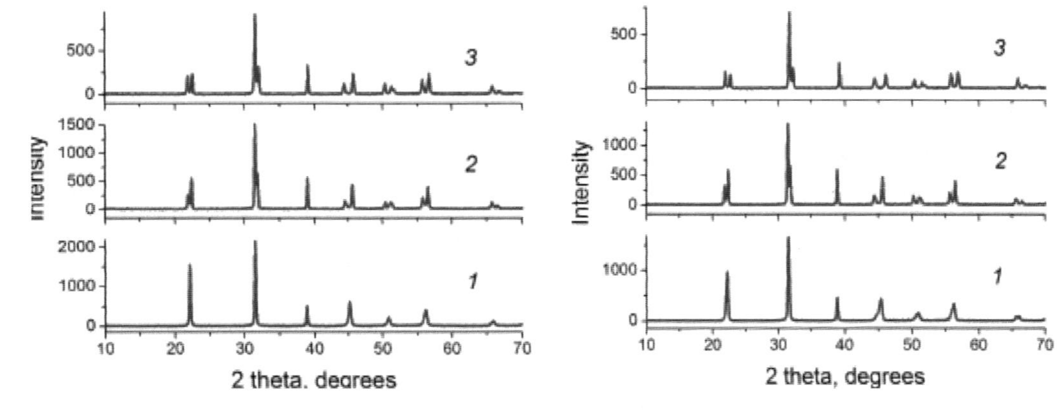

Figure 1. The X-Ray diffraction patterns of the samples (1-x)BiScO3 - xPbTiO3 with x=0.635 (*a*, curve *1*), and with x=0.645 (*b*, curve *1*) modified by the NaCl/LiF in amounts 5 w. % (curves 2), 10w. % (curves 3), sintered at 1323 K (2 h)

The observed changes in the perovskite unit cell parameters (decrease of a, increase of c and increase of c/a ratio in compositions containing £ 10 w. % of additives) may be explained by the substitution of the Bi3+ and/or Pb2+ cations by the Li+ and/or Na+ cations in the A-positions of the perovskite lattice (Figure 3).

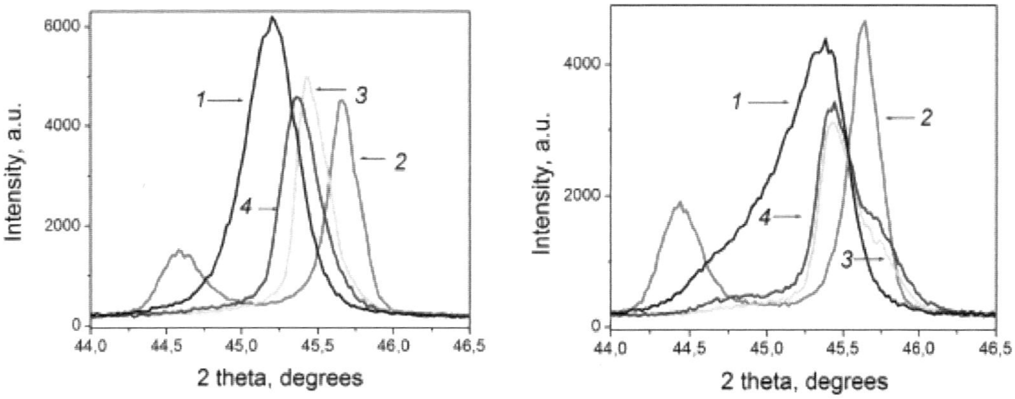

Figure 2. . Parts of the X-Ray diffraction patterns of the samples BSPT1 (*a*, curve *1*) and BSPT2 (*b*, curve *1*) modified by the NaCl/LiF additives in amounts 5 w. % (curves *2*), 10 w. % (curves *3*) and 15w. % (curves *4*)

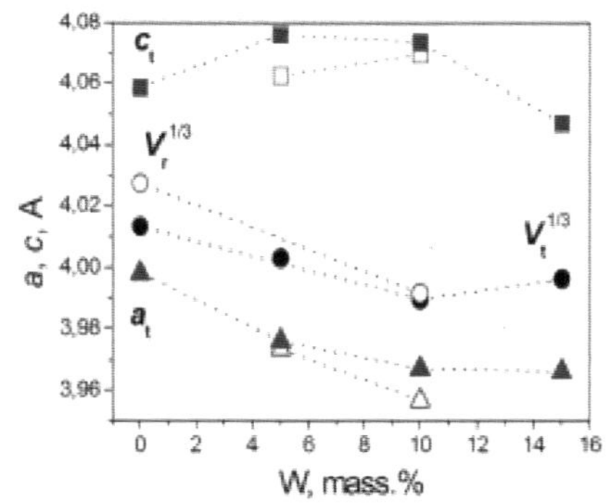

Figure 3. Concentration dependences of the perovskite unit cell parameters *a*, *c* and *V*1/3 for modified samples BSPT1 (empty symbols) and BSPT2 (solid symbols

Improvement of density of ceramics was stimulated by the NaCl/LiF additive introduction. The shrinkage of the samples increased from 15 to 17-19 %. This means that the additives intensified both processes of phase formation and sintering.

The relative intensity of the SHG signal $q=I_{2w}/I_{2w}(SiO_2)$ related to the spontaneous polarization P_s as $q \sim P_s 0.5$ increased from the 1200 (BSPT1) and 1500 (BSPT2) to the 1800 and 2200 values, respectively, with introduction of NaCl/LiF (5 w. %). This indicates to the enhancement of the P_s value with increasing amount of the additive.

Dielectric measurements revealed the 1st order sharp ferroelectric phase transitions marked by peaks in dielectric permittivity and corresponding minima in dielectric loss curves at temperatures 668÷ 730 K (Figure 4). Small increase in the T_C value observed in modified ceramics correlates with the increase in relative content of the tetragonal phase and with changes in the *c/a* ratio proportional to the T_C value. The dielectric loss values of the doped samples are lower than dielectric loss values of the undoped samples at temperatures lower than 550 – 650 K, which are important for poling procedure.

However, the losses in the doped samples are higher than those of the undoped samples at temperatures higher than *TC*. Evidently, such increase in dielectric loss at high temperatures is related to the effects of dielectric relaxation observed at high temperatures in the samples doped by the additives (Figure 5). It is necessary to mark that effects of dielectric relaxation may be related to the presence of oxygen vacancies in the perovskite lattice [6 -8]. Increasing of the T_C value observed for the doped samples indicates to the presence of Li-cations in A- positions of the perovskite structure.

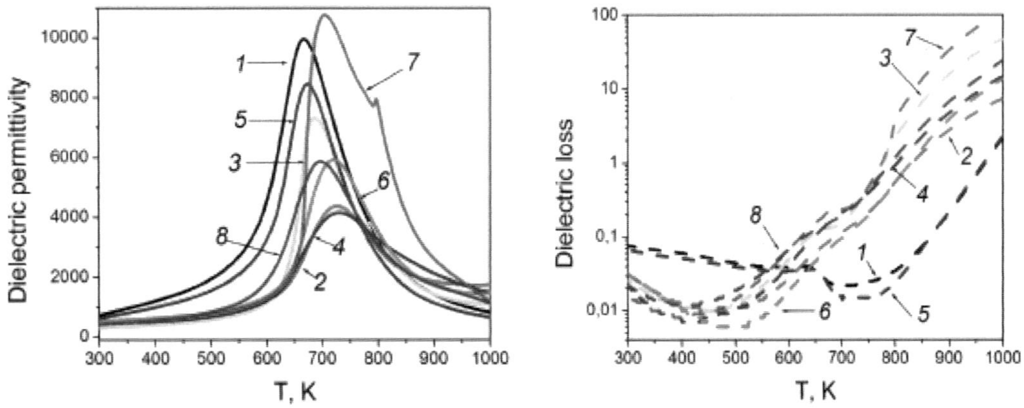

a) **b)**

Figure 4. Temperature dependencies of dielectric permittivity (*a*) and dielectric loss (*b*) of ceramicsBSPT1 and BSPT2 and ceramics modified by the NaCl/LiF additives (initial samples – curves *1*, 5)

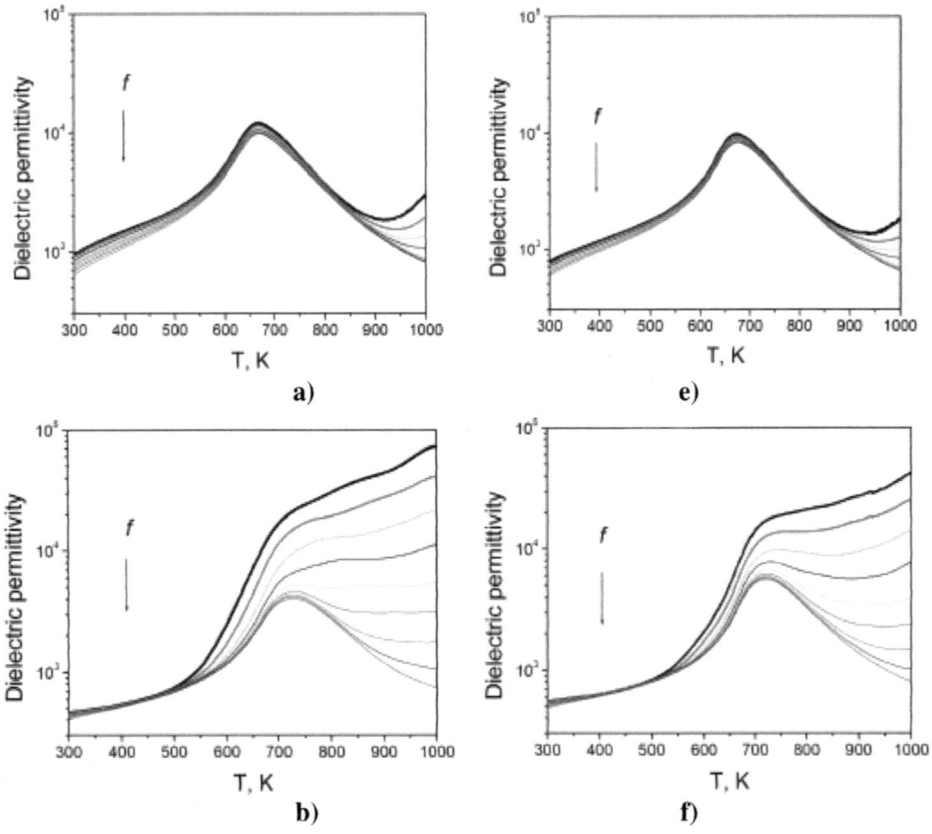

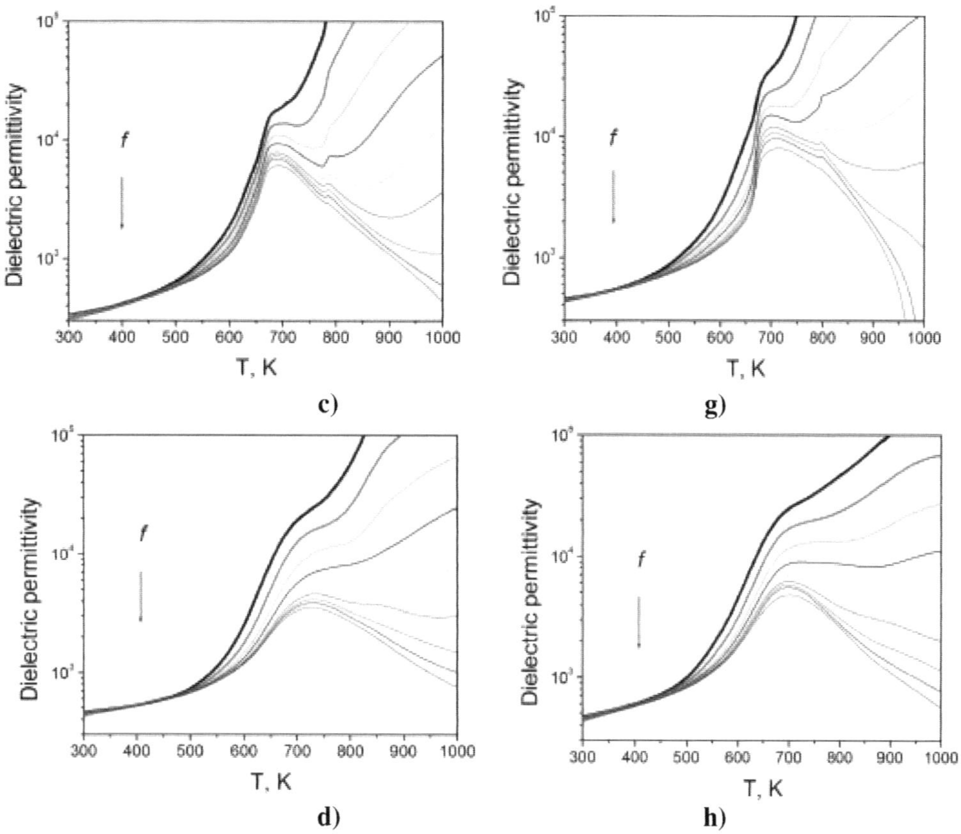

Figure 5. Temperature dependencies of dielectric permittivity (*a-h*) of the BSPT1 (*a-d*) and BSPT2(*e-h*) ceramics modified by the NaCl/LiF additives (initial samples – *a, e*; samples containing 5 w. %– *b, f*; 10 w. % – *c, g*; 15 w. % - *d, h*). Measuring frequencies *f* = 100, 300 Hz; 1, 3, 10, 30, 100, 300kHz, 1 MHz (cooling)

Thus, the BSPT samples doped by NaCl/LiF additives may be promising for the development of new compositions with improved piezoelectric properties.

4. Conclusions

The effects additives NaCl/LiF on the structure parameters, dielectric and ferroelectric properties of compositions from the morphotropic phase boundary in the system (1-*x*)BiScO3 - *x*PbTiO3 with x=0.635and *x*=0.645 have been studied. Positive influence of NaCl/LiF additives on phase formation, sintering process and ferroelectric properties was proved. Small increase in the *T*C value observed in modified ceramics correlates well with the increase in relative content of tetragonal phase and the *c/a* ratio.

Acknowledgements

The authors are thankful to S. Stefanovich for results of the SHG measurements.
The work was supported by the Russian Foundation for Basic Research (Grant 12-03-00388).

References

[1] Eitel, R. E.; Randall, C. A.; Shrout, T. R.; Paul, P. W.; Hackenberger, W.; Park, S.E.2001 Jpn. *J. Appl. Phys.* **40** 5999.

[2] Eitel, R. E.; Randall, C. A.; Shrout, T. R.; Park, S. E. 2002 *Jpn. J. Appl. Phys.* **41** 2099
[3] Zhang, S. J.; Randall, C. A.; Shrout, T. R. 2004 *Jpn. J. Appl. Phys.* **43** 6199
[4] Sehirlioglu, A.; Sayir, A.; Dynys, F. 2010 *J. Am. Ceram. Soc.* **93** 1718
[5] Politova, E. D.; Egorova, B. V, Kaleva, G. M.; Mosunov, A. V.; Stefanovich, S. Yu.;Segalla, A. G.; Zeng, J. 2011 *Bull. Russ. Acad. Sci.* **75** 1166
[6] Politova, E. D.; Egorova, B. V.; Kaleva, G. M.; Mosunov, A. V.; Stefanovich, S. Yu.;Segalla, A. G.; Zeng, J. 2011 *Ferroelectrics* **419** 83
[7] Politova, E. D.; Kaleva, G. M.; Mosunov, A. V.; Segalla, A. G.; Zeng, J. 2012 *KeyEngineering Materials* **512-515** 1363
[8] Kaleva, G. M.; Politova, E. D.; Mosunov, A. V.; Sadovskaya, N. V.; Segalla, A. G.;Zeng, J. 2012 *Inorganic Materials* **48** 953
[9] Politova, E. D.; Kaleva, G. M.; Mosunov, A. V.; Segalla, A. H.; Dosovitskiy, A. E.; Mikhlin, A. L.2013 *J. Adv. Dielectrics* **3** 1350024
[10] Politova, E. D.; Kaleva, G. M.; Mosunov, A. V.; Sadovskaya, N. V.; Segalla, A. H.;Bush, A. A. 2013 *Ferroelectrics* **440** 105
[11] Politova, E. D.; Akinfiev, V. S.; Kaleva G. M.; Mosunov A. V.; Stefanovich, S. Yu.;Segalla, A. H. 2014 *Ferroelectrics* **464** 1
[12] Segalla, A. H.; Nersesov, S. S.; Kaleva, G. M.; Politova, E. D. 2014 *Inorganic materials* **50** 655
[13] Chen, J.; Dong, Sh. 2014 *J. Adv. Dielectrics* **4** 1430002

New method of determination of spot welding-adhesive joint fatigue life using full field strain evolution

T Sadowski[1], M Kneć[2]

[1]Lublin University of Technology, Department of Solid Mechanics, Lublin, Poland
[2]Lublin University of Technology, Civil Engineering Laboratory, Lublin, Poland

E-mail: t.sadowski@pollub.pl

Abstract: Fatigue tests were conducted since more than two hundred years ago. Despite this long period, as fatigue phenomena are very complex, assessment of fatigue response of standard materials or composites still requires a long time. Quite precise way to estimate fatigue parameters is to test at least 30 standardized specimens for the analysed material and further statistical post processing is required. In case of structural elements analysis like hybrid joints (Figure 1), the situation is much more complex as more factors influence the fatigue load capacity due to much more complicated structure of the joint in comparison to standard materials specimen, i.e. occurrence of: welded hot spots or rivets, adhesive layers, local notches creating the stress concentrations, etc.

In order to shorten testing time some rapid methods are known: Locati's method [1] - step by step load increments up to failure, Prot's method [2] - constant increase of the load amplitude up to failure; Lehr's method [2] – seeking for the point during regular fatigue loading when an increase of temperature or strains become non-linear. The present article proposes new method of the fatigue response assessment - combination of the Locati's and Lehr's method.

1. Introduction

Currently, aviation industries use many different complex joining techniques for structural element manufacturing, among other things - hybrid joints, e.g. [3-19]. The most popular are rivet-adhesive [3-8] and hot spot welding (HSW) – adhesive joints [4, 9, 17-18]. The hybrid joints, in assumption, are combination of good properties of each used simple joining techniques. In this article the fatigue process of HSW- adhesive joints will be analysed with application of the new testing methodology. Classically, to estimate the S-N curve (Wöhler curve) for assessment of the fatigue response it is necessary to perform experiments with at least 30 specimens, i.e. for six stress levels with application of five specimen per each stress level. Up to now, some rapid methods are also widely known i.e.: Prott's, Locati's or Lehr's, which shorten the time of testing.

The article presents the new innovative experimental method combining Locati's and Lehr's concepts. The idea rely on strains increase measurement during damage cumulative fatigue process for gradual increase of loading (Figure 2) up to the final failure of specimen. When the strain increase becomes out of linear the damage growth is enough advanced to start unstable part of its sudden uncontrolled increase in the local HSW points. This leads to immediate failure of the samples by HSW shearing.

The fatigue load spectrum, chosen for testing, represents real flight condition previously recorded by the industrial partner. The step by step loading increase during fatigue tests with specially selected number of cycles at each level of loading (Figure 3) allows for effective shortening of the testing time, getting reasonable estimation of fatigue strength.

Obtained results denote that full-field strain measurement requires only few specimens to obtain fatigue load capacity.

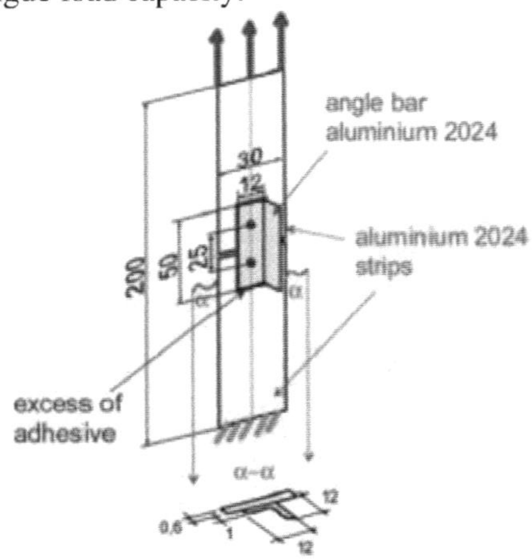

Figure 1. Specimen shape and dimensions

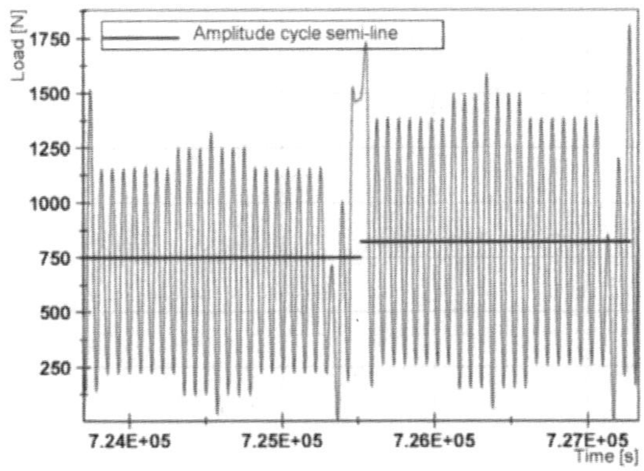

Figure 2. Fatigue load spectrum and load increase moment

2. Experimental setup and new testing methodology.

The experimental setup consists of 3 elements:

1. 25 kN servohydraulic system MTS (USA),
2. 2 Digital Image Correlation (DIC) systems ARAMIS (Germany),
3. National Instrument data acquisition system (USA).

All elements of the testing stand were synchronically triggered during loading program, to get full information about fatigue process of the experimental samples.

The applied load spectrum (Figure 2) corresponds to 2 hour of the airplane flight including starting, ascending, descending and landing. Dynamic loading frequency of servohydraulic system MTS was equal to 23 Hz. The actual values of the carrying load and stroke displacements were recorded with frequency of 1 kHz.

The applied 2 DIC systems ARAMIS allowed for constant monitoring of the deformation process of the specimen from the front and rear sides. After monitoring of displacement fields during fatigue tests the DIC system enables calculation of the major and minor strains. The strains distributions were particularly assessed in the spot areas, where the damage initiation and further propagation takes place, leading to the failure of the hybrid joints.

The whole testing program for estimation of fatigue response of the samples is presented in Figure 3. The fatigue process starts at level 1.5 kN, which denotes about 40% yield force Py, Figure 7. The whole program is divided into 24 loading levels with gradation 75 N. Each loading level consists of 104 cycles, similarly to Locati's method. In the last part of loading program, starting from level 2.775 kN the number of stages increases to 200 (every 50 cycles), as in this part of experiment damage processes initiates and finally leads to failure of the samples. Maximal loading level reached by specimens were equal to 3.25 kN. This value is much below the yield force P_y.

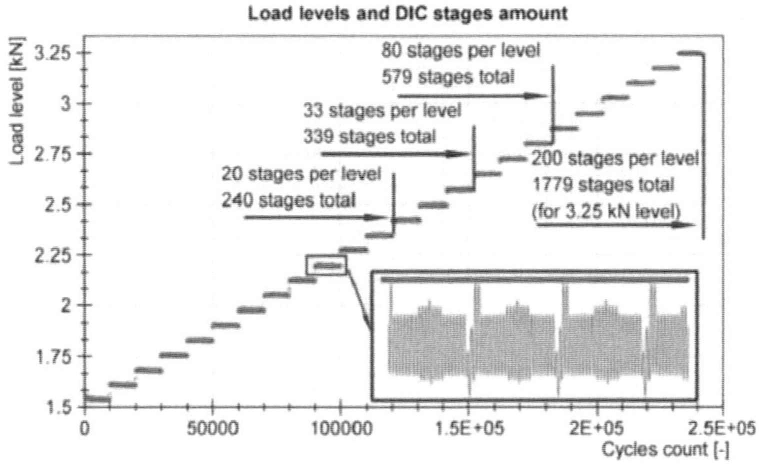

Figure 3. Loading program as a function of cycles amount and DIC stages amount

3. Specimens and HSW microscopic analysis

The specimens, Figure 1, were delivered by industrial partner PZL Mielec (Poland). All specimen were made of aluminium alloy 2024 T352. Each specimen consisted of two sheets and an angle bar, which were initially joined by 2 HSW and further they were strengthened by the Epodur adhesive. Material properties - both for aluminum and an Epodur adhesive - are presented in Table 1.

Figure 4 presents schematically spot weld cross section. The size and shape of the fusion zone (FZ) is crucial for HSW joint. Figure 5 shows the real cross section of the spots. One can observe irregular and not symmetric shapes HSW and the FZ material does not differ of the original material.

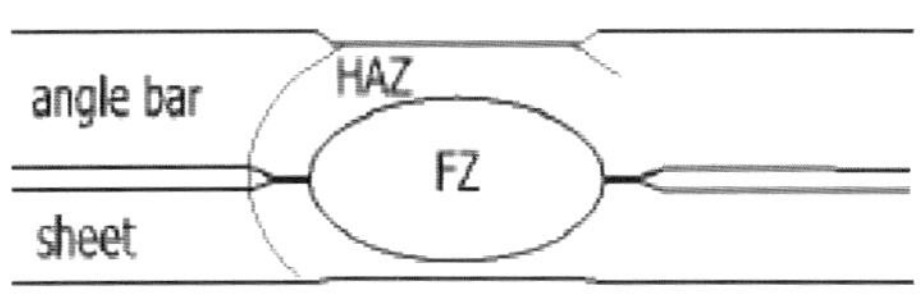

Figure 4. Schematic view of the spot cross section, FZ - Fusion Zone, HAZ - Heat Affected Zone

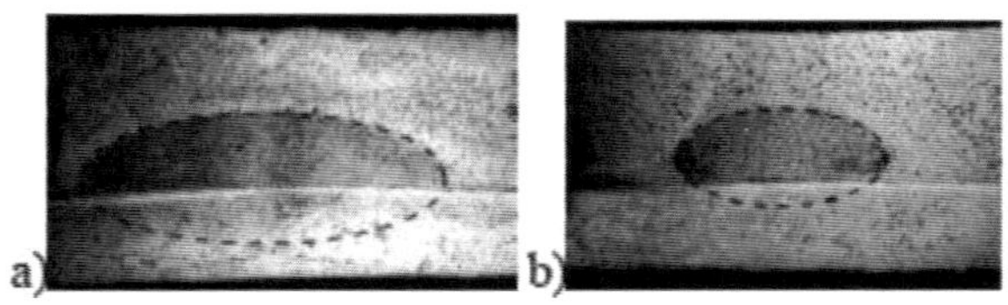

Figure 5. Cross section of two real fusion zones (FZ)

The failure of the hybrid samples starts from the damage of the adhesive layers. The final fatigue failure of the specimen is by shearing of HSW points. Typical fatigue failure mode is presented on Figure 6. Most of the specimens were damaged according to this failure mode.

Table 1. Material properties of the Epodur and aluminum sheets

Parameter	Epodur	2024 T352
E [MPa]	1861	66134
υ [-]	0.39	0.30
Rm [MPa]	40.7	431
A [%]	4.26	20.2

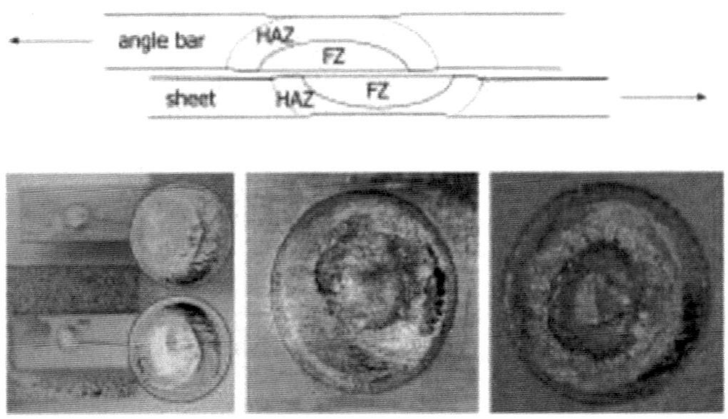

Figure 6. Typical fatigue failure mode

4. Results

Quasi-static response of the hybrid joints (Figure 7) were analyzed by authors in [17]. All fatigue loadings were conducted below yield force P_y point. One can notice that application of the Epodur adhesive leads to increase of the load carrying capacity about 2 times.

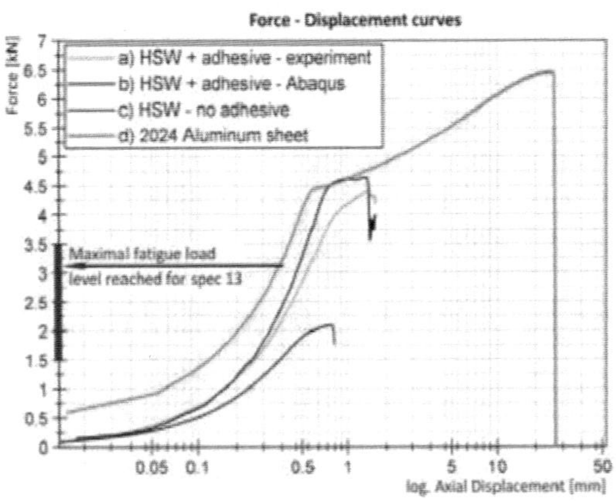

Figure 7. Force - displacement diagram for quasi-static tests

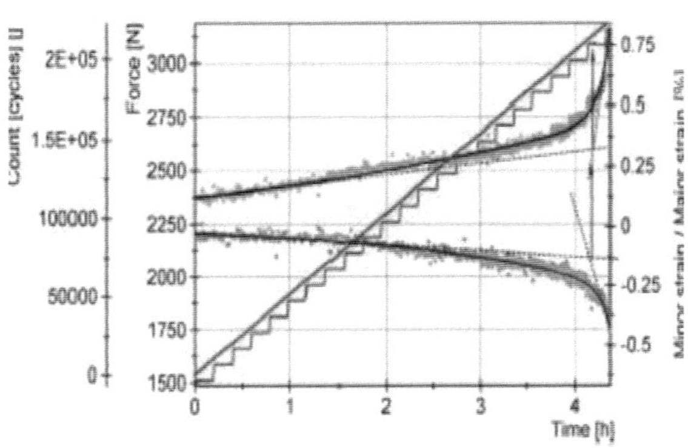

Figure 8. Fatigue test analysis: cycles amount (red), load levels (brown), major strain (blue), minor strain (green) and approximation curves (black)

Experimental observations of the last stage of the fatigue process, just before failure (Figure 8-10), clearly showed that highest strain concentration (higher than 1%) took places in the spots area. Each curve representing average strain change during fatigue process - computed from welded spot area - is very similar to exponential function. For selection of the best approximation function - the coefficient of determination ($R2$) was taken into account, which should be at least 0.85.

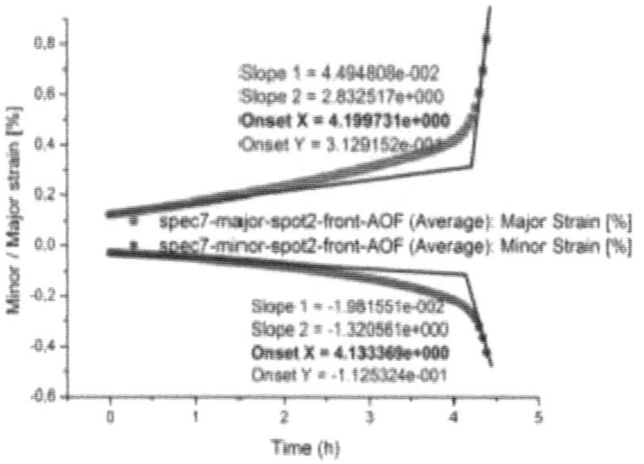

Figure 9. Idea of Lehr's method - two tangent lines analysis

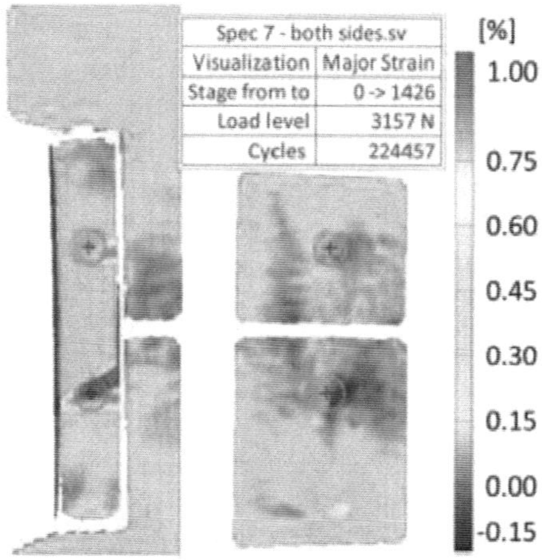

Figure 10. Major strain analysis - last stage of the fatigue test just before failure

5. Conclusions

The new time-saving rapid method for determination of safe levels of loads, cycles amount or displacements was presented in the article. The experimental setup consisting of 2 DIC systems was used to monitor the increase of the local full field strain in the most efforted points of the hybrid joints for the step by step loading increment. The effective application of the method was demonstrated on the analysis of the complex shape hybrid specimens under fatigue.

The proposed approach creates very effective tool for estimation of the fatigue response of different structural elements as well as modern multiphase composite materials, e.g. [19-23]. The analysis comprises a low cycle or even high cycle fatigue problems and can be used in other fields of research. The method is flexible and can be modified to obtain more accurate results with application of more loading levels, if required.

The experimental evidences will be extended to numerical modeling of the gradual fatigue degradation, including different damage models, e.g. [24-32].

Acknowledgment

Financial support of Structural Funds in the Operational Programme - Innovative Economy (IE OP) financed from the European Regional Development Fund Project "Modern material technologies in aerospace industry" No POIG.0101.02-00-015/08 is gratefully acknowledged; task: RT-15: Unconventional technologies of joining elements of aeronautical construction.

This work was financially supported by Ministry of Science and Higher Education within the statutory research number S/20/2015

References

[1] Katarzyński S., Kocańda S., Zakrzewski M. 1967 Testing of metals mechanical properties. WNT, Warsaw (in Polish)

[2] Szala J 1989 Loadings and fatigue life of machine elements. ATR, Bydgoszcz (in Polish)

[3] da Silva, L.F.M.; Pirondi, A.; Oechsner A. 2011 Hybrid adhesive joints. Berlin Heidelberg, *Springer Verlag*;
[4] Monori, F.; Pirondi, A.; Kleiner, F. 2010 *Int. J. Adhesion & Adhesives* **30**, 367
[5] Sadowski, T.; Kneć, M.; Golewski, P. 2010 *Int. J. Adhesion & Adhesives* **30**, 338
[6] Pirondi, A.; Moroni, F. 2009 *J. Adhes. Sci. Technol.* **23,** 1547
[7] Sadowski, T.; Golewski, P.; Zarzeka-Raczkowska, E. 2011 *Comput. Mater. Sci.* **50**, 1256
[8] Sadowski, T.; Zarzeka-Raczkowska E. 2012 *Arch. Metall. Mater.* 57, 1127
[9] Campilho, R.D.S.G.; Pinto, A.M.G.; Banea, M.D.; da Silva, L.F.M. 2012 *Int. J. Adhesion & Adhesives* **27,** 86
[10] Sadowski, T.; Balawender, T. Technology of clinch-adhesive joints. In: da Silva LFM, Pirondi A, Öchsner A, editors. 2011 Hybrid adhesive joints. Berlin Heidelberg: *Springer-Verlag*, 149
[11] Sadowski, T.; Golewski P. 2013 *Arch. Metall. Mater.* **58**, 581
[12] Sadowski, T.; Kneć, M. 2013 *Arch. Metall. Mater.* **58** 119
[13] Balawender, T.; Sadowski, T.; Golewski, P. 2011 *J. Adhes. Sci. Technol.* **25**, 2391
[14] Balawender, T.; Sadowski, T.; Golewski, P. 2012 *Comput. Mater. Sci.* **64,** 270
[15] Sadowski, T.; Balawender, T.; Sliwa, R.; Kneć, M.; Golewski, P. 2013 *Arch. Metall. Mater.* **58**, 163
[16] Pirondi A, Moroni F. Science of clinch-adhesive joints. In: da Silva LFM, Pirondi A, Öchsner A, editors. Hybrid adhesive joints. Berlin Heidelberg: Springer-Verlag, 2011. p. 109-148.
[17] Sadowski, T.; Kneć, M.; Golewski, P. 2014 *J. Adhesion* **90** 346
[18] Sadowski, T.; Golewski, P.; Kneć, M. 2014 *Comp. Struct.* **112** 66
[19] Gömze, L.A.; Gömze, L.N. 2009 *Ěpitöanyag – Journal of Silicate Based and Composite Materials* **61**, 38
[20] Gömze, L.A.; Gömze, L.N. 2009 IOP Conf. Ser.: Mater. Sci. Eng. 47 012033 http://dx.doi.org/10.1088/1757-899X/47/1/012033.
[21] Birman, V.; Bryd, L.V. 2007 *Appl. Mech. Rev.* **60**, 195
[22] Sadowski, T.; Samborski, S. 2008 *Comput. Mater. Sci.* **43**, 75
[23] Postek, E.; Sadowski, T. 2011 *Comp. Interf.* **18**, 57
[24] Campilho, R.D.S.G.; Banea, M.D.; Pchavres, F.J.; da Silva, L.F.M. 2011 *Comput. Mater. Sci.* **50**, 1543
[25] Sadowski T., Marsavina L. 2011 *Comput. Mat. Sci.* **50**, 1336
[26] Sadowski, T.; Golewski, P. 2012 *Comput. Mater. Sci.* **52**, 293
[27] Golewski G., Golewski P., Sadowski T., 2012 *Comput. Mat. Sci.* **62**, 75
[28] Burlayenko, V.; Sadowski, T. 2009 *Comput. Mater. Sci.* **45**, 658
[29] Gajewski, J.; Sadowski, T. 2014 *Comput. Mater. Sci*, **82**, 114
[30] Nakonieczny, K. ; Sadowski, T. 2009 *Comp. Mater. Sci.* **44**, 1307
[31] Petrova, V.; Sadowski, T. 2014 *Meccanica* **49**, 2603
[32] Burlayenko, V.; Sadowski, T. 2014 *Meccanica* **49** 2617

Effect of mineral additives (natural pozzolana and sand of dunes) by substitution of cement on the performance and durability of mortars

M Saidi, B Safi

Research Unit: Materials, Processes and Environment (UR/MPE), University of Boumerdes, Algeria

E-mail: saidimouh@yahoo.fr, safi_b73@umbb.dz

Abstract. The objective of our work consists of the study of the substitution effects of clinker by mineral additions such as: natural pozzolana (PZ) and the sand of dunes (SD) finely crushed on the mechanical properties and the durability of the mortars worked out according to various combinations containing these additions. The results from this research confirm that the substitution of 20% to 30% of cement APC (Artificial Portland Cement) by additions in binary cement (APC + PZ) or ternary (APC + PZ + SD) contributes positively to the mechanical strength of mortars and resistance to the chemical attacks in various corrosive conditions such as: hydrochloric acid, sulfuric acid and nitric acid.

The mechanical strength of the different variants is comparable to those of the APC. The test results of the weight loss and phenolphthalein shows that the chemical resistance of variants (PZ20) and (PZ20 with SD5) are larger compared to the reference mortar APC and other variants.

This study shows that adding value by substituting a part of clinker. This substitution can save20% to 30% of clinker used for the manufacture of cement; this will have a beneficial effect for cement and economically (less energy spent for the clinker burning).

This study contributes to the protection of the environment as to produce one ton of clinker generates about one ton of$CO_2$is harmful to the atmosphere. Based on our results we will reduce from 20% to30% CO_2gasresponsible for the greenhouse effect.

1. Introduction

The cement is required for the manufacture of concrete [1-2]. In terms of energy needed for its production, cement ranks third all materials, second only to steel and aluminum. According to some studies, the production of one ton of cement generates about one ton of CO2. It is responsible for about 5 % of the emissions of this gas on the planet. This situation should be taken seriously because the concrete is expected to play a role increasingly important in the development and maintenance of human activity. Minimizing the removal of these industrial waste and reduction in resource demand usually has duration of longer than concrete "traditional" life. In addition to providing a very high durability, the ternary cement allows the concrete to be much more impervious to chemical attack [3-4]. Another advantage is that it allows the concrete to continue its mechanical performance even after the normal curing period of 28 days, which is why it is used for construction projects to be particularly durable, such as roads, bridges, tunnels, viaducts, dams and oil platforms [5-6-7-8]. The design of concrete is extremely related to the environment in which it will be exposed during the lifetime of a structure. Certain ions in an aqueous environment can be detrimental to the hydrated cement and consequently to the integrity of the concrete material [9-10-11].

The first part of this work is devoted to the development of mortars with additions (pozzolana and dune sand finely crushed) by substituting clinker with different percentages. The second part consists in the characterization of different mortars developed. And finally the third part, we are interested to study the durability of these mortars in various aggressive environments [12-13-14-15].

2. Experimental study

In order to demonstrate the influence of mineral additives on the performance and durability of mortars made based Portland cement with mineral additions (pozzolana and dune sand).

2.1. Materials used

Artificial Portland cement (APC); Washed river sand; Sand of dunes and Pozzolana

2.2. Formulation mortars and Confection specimens

We used prismatic specimens 4 x 4 x 16 cm3 according to standard NF EN 196-1 for testing mechanical strength and durability.

- **With binary cements**: APC + Pouzzolana (PZ) with different percentages (Table1).

Table 1. Different compositions of mortars APC + PZ.

Notation	% of PZ	APC (g)	PZ (g)	River Sand (g)	W/C	Number of test specimens
APC	0%	450	0	1350	0,5	12
PZ10	10%	405	45	1350	0,5	12
PZ20	20%	360	90	1350	0,5	12
PZ30	30%	315	135	1350	0,5	12

- **With ternary cement:** CPA + Pouzzolane (PZ) + sand dune (SD) with different percentages (Table2).

Table 2. Different compositions of mortars APC + PZ + SD

Notation	% of PZ	%SD	APC (g)	PZ (g)	SD (g)	River Sand (g)	W/C	Number of test specimens
$PZ_{15}SD_5$	15%	5%	360	67,5	22,5	1350	0,5	06
$PZ_{20}SD_5$	20%	5%	337,5	90	22,5	1350	0,5	06
$PZ_{20}SD_{10}$	20%	10%	315	90	45	1350	0,5	06

2.3. Confection procedure

After mixing in the automatic mixer (CONTROLAB), test samples were made according to standard NF EN 196-1 in 4x4x16cm^3 prismatic molds and mechanically compacted using a shock table (CONTROLAB) for 60 shocks applied for 60 seconds. Once leveled, the molds containing the samples are stored in the laboratory environment at a temperature of 20° ± 1°C and a relative humidity of about 55 ± 5%. The demolding is effected after a period of 24hours.

2.4. Durability of mortars

To characterize the chemical resistance, prismatic mortar specimens 4x4x16 cm^3 crafted according to NF EN 196-1; demolded after 24 hours, the samples are stored in the open air laboratory at 20 ± 2° C until to 28 days. After 28 days, we proceeded to test different drying by baking (100-105°C) until constant weight (dry weight) before etching. These specimens were then immersed in different solutions to assess the sustainability of each mortar composition:

IOP Conf. Series: Materials Science and Engineering **123** (2016) 012015 doi:10.1088/1757-899X/123/1/012015

Solution of 5% hydrochloric acid (HCl) (PH=2)

- Solution of sulfuric acid (H_2SO_4) diluted to a (PH = 1,45)
- Solution of ammonium nitrate and nitric acid 5 % (PH=2)

The attack due to chemicals specimens immersed in these solutions is evaluated according to ASTM C 267-96.

To assess the degree of degradation by calculating the mass loss of the specimens immersed in various aggressive environments according to time, we adopted the procedure below:

Each 7 days of attack and for a period of 28 days, the samples are cleaned with fresh water to remove the mortar removed altered, dried and stewed until the total elimination of absorbed moisture (the mass becomes constant) at a temperature of 105 °C. The degree of degradation is evaluated by the following formula loss weight:

$$\text{Mass loss (\%)} = [(M1 - M2) / M1)] \times 100$$

With: **M1**: dry mass of the specimens before immersion
M2: dry mass of the specimens after immersion

3. Results and discussions

3.1. Mechanical properties of mortars

The following table presents the results of mechanical strength of test pieces made up of different combinations.

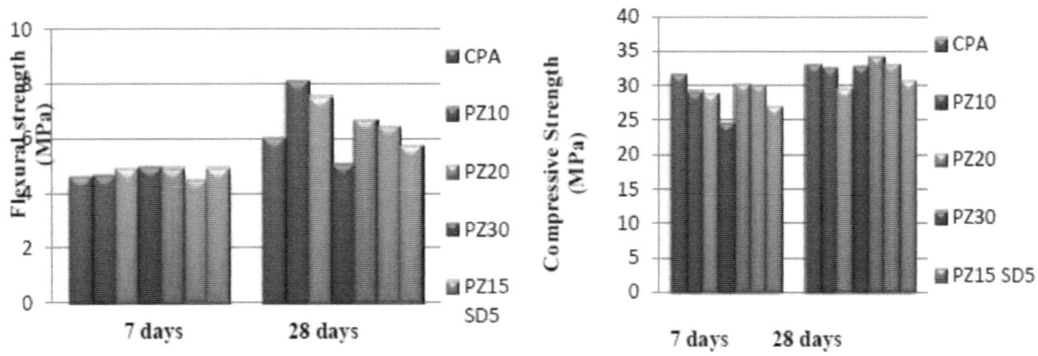

Figure 1. Evaluation of mechanical strengths

The figure shows the effect of substitutions additions rate used on the flexural and compressive strengths of mortars and 7 to 28 days. From this figure we note that the flexural strengths of mortars made with different rates of substitutions pozzolana and sand of dunes are approximate but above resistance control (APC). At 28 days, the resistance has developed except the PZ30. In seven days we observe that the resistances of all mortars of different rates of substitution are approximate resistance mortar control (APC). At 28 days, the resistances are increased especially for mortar (PZ15and SD5).

3.2. Mass loss in different aggressive environments

The following figure shows the histogram comparing weight loss among the three aggressive environments.

IOP Conf. Series: Materials Science and Engineering **123** (2016) 012015 doi:10.1088/1757-899X/123/1/012015

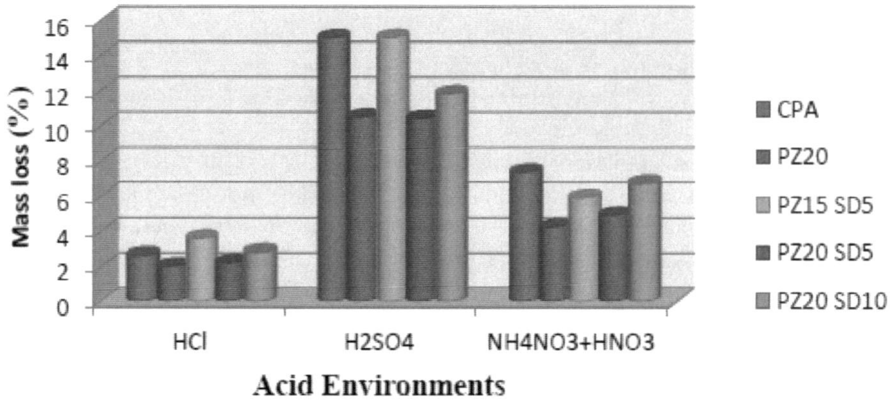

Figure 2. Histogram of mass loss in three different variants aggressive environments

3.3. Determination of the degraded depth

The degraded depths were determined by staining assay with phenolphthalein indicator.

Figure 3. Phenolphthalein indicator staining specimens attacked by HCl, H_2SO_4 and (NH_4NO_3 + HNO_3)

After staining phenolphthalein, one notices the pink color indicates that the basicity of the healthy part (not degraded), so we can determine the depth of degradation by using the measurement of the thickness with a caliper. According to our results, we note that the cement prepared with 20% substitution of pozzolana shows good resistance against the APC and the different variations, this is valid for 3 aggressive environments.

We note that our specimens immersed in various aggressive media for a period of 28 days undergo a much larger degradation in sulfuric acid solution and the depth is greater degradation. As against these specimens are more resistant to the hydrochloric acid solution and the thickness is degraded shallower. It was also noted that the specimens are more resistant (PZ20) and (PZ20 SD5) compared with controls (APC), this is consistent with the results of the resistance mechanically.

4. Conclusion

The study undertaken in this paper indicates that it is possible to exploit deposits of natural resources such as pozzolana and sand dunes to produce the binary and ternary cements in our country. These have compelling interests of technical, economic, environmental and sustainability point of view towards different chemical attacks. The experimental results conducted through this study allow drawing the following general conclusions:

- The mechanical strengths of the different variants are comparable to APC.
- The test results of the mass loss and phenolphthalein show that the chemical resistance of variants (PZ20) and (PZ20SD5) are larger compared to APC and other variants.
- This study adds value by replacing the addition of a portion of clinker.
- This substitution can save 20% to 30 % of clinker used to manufacture cement; this will have a beneficial effect for cement economic (less energy spent for the clinker burning).
- This study contributes to the protection of the environment as to produce 1 ton of clinker generates about 1 ton of CO_2 is harmful to the atmosphere. Based on our findings we will reduce from 20% to 30% CO_2 gas responsible for the greenhouse effect.

References

[1] Baron,J ; Olivier, JP. et .Weiss , J.C. 1997 « Les ciments courants » - « Les bétons, bases et données pour leur formulation » - Edition Eyrolles.
[2] livre : « le ciment et ses application » - école Anglet Cantau (EDC matériaux) - 2006.
[3] D.Achoura,D, Lanos, Ch, Jauberthie, R et Redjel, B. 2004 *Journal Physique IV France* **118** 159
[4] Yeginovali, A and Dilek, F. T. 1995 "Sulfate resistance of mortars containing silica fume as evaluated by different methods " Fly Ash , Silca Fume, Slag and Natural Pouzzolans Concrete, Proceedings, Fifth International Conference, Milwaukee, Wisconsin, USA- Vol 2 , pp 795-813.
[5] Turanli, L; Uzal, B; Bektas, F. 2004 "Effect of large amounts pozzolana addition on properties of blended cement" *Cement and Concrete Research.*
[6] Lee , S. T; Moun, H. Y; Swamy R. N. 2005 *Cement & Concrete Composites* **27** 65
[7] Mangat, P.S.and Khatib, J. 1995 *ACI Materials Journal* **92** 587
[8] Le Bellégo, C., Gérard, B. et Pijaudier-Cabot, G. 2001 Mechanical analysis of concrete structures submitted to an aggressive water. Fracture Mechanics of Concrete Structures. Edited by D. B. E AL. R. Swets and Zeitlinger, Lisse, pp. 239-246.
[9] Sebsadji, S. 2005 Les dégradations dans les ouvrages en béton armée : causes et prévention. Colloque CREDIMAT.
[10] Leroux, A. Orsetti, S. 2000 " les réactions sulfatiques : conditions de formation, structure et expansion des minéraux secondaires sulfatés " Bulletin de liaison des Laboratoires Centrales des Ponts et Chaussées, N° 225, Réf 4316, pp 41-50 Mars-Avril.
[11] Janotka, I and Krajci, L. 2000 "Utilization of natural Zeolite in Portland Pozzolana cement of increased sulphate resistance"In"Durability of Concrete Proceedings "Fifth International Conference, Barcelona, Spain-Vol I, pp 223-236.
[12] Ramezanianpour, A. K. 1987 « Engineering properties and morphology of pouzzolanic cement concrete», PhD Thesis, University of Leeds, 310 pages-April.
[13] Roy,D. M; Arjunan, P & Silsbee, M. R 2001 *Cement and Concrete Research* **2** 1809
[14] Mansour, S. 2003 " Etude des propriétés des mortiers et bétons à base de ciments ternaires " Mémoire de Magister – Université SAAD DAHLEB – BLIDA.
[15] Senhadji, Y 2006" L'influence de la nature du ciment sur le comportement de mortiers vis-à-vis des attaques chimiques (acides et sulfatiques) " Mémoire de Magister – I.G.C.M.O – ORAN

Physics and chemistry of minerals under laser processing

O Kotova[1] and N Leonenko[2]

[1]Institute of Geology Komi SC UB RAS, Syktyvkar, Russia
[2]Institute of Mining FEB RAS, Khabarovsk, Russia

E-mail: kotova@geo.komisc.ru

Abstract. New experimental data of the influence of laser irradiation on the phase composition minerals of bauxite and red mud and their technological features have been represented. The mechanisms of formation of micro- and nanophases on mineral surfaces under the influence of laser irradiation were shown. It is underlined that changes of condition of surface atoms, concentration of metals, agglomeration of titanium minerals and new phase formation occurred under laser irradiation on bauxites and red mud.

1. Introduction

The interaction of laser irradiation with mineral substance is characterized by many important factors compared to, e.g., mechanical processing. The degree of influence of laser irradiation on mineral (ore) depends on the ability of the latter to absorb the irradiation on of the given laser wavelength. It can be compared to mechanical processing based on such features as hardness and viscosity. It is considered that main energy output during utilization of mineral resources is connected with ore degradation to expose minerals. Such an approach is not sufficient enough due to variety of entry of metals into host minerals, besides the excessive crushing is negative. The laser processing allows avoiding excessive crushing of the material, new 0.1 mm phase formation and metal agglomeration are observed on mineral surface.

The main task of the paper - to show mechanisms of laser processing of minerals (ore), which open the perspective of new material creation, utilization of "invisible" gold and other valuable metals stimulating transition to the industry of laser technologies.

2. Materials and Methods

The objects of studies – natural and technogenic mineral systems of bauxite and iron-containing concentrate from red mud (Russia). The sample processing was conducted with the use of ytterbium fiberoptical continuous wave laser LS-06 with wavelength λ = 1060 nm. The power of output irradiation varied from 100 to 400 W. The samples were studied by scanning electron microscopy. Element analysis of initial composition of the studied sample, study of structural and element chemical changes under laser irradiation were conducted with LEO EVO 40HV scanning electron microscope (Carl Zeiss, Germany) with INCA-ENERGY energy dispersive analyzer. The imaging of the studied objects was supported by SE-detector, which gives information about sample topography. Additionally QBS-detector was used. Localization of sensing beam to the spot with diameter approximately $3 \cdot 10^{-5}$ mm with the depth of penetration of beam ~ 10^{-3} mm allowed to obtain data on qualitative and quantitative distribution of chemical elements in the studied mineralogical objects.

3. Results and discussions

Low-iron bauxites. The method of low-iron bauxites treatment for processing is suggested. The discovery is related to the sphere of low-iron bauxite concentration and can be applied for preliminary preparation for processing, e.g., during bauxite processing for titanium, which allows increasing integrity and efficiency of bauxite processing. The integrity of bauxite processing is realized not to full extent, mainly for rare earth elements. The technical result is achieved due to the method of preparation of low-iron bauxites includes rough preparation of the surface and subsequent irradiation.

Laser irradiation will be absorbed by mineral phases with greater absorption capacity (absorption depends also on melting and evaporation temperature, volatility etc.). The absorption capacity of mineral increases during laser impulse due to thermal emission and photoeffect. These processes result in electron emission, which increases absorption.

Before processing the sample represented rocks with boehmite predomination, which was seen on the diffraction pattern of non-irradiated sample (Fig.1).

Initially the sample surface represented grained boehmite matrix (Fig.2(a)) with titanium in the form of rutile and ilmenite. However titanium can be isomorphically included in lattices of other minerals. According to Belyaev with co-authors [1] up to 3.5 % TiO_2 is included in the lattice of boehmite. During our micro X-ray studies of bauxites, titanium is absent (or beyond sensitivity) in boehmite matrix.

The conducted studies showed that redistribution of bauxite substance is observed with titanium concentration in sinter during laser processing of kaolinite-boehmite bauxites (according to statistical interpretation of microprobe scanning the content of titanium increases in average at 1.7 times), at that titanium concentration is observed at certain values of energy density of laser irradiation. During the experiment the optimal parameters of energy density of laser irradiation were determined, at which a new phase crystallization was observed.

Iron-containing concentrate from red mud. Figure 3 shows the surface of initial sample (-0.071 mm fraction) and element microanalysis of local zone of iron-containing concentrate from red mud.

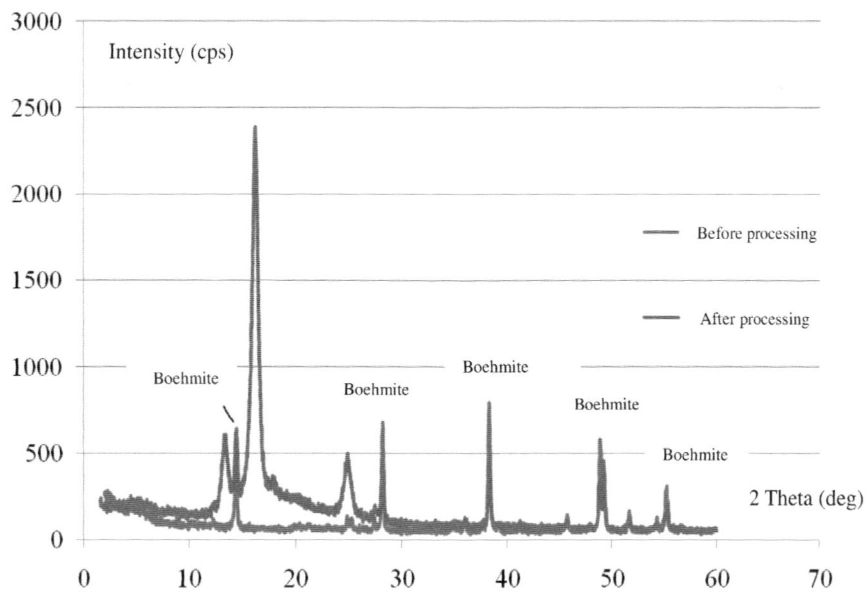

Figure 1. Diffraction patterns of initial and laser-processed bauxite

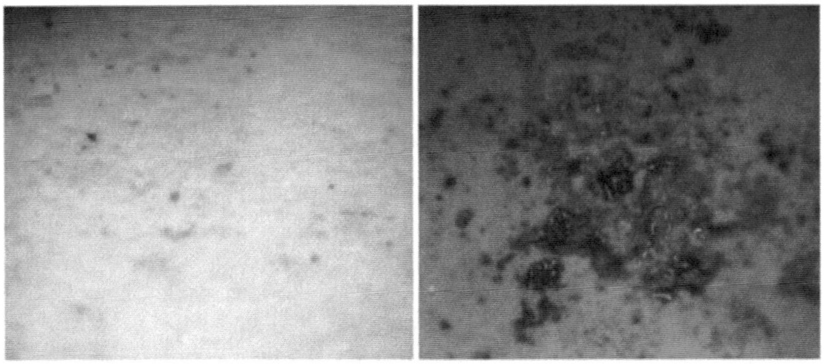

Figure 2. Surface of low-iron bauxite from Vezhayu-Vorykva deposit before (a) and after (b) laser processing

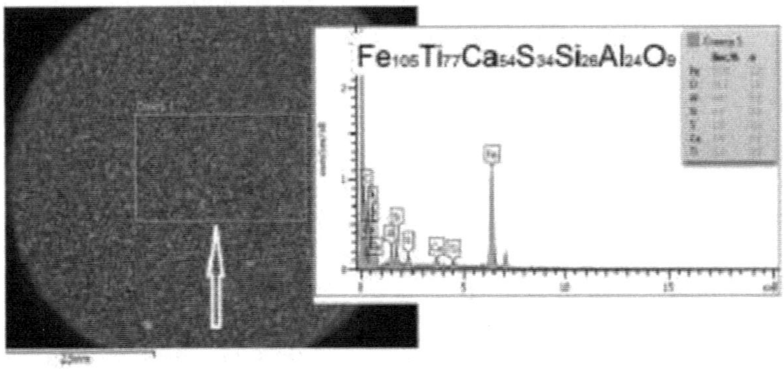

Figure 3. Surface of initial sample of iron-containing concentrate from red mud with -0.071 mm fraction and element microanalysis in local zone

Iron content was 50 % (wt%) according to the chemical analysis. The element analysis of iron-containing concentrate from red mud showed the presence of iron, titanium, calcium, silica, aluminum and oxygen in the following ratio: $Fe_{105}Ti_{77}Ca_{54}S_{34}Si_{26}Al_{24}O_9$.

During the experiments a sample of iron-containing concentrate from red mud (fraction - 0.071 mm) was placed on a special graphite base. The rate of laser process was continuous up to 1 mm/sec. As a result of laser processing the obtained samples were of spherical shape with chemical non-homogeneity. The laser processing with power from 100 to 400 W resulted in the agglomerated pellets of iron-containing concentrate from red mud (Fig.4).

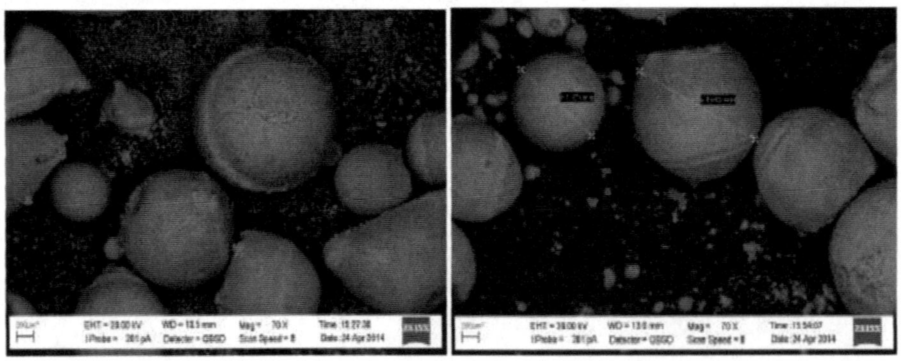

Figure 4. Agglomerated pellets of iron-containing concentrate from red mud

The size of the pellets varied form 0.5 mm to 2.0 mm. The pellets were represented by glass-like spherical structures with a high degree of hardness, which were very difficult to crush in agate mortar. The hardness was 6-7 by Mohs scale. The surface of the sinters (Fig.2) showed local zones of gold and platinum concentration: $Au_{547}Pt_{536}Rb_{103}Fe_{44}Ca_{22}O_4$ and $Au_{753}Pt_{738}Fe_{61}Ca_{31}O_5$ (Fig.5).

Figure 6 represents the element analysis of the following composition: a - $Au_{434}Pt_{426}Fe_{35}Ca_{18}O_3$, b - $Au_{336}Pt_{330}Fe_{27}O_2$.

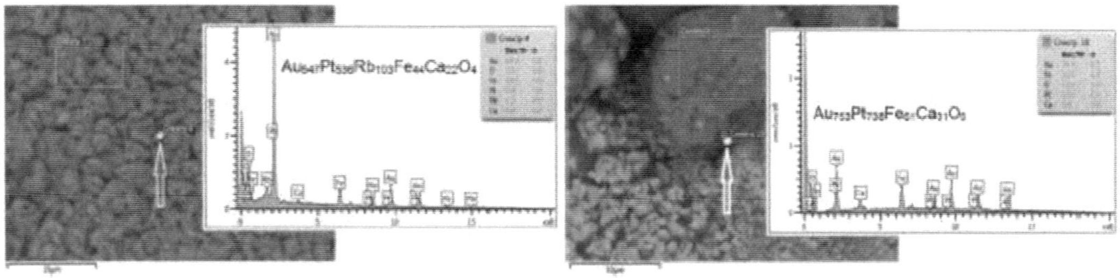

Figure 5. Gold and platinum on the agglomerated pellet surface of iron-containing concentrate from red mud (laser power 100 W)

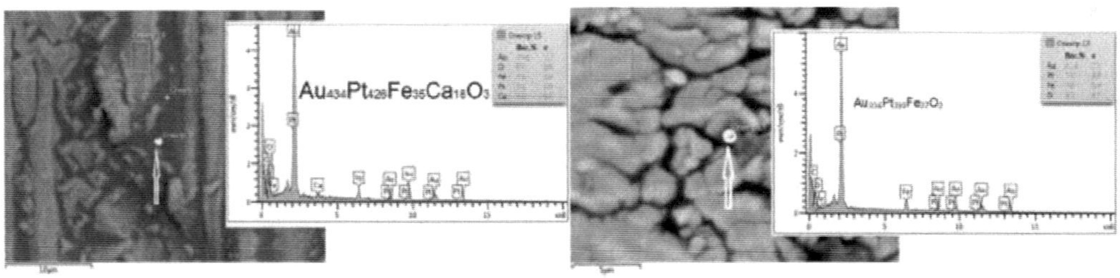

Figure 6. Gold and platinum on the agglomerated pellet surface of iron-containing concentrate from red mud (laser power: a – 125 W and b -150 W)

We observed various surface structures depending on the laser power and irradiation density at given parameters. E.g., figure 7 represents images of modified surface structures with element analysis and identification of gold, platinum, zircon, tungsten at laser processing from 100 to 400 W.

4. Conclusion

The method of bauxite laser processing has been developed, which resulted in redistribution of substance with agglomeration of titanium minerals and new phase formation, which are recoverable by conventional methods of concentration.

The laser processing results in the thermal disintegration of crystal lattices of finely disseminated minerals of iron-containing concentrate from red mud, then fast thermal recrystallization, defragmentation and sintering due to laser exposition of second duration. This scenario leads to increasing chemical homogeneity of dispersive mineral objects. The concentration and agglomeration of metals, particularly gold and platinum, into larger formations occurs under the influence of laser irradiation. Apart from gold and platinum, under laser processing at 125 W hafnium, zircon, and at 200 W – bismuth, at 400 W – tungsten are identified on the surface of sinters of iron-containing concentrates from red mud. However it should be noted that we did not find platinum on the surface at power more than 200 W.

Studies presented above can be a scientific base to develop new technologies of extraction of submicron and nanometer forms of noble metals and other valuable components.

Figure 7. Images of modified surface structures with element analysis and identification of gold, platinum, zircon, tungsten (laser power from 100 to 400 W)

5. Acknowledgements
The work was done under financial support of RAS programs (project 15-11-5-33).

References

[1] Belyaev et al 1975 Regularities of formation and placement of Timan bauxite (Syktyvkar, Institute of Geology AS USSR) (in Russian)
[2] Vakhrushev A et al 2012 Vestnik of Institute of Geology of Komi SC UB RAS **10** 14. (in Russian)
[3] Kotova O et al 2012 Vestnik of Institute of Geology of Komi SC UB RAS **10** 32. (in Russian)
[4] Leonenko N et al 2007 Notes of mining institute **171** 113. (in Russian)
[5] Leonenko N et al 2013 Development of sceintific basis of recovery of submicron forms of gold from technogenic raw (Saint-Petersburg, Saint-Petersburg state technological university) (in Russian)

The use of experimental bending tests to more accurate numerical description of TBC damage process

T Sadowski, P Golewski

Faculty of Civil Engineering and Architecture, Lublin University of Technology Lublin, Poland

E-mail: sadowski.t@gmail.com

Abstract. Thermal barrier coatings (TBCs) have been extensively used in aircraft engines to protect critical engine parts such as blades and combustion chambers, which are exposed to high temperatures and corrosive environment. The blades of turbine engines are additionally exposed to high mechanical loads. These loads are created by the high rotational speed of the rotor (30 000 rot/min), causing the tensile and bending stresses. Therefore, experimental testing of coated samples is necessary in order to determine strength properties of TBCs. Beam samples with dimensions 50x10x2 mm were used in those studies. The TBC system consisted of 150 μm thick bond coat (NiCoCrAlY) and 300 μm thick top coat (YSZ) made by APS (air plasma spray) process. Samples were tested by three-point bending test with various loads. After bending tests, the samples were subjected to microscopic observation to determine the quantity of cracks and their depth.The above mentioned results were used to build numerical model and calibrate material data in Abaqus program. Brittle cracking damage model was applied for the TBC layer, which allows to remove elements after reaching criterion. Surface based cohesive behavior was used to model the delamination which may occur at the boundary between bond coat and top coat.

1. Introduction

The novel materials are different types of composites obtained as mixtures of various phases and further subjected to specific technological process in the manufacturing, e.g. [1, 2]. The materials engineering allows for designing of almost arbitrary internal structure of the composites, particularly important for industrial demands, e.g. 2-phase ceramic materials [3-9]. A very important example are functionally graded materials, possessing gradation of the material properties, (e.g. [10-20]), for possible applications in thermal barrier coating (TBC) of engines critical parts.

The study of TBC protective coatings is a very complex and difficult issue. The mere fact, that the protective coating is applied on the turbine blades with complex shapes, makes the stress and strain fields are non uniform [21, 22] in comparison to simple layered TBC structure. In addition, the blades also have internal cooling channels, which introduce variable stiffness of the substrate and are a source of thermal stresses formation. Therefore, it is necessary to strive for their most optimal distribution, also taking into account the cooling efficiency ratio [23].

In addition to thermal loads, turbine blades are also carry the considerable mechanical loads. As a result of not overlapping the direction of centrifugal force with the centers of gravity of blade cross - sections, tends to form a complex stress state caused mainly by stretching and bending. Such a complex case was analyzed at work [24].

A very important issue that will have an influence on the durability and strength of protective layer are environmental loads. Due to the fact that the TC layer (top coat) is porous it does not constitute a barrier to oxygen contained in the exhaust gas. Therefore, during blade operation, it comes to gradual increase of TGO (thermally grown oxide) layer on the interface between TC and BC (bond coat). The growth process of this layer and microscopic measurements are presented in [25].

The difficulty, that always appears during TBC coating research is the strain measure. The reason for this is that, the coating thickness is about 0,5mm, and would be difficult to use eg. tensometric gauges. Therefore, in this paper, a solution to solve this problem by using optical displacement measurement system Aramis was proposed.

2. Description of test method

The method presented in this work consisted of three steps: bending test with strain observation by Aramis system, microscopic observations and damage model calibration in Abaqus program.

Experimental studies were carried out on the MTS 25kN testing machine. Due to the small sample sizes b x h x w (10mmx2mmx50mm) previously designed handle was used for the bending test, which allows to adjust the supports and conduct research for different loads.

Due to the fact the Aramis system was used, side surfaces of the samples had to be first carefully sanded and polished (Figure 1).

Figure 1. Samples preparation

Figure 2. Bending test

The three point bending test was carried out in a static way (Figure 2). During the test, both the force, displacement and strains of the side surface of the sample with TBC layer were registered. The research was carried out for 10 load levels from 50N to 500N with the increment of 50N.

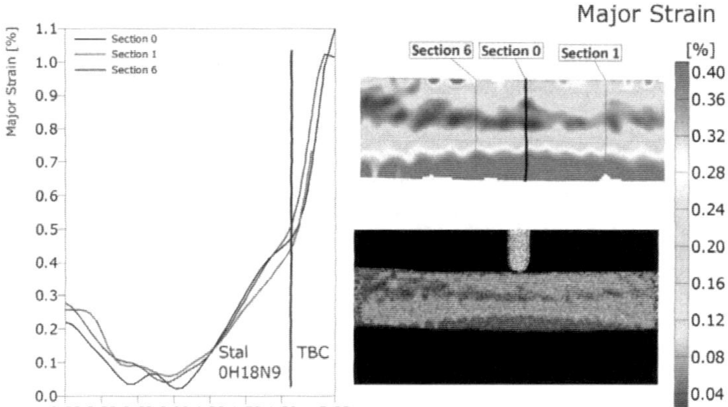

Figure 3. The results from Aramis system

A high noise occurred for small loads from 50N to 250N and in strain fields it was difficult to distinguish the individual layers: substrate and TBC. From 300N to 400N a clear distinction of individual layers already occurred, but just at the edge there were still areas where the system could not determine the correct values. The best effect was obtained for the largest load. In Figure 3 we can clearly read the position of the neutral axis, and the maximum strain values. Making cross - sections (section 0, 1 and 6), we can read the values for both the BC and TC layers.

3. FEM model description

The numerical model was built with elements of type CPS4R (A 4-node bilinear plane stress quarrilateral, reduced integration, hourglass control). For a beam model 12500 elements were used and for a single support 289. The substrate was made from stainless steel OH18N9 with 2mm thickness (Figure 4) for which a separate strength tests were carried out. The following values were assumed for the substrate: E = 200 GPa, n = 0.3, R_e = 190 MPa, R_m = 700 MPa, A = 0.45.

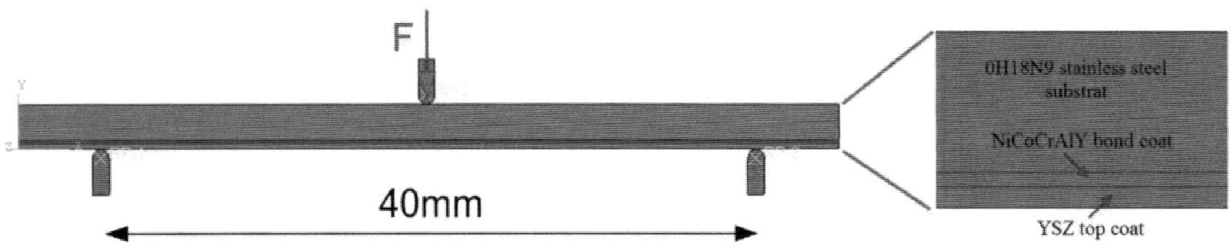

Figure 4. Boundary conditions and structure of FEM model.

The material data for bond coat (BC) and top coat (TC) were partially taken from the literature and confirmed in our own tests using microhardness testing. The samples were made by air plasma spraying method (APS) by PZL Rzeszów and the same covering was applied also on guide turbine blades of PZL 10W engine. In order to make the numerical model closer to reality, the material model for ceramic coating also includes damage by using brittle cracking model. This is a fuzzy type model representing discontinuities in the brittle material. Therefore, a single crack is not tracked like in the case of X-FEM method [26]. Crack initiation is based on a simple Rankine hypothesis, which assumes that a crack arises when the maximum tensile stresses reach the tensile strength σ_t^I.

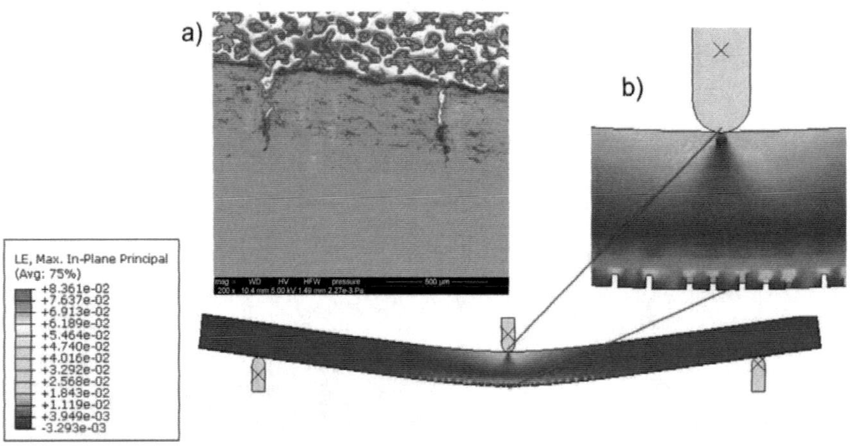

Figure 5. Damage of TBC coating:
a) results from microscopic observations, b) FEM simulation

Additionally, the brittle cracking model requires also the definition of damage evolution. The following criteria are used to describe the damage evolution: stress - strain criterion, fracture energy criterion and displacement criterion. Figure 5 summarizes two results: a) the experimental test and b) FEM simulation. Microscopic observations were carried out on Quanta FEG scanning electron microscope of FEI company. The coating thickness (TC) was about 300μm, and the thickness of bond coat 150μm, these values were also adopted in the simulation. In Figure 5a there is also visible the folding of the individual layers. It should also be pointed out that the crack does not always occur only in the ceramic coating (TC) and may also propagate to the bond coat (BC). Therefore, further works in research and FEA simulations are needed in order to describe the damage process in as much detail as possible, including delamination with application of the cohesive model, e.g. [27-33] . The example has confirmed that it is possible to use strain measurements of Aramis system and microscopic observations for a better description of damage in Abaqus program.

4. Conlusions

In this paper three - point bending tests were performed using an innovative method of system Aramis application for strain measurement in TBC layer. The tests were performed for 10 load levels from 100N to 500N with increments of 50N. For each load level microscopic observations were carried out in search of places where a crack or delaminating occurred. The results of deformation and damage information allowed for more detailed description of brittle cracking model for TBC in Abaqus program.

References

[1] Gömze,L.A.; Gömze L.N. 2009 *Ěpitöanyag – J. Silicate Based and Comp. Mat.* **61** 38
[2] Gömze,L.A.; Gömze L.N. 2013, *IOP Conf. Ser.: Mater. Sci. Eng.* 47 012033 http://dx.doi.org/10.1088/1757-899X/47/1/012033.
[3] Sadowski, T. 2012, *Comput. Mat. Sci.*, **64**, 209
[4] Sadowski, T.; Marsavina, L. 2011, *Comput. Mat. Sci.,* **50**, 1336
[5] Sadowski, T.; Samborski, S. 2008, *Comput. Mat. Sci.*, **43**, 75
[6] Marsavina, L.; Sadowski, T. 2009 *Comput. Mat. Sci.* **44**, 941
[7] Marsavina, L.; Sadowski, T. 2009 *Comput. Mat. Sci.* **45**, 693
[8] Marsavina, L.; Sadowski, T. 2007 *Int. J. Frac.* **145**, 237
[9] Postek, E.; Sadowski, T. 2011 *Comp. Interf.* **18**, 57
[10] Birman, V. ; Byrd, L. V. 2000 *Appl. Mech. Rev.* **53**, 147
[11] Sadowski, T. ; Neubrand, A. 2004 *Int. J. Fract.* **127**, 135
[12] Ivanov, I.; Sadowski, T.; Pietras, D. 2013 Eur. *Phys. J. Special Topics* **222** 1587
[13] Petrova, V.; Sadowski, T. 2014 *Meccanica* **49**, 2603
[14] Sladek, J.; Sladek, V.; Zhang, Ch. 2005 *Comp. Mater. Sci.* **32**, 532
[15] Hosseini, S.M.; Abolbashari, M.H. 2012 *Heat Transf. Eng.* **33**, 261
[16] Sadowski, T.; Nakonieczny, K. 2008 *Comput. Mat. Sci.* **43**, 171
[17] Sadowski, T.; Ataya, S.; Nakonieczny, K. 2009 *Comp. Mater. Sci.* **45**, 624
[18] Nakonieczny, K. ; Sadowski, T. 2009 *Comp. Mater. Sci.* **44**, 1307
[19] Birsan, M.; Altenbach, H.; Sadowski, T.; Eremeyev, V.; Pietras, D. 2012 *Comp. Part B* **43**, 1315
[20] Birsan, M.; Sadowski T.; Marsavina L.; Linul E.; Pietras D. 2013 *Int. J. Sol. Str.* **50** 519
[21] Sadowski, T.; Golewski, P. 2011 *Comp. Mat. Sci.* **50** 1326
[22] Sadowski, T.; Golewski, P. 2012 *Def. and Diff. Forum* **326-328** 530
[23] Sadowski, T.; Golewski, P. 2012 *Comp. Mat. Sci.* **52** 293
[24] Sadowski, T.; Golewski, P. 2012 *Comp. Mat, Sci.* **64** 285
[25] Sadowski, T.; Golewski, P. 2014 *Sol. St. Phenom.* **216** 128

[26] Golewski, G. L.; Sadowski, T.; Golewski P. 2012 Comp. Mat. Sci. **62** 75
[27] Burlayenko, V.; Sadowski, T. 2009 *Comp. Mater. Sci.* **45** 658
[28] Burlayenko, V.; Sadowski, T. 2014 *Meccanica* **49** 2617
[29] Sadowski, T. ; Golewski, P. ; Zarzeka-Raczkowska, E. 2011 *Comp. Mat. Sci.* **50** 1256
[30] Burlayenko, V.; Sadowski, T. 2014 Comp. Struct. **108** 354
[31] Sadowski, T.; Golewski, P.; Kneć, M. 2014 *Comp. Struct.* **112** 66
[32] Sadowski, T.; Marsavina, L.; Peride, N.; Craciun, E.-M. 2009 *Comput. Mat. Sci.* **46** 687
[33] Balawender, T.; Sadowski, T.; Golewski, P. 2011 *J. Adhes. Sci. Technol.* **25**, 2391

IOP Conf. Series: Materials Science and Engineering **123** (2016) 012018 doi:10.1088/1757-899X/123/1/012018

Thermal Insulation Properties Research of the Composite Material "Water Glass - Graphite Microparticles"

V A Gostev[1], E A Pitukhin[2], A S Ustinov[1] and A S Shelestov[1]

[1]Physical Engineering Department, Petrozavodsk State University, Petrozavodsk, Russia

[2]Mathematical Department, Petrozavodsk State University, Petrozavodsk, Russia

E-mail: vgostev@psu.karelia.ru, Eugene@petrsu.ru, anton-ustinov@petrsu.ru and shelestov@psu.karelia.ru

Abstract. Research results for the composite material (CM) "water glass - graphite microparticles" with high thermal stability and thermal insulation properties are given. A composition is proposed consisting of graphite (42 % by weight), water glass $Na_2O(SiO_2)_n$ (50% by weight) and the hardener - sodium silicofluoride Na_2SiF_6 (8% by weight). Processing technology of such composition is suggested. Experimental samples of the CM with filler particles (graphite) of a few microns in size were obtained. This is confirmed by a study of samples using X-ray diffraction analysis and electron microscopy. The qualitative and quantitative phase analysis of the CM structure was done. Values of limit load causing destruction of the CM were identified. The character of the rupture surface was detected. Numerical values of the specific heat and thermal conductivity were defined. Dependence of the specific heat capacity and thermal conductivity on temperature during monotonic heating was obtained experimentally. Studies have confirmed the increased thermal insulation properties of the proposed composition. The CM with such properties can be recommended as a coating designed to reduce heat losses and resistant to high temperatures. Due to accessibility and low cost of its components the proposed material can be produced on an industrial scale.

1. Introduction

Currently one of the most important problems is the creation of materials and coatings designed to reduce heat loss and resistant to high temperatures. One way to solve this issue is to develop a technology of preparation and application of composites, which when applied to a surface slow down its heating [1]. Over the past decades, composites have been extensively put into practice and replaced traditional materials in energy, transportation, electronics and other fields of endeavor. The difference between the traditional materials and most of the composite materials is that the production process of the latter can be combined with manufacturing process of a product [1]. The use of graphite as a filler [2] is substantiated by its high temperature and chemical resistance. There are widely known composite materials (CM) based on magnesium and aluminum oxides [3-6] and other alumina-components that are similar in composition but have different reactivity with respect to water glass. There is also international experience with composite technology based on carbon fibers [7-12].

The authors of this paper designed, produced and experimentally investigated a composite material of a new structure with high thermal stability and thermal insulation properties.

2. Investigation of the source filler material

The phase composition of the original graphite was investigated using X-ray diffraction analysis performed on X-ray diffractometer DRON-3 using FeKa radiation. The intensity at each fixed angle was recorded every 20 seconds. The images were taken in reflection geometry within the angular range 3-145° with the step size of 0,05°. From the analysis (figure 1) it was concluded that the modification of graphite is hexagonal. The material revealed the presence of SiO_2 impurity up to 3% by weight.

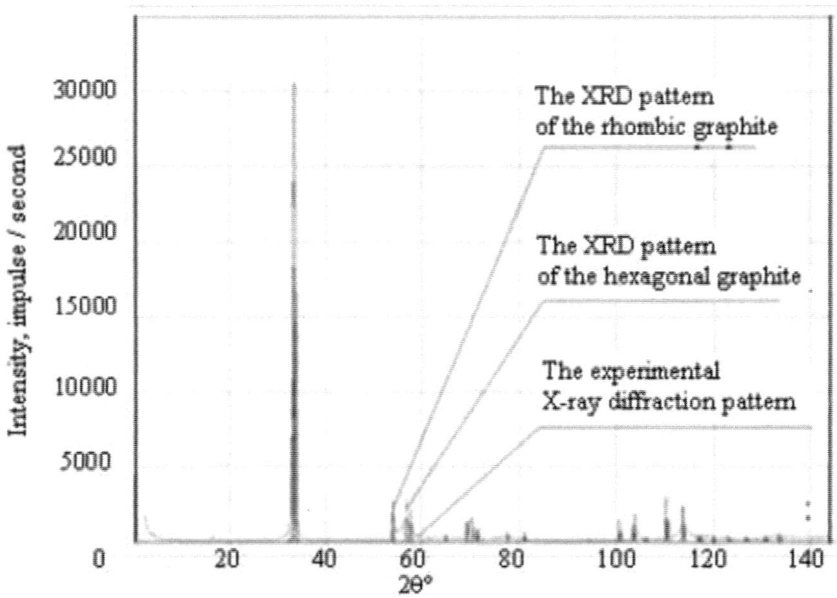

Figure 1. General view of the theoretically calculated X-ray diffraction of the rhombic and hexagonal graphite, as well as the experimental X-ray diffraction pattern of the graphite sample in the range of 2θ=40°-80°

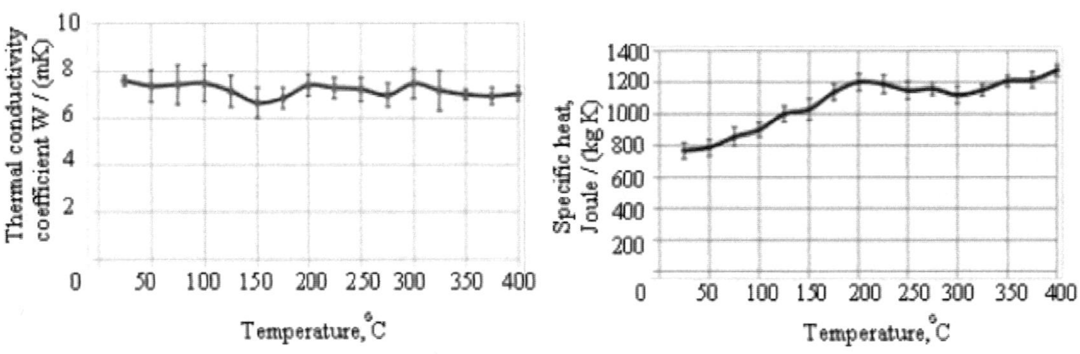

Figure 2. The values of the thermal conductivity coefficient X of graphite (a) and the values of the specific heat *c* of graphite (b) when the temperature changes from 25°C to 400°C

In order to determine the thermal stability of the initial samples and evaluate thermal insulation properties, the thermal conductivity coefficient (Figure 2a) and the specific heat (Figure 2, b) were measured at temperature ranging from 25°C to 400°C using the IT-X-400 and the IT-c-400 in monotonic heating mode [13-15]. The results of the measurement were processed using the methods of mathematical statistics. Random error in measurements with Student's coefficient equal to 2.9 and probability of 0.90 is shown in the graphs (Figure 2).

The Fig. 2a shows that the thermal conductivity coefficient of the CM when temperature is increased from 25°C to 400°C does not change significantly and remains in the range of 6-8 W/(m-K), taking into account random error in measurement. Fig. 2b illustrates the piecewise linear nature of the increase in the specific heat. With temperature changing from 25°C to 200°C there is a significant increase in the specific heat from 800 Joule/(kg-K) to 1200 Joule/(kg-K), compared with heating at temperature changing from 200°C to 400°C, when the specific heat increases slightly from 1200 Joule/(kg-K) to 1270 Joule/(kg-K). Such dependence can be explained in particular by evaporation of the internal moisture in the first section of heating of the CM.

3. Preparation of composite material based on water glass with graphite filler

At the first stage of production of the micro composition, micro powder of graphite with a particle size of 1-10 microns was obtained by milling the initial samples, which, in our opinion, provides a larger wetting surface, increases adhesion and expands the area of phase contact. Milling was performed on a laboratory planetary centrifugal mill - Hephaestus activator-2 (AGO-2U).

At the second stage the composite material was prepared. For the binder component we selected soda water glass Na_2O (SiO_2) n with the silicate modulus 2, 8. Water glass composition was prepared at ambient temperature by manually mixing in a mortar graphite powder (42 wt%), water glass (50% by weight) and hardener - sodium fluorosilicate Na_2SiF_6 (8% by weight). The ratio of components was chosen by the authors experimentally upon condition that there is no destruction of the samples when heated to 100 °C, making it possible to conduct studies at elevated temperature. Percentage ratio by weight was determined by weighing on an electronic balance. The process of wetting the particles with the binder component lasted 7 days. Pictures of the surfaces of the samples taken with electron microscope Hitachi SU 1510 (Figure 3) confirmed that the graphite particles in the composition have a size of several micrometers. The surface structure is clearly visible.

a b

Figure 3. Photomicrographs of the surface of the water glass composition with graphite filler with the horizontal picture size: 10 microns (a); 50 microns (b)

4. Mechanical testing of the composite material

An important characteristic that determines the mechanical properties of the composite material is adhesion [6] This characteristic greatly affects reaching the limit state under loads. In order to investigate the processes that occur during mixing graphite with water glass in the presence of a hardener, we conducted mechanical tests of the composite material [16, 17] to determine the load required to tear the test coverage off the base surface. The finished composition was applied on wood samples (thickness 2 mm). After thorough drying, the samples were torn off measuring the load at the breaking point with accuracy up to 10 N/m^2.

IOP Conf. Series: Materials Science and Engineering **123** (2016) 012018 doi:10.1088/1757-899X/123/1/012018

Figure 4. Tools to measure the load required to tear off the coating and pictures of the rupture surface after the experiment (Electron Microscope Hitachi SU 1510): wooden tool (general view) (a); wooden tool (side view) (b); type of the rupture surface (c); micrographs of the rupture surface (scale 10 microns) (d); micrographs of the rupture surface (scale 50 microns) (f)

The load was applied to the sample by hanging weights on hook tools (Figure 4a, b). The destruction of the samples (Figure 4,c-f) occurred along the boundary of the composite material and the base surface at the structure in the longitudinal section. Cracks in the samples were not formed. Limit fixed load value was 1,22 MPa.

5. Investigation of structure and physical properties of the composite material based on water glass with graphite filler

In order to ensure reliability of the results of the experiment four identical samples were investigated. To determine the thermal stability of the composite two samples were monotonically heated from 25° C to 400° C. The other two samples were control samples. We have performed X-ray analysis of the samples on DRON-3 diffractometer (FeKa radiation, wavelength $1{,}93597 \cdot 10^{-10}$ m) (Figure 5).

Comparison of the radiographs of all four samples - two heated up to 400° C (Fig. 5, a, b), and two that have not been heated (Figure 5, c, d) - makes it clear that the position of the diffraction peaks coincide. Qualitative analysis showed that the samples are multiphase. Graphite phase belongs to hexagonal graphite. Silicon dioxide SiO_2 is in the crystalline state of a-quartz. Change of the background, when 2θ° = 27, indicates that the composition also contains amorphous silicon dioxide SiO_2, and sodium fluoride NaF in the form of crystals. The presence of sodium silicofluoride Na_2SiF_6 phase in crystalline form shows that the chemical reaction of silica gel, which binds the filler phase, did not go completely.

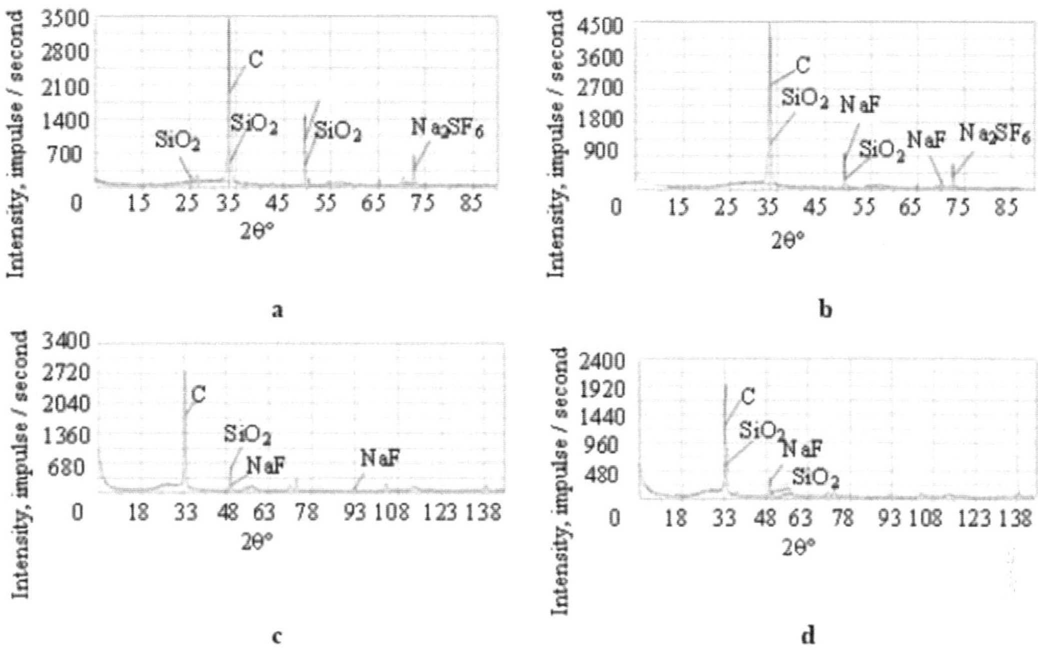

Figure 5. XRD patterns of the samples: No. 1 (a); No. 2 (b); No. 3 (c); No. 4 (d)

Quantitative analysis of the phase composition was performed on electron microscope Hitachi SU 1510 and DRON-3 diffractometer. The analysis is based on the exact position of the diffraction peaks. The results of the CM research are shown in the Table. The emission spectra obtained with the electron microscope are shown in Figure 6.

Table. The values of mass and mole fractions of the components of the CM

Phase	Mass fractions of the components in the samples,%				Mean,%	Molar mass of, g / mol	Mole fraction, x
	No. 1	No. 2	No. 3	No. 4			
C	42,1	42	44,2	42,4	42,6	12	0,76
NaF	22,8	21	22,8	21,7	22,1	41,98	0,11
SiO2	33	32,7	30,9	34,2	32,8	60,08	0,12
Na2SiF6	2,1	4,3	2,1	1,7	2,5	188	0,01

Thus, the authors demonstrated that the investigated material has heat resistance and is able to maintain substantially unchanged chemical structure in the temperature range from 25°C to 400°C.

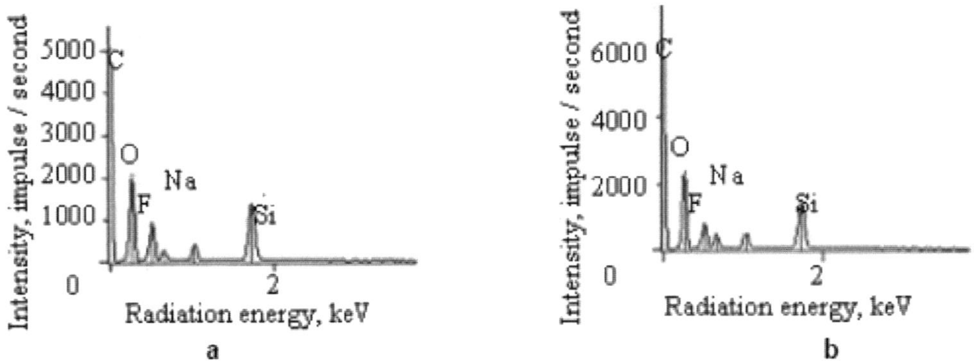

Figure 6. The emission spectrum of the CM samples not heated (sample No. 1) (a) and heated (sample No. 2) (b)

To evaluate the thermal insulation properties of the samples we measured the CM thermal conductivity coefficient X, W/(m-K) (Fig. 7a) and the specific heat c, Joule/(kg-K) (Fig. 7, b) at the temperature ranging from 25 ° C to 400 ° C by using the IT-X- 400 and IT-c-400 in monotonic heating mode [7-9]. The measurement results were processed by the methods of mathematical statistics. Random error in measurements with Student's coefficient equal to 2.9 and probability of 0.90 is shown in the graphs (Figure 7).

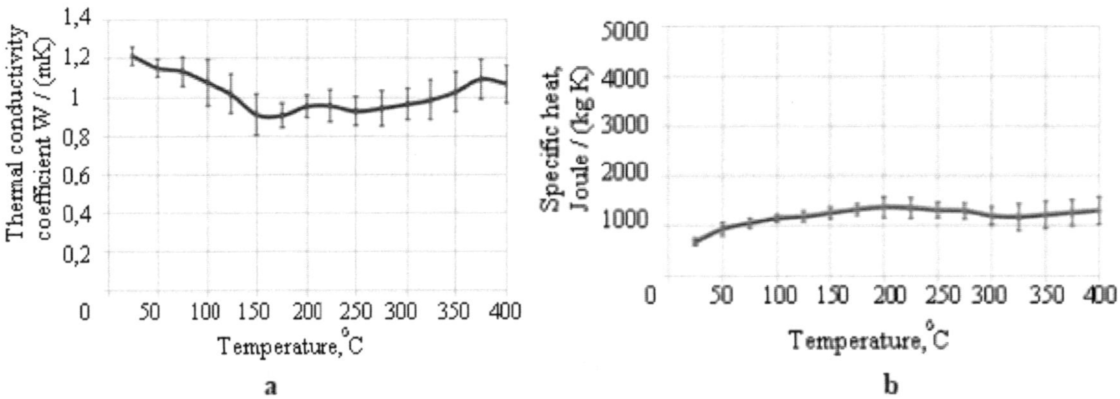

Figure 7. Temperature dependence of thermal conductivity coefficient λ of the CM samples (a) and specific heat ccm (b) in the range of 25°C to 400°C

6. Conclusion

The composite material with high heat resistance and insulating properties was studied [6, 18]. XRD analysis of the samples subjected to heating to 400 ° C showed that the material is able to maintain chemical structure unchanged, which confirms its higher thermal stability. The experimentally obtained dependence of the specific heat capacity and thermal conductivity coefficient on temperature showed good thermal insulation properties of the composite material.

Composite material with these characteristics can be used as a material or coating designed to reduce heat loss and resistant to high temperatures. Due to general availability and low cost of its components, the proposed material can be produced on an industrial scale.

7. Acknowledgment
This work was supported by the RF Ministry of Education and Science as a part of state program in the scientific field, projects N° 3.757.2014/K, program of strategic development of the Petrozavodsk State University 2012-2016.

References
[1] Karpov Y S, Ivanovskaya O V 2001*Composite materials: components, structure, processing intoproducts. Kharkov National Aero pace University Publishing House,* 153
[2] *Chemical Encyclopedia: In 5 volumes.* 1998 / Ch. Ed. Knuniants I L [until 1992], Zephirov N S (from1995) Moscow, Soviet Encyclopedia; Russian Encyclopedia
[3] Korenev V I, Danilov V V 1996 *Soluble and water glass. St. Petersburg: Construction Publishing* 213
[4] Romanenkov I G, Levites F A 1991 *Fire protection of building constructions. Moscow: Construction* Publishing, 320
[5] Meshalkin A B, Kaplun, A B 2001 *Journal of Inorganic Chemistry* **48** (10) 1712
[6] Eremin N.V. 2007 *Fire-retardant compositions based on liquid glass and mechanically activated aluminum and magnesium oxides: PhD Thesis in Engineering Science. Tomsk* 156
[7] Murariu M., Dechief A.L., Bonnaud L., Paint Y., Gallos A., Fontaine G., Bourbigot S., Dubois P. 2010 *Polymer Degradation and Stability* **95** (5) 889
[8] J Huang, D G Baird, J E McGrath. 2005 *Journal of Power Sources* **150** 110
[9] S Kishor Kumar, P Deepalekshmi, T Sabu, G Yves 2013.*Progress in Polymer Science* Manuscript, August (1) 107
[10] A Das, Gaurav R Kasaliwal, R Jurk, R Boldt, D Fischer, K Werner Stockelhuber, G Heinrich 2012 *Composites Science and Technology* **72** (16) 1961
[11] C Barreto, J Proppe, S Fredriksen, E Hansen, R W Rychwalski 2013 *Polymer* **54** (21) 3574
[12] B Wang, Y Jiao, A Gu, GLiang, L Yuan 2014 *Composites Science and Technology* **91** (31) 8
[13] GOST 7076-99. Building materials and products. Test method for thermal conductivity. Introduced 01.04.2000. Moscow, 2000, 27 p.
[14] GOST 23250-78. Building materials. Test methods for specific heat. Introduced 01.01.1979. Moscow: Publishing House of Standards, 1979, 11 p.
[15] Volkov *DP.* Devices and methods for measurement of thermo physical properties of substances. St.-Petersburg, 2006, 71 p.
[16] GOST 15140-78. Paints and varnishes. Test methods for adhesion. Introduced 01.01.1979. oscow: Publishing House of Standards, 1979, 12 p.
[17] GOST 28574-90. Corrosion protection in construction. Concrete and reinforced concrete. Test methods for adhesion of protective coatings. Introduced 01.01.1991. Moscow: Publishing House of Standards, 1990, 7 p.
[18] Volkova V K 2011 Thermophysical properties of composite materials with polymeric matrix and solid solutions. Publishing House of Science Education 101 p.

IOP Conf. Series: Materials Science and Engineering **123** (2016) 012019 doi:10.1088/1757-899X/123/1/012019

Evolution of microstructure and precipitates in 2xxx aluminum alloy after severe plastic deformation

B Adamczyk-Cieslak[1], J Zdunek[1], and J Mizera[1]

[1]Materials Science and Engineering Faculty, Warsaw University of Technology, Woloska, Warsaw, Poland

E-mail: badamczyk@inmat.pw.edu.pl

Abstract: This paper investigates the influence of precipitation on the microstructure development in a 2xxx aluminum alloy subjected to hydrostatic extrusion. A three step reduction of the diameter was performed using hydrostatic extrusion (HE) process: from 20mm (initial state) to 10 mm, 5 mm and 3 mm, which corresponds to the logarithmic deformations $\varepsilon = 1.4$, $\varepsilon = 2.8$ and $\varepsilon = 3.8$ respectively. The microstructure and precipitation analysis before and after deformation was performed using transmission electron microscope (TEM), and scanning electron microscopy (SEM). As a result of the tests, a very significant influence of precipitation on the degree of refinement and mechanism of microstructure transformation was stated.

1. Introduction

Alloys of the 2xxx series contain, as additives, beside copper (up to 6 % wt.), also magnesium, silicon and manganese (up to 1,2 %) as well as small amounts of iron, nickel, titanium, zirconium and lithium. They characterize in good mechanical properties and they also have the properties of heat-resisting materials, owing to the formation of phases rich in Fe, Mn and Ti [1,2]. Many alloy additives and impurities enable the formation of a wide range of intermetallic phases of a diversified composition. The basic reinforcing phases are Al2Cu and Al2CuMg, as well as the metastable elements of the precipitation sequence [3-5]. What is more, the literature [6-11] identifies the following phases: heterogeneous precipitates Al(CuMnFeSi), Al7Cu2Fe, Al6MnFe2, (Al,Cu)6Mn, Al8Fe2Si, Al10Fe2Si, Al6(CuFeMn) as well as particles of the following types: AlCuFeMn(Si), Al7CuFe2, Al12(FeMn)3Si, Mg2Si, Al6Cu2(MnFe), Al6Mn, Al4(CuMgSi), Al10Cu3Mg3, (AlCu)6Mn. The literature [12-16] suggests that, for many commercial aluminium alloys (also of the 2xxx seires), an improvement of the mechanical properties can be achieved by means of modern plastic treatment methods which apply such techniques as Severe Plastic Deformation (SPD), chemical composition modification, as well as modern methods of multistage thermal treatment. The aim of this work is an analysis of the effect of the SPD technique, i.e. hydrostatic extrusion, on the grain and second phase particle refinement.

2. Experimental procedure

The investigated material was a commercial aluminum cooper alloy of the chemical composition given in table 1.

Table 1. Chemical composition of aluminium-cooper alloy

Alloy	Element (wt. %)						
	Mg	Mn	Cu	Si	Fe	Zn	Cr
2014	0,2-0,8	0,4-1,2	3,9-5	0,5-1,2	0,7	0,25	0,1

The alloy was obtained in the process of semi-continuous casting of ø 215 mm ingots. Next, the ingots without additional annealing treatments were subjected to the extrusion process, followed by annealing: 450oC /6h and cooling with furnace 30oC /1h down to the temperature of 200oC. Then the alloy was air cooled. The microstructure of the alloy in the initial state was built of coarse, elongated grains of the average diameter of 100 μm. In the next step, the sample was deformed at room temperature with the use of the hydrostatic extrusion (HE) method. Rods of the initial diameter of 20 mm were subjected to HE in three consecutive steps: 10 mm, 5 mm and 3 mm, which corresponds to the logarithmic deformations $\varepsilon = 1.4$, $\varepsilon = 2.8$ and $\varepsilon = 3.8$, respectively The extruded specimens were water cooled at the die outlet.

The microstructure of the deformed samples was examined on their transverse section. The alloy microstructures were examined using transmission electron microscope (TEM) JEOL 1200 with the accelerating voltage of 120kV. The mean sizes of the grain deq and the intermetallic particles (defined as the diameter of a circle which has the surface area equal to the surface area of a given grain) were determined by TEM (grains) and SEM (particles) for the minimum of 200 randomly selected objects.

3. Results

The 2014 alloy contains intermetallic particles distributed randomly in the α-Al matrix. The results of the SEM/EDS analysis showed three visible types of particles (figures 1, 2): a mainly white, spherical Al2Cu (θ-phase), an irregularly shaped Al-Cu-Mn-Fe-Si phase and small, spherical gray particles Al-Cu-Mg-Si.

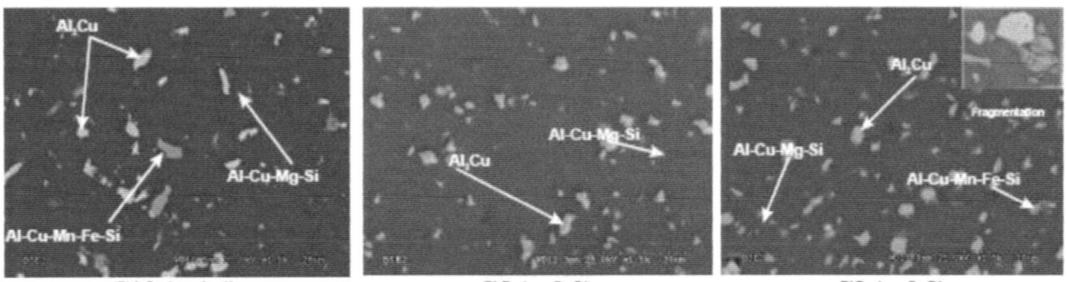

Ø10 (ε=1.4) Ø5 (ε=2.8) Ø3 (ε=3.8)

Figure 1. BSE-SEM images of the surface of the 2024 alloy after different step of extrusion

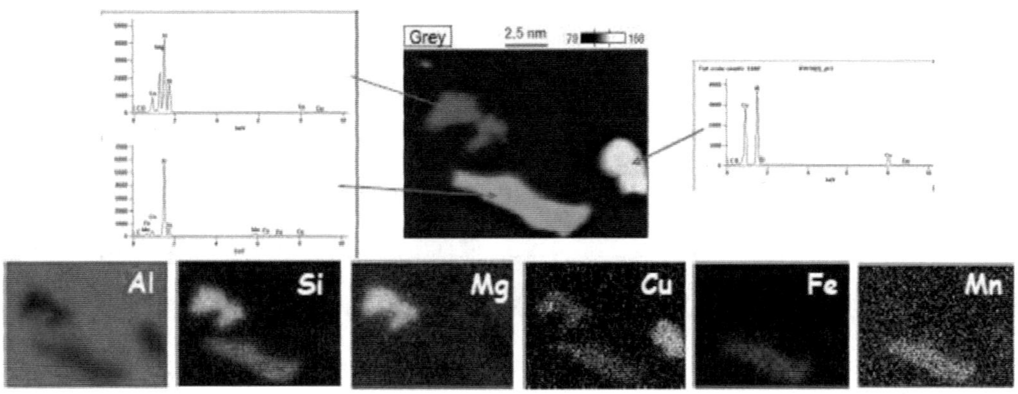

Figure 2. The EDS spectrum and mapping of the surface elements in 2024 alloy

The particles of Al2Cu and Al-Cu-Mn-Fe-Si broke up as a result of the HE, and this was accompanied by a re–distribution, which assured a more homogeneous microstructure.

The Al-Cu-Mg-Si particles, after deformation, only slightly changed their dimensions, in comparison to the initial state. The highest fragmentation was observed after the first step of extrusion (figure 3). The microscopic observations showed that the grain size decreases from about 100 μm in the initial state to 0,2 μm after the last step of extrusion (figure 4).

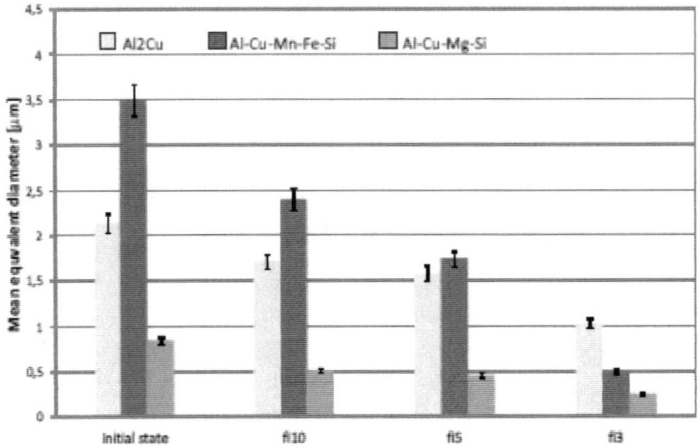

Figure 3. Mean equivalent diameter of the particles as a function of deformation mode

Figure 4. The microstructures of the 2024 alloy after different step of extrusion

The analysis of the microstructure of the investigated alloy before deformation indicates that most of the dislocations are present near the particles. In the areas with a higher density of dislocations, the recovery and recrystalization proceed easier, and thus they become places where a fine microstructure is created. The observations using transmission electron microscopy showed that, as a result of the deformation, the grain size decreases near the large particles (Fig. 5).

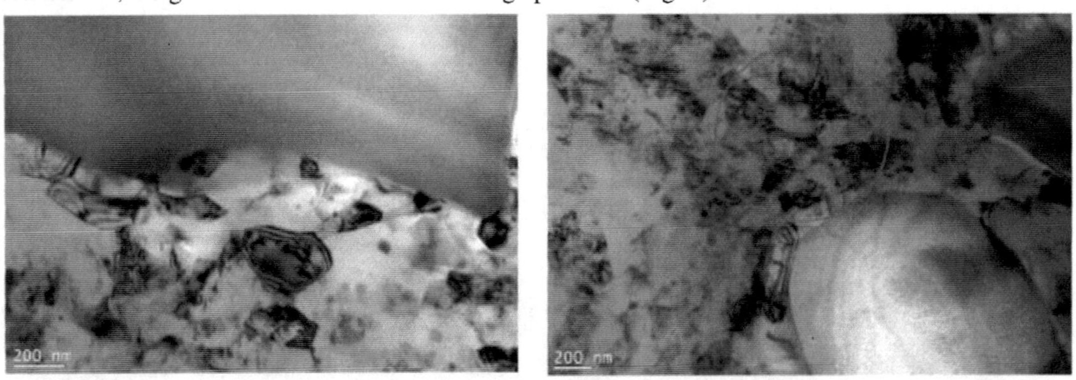

Figure 5.TEM images of the microstructure near to large particles in the samples after HE

The grains are equi – axial and smaller than those where the zone of particles interactions does not occur. The results indicate, that the HE method is most effective in reducing the grain size near the particles – in this case, the smallest grain size (deq = 80-100 nm) was obtained.

4. Conclusions
The results obtained in the present study lead to the following conclusions:

- In the 2024 alloy, three types of particles were visible: Al2Cu (θ-phase), Al-Cu-Mn-Fe-Si and Al-Cu-Mg-Si.
- The use of the HE method results in a significant increase of the volume fraction of the small particles and a decrease of the fraction of the large ones.
- The HE process brings about equi-axial nanograins in the vicinity of the large primary particles and leads to a significant refinement of microstructure.

Acknowledgments
The studies were financially supported by the research project of the National Science Centre No 2013/09/B/ST8/03520.

References
[1] Zeren M 2005 *J. Mater. Proc. Tech.* **169** p 292
[2] Fuller Ch. B, Seidman D N, Dunand D C 2003 *Acta Mater.* **51** p 4803
[3] Sha G, Marcelu RKW, Gao X, Muddle B C, Ringer S P 2011 *Acta Mater.* **59** p 1659
[4] Meyveci A, Karacan I, Ugur C, Durmus H 2011 *J. Alloy Compd* **491** p 278
[5] Lochte L, Gitt A, Gottstein G, Hurtado I 2000 *Acta Mater.* **48** 2969
[6] Scholes F H, Furman S A, Hughes A E, Markley T A 2006 *Corros. Sci.* **48** p 1812
[7] Gaziziv M, Kaibyshev R 2012 *J. Alloy Compd* **527** p 163
[8] Guillaumi V, Mankowski G 1999 *Corros. Sci.* **30** p 310
[9] Sharifitabar M, Nami H 2011 *Composites* Part B **42** p 2004
[10] Kumaran S M 2012 *J. Alloy Compd* **539** p 179
[11] Wang S C, Starink M J 2005 *Int. Mater.* **50** p 193
[12 Gao N, Starink M J, Furukawa M, Horsta Z, Xu C, Langdon T G 2005 *Mater. Sci. Eng.* A **410-411** p 303
[13] Buha J, Lumley R N, Crosky A G, Hono K 2007 *Acta Mater.* **55** p 3015
[14] Karnesky R A, Dunand D C, Seidman D N 2009 *Acta Mater.* **57** p 4022
[15] Ma Y, Zhou X, Thompson G E, Hashimoto T, Thompson P, Fowles M 2011 *Mater. Chem. Ph.* **126** p 46
[16] Huang L P, Chen K H, Li S, Song M 2007 *Scripta Mater.* **56** p 305

Hydrogen motion in Zircaloy-4 cladding during a LOCA transient

T Elodie[1], D Jean[1], G Séverine[1], B M-Christine[2], C Michel[3], P Berger[4], B Martine[5] and A Antoine[5]

[1]Institut de Radioprotection et de Sûreté Nucléaire (IRSN), Centre d'Etudes de Cadarache, PSN-RES/SEREX/LE2M, Saint-Paul lez Durance Cedex, France
[2]INSA de Lyon, Laboratoire de Mécanique des Contacts et des Structures (LaMCoS), Villeurbanne cedex, France
[3]École Centrale de Nantes, Institut de recherche en génie civil et mécanique (GEM), Nantes, France
[4]Laboratoire d'Etude des Eléments Légers (LEEL), SIS2M, CEA-Saclay, Gif sur Yvette, France
[5]EDF-R&D, Les Renardières, département MMC, Moret sur Loing, France

E-mail: elodie.torres@irsn.fr

Abstract. Hydrogen and oxygen are key elements influencing the embrittlement of zirconium-based nuclear fuel cladding during the quench phase following a Loss Of Coolant Accident (LOCA). The understanding of the mechanisms influencing the motion of these two chemical elements in the metal is required to fully describe the material embrittlement. High temperature steam oxidation tests were performed on pre-hydrided Zircaloy-4 samples with hydrogen contents ranging between 11 and 400 wppm prior to LOCA transient. Thanks to the use of both Electron Probe Micro-Analysis (EPMA) and Elastic Recoil Detection Analysis (μ-ERDA), the chemical elements partitioning has been systematically quantified inside the prior-β phase. Image analysis and metallographic examinations were combined to provide an average oxygen profile as well as hydrogen profile within the cladding thickness after LOCA transient. The measured hydrogen profile is far from homogeneous. Experimental distributions are compared to those predicted numerically using calculations derived from a finite difference thermo-diffusion code (DIFFOX) developed at IRSN.

1. Introduction

During a LOCA transient, the fuel cladding tubes are subjected to high temperature steam oxidation before core reflooding. During this type of accident, metallurgical evolutions due to α-β phase transformations and partitioning of oxygen/hydrogen but also of the main alloying elements is observed in the material. These phenomena have a strong influence on the material mechanical properties [1]. The microstructure results in a complex distribution of oxygen and hydrogen within the cladding thickness. At 1200°C, the hydrogen has low affinity for α(O) and an excellent solubility in the β-phase. During high temperature oxidation, most of the hydrogen moves towards the β-phase and is expected to remain in the prior β-phase after quenching [1, 2, 3]. The understanding of the mechanisms influencing the motion of these two chemical elements is required to fully describe the post-quenched embrittlement.

This study aims at determining oxygen and hydrogen concentrations in cladding materials after a LOCA transient. Several 70 mm long Zircaloy-4 samples are pre-hydrided with hydrogen contents ranging between the as-received content (~ 10 wppm) and 400 wppm. The samples are then steam oxidized at high temperature (1200°C), to simulate LOCA conditions, and quenched in a water bath at room temperature. The microstructure of the material, especially the phase distributions within the cladding thickness, is systematically characterized by metallography. Local oxygen and hydrogen distributions are specifically investigated using Electron Probe Micro-Analysis (EPMA) and micro Elastic Recoil Detection Analysis (μ-ERDA). Image analysis and metallographic examinations are combined to provide average oxygen profile as well as hydrogen profile within the cladding thickness after LOCA transient. Experimental average radial distributions are then compared to those predicted numerically using calculations derived from the DIFFOX code developed at IRSN [4].

2. Materials and methods

2.1. Materials

The specimens investigated in this study have been cut from low tin 17x17 SRA Zircaloy-4 fuel cladding tubes provided by AREVA-NP. The tubular samples are 70 mm long, have a 9.5 mm outer diameter and a 570 μm thickness respectively. Chemical composition of the tested material is detailed in Table 1.

Table 1. Chemical composition of the tested Zircaloy-4

Sn (wt %)	Fe (wt %)	Cr (wt %)	O (wt %)	H (wppm)
1.3	0.21	0.10	0.13	10

2.2. Experimental process

2.2.1. Hydrogen gaseous charging. The hydrogen gaseous charging of the 70 mm long cladding tubes is performed at the École Centrale de Paris. Each Zircaloy-4 sample is inserted in a quartz tube under high vacuum heated by a furnace at 420°C. Pure hydrogen is injected into the quartz tube. The charging temperature is chosen sufficiently low to avoid cladding recrystallization while permitting homogenous distribution of hydrogen in the sample. The sample is then slowly cooled down at 20°C/h to obtain large circumferential hydrides and uniformly distributed across the sample thickness. After gaseous charging, tubes are cut into 20 mm long segments for the oxidation. Thin samples are also cut for hydrogen measurements and metallographic analysis.

2.2.2. High temperature oxidations. Two-sided steam oxidations are performed in a vertical furnace at 1200°C in a mixture of argon and steam. The argon and steam flow rates are respectively 10Nl.min-1 and 500g.h-1 (50 vol% argon and 50 vol% steam) in order to prevent steam starvation and limit hydrogen pick-up. Oxidation durations are adjusted to reach ECR (Equivalent Clad Reacted) values of 15, 20 and 25%. The ECR is assessed using Cathcart-Pawel correlation assuming inner and outer wall oxidation [5]. Oxidation is interrupted by dropping the sample into a water bath at room temperature. The hydrogen content is also measured after oxidation for each sample by hot extraction. The microstructure of the material, especially the phase distributions within the cladding thickness, is characterized by metallography. Oxygen profiles are systematically measured using EPMA on each radial cut.

2.3. Samples characterizations

2.3.1. Hydrogen content measurement

The hydrogen content of the samples is measured using a Brücker ONH mat 286 by melting the sample at 1500°C in the presence of a carrier gas (argon). The hydrogen is then released in the device. The hydrogen content is determined by catharometry comparing the conductivity of the mixture with the one of pure argon. For 70 mm long tube samples, four axial locations are systematically characterized. A minimum of four analyses at each axial location are performed to check the homogeneity of hydrogen content.

2.3.2. Metallography examination
After oxidation, two cross sections are cut are at each end for all the tested samples for metallography. The thicknesses of the oxide and □(O) layers are systematically measured at four different angular locations at both outer and inner surfaces on each metallographic radial cut. The phase distribution of the samples is estimated by image analysis based on a half ring metallography reconstruction. The metallography is first transformed into a binary image of the two phases in the metal. This binary image is then used to determine the radial profile of α (O) phase fraction. An illustration is shown in Figure 1.

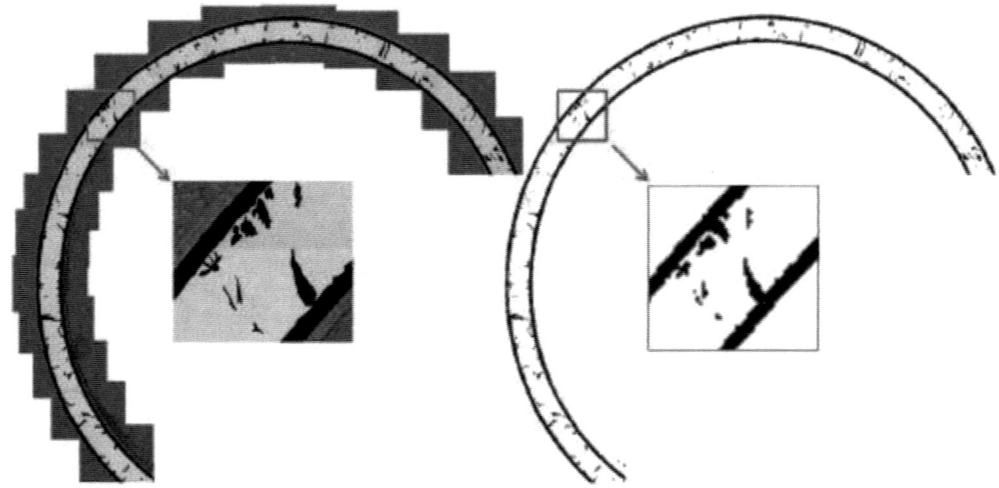

Figure 1. Example of the reconstruction and the binarisation of the optical microscopies (H66-7)

2.3.3. Electron Probe Micro-Analysis (EPMA)
Each radial cross section of the samples is embedded in a conductive resin with an as-received Zircaloy-4 sample. A mechanical polishing and then a chemical etching with a fluohydric acid solution is performed before introduction in the microprobe vacuum chamber. Oxygen K□ line is measured with a W/Si multi-layer synthetic crystal. Zircaloy-4 standards with homogeneous and known oxygen contents are used to perform oxygen calibration. The oxygen measurement uncertainty is estimated to ± 0.2 wt%. All the profiles are measured within the thickness with 1 μm stage displacement step in each phases and 250 nm stage displacement step at the zirconia/α(O) and α/prior-β interfaces. Points analysis are also performed in the α(O) inclusions, the α(O) layer and in the prior-β-phase.

2.3.4. Elastic Recoil Detection Analysis (ERDA)
The hydrogen distribution and quantification within each phase of the material are quantitatively measured at a microscopic scale using micro Elastic Recoil Detection Analysis (μ-ERDA). A 3 MeV 4He+ incident beam with an intensity of 500 pA is focused on the sample at a glancing angle of 15° with respect to the sample surface. Hydrogen atoms ejected by elastic collision are collected with a Si surface barrier detector placed symmetrically, with a 15° grazing angle. The energy of recoil hydrogen atoms is related to the depth from which they are scattered to the surface (the deeper hydrogen is located, the higher is the energy loss). Measurements are performed using a 3x2 μm^2 beam size and a 1x1 μm surface scanned step. Due to the grazing incidence, the actual beam size increases by a factor 4 according to the horizontal axis. Scan durations are about 4 to 5 hours to obtain 400x100 μm^2 maps.

The hydrogen measurement uncertainty is estimated to ± 10%. A homogeneously hydrided Zircaloy-4 sample is used to determine the accuracy of ERDA analysis. The hydrogen contents measured by sample fusion and by ERDA of this sample are respectively 190 ± 43 wppm and 207 ± 20 wppm.

3. Results and discussions

3.1. Chemical elements partitioning

Alloying element segregation within the cladding samples are systematically analyzed using EPMA. During the high temperature oxidation, some strong partitioning of the chemical elements occurs within the thickness of the cladding tube due to the oxygen diffusion and the associated progressive transformation of the β-phase into an oxygen stabilized α(O) layer [6, 7]. Typical illustration of the concentration profiles for alloying compounds obtained using EPMA are presented in Figure 2. The residual prior-β phase is enriched in iron and chromium consistently with their β-stabilizing properties. On the contrary, oxygen is segregated into the α phase. The chromium concentration follows a diffusion-like profile whereas tin is uniformly distributed into the two phase region without any noticeable segregation.

Experimental oxygen distributions are compared to the predictions of a 1-D finite-difference code developed by IRSN to model the oxygen diffusion equations [4]. The calculations accurately determine the α-layer thickness but strongly underestimate the oxygen content within the α(O) inclusions in the α(O)+β regions. EPMA indicates that in the α(O) inclusions and in the prior-β phase, the oxygen content is constant but different in each of the two phases. Consequently, using the α(O)-phase radial distribution, it is possible to determine the equivalent averaged oxygen radial profile. A rather continuous oxygen profile is obtained between the α(O) layer and the two-phase region. Experimental and calculated profiles are illustrated in Figure 3.

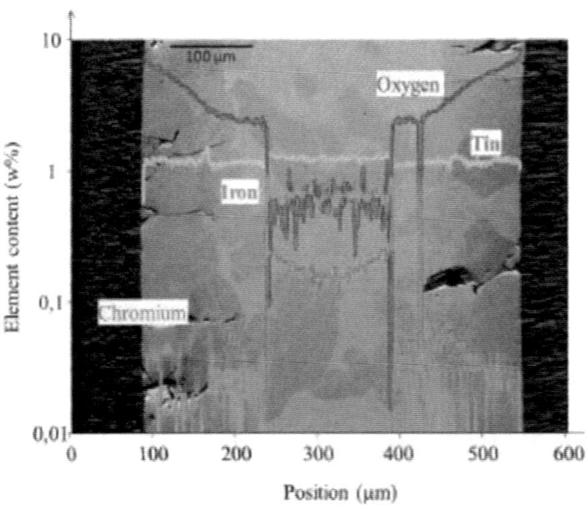

Figure 2. Typical concentration profiles in chemical elements alloy obtained by EPMA after oxidation at 1200°C during 624s

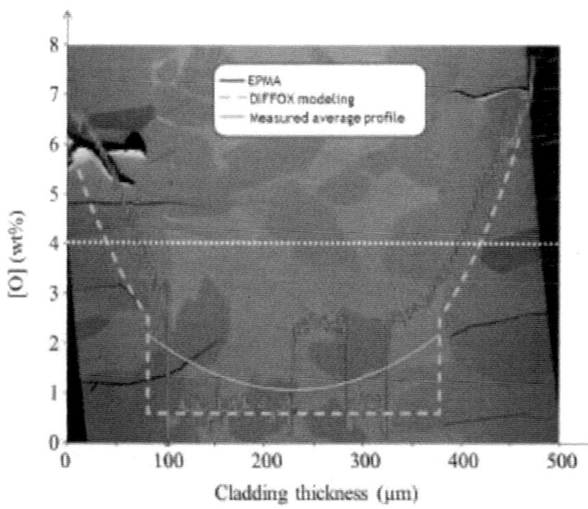

Figure 3. EPMA oxygen profiles versus Diffox modeling and measured average profile taking into account phases distribution

3.2. Local hydrogen quantification

Electron Recoil Detection Analysis (ERDA) has been performed on some oxidized samples to determine the hydrogen distribution within the cladding material. A typical map obtained by ERDA is illustrated in Figure 4. A hydrogen enrichment of the β-phase and depletion of the α(O) phase is evidenced. A local hydrogen enrichment is also observed around the α(O) grains in the beta side in agreement with Rapsaet study [2]. Far from α(O)-prior β grain boundaries, the hydrogen is homogenously distributed and its local content is close to equilibrium concentration. ERDA measurement was further used to determine hydrogen diffusion profile normal to grain boundaries from α(O) layer to the homogeneous prior-β phase. The hydrogen enriched layer is about 25 μm thick in the beta phase. A typical hydrogen diffusion profile derived from the hydrogen content analyses normal to a grain boundary is illustrated in Figure 5.

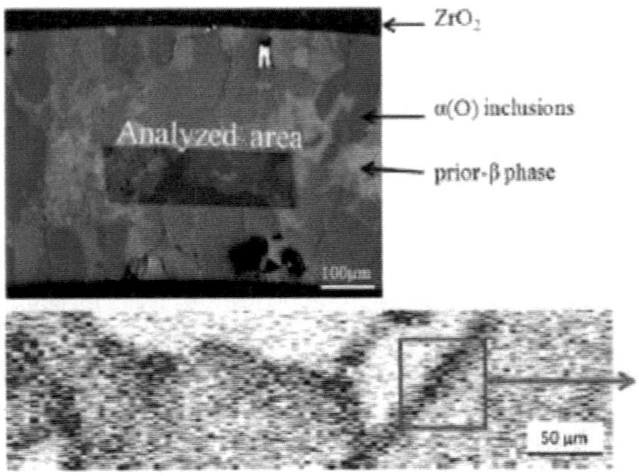

Figure 4. Hydrogen distribution measured by ERDA after oxidation at 1200°C during 624s

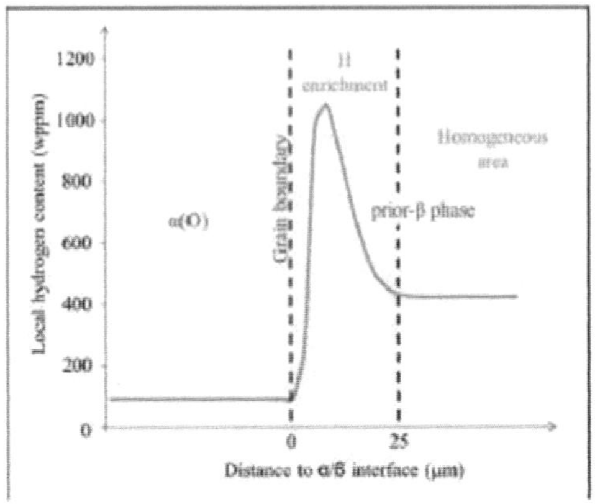

Figure 5. Hydrogen diffusion profile measured by ERDA in the red area

As for oxygen content, an assessment of the averaged hydrogen content along the radial direction is obtained by combining the α(O)-phase radial distribution assuming equilibrium concentrations in each of the two co-existing phases. The measured average hydrogen profile shows continuity at α(O) layer and two-phase region. This profile is plotted in Figure 6. Hydrogen follows a complex profile within the cladding thickness controlled by the inclusion distribution in the two-phase domain. Hydrogen migrates from hydrogen depleted regions towards hydrogen rich regions implying that its motion is not governed by usual concentration gradients but most probably by more complex processes.

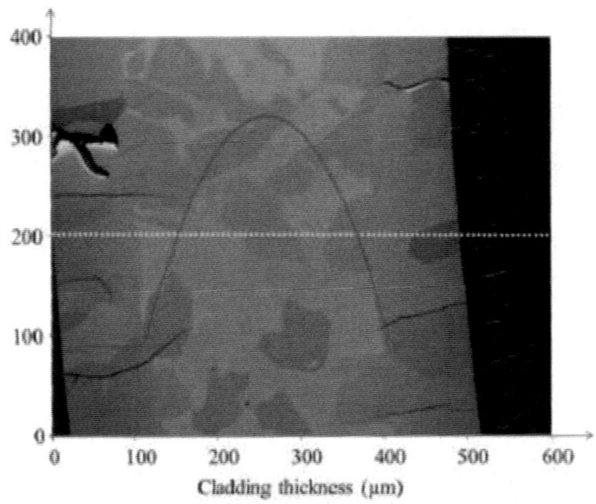

Figure 6. Hydrogen measured average profile taking into account phases distribution

4. Conclusions and perspectives

The present study addresses oxygen and hydrogen average contents in a Zircaloy-4 cladding after a LOCA steam oxidation transient. Within the samples, special attention was paid to the region with co-existing α(O)+βphases. This region is not homogenous and considering 1-D diffusion modeling, a special data treatment has to be performed to compare experimental data to modeling results.

Using ERDA and EPMA measurements, the distribution of the main alloying elements within the steam-oxidized cladding samples was thoroughly characterized. The main chemical compounds are close to equilibrium except far from α/β grain boundaries. At these locations, Cr and H are locally not homogeneously distributed whereas other elements like oxygen and iron are rather homogeneously distributed in each phase, without any measurable local enrichment for these two last elements. Additionally, the α(O)-phase radial profile was derived from metallographic analysis.

Considering these experimental observations and assuming that the phase contents are uniform, radial profiles of oxygen and hydrogen have been determined. The oxygen profile is compared to diffusion simulations using the DIFFOX code. The code prediction shows that improvements are necessary to better model the diffusion of both oxygen and hydrogen in the mixed phase region.

5. Acknowledgment

Authors are grateful to P. Lacote and G. Montigny for their technical assistance.

References

[1] J-C Brachet, V Vandenbergue-Maillot, L Portier, D Gibon, A Lesbros, N Waeckel, J-P Mardon, 2008 *Journal of ASTM International*, **5** (5), Paper ID JAI101116

[2] C Raepsaet, P Bossis, D Hamon , J L Béchade, J C Brachet 2008 *Nuclear Instruments and Methods in Physics Research section B* **266** 2424

[3] J-C Brachet, C Toffolon-Masclet, D Hamon, T Guilbert, G Trego, J Jourdan, A Stern, C Rapsaet 2011 *Solid State Phenomena* **172-174** 753

[4] C Duriez, S Guilbert, A Stern, C Grandjean, L Belovsky, J Desquines, 2010 *Journal of ASTM International* **8** 2

[5] J V Cathcart, R E Pawel, R A McKee, R E Druschel, G J Yurek, J J Campbell, S H Jury 1977 "Zirconium metal-water oxidation kinetics IV. Reaction rate studies" NRC report ORNL/NUREG-17

[6] C Desgranges, C Toffolon-Masclet et al. 2011 *Solid State Phenomena* **172-174**

[7] J-C Brachet et al. 2001 "Mechanical behavior at Room Temperature and Metallurgical study of low-Tin Zy-4 and M5TM after oxidation at 1100°C and quenching", Proc. of TCM on "Fuel behavior under transient and LOCA conditions", IAEA, Halden-Norway, Sept. 10-14

Formation of Catalyst Model Dispersed of Pd on a thin MgO(100)

F Baara[1], A Chemam[2]

[1]Laboratory of surfaces and interfaces of solids studies - (LESIMS) –Department of Physics, Faculty of Sciences, Badji Mokhtar University-Annaba-Algeria
[2]Preparatory School for Science and Technology of Annaba, Annaba, Algeria

E-mail: a.chemam@epst-annaba.dz

Abstract. The nucleation kinetics or the formation of a catalyst model dispersed for the system Pd /thin MgO (100) are calculated by developing many programs using Fortran software. This simulation is based upon parameters studied in situ by transmission electron microscopy (TEM), related to the first quantitative study on the nucleation and the growth. Palladium nanoparticles deposited on thin MgO are tested in the temperature range 573–1073 K and deposition time of 1000 s. The nucleation kinetics are interpreted according to the theory of random nucleation. The general scheme is consisting of three stages namely, nucleation, growth and coalescence. The saturation density of clusters decreases when the substrate temperature increases following Arrhenius law. This behavior is in agreement with a recent AFM study for Ag/MgO and Au/MgO. The phenomenon of coalescence is explained via island migration process. It is shown that the coalescence occurs more rapidly when the substrate temperature is high.

1. Introduction

Nowadays, nano-objects provide a promising research for the identification of new fundamental properties of the materials and their potential technological applications. Much effort is devoted to understand the physical and chemical properties of materials, which can serve as model catalyst systems. Consequently, fundamental studies have been carried out on a range of heterogeneous catalyst, for example, metal islands grown on thin films [1-4] or on single-crystals surfaces [5, 6]. Palladium deposits on the MgO(1 0 0) surface have become one of the most widely used model systems, and have given rise to many detailed experimental studies [7-9]. Although the main microscopic steps governing nucleation and growth of the films are now understood, detailed characterization of these processes has proven difficult. Earlier, empirical and theoretical studies of Pd over single crystals MgO, investigated defect nucleation [10,11] when nucleation centres occupy minority of sites .On the other hand , the results of nucleation kinetics over thin films governed by random nucleation [1,12],each atomic site is potentially a nucleation centre . In this study, we build upon many experimental and theoretical studies [1, 13] have been carried out to understand these processes. The aim of this work is to investigate the microscopic mechanisms focusing on the nucleation, growth and coalescence of Pd / thin MgO (100) using Fortran software.

1.1. Transmission electron microscopy experiments
To understand the first quantitative study of nucleation and growth of Pd on thin layer of MgO(100),we exploited the experimental work of C.Henry et al [1],who used transmission electron microscopy and electron diffraction at high energy to measure the Pd island density as a function of time a given temperature and a constant flux.

1.2. Over view of nucleation and growth theories
Nucleation on a surface has been discussed in both classical thermodynamic and in atomistic terms, and both have a long history. The growth process occurs by accretion of adatoms. It is described by the atomistic nucleation theory, which has been developed by Zinsmeister [15].

1.2.1. Nucleation kinetics. The rate equations given by Zinsmeister express the variation with time of the number of clusters of size i:

$$\frac{dn_i}{dt} = \omega_{i-1}\, n_{i-1} - \omega_i\, n_i \qquad \text{for } i = 2,3\ldots\infty \qquad (1)$$

1.2.2. ω_i is the attachment frequency of an adatom to a cluster containing i atoms which is expressed by:

$$\omega_i = \sigma_i\, D n_i \qquad (2)$$

Where σ_i is the capture number for a cluster of size i. Forsake of simplicity Zinsmeister assumes that it is a constant between 1 and 4. From this scheme the nucleation frequency is:

$$J = 2\,\omega_1\, n_1 \qquad (3)$$

Supposing that the growth is negligible, the density of adatoms is equal to the stationary value:

$$n_1 = F\tau \qquad (4)$$

Where F is the flux of atoms impinging on the substrate and τ the mean life time of an adatom before desorption. Then, combining equations (2) and (4) the nucleation rate becomes:

$$J = 2\sigma_1\, D\, F^2\tau^2 \qquad (5)$$

The nucleation rate is proportional to the square of the impinging flux for a homogeneous substrate without defects.

1.2.3. Growth kinetics. Zinsmeister has solved the system of differential equation assuming a constant value for the attachment frequency (ω_i) [28]. However by this treatment several characteristics of the growth of clusters are not taken into explanation. Several researches have tried to treat more precisely the calculation of the attachment frequencies [16–18]. In the typical growth, the diffusion of adatoms is limited by desorption and the diffusion length X_s of an adatom is:

$$X_s = (D\tau)^{1/2} \qquad (6)$$

1.2.4. Where D is the surface diffusion coefficient. It is expressed by:

$$D_s = (a_0^2\, \nu_d)\, \exp\left(-\frac{E_d}{KT_s}\right) \qquad (7)$$

Then the mean life time of a physisorbed molecule as explain by:

$$\tau = (1/\nu_a)\, exp\left(\frac{E_a}{KT_s}\right) \qquad (8)$$

1.2.5. where E_a and E_d is the adsorption and the diffusion energy, and a_0 the jump distance , v_d and v_a the frequency factors for the diffusion and the adsorption process, K the Boltzmann constant and T_s the substrate temperature.

1.2.6. Coalescence. The transformation of two touching nuclei into one nucleus can be described by a time constant shown by Nichols and Mullins [19]:

$$\tau_c = 0.2 \left(\frac{R^4}{B}\right) \quad with\ B = \frac{\gamma\ \Omega^{4/3} D_s}{K\ T_s} \qquad (9)$$

Where R is the radius of the coalescing spheres, γ the surface free energy and Ω the atomic volume of Pd.

1.2.7. Algorithm. We have been developed many programs exploiting Fortran software. The following list details this mapping.

1. Pd deposition flux rate on MgO (100) is $1.13x1O^{13}$ atoms cm^{-2}. s^{-1}.
2. Pd atoms are deposited randomly onto the surface with activation energy of about 0.22 eV.
3. Pd nanoparticles deposited on thin MgO are tested in the temperature range 573–1073 K and deposition time of 1000 s.
4. Pd islands are approximated to be three-dimensional clusters.
5. The entry parameters are: the velocity of nucleation, velocity of growth, the average mean life time, the surface repeat distance, the diffusion length, the surface free energy, the atomic volume of Pd, the activation and the diffusion energies.

2. Result

Table 1. Calculated of the cluster density (cm^{-2}) at different substrate temperatures (K) and deposition times (s)

t (s)	573 K	*673 K*	773 K	873 K	973 K	*1073 K*
10	9.39E11	4.85E11	2.97E11	2.03E11	1.50E11	1.18E11
70	3.02E12	1.56E12	9.58E11	6.56E11	4.86E11	3.81E11
τ_c	No	No	9.80E11	6.71E11	4.98E11	3.90E11
1000	3.02E12	1.60E12	6.77E11	2.59E11	1.64E11	1.19E11

'Figure1'shows the variation of cluster density as a function of exposure time at different substrate temperatures ranging from 573K to 1073K and a constant palladium flux $1x1O^{13}$ atoms cm^{-2}. S^{-1}. For Ts = 573K and 673 K, we can see that the density of clusters is increasing rapidly after 10 to 70 s (see table 1) due to the large adsorption energy for the Pd adatom confirming the nucleation stage, up to a plateau (saturation density) corresponding to n_s = 3.10^{12} cm^{-2} and $1.6x10^{12}$ cm^{-2} respectively. A similar behavior is observed for the remaining substrate temperatures till the coalescence occurrence, were the cluster density decreases. It is worth to note that the cluster density decreases when the temperature increases.

In 'Figure 2', the saturation density is plotted in an Arrhenius diagram. It is represented by the equation: $n_s = B_0 .exp (E / K .T_s)$. When the activation energy E is equal to 0.22 ± 0.05 eV, B_0 (preexponential factor) is $3.63.10^{10}$ cm^{-2}. We show that it increases when the substrate temperature decreases. This behavior is in agreement with a recent AFM study for Ag/MgO [20], Au/MgO (100) [21] and our previous studies by TEM [1] on the same system.'Figures 3' and '4 'show the variations of the covered area (A) and the coalescence parameter (B) as a function of substrate temperatures obtained under the same conditions. We see that the fraction of covered area decreases when deposition temperature rises. An opposite behavior is observed for the coalescence parameter.

'Figures 3' and '4 'show the variations of the covered area (A) and the coalescence parameter (B) as a function of substrate temperatures obtained under the same conditions. We see that the fraction of covered area decreases when deposition temperature rises. An opposite behavior is observed for the coalescence parameter.

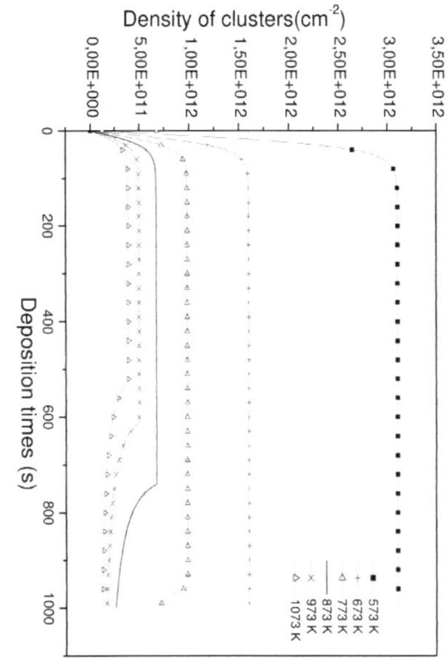

Figure 1. Nucleation kinetics of Pd on MgO (100) at different substrate temperatures for a pd flux 1x1O[13] atoms cm^{-2}. s^{-1}.

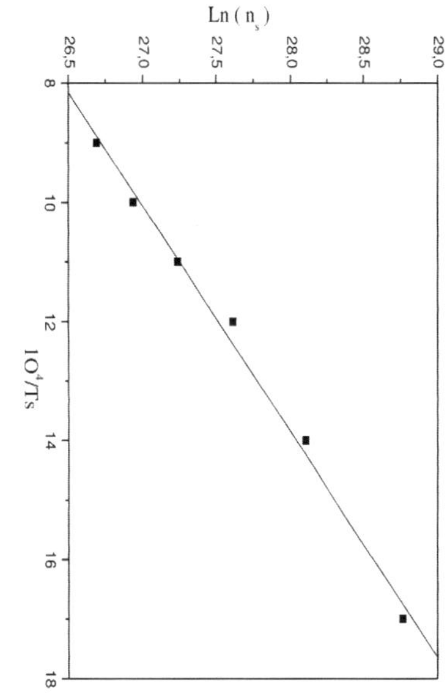

Figure 2. Arrhenius plot of the saturation density of palladium clusters on MgO (100).

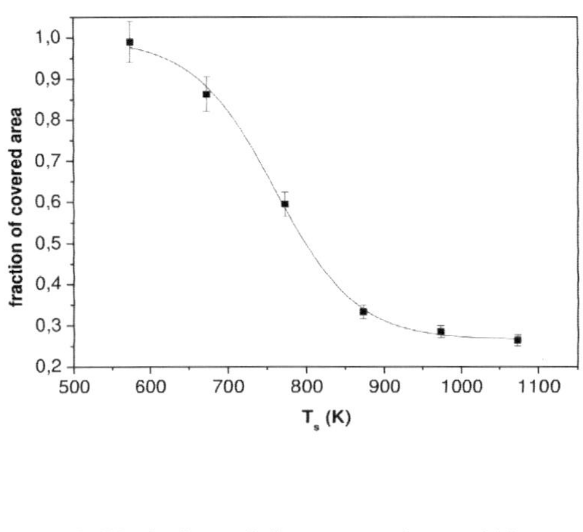

Figure 3. Variation of the covered area (A) as a function of substrate temperatures

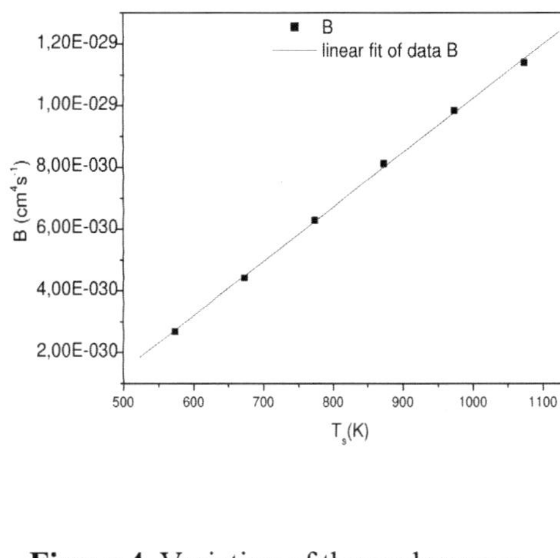

Figure 4. Variation of the coalescence parameter (B) as a function of substrate temperatures

3. Discussion

From our previous empirical result for Pd/MgO thin film, the initial nucleation curves suggest the occurrence of a random nucleation mechanism. The theory of this nucleation process is expressed by the kinetic equations of Zinsmeister [15] . Robinson and Robins [29] have given analytical solutions in two limit cases, namely at low and high temperature. In this work ,we focus on the high substrate temperatures(regime of incomplete condensation) ,where the particle number density is given by :

$$n = n_s \tanh\left({}^{t}/{}_{\tau}\right) \tag{10}$$

When the random nucleation model is used ,the calculated curves in the first stage agree well with the experimentally measured [1] time dependencies of island density (see nucleation regime in fig. 1).Hence, the curves show a plateau as a maximum particle density reached at the end of the nucleation regime characterizing the Volmer Weber growth .The value of saturation of the island density is an important parameter which determine the mode of thin film growth. The linear behavior of the Arrhenius plot observed for the temperature dependence of the saturation density of clusters has been also found in the case of Ag / Ar-cleaved MgO(1 0 0) [20], Au /MgO(1 0 0) [21] and Pd/ UHV-cleaved MgO(1 0 0) [22,23] .

$$n = n_s\ (1 + t/\ \tau)^{-3/m}\text{, with m = 7} \tag{11}$$

The coalescence curves were better fitted with a Cluster diffusion model [14] rather than Ostwald ripening model. The most crucial parameter in our results is the coalescence time. It is defined as the mean time for two clusters that come into contact to coalesce. From equation (9), we can note that two parameters are important to determine the duration of the coalescence stage, which are R and B. One can also notice the influence of the deposition temperature, that modifies the clusters coalescence time. It is clearly seen that clusters coalesce more rapidly at high temperature [24]. This phenomenon is explained by the process of island migration. The process of island migration in this calculation is essentially described by the parameter B. The derived B values from 'Figure.4 are not high enough meaning that the process of island density coalescence is not ignored even at the initial stages of deposition [25]. The mechanism of coalescence which can be expected at such a low value of surface coverage is the migration of islands on the surface, prior to the mechanism of immobile islands in which the coalescence occurs at high values of surface coverage when the islands touch each other [26].

The surface coverage does not depend on B parameter, but it strongly depends on the cluster density and the shape of the particles. The fraction of the substrate covered by the clusters, which is considered as the contact surface of the half sphere is a circle can be written as:

$$A = \frac{\pi}{4}\ n\, D^2 \tag{12}$$

In this range of temperatures, the diameter (D) of the clusters follows a power law: $D_o\, t^P$ with 0.33 <P<0.55. At 573K and 673K, our previous experimental work yielded values of D_o; P equals 0.024 nm, 0.8 and 0.372 nm, 0.39 respectively [1]. The obtained values are used for the calculated curves of the covered area of the substrate surface.

As indicated in the last section the cluster density decreases when the substrate temperature increases due to the increased ad-atom mobility [16]. In addition, we assumed that D is relatively constant with a low error (ΔD / D=11%).Accordingly, the fraction of the substrate covered by the particles decreases and the decrease is more pronounced when the coalescence occurs. This behavior is in good agreement with growth rate of the particles in the case of the random nucleation theory [27].

4. Conclusion

In the present study, we have simulated the microscopic mechanisms, which can calculate various parameters related to the formation of Pd / thin MgO (100) using FORTRAN software. The formation kinetics follows a general scheme consisting of three stages: nucleation, growth and coalescence. It is determined that saturation density obeys an Arrhenius law with activation energy of 0.22 eV as a fit parameter. It is observed that the variation of island density upon time reaches the saturation. In some cases the cluster density decreases slowly after a saturation regime. The latter decrease is interpreted via the processes of islands migration .It is shown that the coalescence time and the fraction of the substrate covered by the clusters decrease when the temperature increases. Furthermore, the clusters coalesce more rapidly when the temperature increases.

Acknowledgment

R. Chemam, A. Amara and M.Ghares are gratefully acknowledged.

References

[1] C Chapon, C R Henry, A Chemam 1985 *Surface Sci.* **162** 747
[2] C Stiehler, Y Pan, W-D Schneider, P Koskinen, H Hakkinen, N Nilius, and H-J Freund 2013 *Phys Rev* B **88** 115415
[03] X Shao, Y Cui, W-D Schneider, N Nilius, and H-J Freund, C 2012 *Phys. Chem.* **116** 17980
[04] H-F Wang, H Ariga, R Dowler, M Sterrer, H-J Freund 2012 *Journal of Catalysis* **286** 1
[05] P Nolte, A Stierle, N Kasper, N Y Jin-Phillipp, N Jeutter, H Dosch 2011 *Nano Lett* **11** 4697
[06] P Nolte, A Stierle, N Kasper, N Y Jin-Phillipp, H Reichert, A Rühm, J Okasinski, H Dosch, S Schöder 2008 *Phys. Rev.* B **77**
[07] C Xu, W S Oh, . Liu, D Y Kim, D W Goodman 1997 *J. Vac. Sci. Technol.* A **15** 1261
[08] C R Henry, C Chapon, C Duriez, S Giorgio 1991 *Surf. Sci* **253** 177
[09] H Fornander, L Hultman, J Birch, J-E Sundgren 1998 *Journal of Crystal Growth* **186** 189
[10] L Xu, C T Campbell, H Jo'nsson, G Henkelmana 2007 *Surface Science* **601** 3133
[11] J A Venables, J H Harding 2000 *Journal of Crystal Growth* **211** 27
[12] G Renaud 2005 *American Institute of Physics* **748** 63
[13] V N E Robinson and J I Robins 1974 *Thin Solid Films* **20** 155
[14] D Kashchiev 1976 *Surf. Sci.* **55** 477
[15] Zinsmeister 1966 *G.Vacuum* **16** 529
[16]. Halpern 1969 *J. Appl. Phys.* **40**, 4627
[17] Venables 1973 *J.A. Phil. Mag.* **27**, 697
[18] Lewis B. 1974 *Phil. Mag.* **29** 1253
[19] F A Nichols and W W Mullins 1965 *J. Appl. Phys,* **36** 1826
[20] A Menck, Ph.D. thesis, Lausanne, 1998
[21] K Højrup-Hansen, S Ferrero, C R Henry 2004 *Applied Surface Science* **226** 167
[22] M Meunier, Ph.D. thesis, Marseille, 1995
[23] C R Henry 1998 *Surf. Sci.* Rep. **31** 231
[24] J Carrey, J-L Maurice, F Petro, A Vaurès 2002 *Surface Science* **504** 75
[25] J Carrey and J-L Maurice 2002 *physical review B*, volume **65**, 205401
[26] M C Bartelt, J W Evans 1994 *Surf. Sci.* **314** L229
[27] M Harsdorff 1982 *Thin Solid Films* **90** 1
[28] C R Henry, H Poppa 1988 *J. Vac. Sci. Technol.* A **6** 1113
[29] V N E.Robinson and J I Robins 1974 *Thin Solid Films* **20** 155

Influence of Hemp Fibers Pre-processing on Low Density Polyethylene Matrix Composites Properties

S Kukle[1], R Vidzickis[2], Z Zelca[3], D Belakova[4] and J Kajaks[5]

[1, 2, 3, 4]Institute of Design Technologies, Riga Technical University, Latvia
[5]Department of Polymer Material Technologies, Riga Technical University, Latvia

E-mail: skukle@latnet.lv

Abstract. In present research with short hemp fibres reinforced LLDPE matrix composites with fibres content in a range from 30 to 50 wt% subjected to four different pre-processing technologies were produced and such their properties as tensile strength and elongation at break, tensile modulus, melt flow index, micro hardness and water absorption dynamics were investigated. Capillary viscosimetry was used for fluidity evaluation and melt flow index (MFI) evaluated for all variants. MFI of fibres of two pre-processing variants were high enough to increase hemp fibres content from 30 to 50 wt% with moderate increase of water sorption capability.

1. Introduction

The plant fiber reinforced plastics (PFRPs) manufacturing technique is interrelated with such key composite parameters as volumetric composition, reinforcement type and form, and matrix type. The selection of fibers type, fibers extraction process and fibers surface modification technique, fibers volume fraction and interfacial properties, reinforcement packing arrangement and orientation are matters of greatest importance. The interactive effect of all these parameters can influence the variation in mechanical properties of PFRPs produced through different manufacturing processes. [1]. Hemp as a yearly renewable lignocellulose fibers source with a low ecological footprint evaluated by 26 criteria [2] and with a high yields in temperate climate zone [3] are an attractive source of reinforcement for PFRPs.

However, hemp fibers properties can differ greatly depending on variety, agricultural complex applied, weather of growing season, type, grade, harvest quality, preprocessing methods, yield, etc. [4]. Only natural fibers of high technical quality guarantee sufficient reproducibility of the mechanical characteristics of bio composites [5]. Investigation of non-controllable (weather) and controllable factors (variety, agricultural complex, stem and fibers processing parameters) influences on fibers and composite properties will allow to discuss and to evaluate anticipated unavoidable variation intervals of composite properties [6] in the same way as in textile industry.

The latest development, due to the environmental and ecological factors, biobased polymers from renewable resources have found increasing attention as potential alternatives to currently dominating petroleum based polymers [7]. Study shows that current producers of bio-based polymers had estimated production capacity nearly 12 million tons by 2020. With an expected total polymers production of about 400 million tons in 2020, the bio-based share should increase from 1.5% in 2011 to 3% in 2020, meaning that bio-based production capacity will grow faster than overall production [8].

Biobased linear low density polyethylene (LLDPE) from renewable raw materials behaves the same as PE based on petroleum. Projections for year 2016 evidence that biodegradable bioplastics share will reach only 13.4% from anticipated bioplastics production volume 5 779 000 t/per year, rest 86.6% will refer to biobased / non-biodegradable bioplastics [11] due to the fact that they can easily be recycled and can thus be included in the current waste separation process, and processed into new bioPE products using conventional technologies without requiring extra investments [12].

2. Experimental methods

Research focused on improving the mechanical properties of LLDPE composites reinforced with hemp fibres. As show previous studies, due to their polar and hydrophilic nature, natural fibres are incompatible with the hydrophobic polymers. This incompatibility could leads to the weak interfacial adhesion as well as to the non-uniform dispersion of fibres within the composite. To overcome incompatibility several strategies have been tested to enhance the adhesion between the lignocellulosic fillers and the polymer matrix. These strategies generally involve fibres modifications by physical or chemical methods [13, 14, 15 and 16]. Chemical modification of natural fibres involves various chemical treatments in order to reduce the content of hydroxyl groups or introduce cross-linking between the filler fibres and the polymeric matrix. Water boiling, sodium hydroxide treatment, esterification with acetic anhydride and grafting with maleic anhydride (MAH) were used (Table1).

Table 1. Plan of experiments

Without pre-processing ***WP***	**Pre-treatment**			**MAH coupling,** ***5%MAH***
	Water boiling ***WB***	**Alcali, 2%*Alc***	**Acetilation, 11% Acetil**	
Decortification, scutching				
	2 hours, 100°C	1 hour, 20°C	4 hours, 20°C	
Washing with destilled water, drying				
Cutting, grinding, sifting, fibers length < 1mm				
Camera drying, 12 hours 105oC				
Two step roller mixing-compounding (LabTechScientific) temperature 155°C and 120°C, 10 minutes				

Fibres of hemp variety Bialobrzeskie grown in the East part of Latvia and linear density 3.62 tex were used in this study to reinforce linear low density polyethylene (LLDPE) grade LL 6201 matrix.
Fibers of three variants marked as WB, Alc and Acetil (Table 1) were subjected to the pretreatments discussed above. Composites were prepared by components mixing on two rolls mills, then cooled, granulated and pressed in 1 mm thick sheets from which were cut off specimens for tensile tests following standard ASTM D 638 M-93.

Tensile tests were conducted on a universal material testing equipment UTS-100 (standard ASTM D 638 M – 93). Fluidity was estimated by melt flow index method MFI (T=1900 C, P=2.16 kg) according to the standard ASTM D 1238-90b. Microhardness (MH) was examined by Vickers M41 at load 200g. Water exposure experiments (standard ASTM D 570-98) were done at room temperature (+23°C).

3. Results and Discussion

Properties of produced composites vary in vide ranges: mean tensile strength in a range from 7.7 to 22.5 MPa, tensile modulus in a range from 647 to 11 604 MPa, microhardness in a range from 55.4 to 93.4 (Fig. 1, 2).

Tensile strength 22 MPa and modulus 1.31 GPa of the first samples group with MAH coupling (Fig.1) are quite close to the corresponding properties of Hemp/PP 30 wt %: 30 MPa, 1.56 GPa [1].

Samples of II group without pretreatment (WP) and with alcali pretreatment (Alc) shows light difference in position and could not proposed that alcali pretreatment applied in experiment improve substantially hemp/LLDPE composites mechanical properties.

III group with poor mechanical properties evidence that water boiling and acetilation with technologies used in experiment influence fibers structure.

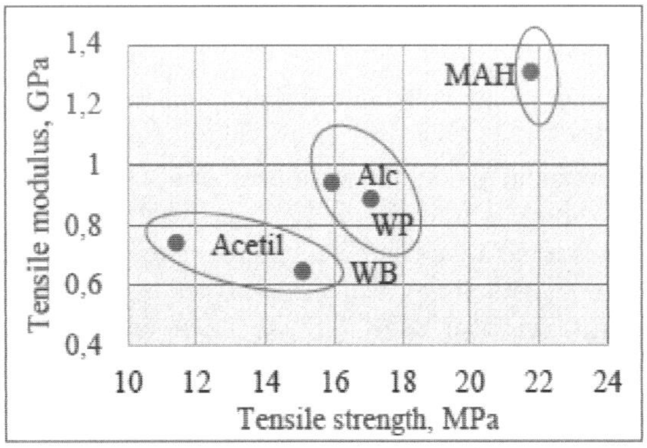

Figure 1. Tensile properties of Hemp/LLDPE 30 wt% composites

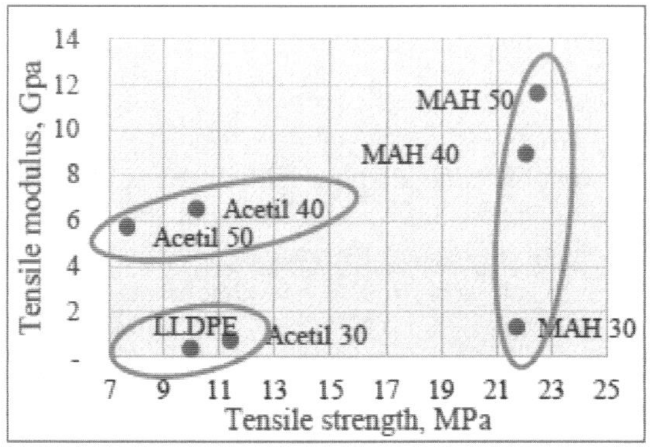

Figure 2. Tensile properties of Hemp/LLDPE composites as function of fibers content

As seen from Fig.2 (right) tensile modulus grow fast with fibers content increase from 30 to 50 wt% for MAH coupling, at the same time tensile strength increase is moderate. Acetilation of fibers doesn't change substantially tensile strength of composite compare to LLDPE (Fig.1) and drop below this level with the increase of fibers content fro 40 to50 wt%. Tensile modulus of acetilated fibers composite increases from 2 times to 17.5 times compare to LLDPE modulus if fibers content increase from 30 to 40 wt % (Figure 2, right).

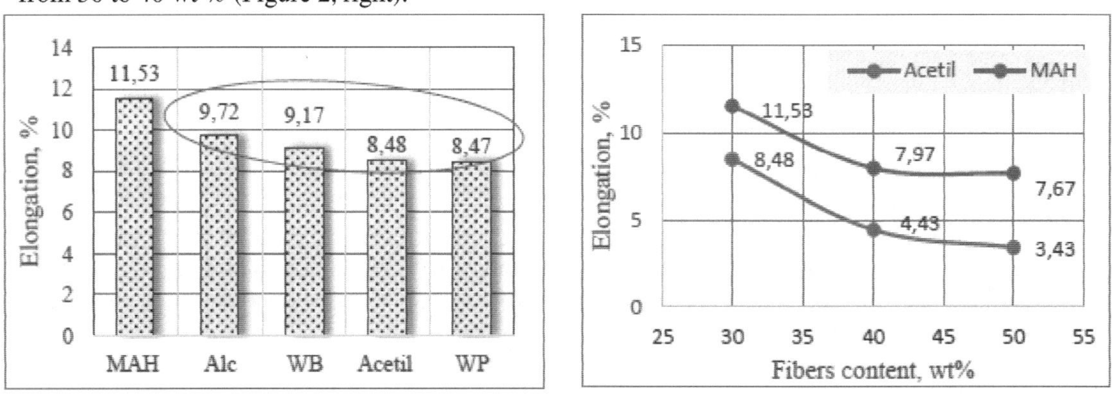

Figure 3. Composites elongation at break. Hemp/LLDPE 30 wt% (left) and 30-50 wt% (right)

Elongation at break of variants with 30 wt% fibers vary in range 1.25 % exept variant with MAH having the highest elongation value 11.53 % (Figure 3, left). In result of hemp fibers acetilation pretreatment elongation at break of LLDPE matrix composite drops down fast with fibers content increase reaching minimum 3.43 % for Hemp/LLDPE 50 wt%. Elongation of composites grafted with MAH are almost 2% higher for all three variants (Figure 3, right).

Microhardness quantify the resistance of a material to the plastic deformation. The highest values of microhardness show composite without fibers pretreatment, microhardness gradually decrease starting with NaOH pretreated (Alc) to the acetilated fibers (Acetil) composite (Figure 4, left). Microhardness increase 1.35 times when fibers content with acetilation pretreatment increse from 30 to 40 wt% and after decrease to the level of the composite microhardness with the fibers content 30 wt%.

Microhardness increases dramatically when fibers content of MAH grafted PE increase fro 40 to 50 wt%.

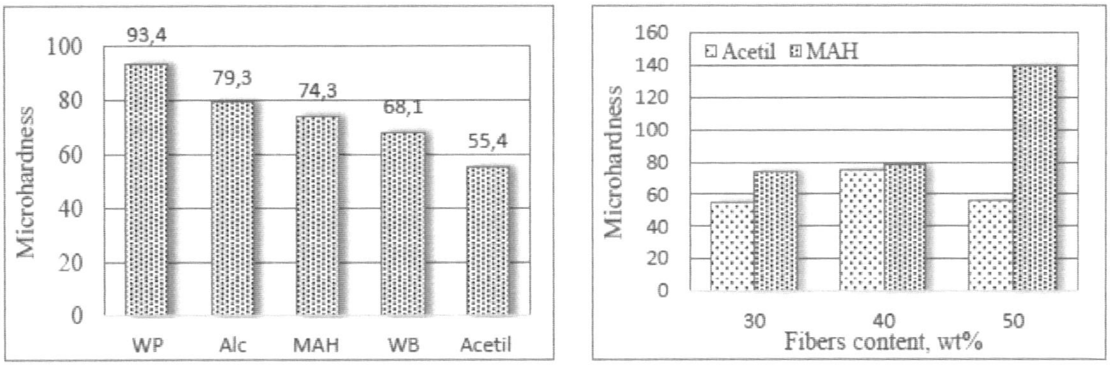

Figure 4. Microhardness, MPa. Hemp/LLDPE 30 wt% (left) and 30-50 wt% (right)

Acetic anhydride pretreatment of hemp fibers ensure good fluidity for all three Acetil variants. Melt flow index decreases 9.3 times when content of fibers increases from 30 to 50 wt% (Figure 5.) Melt flow index of compositions with 30 wt% fibers content decreases from 1.7 g/10 min (water boiled fibers) to 0.4 g/10 min (alcali treated fibres) with medium for this range 0.8 g/10 min (MAH 30).

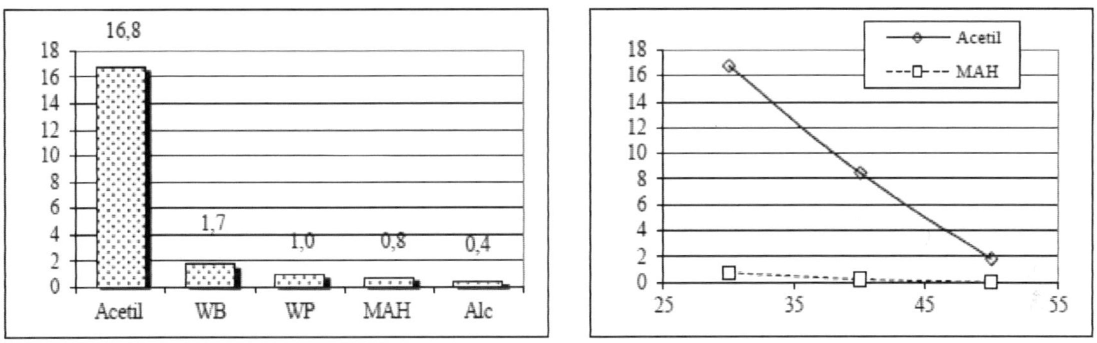

Figure 5. Melt Flow Index, g/10min

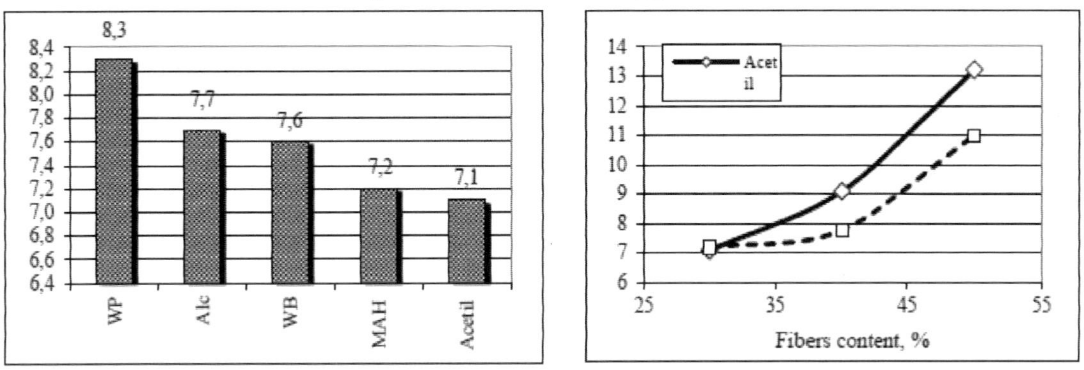

Figure 6. Water sorption % after 240 h. Hemp/LLDPE 30 wt% (left) and 30-50 wt% (right)

Water sorption of variants with fibers content 30 wt% vary from 7.1 to 8.3% (Figure 6, left.). Water sorption capability grows fast following acetilated and MAH grafted fibers content increase from 30 to 50 wt% (Fig. 6, right).

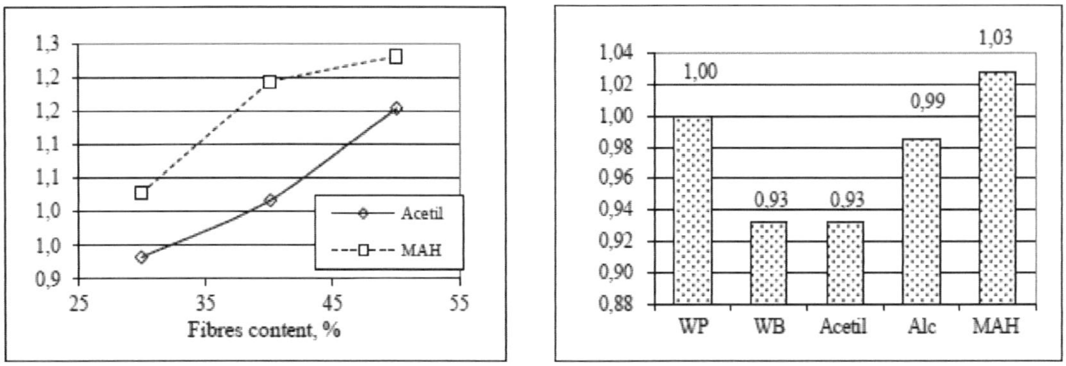

Figure 7. Composites density

Composite density increase following increase of fiber content from 30 to 50 wt% (Fig. 7, left).

4. Conclusions

Hemp/LLDPE composites with different properties are produced:

- with density in a range from 0.93 to 1.23 G/cm3; increase with fibers content increase;
- tensile strength vary in a range from 7.7 to 22.5 MPa;
- elongation at break vary in a range from 3.4 to 11.5 %; decrease with fibers content increase;
- tensile modulus vary in a range from 647 to 11 604 MPa; increase substantially with fibers content increase;
- microhardness vary in a range from 55.4 to 93.4; decrease substantially if fibers are cubjected to preprocessing;
- water sorption after 240 h vary in a range 7.1 to 13.2 % and increase substentially with fibers content increase.

In result of acetilation and water boiling tensile strength decrease and increase in result of coupling with 5 wt% MAH. Elongation at break increases after hemp fibers alcali treatment and coupling with 5 wt% MAH. Tensile modulus substentially increase after coupling with 5 wt% MAH and decrease if hemp fibers are subjected to the water boiling and acelilation. Microhardness drops down and water sorption increase for all apllied fiber pretreatments.

Grafting with a such coupling agent as maleic anhydride allows produced hemp/LLDPE 30 wt% composites with mechanical properties close to the corresponding properties of Hemp/PP and Flax/PP 30 wt% composites. Although water boiling improve fluidity, other composite properties do not convince that water boiling pretreatment could improve hemp fibers compatibility with LLDPE matrix as well the same coresponds to the alcali 2wt% pretreatment. 11wt% acetilation pretreatment offer good fluidity and low elongation at break, but substantially lessen composite tensile strength and microhardness.

Short hemp fibres reinforced LLDPE matrix composites with fibres content in a range from 30 wt% to 50 wt% could be used in technical applications where elastomeric materials that resist crack growth under strain are needed.

Acknowledgement
This research is co-financed by the ESF within the project "Development of innovative technologies for the accumulation and production of heating and cooling", project implementation agreement No. 2013/0064/1DP/1.1.1.2.0/13/APIA/VIAA/050.

References

[1] D U Shah 2013 *J. Mater. Sc.* **48** 6083 DOI 10.1007/s10853-013-7458-7

[2] Small E, Marcus D 2002 Hemp: A New Crop with New Uses for North America* [book auth.] Whipkey J, No: Janick. [ed.] J., Whipkey, A. Janick. *Trends in new crops and new uses.* (Alexandria : ASHS Press) pp 284-326

[3] Freivalde L, Kukle S, Puriņa B and Stramkale V 2010 Durability Properties of Hemp Fibers.// Proc. of the 5th Int. Textile, Clothing & Design Conf., Dragcevic (Ed.), ISSN 1847-7275, (Dubrovnik, Croatia) pp 83-86

[4] Kukle S, Stramkale V, Kalniņa D and Solizenko R 2012 Comparision of Hemp Fiber Properties. Proc. of the 6th Int. Textile, Clothing & Design Conf., ISSN 1847-7275, (Dubrovnik, Croatia) pp 76-80

[5] Pickering K, 2008 (Ed.), Properties and performance of natural-fiber composites. (Boca Raton: CRC Press LL)

[6] Harmsen P, Hackmann M, 2014 *Biofuels, Bioproducts and Biorefining* **8** (3) 306

[7] Production Capacities for Bio-based Polymers in Europe - Status Quo and Trends towards 2020. Available from: http://www.bio-based.eu/market_study/pressrelease

[8] Mülhaupt R, 2013 *Macromolecular Chemistry and Physics.* **214** (2) 159 Available from: http://www.braskem.com/site.aspx/green-products-USA

[9] Ed de Jong. Overview of biorefineries and current status. Tomorrow's biorefineries in Europe. 2014 IEA Bioenergy Task 42. Brussels, Available from: http://www.iea-bioenergy.task42-biorefineries.com

[10] Global production capacity of bioplastics 2009-2016 Available from: http://en.european-bioplastics.org/

[11] Braskem. Green Products. 2011 Available from: http://www.braskem.com/site.aspx/chemistry-sustainable-USA Small, E. and D. Marcus. 2002 Hemp: A new crop with new uses for North America. p 284–326

[12] Mwaikambo L Y, Ansell M P 2002. *J. Appl. Pol. Sc.* **84** 2222 © Wiley Periodicals, Inc

[13] Álvarez-Chávez C R, Edwards S, Moure-Eraso R, Geiser K 2012 *J. Cleaner Production* **23** 47

[14] Lee Y, Delille K, Bismarck A 2011 Greener Surface Treatments of Natural Fibres for the Production of Renewable Composite Materials *Cellulose Fibers: Bio – and Nano – Polymer Composites: Green Chemistry and Technology* Kalia S, Kaith B S, Kaur I, – Berlin: Springer, pp 156 – 158, ISBN 978-3-642-17369-1

[15] Espert A 2005 Strategies for improving mechanical properties of polypropylene/cellulose composites. Doctoral thesis, comprehensive summary. ISBN 91-7283-998-8

[16] Martin L W, Addcomp North America Inc. Overview of Maleic-Anhydride-Grafted Polyolefin Coupling Agents. A guide to understanding their uses, benefits, functions, selection, and developments.
Available from: http://www.speautomotive.com/SPEA_CD/SPEA2012/pdf/TP/TP9.pdf

Changeing of fly ash leachability after grinding

J Lakatos[1], R Szabo[2], A Racz[2], O Banhidi[1],G Mucsi[2]

[1]Institute of Chemistry, Faculty of Material Science and Engineering, University of Miskolc, Hungary
[2]Institute of Raw Materials Preparation and Environmental Processing, Faculty of Earth Science and Engineering, University of Miskolc, Hungary

E-mail: mtasotak@uni-miskolc.hu

Abstract. Effect of grinding on the reactivity of fly ash used for geopolymer production was tested. Extraction technique using different alkaline and acidic solutions were used for detect the change of the solubility of elements due to the physical and mechano-chemical transformation of minerals in function of grinding time. Both, the extraction with alkaline and acidic solution have detected improvement in solubility in function of grinding time. The enhancement in alkaline solution was approx. 100 % in case of Si and Al. The acidic medium able to dissolve the fly ash higher manner than the alkaline, therefore the effect of grinding was found less pronounced.

Introduction

Geopolymer production is one of the promising way of recycling the fly ash which large amounts has accumulated in deposited form all over the world where coal are used as fuels in power plants. Geopolymers can be produced by generate reactions between the fly as particles which are activated during the pre-treatment. Activation of fly ash can be done by mechanical and/or chemical way. The mechanical activation can be carried out by grinding in conventional ball mills or in high energy density mills (vibratory, stirred media mills) etc.. The chemical activation generally means a treatment with concentrated alkaline solutions which able to dissolve appropriate amount of network forming component of the alumino-silicate matrix. The efficiency of this process depends on the type of fly ash which has influenced by the origin of the coal used and the combustion technology applied. The efficiency and rate of chemical activation can modified by grinding as a mechanical preactivation procedure. This process breaking up the spherules and improves the accessibility of the soluble minerals or can generate such kind of mechano-chemical processes which can improve the manner of structural defects, enhancing the concentration of the more reactive phases. In this study the changing of leaching characteristics of the grinded fly ash are demonstrated. Based on this the effect of grinding on activation of fly ash before geopolymerisation was tested.

Many papers dealt with the effect of mechanical activation on the properties of the formed geopolymers [1-6]. These works has demonstrated that applying mechanical activation the formed geopolymer properties will enhanced. However little information gathered about the details, the effects of the mechanical pre-treatments on the efficiency of the subsequent chemical activation. Since the chemical activation based on the dissolution of the network forming elements (Si, Al, Fe) the main consideration in this paper is how the grinding can modify of these elements mobility.

Leaching characteristics of ash and the ash base geopolymers are very important in environmental issues. Expecting low environmental risk, or impact the low solubility is favourable. The opposite r is the demand in case of geopolymer production since large concentration of the mobile multivalent ions required to fix to each other properly the particles or expected be able to form strong inorganic polymer network. The leaching tests, since it based on same process a the chemical activation, can provide direct information about that kind of raw material transformations which determine the efficiency of the forthcoming chemical activation process.

According to Provis et al. [7] and Izquierdo et al. [8] many elements are present in the fly ash can fixed in geopolymer framework. However there are some which mobility change in the undesired direction during the geopolymerisation. They concluded that geopolymers suitable for immobilization of many hazardous elements from fly ash like Be, Bi, Cd, Co, Cr, Cu, Nb, Ni, Pb, Sn, Th, U, Y, Zr and rare earth elements, but like As, B, Se, V, W which has oxianionic forms can mobilizing in high alkaline conditions.

2. Materials and methods

The fly ash used in this investigation originated from Matra power plant, Hungary. It has the following physical and chemical properties. Moisture content was 0.27 m/m%. The bulk density of raw fly ash is 0.73 g/cm^3, particle density 1.93 g/cm^3. Particle size distribution and specific surface area measured by Horiba LA950 V2 type laser particle size analyzer; x_{10}=10.77 μm; x_{50}=52 μm; x_{80}=119 μm, specific surface area: 1152 cm^2/g. Based on XRF analysis the measured chemical composition of fly ash are given in Table 1. From the analysis the SiO_2/Al_2O_3 ratio was found to be 3.34 and the sum of SiO_2, Al_2O_3 and Fe_2O_3 content of the fly ash was 73.49 wt%.

Table 1. Chemical composition of the fresh lignite type fly ash (Visonta)

Component	Mass concentration, m/m %
SiO_2	48.1
Al_2O_3	14.4
Fe_2O_3	11.0
Na_2O	0.37
K_2O	1.76
CaO	11.8
MgO	3.3
TiO2	0.49
P_2O_5	0.26
MnO	0.17

Fly ash was ground in tumbling mill for 5, 10, 20, 30 and 60 minutes, with 80% of critical revolution number and 110 % mill filling ratio.

Changing of the characteristic particle size and the surface are in function of the grinding time are given in the Table 2.

Table 2. Changing of the lignite type fly ash (Visonta) particle size and surface area in function of grinding time.

Grinding time, min.	Characteristic particle size, d_{50}, μm	Characteristic particle size, d_{80}, μm	Specific surface area, cm^2/g
0	52	119	1150
5	45	92	1250
10	30	70	1500
20	25	50	1700
30	20	45	1850
60	11	28	5400

Alkaline extractions were performed using 1, 5 an 10 mol/l NaOH solution. The solid liquid ratio was 1:10. After the desired contact time an appropriate volume of the solution was taken, acidified with 1 mol/l HCl and after dilution the concentrations of the selected elements were measured by ICP.

In the second set of the leaching tests extractions with distilled water, 1 mol/l HCl and 2 mol/l HNO3 was taken at 1:50 solid: liquid ratio. The samples were analysed as the pervious ones. All leaching tests were performed at ambient temperature.

3. Results and discussion

The figure 1-3. show the effect of grinding on the leachability of main component elements in 1, 5 and 10 mol/l NaOH solutions. The Si, Al and K show larger solubility in case of longer grinded samples.

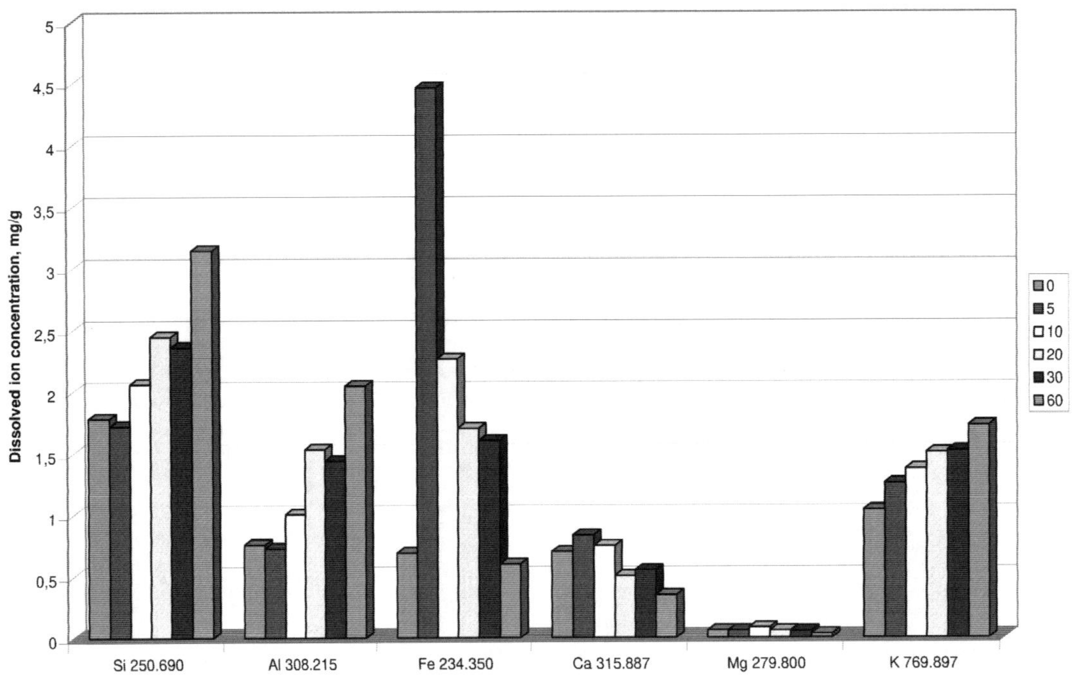

Figure 1. Leachability of elements in function of grinding time. Solution 1 mol/ l NaOH, solid/liqid rato: 1/10. Contact time: 1 day

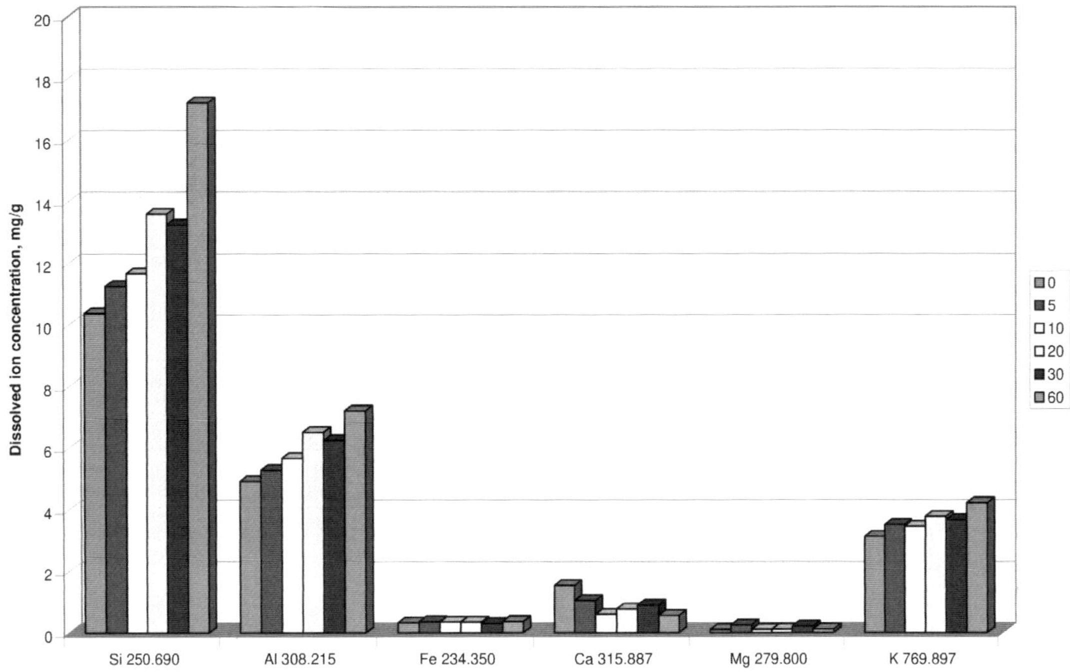

Figure 2. Leachability of elements in function of grinding time. Solution 5 mol/ l NaOH, solid/liqid rato: 1/10. Contact time: 1 day

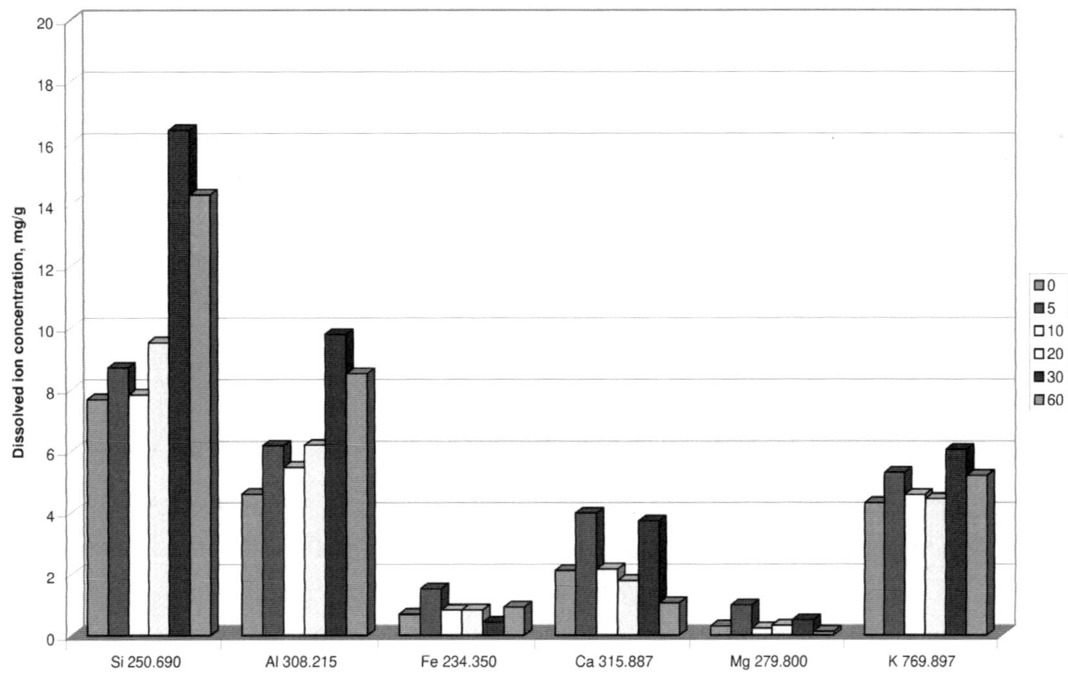

Figure 3. Leachability of elements in function of grinding time. Solution 10 mol/ l NaOH, solid/liqid rato: 1/10. Contact time: 1 day

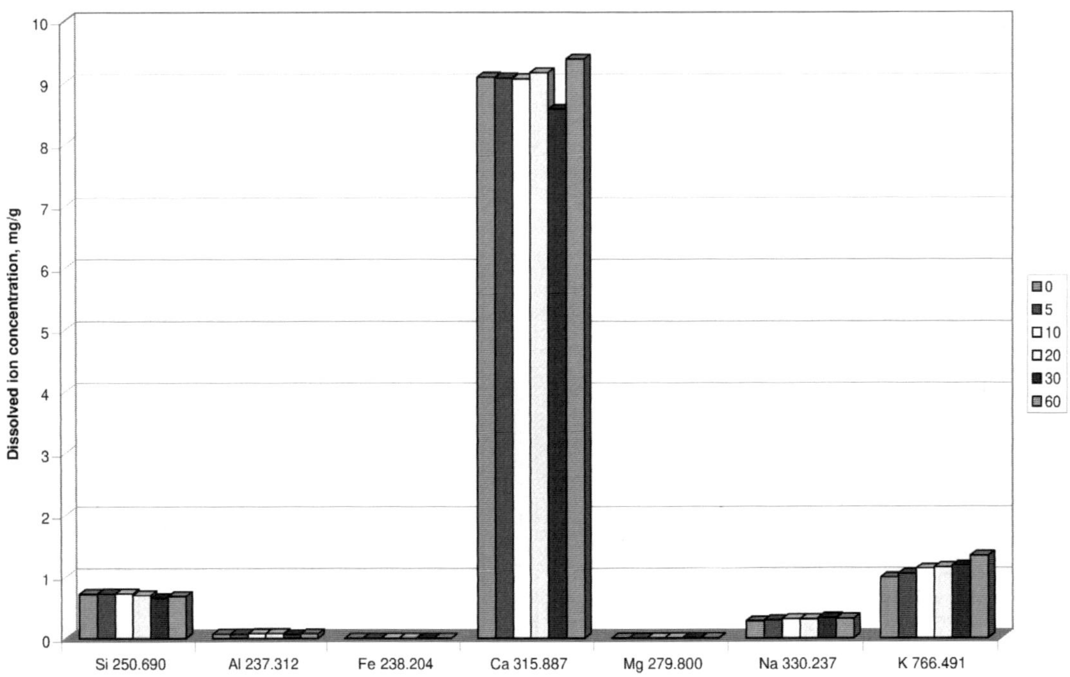

Figure 4. Leachability of elements in function of grinding time. Solution: distilled water, solid/liqid rato: 1/50. Contact time: 36 days

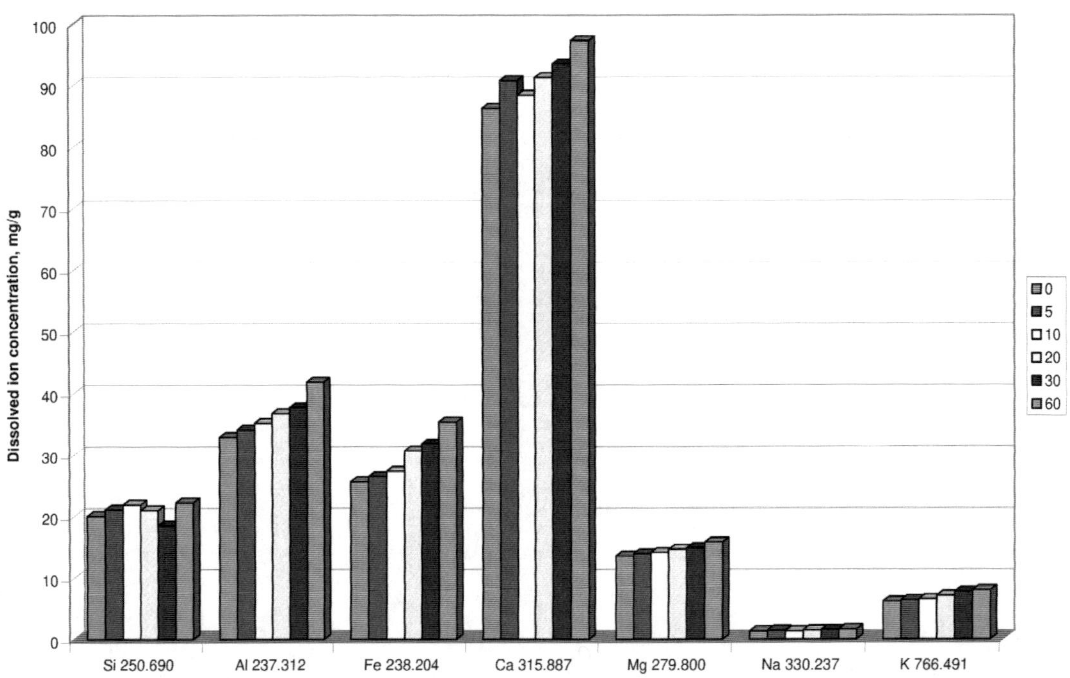

Figure 5. Leachability of elements in function of grinding time. Solution: 1 mol/l HCl, solid/liqid rato: 1/50. Contact time: 36 days

The complex effect of grinding: improving accessibility of elements braking the spherules and the mechano-chemical transformation of the minerals is probable the cause of the different tendency have got for Fe and Ca. compared to Si, Al and K. In the 10 mol/l leaching solution the concentration of the dissolved Si was found less than in the 5 mol/l solution case, however the K which probable liberated from the dissolved silicates further improved. One explanation of this finding is the consumption of the dissolved Si for network formation parallel with the dissolution. It was directly proved by the consolidated deposit formation in the alkaline solution- fly ash slurries (3 months) of the 5 and 10 mol/l NaOH systems which were let standing as long as duration of 3 months.

The second set of leaching were done using distilled water, 1 mol/l HCl and 2 mol/l HNO_3. In these leachates the tested elements dissolved differently than in alkaline medium, see figs.4,5. The Si was less soluble in acidic medium. The Al, Fe had higher solubility, the Ca was completely dissolved. Each tested elements solubility (except Si) increased in function of grinding time, however the manner of the enhancement was less than in case of the alkaline leachates. The dissolution of the tested elements in 2 mol/l HNO3 was similar as it was in the 1 mol/l HCl case.

4. Conclusions

Both, extraction with alkaline and acidic solution have detected improvement in solubility in function of grinding time. The enhancement in alkaline solution was approx. 100% in case of Si and Al. The acidic medium able to dissolve the fly ash higher manner than the alkaline, therefore the effect of grinding was found less pronounced.

Acknowledgement

The described work was carried out in the frame of Center of Excellence of Sustainable Resource Management, University of Miskolc, Hungary.

References

[1] Kumar S, Kumar R, Alex T C, Bandopadhyay A, Mehrotra S P 2007 *Adv Appl Ceram* **106** (3) 120
[2] Álvarez-Ayuso É, Querol X, Plana F, Alastuey A, Moreno N, Izquierdo M, Font O, Moreno T, Diez S, V´ázquez É, Barra M 2008 *J Hazard Mater* **154** 175
[3] Mucsi G, Lakatos J, Molnár Z, Szabó R 2014 *9th International Conference "Environmental Engineering"*, Vilnius, Lithuania, 22–23 May 2014
[4] Mucsi G, Molnár Z, Papp A 2014 *XXVIII. microCAD International Multidisciplinary Scientific Conference*, University of Miskolc, 10-11 April 2014
[5] Mucsi G, Csőke B, Mádai F, Debreczeni Á, Rácz Á, Molnár Z 2013 *XV Balkan Mineral Processing Congress* Sozopol, Bulgária, 12.06.2013-16.06.2013. Sofia, pp. 864-868. (ISBN: 978-954-353-218-6)
[6] Mucsi G, Csőke B, Molnár Z 2010 *Proceedings of the XIIth International Mineral Processing Symposium* 6-8. October 2010. Cappadocia Nevsehir, Turkey ISBN: 978-975-491-295-1, 1235
[7] Provis J L, van Deventer J S J 2009 *Geopolymers, structure, processing, properties, and industrial applications*. (Woodhead publishing in materials, New York), ISBN: 978-1-84569-638-2 (e-book) 421
[8] Izquierdo M, Querol X, Davidovits J, Antenucci D, Nugteren H, Pereira C F 2009 *J Hazard Mater* **166** 561

Mineralogy and crystal chemistry of iron in the Timan bauxite and products of their technological processing

O Kotova, V Silaev, V Lutoev and A Vakhrushev

Institute of Geology Komi SC UB RAS, Syktyvkar, Russia

E-mail: kotova@geo.komisc.ru

Abstract. Mineralogical and geochemical features of two series of samples of typical bauxites from two deposits of Middle Timan mining area (Vezhayu-Vorykva and Svetlinskoe) were studied. The phase composition of ferrous bauxites generally is boehmite, hematite, ultradisperse low-ordered goethite and berthierine. In a boehmite and kaolinite structural impurity of iron to 10%, and in the iron oxidehydroxides aluminum impurity is revealed. On iron content bauxites are subdivided into three mineral types for which quantitative data on valence states of ions of iron and proportions of their distribution last on nonequivalent structural positions in hematite, goethite and berthierine are obtained. Noble metals (Ag, Au, Ir, Rh, Pd) concentrating in bauxites are revealed for the first time. Obtained data can lead to decrease of power consumption during aluminum production and high quality ceramics, to provide production of valuable iron oxide, and also to minimize the ecological harm from accumulation of bauxite wastes.

1. Introduction

Russia is one of the world's largest aluminum producers, however it lacks own aluminous raw due to limitation of the explored reserves not exceeding 3 % from the world reserves. Besides a negative role is played by relatively poor quality of bauxites, in particular their iron content. It is obvious that in such conditions it is possible to meet the needs of Russian aluminum industry from domestic raw only by application of new processing technologies, which allow, first, to utilize ferrous varieties of bauxites stored in dumps, and secondly, to utilize natural raw more efficiently. The major precondition of success is the results of mineralogical-geochemical and crystallochemical studies of bauxite substance received by means of advanced analytical methods, and use of more effective technological methods. This has a special urgency in application to developing Timan bauxite province, where about 30 % of Russian reserves of aluminous raw are concentrated, which is characterized by heterogeneous composition and iron content.

2. Subjects and Methods

Objects of study are represented by two series of samples of typical bauxites from two most promising deposits of Middle Timan mining area - Vezhayu-Vorykva and Svetlinskoe. Corresponding bauxitea formation (bauxites, allites, aleurites and kaolinite-hydromicaceous clays) lies unconformably on Upper Riphean dolomitized limestones and argillaceous shales overlapped by sandstone-argillaceous and igneous-terigenous coastal-marine Devonian-Carboniferous sediments and Later Devonian platobasalts [1]. By its mineral composition the Middle Timan bauxites are traditionally subdivided on kaolinite-boehmite, chamosite-boehmite, hematite-boehmite, hematite-chamosite-boehmite types. By chemical composition they are subdivided on low-ferrous fire-resistant (Al_2O_3 56–75 wt %, Fe_2O_3 2–8 wt %) and ferrous (Al_2O_3 42–48 wt %, Fe_2O_3 12–36 wt %) varieties. As it is known at present in Russia and abroad bauxite raw is processed from contamination by chemical Bayer process. However this method is profitable only for relatively high-quality bauxites. But Timan bauxites along with oxide minerals of aluminum include oxide and silicate forms of mixed alumoferrous and ferrous-aluminum composition, which is difficult to enrich and process. Due to this reason brown bauxite varieties (Fig.1) with total Fe_2O_3 content from 25 wt% are sorted out already in quarries, which leads to significant losses of not only aluminum, but also valuable small elements such as V, Ga, Li, some rare lantanoids. Our studies and experiments have shown that such losses can be minimized by the use of new processing methods.

Modern physical and physical-chemical methods were used: computerized reflective and transmission optical microscopy (POLAM P-312 and OLYMPUS BX51); analytical scanning electron microscopy (Jeol JSM-6400); X-ray fluorescent analysis (Horiba MESA-500W); ICPMS (Perkin Elmer ELAN 9000); thermal analysis (derivatograph Shimadzu DTG-60A/60AH); x-ray diffractometry (Shimadzu XRD-6000); IR-spectrometry (Fourier spectrometer Lumex FT-02 Intralum); 57Fe Moesbauer Spectroscopy (MS-1104Em); magnetometry; ferromagnetic and nuclear magnetic resonance; magnetic separation in weak magnetic field (less 0.5 T) at 0.01–0.4 A (SIM-1).

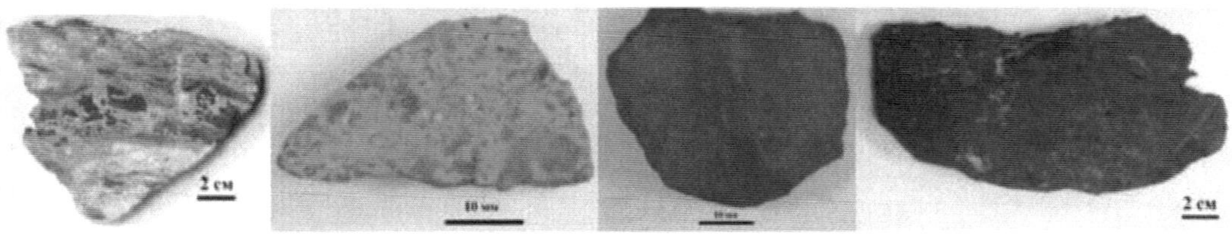

Figure 1. Change of color of Timan bauxites due to iron content from light low-ferrous variety toward brown highly ferrous variety

3. Experimental

Objects of study are bauxite samples with a wide variety of iron content collected from two main deposits of Middle Timan bauxite area – Vezhayu-Vorykva and Svetlinskoe. These samples represent two most enrichment-contrasting types of bauxites with considerable difference in chemical composition (wt%): 1) **low-ferrous** – SiO2 8–32, Al_2O_3 45–85, Fe_2O_3 2–18, silicon module (Al_2O_3/SiO_2) 1.5–11; 2) **ferrous** – SiO2 4.5–38, Al_2O_3 25–63, Fe_2O_3 12–38, silicon module 1–13. According to Yudovich's lithochemical classification [4] Timan bauxites vary from hypo-normoferrous normal superhydrolysates (low-ferrous variety) to normosuperferrous normosuperhydrolysates (ferrous variety). Their composition contains 54 microelements including all 15 lantanoids. By their chemical and geochemical features these elements can be subdivided into 5 groups: 1) low charge lithophites (alkaline, alkaline-earth, rare earth and rare elements) - Li, Rb, Be, Cs, Ba, Sr, Y, Ln, Mo, W, Sn,Tl; 2) multicharge lithophiles (elements-hydrolysates) - Ca, Ge, Sc, Zr, Hf, Nb, Ta, Th, U; 3) siderophilous and chalcophylous elements – Cr, Ni, Co, Cu, Zn, Cd; 4) noble metals – Ag, Pd, Pt, Rh, Ir; 5) semimetals and non-metals – Bi, Sb, As, Te, Se, B. The analysis showed that by relative content (normalized by the average composition of terrigenous sedimentary rocks in earth crust – Clark concentration) and distribution pattern the elements are subdivided into three groups (Fig.2): 1) excessive, i.e. concentrated in bauxites - Sc, Ga, Ge, Zr, Hf, Nb, W, Ln, U, Cd, Pb, Ag, Pd, Rh, Ir, Bi, Sb, As, Te, Se; 2) multidirected – Li, Be, Ba, Sr, Y, Cr, Cu, Ni, Co, Zn, Mo, Sn, Ta, Th, Tl, B; 3) deficit, i.e. decreasing at bauxite formation – Rb, Cs, Pt.

These data testify to that low charge lithophylous elements tend to decrease in bauxites, and multicharge elements-hydrolysates tend to accumulate, including metals, semimetals and nonmetals. Siderophylous and chalcophylous elements are characterized by intermediary geochemical features. It is obvious that such geochemical differentiation fully agrees with hypergene origin of the studied rocks. It should be considered that a number of noble metals is included in the group of excessive elements forming the following succession of relative accumulation: Ir < Rh < Ag < Pd. This specifies the necessity of more careful relation to bauxites as the source of not only aluminum, but some rare highly valuable metals. It is also determined that ferrous variety of bauxites is characterized by higher content of excessive microelements.

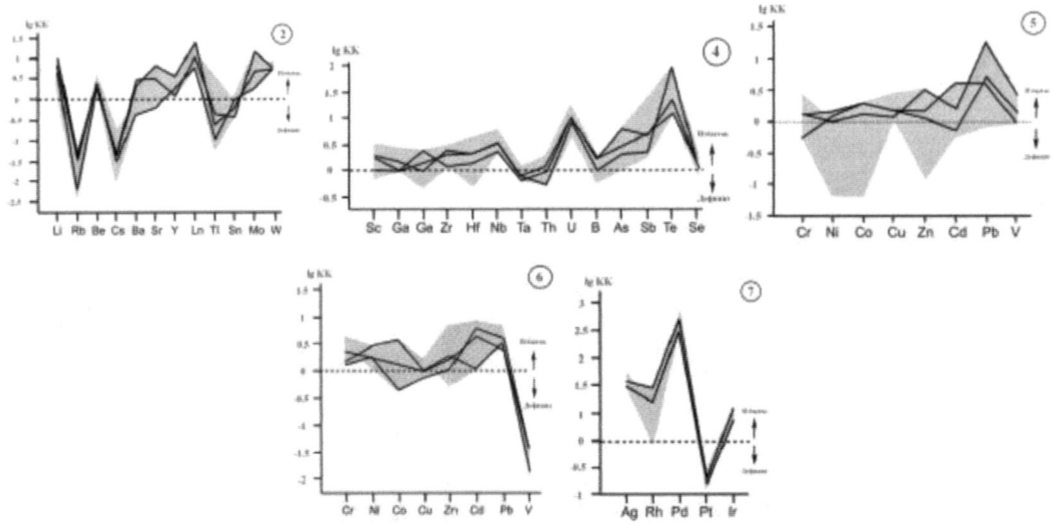

Figure 2. Normalized concentrations of microelements in Timan bauxites on example of Svetlinskoe deposit. Fill – assemblages of trends, lines – trends in ferrous bauxites

Phase composition of bauxites was determined by X-ray difractometric, IR spectroscopic and thermographic methods. The received patterns showed reflections from boehmite, kaolinite, 7 Å septechamosite phase, hematite, crystal-imperfect goethite, rutile and and alusite. IR absorption spectra well presented main lines of deformation and valency fluctuations of chemical bonds Al–O in boehmite and septechamosite phase. Besides these spectra show lines conditioned by chemically unbounded sorbed water (80–100°C) associated with X-ray amorphous phases, dehydration of boehmite (500–600°C) and septechamosite phase (650–900°C). Some thermograms in area 300–350°C represent endothermal peak corresponding to hydrogoethite dehydration.

As is known, the ferrous silicate phase in exogenic and hypergene rocks is treated either as chamosite - 14 Å phylosilicate with chlorite structure (three-layer package with T-O-T alternation) and composition $(Fe^{2+}\cdot Fe^{3+},Al)_5Al[(Si_3Al)_4O_{10}](OH)_8$, or as septechamosite - 7 Å phylosilicate with kaolinite structure (two-layer package with T-O alternation) and composition $(Fe^{2+}\cdot Fe^{3+},Mg)_{2-3}[(Si,Al)_2O_5](OH)_6$. The latter mineral is determined as berthierine. Namely this mineral occurs in the studied Timan bauxites

The mineral composition of Timan ferrous bauxites, estimated by the phase analysis, is predominated by boehmite with its content above 60 mol%. This testifies to potentially high quality of the raw. Admixtures to boehmite is berthierine (to 30–35 mol %), hematite (to 15–20 mol%), goethite (to 6–7 mol %). Besides ferruginous substance is determined, which content can reach 10 mol% and it seems to be amorphous by X-ray data. It is also determined that the growth of content of oxide and oxidehydroxide phases in ferrous bauxites is accompanied by the rapid decrease of berthierine and increase the disproportion between lithophylous (decreasing) and siderochalcophylous (increasing) microelements.

The crystal-chemical features of ferrous bauxites were studied by Moessbauer spectroscopy. The imaging was conducted in the range (–11 … +11)·10–3 m/s at normal conditions. The isomeric shift was determined in relation to α-Fe. Time of spectra accumulation was from 16 to 180 hours in relation to iron content in the samples. The conducted experiments resulted in two variants of multicomponent spectra. The first variant – practically completely sextet – is characteristic for hematite-containing bauxites. The second variant – doublet – is characteristic of silicate-ferrous bauxites. The obtained data testify to that iron in Timan bauxites is not only present in various valency conditions, but also included in the structure of several crystal-chemically heterogeneous minerals – oxides, oxidehydroxides, silicates – distributing to non-equivalent structural positions in them as well.

The oxide phases in Timan bauxites are represented by hematite and ultradisperse (with sizes up to 20 nm and less) low-ordered goethite. In Moessbauer spectra these phases are corresponded by inherent line sextets with various values of ultrathin field intensity on iron nuclei (see Table 1). These values for hematite in all cases appeared lower than reference, which can be explained by the presence of aluminum admixture, which decreases the efficient value of ultrathin magnetic field intensity. The estimated concentration of aluminum admixture makes about 10 at% in hematite of Timan bauxites. The sextet components of goethite phase in spectra are partly relaxed due to ultradispersity of this phase.

Some varieties of ferrous bauxites contain the admixture of disperse x-ray-amorphous oxide alumoferrous and silicious substance. The size of particles, according to X-ray and synchrotron small-angle scattering methods, varies from 35–40 to 20–25 nm. The fluctuation of X-ray amorphous degree of this part of bauxite substance was estimated by the level of heat fluctuations.

Table 1. Parameters of Moessbauer spectra of Timan ferrouos bauxites

Mineral type	QS, mm/s	IS, mm/s	H, kE	S, %	Fe position	Fe^{2+}/Fe^{3+}
Hematite-boehmite	-0.21	0.37	500, 420–511	72.5	Hematite -Al	0
	-0.28	0.36	330, 200–380	7.8	goethite -Al	
	2.42	1.19	–	2.8	$Fe^{2+}{}_{VI}$, kaolinite	0.16
	0.84	0.34		11.6	$Fe^{3+}{}_{VI}$, kaolinite	
	0.47	0.37		5.3	$Fe^{3+}{}_{IV}$, kaolinite	
Hematite - berthierine-boehmite	-0.21	0.37	500, 410–511	52.4	Hematite -Al	0
	-0.28	0.36	314, 200390	13.4	goethite-Al	
	2.64	1.13	–	23.3	$Fe^{2+}{}_{VI}$(M2), berthierine	2.53
	2.28	1.08		1.2	$Fe^{2+}{}_{VI}$(M1), berthierine	
	0.83	0.44		7.0	$Fe^{3+}{}_{VI}$, berthierine	
	0.5	0.31		2.7	$Fe^{3+}{}_{IV}$, berthierine	
Berthierine-boehmite	2.654	1.124	–	62.8	$Fe^{2+}{}_{VI}$(M2), berthierine	2.15
	2.28	1.06		5.5	$Fe^{2+}{}_{VI}$(M1), berthierine	
	0.71	0.52		12.2	$Fe^{3+}{}_{VI}$, berthierine	
	0.57	0.34		19.5	$Fe^{3+}{}_{IV}$, berthierine	

Note. QS – quadrupole splitting, IS – isomeric shift, HHF – effective value of ultrathin magnetic field intensity (average and variation range), S – area for relevant component

According to crystal-chemical features of iron distribution ferrous bauxites are subdivided into three mineral types. The first type – **hematite-boehmite** – iron bonds with Al-hematite at more than 70%, it is determined by goethite admixture at about 8%, about 20% of iron is related to the alumosilicate phase, probably kaolinite. In kaolinite the main part of iron is in tervalent state (Fe3+/Fe2+ = 6.03), occupying mainly octahedral and in lesser extent tetrahedral positions. Insignificant amount of iron is present in kaolinite lattice in bivalent state, also occupying octahedral positions. In the second type – **hematite-bethierine-boehmite** – more than 50% of iron is confined to hematite, about 13% relates to goethite, and about 34% is included to berthierine. In berthierine, as in other phylosilicates with kaolinite structure type, Fe2+ ions fill M1 and M2 octahedral positions characterized by *trans* and *cis* configurations of OH-group position. The occupation of M2 positions by these ions results in great deformations of lattice, therefore these positions are assigned by the most intense doublet with greater quadrupole splitting (QS). Accordingly doublet with smaller QS is related to Fe2+ ions in M1 positions. The doublets, related to Fe3+ ions in berthierine, are also assigned to different positions: with greater QS – top octahedral M2, and with smaller QS – to octahedral M1 or tetrahedral positions. In the studied case 68 % of iron in berthierine is represented by Fe2+ ions in M2 octahedral positions. Fe3+ part makes 28 %, and Fe3+(IV)/Fe3+(VI) ratio reaches 0.39. Total Fe3+/Fe2+ ratio in berthierine is 0.40. In the third type – **berthierine-boehmite** – practically all iron bonds with berthierine. The doublet Moessbauer spectrum for such bauxites contains only components corresponding to irono ions in phylosilicate lattice. The parameters of doublets, corresponding to M1 and M2 positions, are very close to the parameters for berthierine in literature. Fe2+ part in octahedral M2 positions of berthierine in this case is 63%, Fe3+ ions – 32%, Fe3+(IV)/Fe3+(VI) ratio is estimated at 1.6. Total Fe3+/Fe2+ ratio reaches 0.46.

4. Modification of technological properties

The experiments with the bauxites revealed considerable increase of magnetic product yield in size class –315+250 μm after drying at 150 °C. This means that the considerable increase of efficiency of utilization of ferrous bauxite raw can be reached already by the change of surface state, in particular by decreasing degree of its hydroxylation. The experiments with red mud revealed that it is characterized by much greater (4-5 times) particle dispersity in comparison with ferrous bauxites. Particles of class –71 μm make 81% which are divided equally between classes –71+45 and –45 μm. The division of red mud into non-magnetic and magnetic fractions occurs in the conditions of weak magnetic field. Non-magnetic material dominates in red mud at the initial sample. At that magnetic fraction yield decreases with the decrease of particle size, however in –45 μm class this yield slightly increases due to increasing content of hematite and goethite. After two hours drying at 150 °C the trend of yield decrease is preserved and covers both size classes. After drying at 300 °C the magnetic product yield increases considerably and grows with decreasing particle size. Above data testify to that the efficiency of magnetic separation of initial ferrous bauxites and red mud is influenced by two factors: content of weak magnetic oxides and oxydehydroxides (favorable factor) and hydroxylation of particle surface (unfavorable factor). The influence of the last factor can be minimized by the way of drying geomaterials and their products to 150–300 °C. However this procedure only partly improves separation of aluminum raw without a real success. To achieve greater degree of separation of ferrous and aluminum components of the studied bauxites we, together with the colleagues from the Ukraine, conducted the pioneer program of magnetization experiments. During these experiments we achieved transformation of initially non-magnetic and low-magnetic oxides and oxihydroxydes of iron into strong magnetic oxides [2, 3]. The experiments were conducted in water environment in the conditions of electromagnetic field in the presence of chemical catalyst with reduction characteristics. The physical essence of the transformation is in recharge of iron ions in the structure of relevant minerals. The objects of our studies were samples of high-ferrous bauxites of hematite-boehmite and hematite-berthierine-boehmite types, average mineral composition is within the range (mol%): boehmite 62.9–65.6; berthierine 0–12.6; hematite 17.4–11.9; goethite 6.3–6.5; rutile + anatase 2.6–3.4; X-ray amorphous phase 10.8–0. It was determined that at the microwave influence on ferrous bauxite the phase composition and magnetic characteristics were considerably changed. Low-magnetic oxides and oxihydroxides - hematite and goethite already at low temperature 95°C transit to strong magnetic state due to the transformation of these minerals into maghemite and magnetite. This is confirmed by X-ray diffraction analysis, which reveals that initial hematite is decreased (reflexes d/n = 2.702 and 2.518 Å), but instead a newly formed magnetite phase appeared (d/n = 2.969, 2.525, 2.096, 1.612 Å) in the modified samples. The degree of transformation directly depends on the conditions of magnetization. The magnetic characteristics of bauxites act accordingly with phase-crystal-chemical changes. In particular, the value of saturation magnetization increases at about an order in comparison to the initial stage (Fig.3).

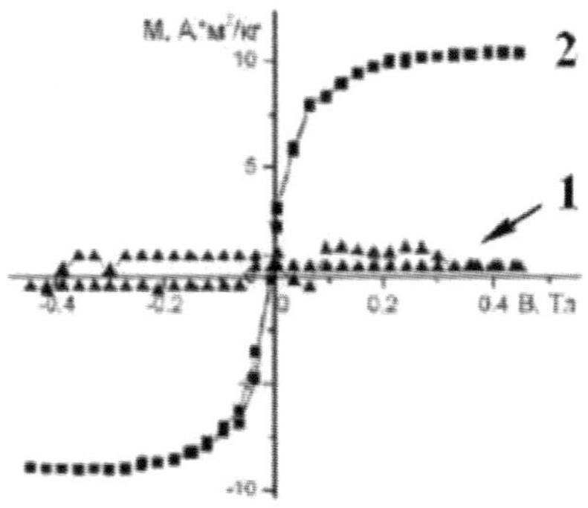

Figure 3. Dependence of magnetization degree (M) of typical sample of hematite-boehmite bauxite, conditioned by outer magnetic field (B). 1, 2 – magnetization curves for initial and modified samples accordingly

5. Conclusion

Results of the experiments convince that it is possible to considerably improve the technological features of ferrous bauxites by a certain influence, namely, to increase the efficiency of magnetic separation by directed phase-crystal-chemical transformation of ferrous minerals in bauxite composition. This transformation can be characterized by the following equations: 1) $12FeO(OH) \rightarrow 6Fe_2O_3 + 6H_2O\uparrow \rightarrow 4Fe_3O_4 + O_2\uparrow$; 2) $6Fe_2O_3 \rightarrow 4Fe_3O_4 + O_2\uparrow$. It is obvious that this transition is carried out gradually with intermediary phase of maghemite and can have various degrees of completion. Therefore to achieve maximal efficiency of separation it is necessary to vary the mode of magnetic separation. We consider the achieved result as the scientific precondition to development of a new method of bauxite separation without power-consuming chemical processing. This method can lead to decrease of power consumption during aluminum production and high quality ceramics, to provide production of valuable iron oxide, and also to minimize the ecological harm from accumulation of bauxite wastes.

6. Acknowledgements

The studies are financed by grant of RFBR No 14-05-90403 and RAS programs (project 15-11-5-33).

Reference

[1] Vakhrushev A et al 2012 *Vestnik of Institute of Geology of Komi* SC UB RAS **10** 14
[2] Lyutoev V et al 2013 *Vestnik of Institute of Geology of Komi* SC UB RAS **1** 20
[3] Ponomarenko A et al 2013 *Modern problems of theoretical, experimental and applied mineralogy* (Yushkin's Readings-2013) (Syktyvkar, Geoprint)
[4] Yudovich Ya and Ketris M 2000 *Basis of lithochemistry* (Saint-Petersburg, Nauka

IOP Conf. Series: Materials Science and Engineering **123** (2016) 012025 doi:10.1088/1757-899X/123/1/012025

Titanium minerals for new materials

O Kotova[1], E Ozhogina[2], A Ponaryadov[1] and I Golubeva[1]

[1]Institute of Geology Komi SC UB RAS, Syktyvkar, Russia
[2]All-russian scientific-research institute of mineral resources, Moscow, Russia

E-mail: alex401@rambler.ru

Abstract. The mineral composition of titanium minerals of modern coastal-marine placer in Stradbroke Island (Australia) and Pizhma paleoplacer in Middle Timan (Russia) has been presented. The physical features of titanium minerals and their modification methods were shown. Photocatalysts on the basis of the Pizhma leucoxene were developed for water purification.

1. Introduction

Titanium is related to strategic mineral resources. The mineral composition of placer titanium ores and localization pattern of ore minerals determine their processing and enriching technologies. The paper presents new data on the mineralogy of placer titanium ores on an example of modern coastal-marine placer in Stradbroke Island, Eastern Australia, and Pizhma paleoplacer in Middle Timan, Russia, and materials on their basis.

2. Experimental

The ore sand samples of modern coastal-marine placer originate from the Enterprise Mine related to East-Australian ore province stretching to 13 km with average width 3 km along the eastern coast of Queensland.

Titanium-containing mineral samples were collected from the Pizhma placer in Middle Timan. The Pizhma placer is one of the greatest titanium deposits. The reserves are evaluated at 2.5 billion tons. The open mining of titanium can function during 150 years. The resources of the deposit can support titanium mining during 500 years.

The samples were studied by the following methods: optical-mineralogical (stereomicroscope MBS-10, polarizing microscope POLAM L-311), semiquantitative x-ray phase analysis (x-ray difractometer X'Pert PRO MPD). Besides microprobe (VEGA 3 TESCAN) and x-ray fluorescent analysis (XRF-1800 Shimadzu) were used.

3. Results and discussion

Earlier on an example of Australian titanium placer the process of transformation of ilmenite into rutile was shown. By the mineralogical composition ores of the both deposits are complex: enriched by valuable minerals [1, 2]. Apart from main ore concentrates it is possible to obtain accompanying nonmetallic products. This will increase the efficiency of deposit exploitation.

The mineral composition and occurrence structures of minerals in ore, products, concentrates, wastes are changing during utilization of mineral raw. The main method of monitoring of composition and features of titanium ores and technological products is x-ray phase analysis. The mineral compositions of the samples (Fig.1) are represented in Table 1.

Ilmenite dominates in ore sands of Stradbroke Island, and leucoxene dominates in the ores of the Pizhma titanium deposit. Australian ilmenite and its altered varieties are mainly characterized by a very high MnO content (from 5.24 to 11.08 %). The irregular distribution of iron oxides, titanium and manganese in the altered ilmenite was shown in the paper [3]. E.g., in the areas of substitution of ilmenite by pseudorutile the concentrations of the given elements are greatly various due to various ratios of basic components in each grain. Their ratios are equal in the area of rutile evolution.

Titanium ores and quartz sandstones are two basic types of raw of the Pizhma deposit. Moreover, the high content of gold, diamonds and also rare earth elements (REE) and rare metals (their forms are not determined) were studied. We found native copper on the surface of minerals composing titanium-bearing sandstones of the Pizhma placer.

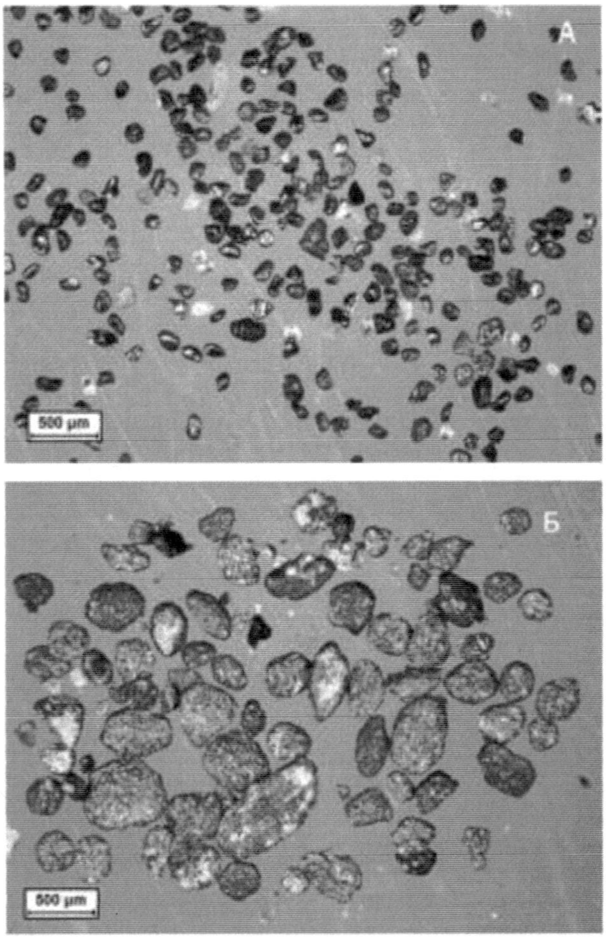

Figure 1. Titanium minerals of modern coastal-marine placer in Stradbroke Island, Eastern Australia (A) and Pizhma paleoplacer in Middle Timan (B)

Table 1. Mineral composition of ore sands

Mineral	Theoretical formula	Content weight %
Stradbroke Island		
Ilmenite	$FeTiO_3$	39
Brookite	TiO_2	3
Rutile	TiO_2	8
Pseudorutile	$Fe_2Ti_3O_9$	19
Zircon	$Zr(SiO_4)$	6
Quartz	SiO_2	2
Kyanite	$Al_2O(SiO_4)$	1
Epidote	$Ca_2Al_2Fe_3+(SiO_4)_3OH$	1.5
Tourmaline	$NaFe^{2+}{}_3Al_6Si_6O_{18}(BO_3)_3(OH)_4$	4
Garnets	$R^{2+}{}_3R^{3+}{}_2[SiO_4]_3$, R^{2+} – Mg, Fe, Mn, Ca; R^{3+} – Al, Fe, Cr	1.5
Chromespinelides	$(Mg, Fe)Cr_2O_4$	6
Amphibole	$R_7[Si_4O_{11}]_2(OH)_2$, R – Ca, Mg, Fe	5
	Crystal phase total:	96
Pizhma placer		
Ilmenite	$FeTiO_3$	7.5
Anatase	TiO_2	1
Rutile	TiO_2	18
Pseudorutile	$Fe_2Ti_3O_9$	45
Hematite	Fe_2O_3	1.5
Quartz	SiO_2	10
Pyrite	$Al_2O(SiO_4)$	1
Mica	$K_{<1}Al_2(OH)_2(AlSi_3O_{10})*nH_2O$	4
	X-ray amorphous phase	12
	Total:	100

The native copper up to 12 μm occurs both on the surface of quartz, covered by iron hydroxides, and on leucoxene. Two copper occurrence forms found are as:

1) inclusions of monocrystals of isometric (Fig.2A) or angular irregular shape (Fig.2B) in cavities of quartz grains covered by iron hydroxides;
2) intergrowths of plate (Fig.2C) or dendrite-like (Fig.2D) polycrystals.

Apart from native copper there is cuprite.

The treatment methods and mineral processing depends on the mineral composition of the sands. According to the technological features of rocks (density and magnetic) studied placers are close. The obtained results of physical studies, mineral composition features, morphostructural characteristics and degree of alteration of titanium minerals from the placers specify a high potential of physical methods of processing (gravitational and magnetic separation, flotation) and possible application of combined methods of processing [3].

Production of pigment titanium dioxide for further production of titanium white, paper, plastics etc is the usual application area of titanium concentrates. Titanium dioxide of high chemical purity is used to produce optically transparent glass, fiber optics, electronics (iPad), piezoceramics, in medical and food industry.

We designed photocatalysts based on leucoxene from Pizhma placer [5]. The results showed that the photocatalysts based on rutile, synthesized from leucoxene from Pizhma deposit, can be applied to decay phenols in water. The dependence of photocatalyst activity from specific surface area and presence of noble metal atoms (Au, Pt) was shown.

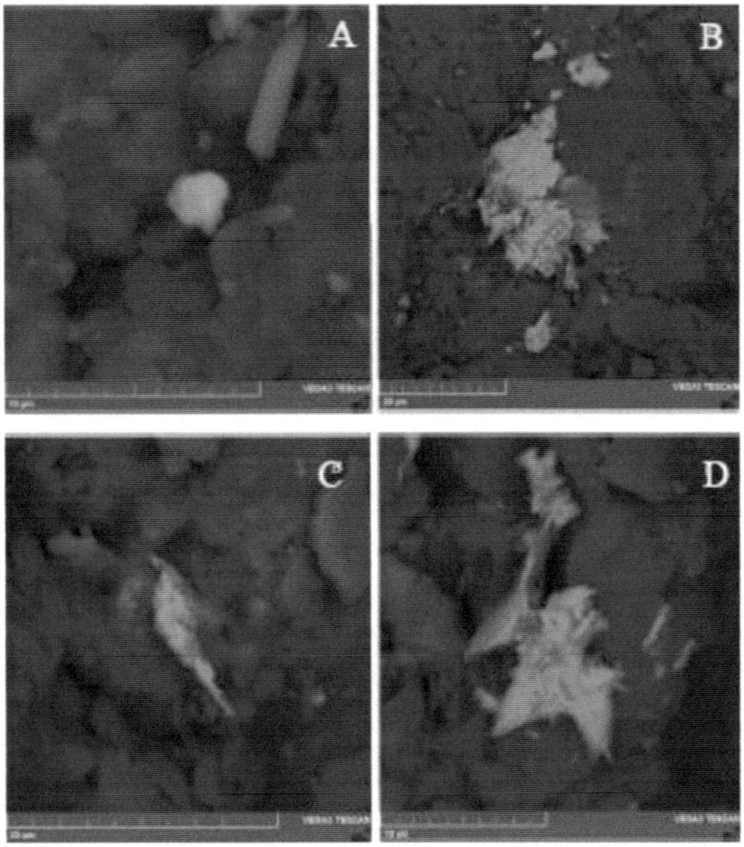

Figure 2. Inclusion of copper of isometric (A), angular irregular (B) shapes, intergrowth of plate (C) and dendrite-like copper with quartz crystal (D)

4. Conclusion

The mineralogical features of titanium minerals from modern coastal-marine placer in Stradbroke Island, Eastern Australia and Pizhma paleoplacer, Russia, have been studied. Their physical and chemical features (magnetic, electrical, sorption etc.) and their modification methods were studied. New areas of application of leucoxene concentrate were suggested: photocatalysts for water purification.

5. Acknowledgments

The work was done under financial support of RAS programs (project 15-18-5-45).

References

[1] Askhabov A et al 2012 Vestnik of Institute of Geology of Komi SC UB RAS **11** 39. (in Russian)

[2] Dyadchenko M and Khatuntseva A 1961 Stages of ilmenite alteration in hypergeny conditions in Problems of Mineralogy of sedimentary structures (Lviv, Lviv University) (in Russian)

[3] Golubeva I et al 2013 Vestnik of Institute of Geology of Komi SC UB RAS **9** 24. (in Russian)

[4] Ignatiev V D and Burtsev I N 1997 Timan Leucoxene: mineralogy and technology (Saint-Petersburg, Nauka)

[5] Ponaryadov A and Kotova O 2013 *Materials Science and Engineering* **47**

Wet gringing of zeolite in stirred media mill

G Mucsi[1], K Bohács[1]

[1]Institute of Raw Material Preparation and Environmental Processing, University of Miskolc, 3515 Miskolc, Hungary

E-mail: ejtbohacs@uni-miskolc.hu

Abstract. In the present study the results of systematic experimental series are presented with the specific goal of optimizing the zeolite nanoparticles' production using a wet stirred media mill. The diameter of the grinding media as well as the rotor velocity were varied in the experiments. Particle size distribution and „outer" specific surface area of the ground samples were measured by a laser particle size analyser. Additionally, BET, XRD and FT-IR analyses were performed for the characterization of the „total" specific surface area as well as the crystalline and material structure, respectively. Based on the results of the laboratory experiments it was found that wet stirred media milling provided significant reductions in the particle size of zeolite. Furthermore, the crystallinity of the samples also decreased, so not only the physical but the mineralogical characteristics of zeolite can be controlled by stirred media milling.

1. Introduction

Zeolites are natural crystalline aluminosilicates. Structurally, zeolites are tectosilicates exhibiting an open three-dimensional structure containing cations needed to balance the electrostatic charge of the framework of silica and alumina tetrahedral. To collapse of the crystal structure can be desired in some cases since different strength of Bronsted and Lewis acid and base sites can thus be obtained [1, 2]. This, in turn, can provide a better catalytic performance [3].

Ion-exchange properties of zeolites can be utilized in agriculture because of their large porosity and high cation-exchange capacity[4]. They can be used as both carriers of nutrients and a medium to free nutrients [5]. Zeolite incorporation in soil was found to increase crop yields and to promote nutrient use efficiency [6]. Other possible uses being investigated include applications as a carrier of slow-release fertilizers, insecticides, fungicides, and herbicides, and as a trap for heavy metals in soils. [7]

Stirred media mill is a high energy density mill belonging to the group of drum mills, which is used for industrial nanosize particle production in wet mode. Stirred media mill is an important milestone in the mill development, because while before the movement of the grinding media was carried out by the rotation (ball mill) or vibration (vibration mill) of the milling chamber, in this case the rotation of the concentrically or eccentrically placed rotor makes the grinding media moved. As a result, the grinding balls colliding with each other; are pressed to the mill liners; and colliding with the rotor. The particles of the feed material are fractured by the impact, pressure, collide and friction stresses [8].

Stress model are widely used for the description of the stirred media milling. Using the product related stress model described by Kwade et al. [9], the specific surface area or the particle size can be related to the operation parameters, thus creating the purposeful control of the product of the stirred media mill. The degree of stress energy has an optimum. When the energy input is high enough to fracture the agglomerates or particles into smaller ones by only one stress event, the optimum stress energy is achieved.

Stirred media mills can be used for a lot of various aims besides the ultrafine grinding. One is the mechanochemical activation of the materials due to the high energy density in the milling chamber, this was presented by Mucsi et al [10]. Stirred media mills can also be used in dry or wet mode. In dry mode, fundamentals of particle rounding and surface roughness reduction was revealed by Racz [11].

2. Materials

Natural zeolite used in this study originated from Mád deposit, Hungary. Its main mineral components were identified as clinoptilolite, cristobalite, smectit by XRD pattern. The natural zeolite was preliminary crushed and ground resulting in a maximum particle size less than 150 μm.

Table 1. Chemical composition of natural zeolite used in the experiment (in w/w%, dry basis)

Oxides	w/w%	Elements	ppm
SiO_2	74,7	Cu	6
Al_2O_3	13,2	Zn	67
MgO	0,96	Pb	44
CaO	1,54	Rb	194
Na_2O	0,19	Sr	437
K_2O	1,53	Ba	143
Fe_2O_3	1,54	Cr	>5
MnO	0,035	Co	>5
TiO_2	0,092	Ni	>5
P_2O_5	0,017	Zr	182
$\sum$	93,834		

From the chemical composition of natural zeolite (measured by Rigaku XRF apparatus) used in the experiment can be seen the main components are SiO_2, Al_2O_3 and MgO (Table 1). The SiO_2/Al_2O_3 ratio is 5,66.

3. Methods

The experiments were carried out in a stirred media mill (Netzsch MiniCer) in circular mode. During the experiments the particle size distribution F(x) and the „outer" specific surface area of the ground samples were measured and calculated by a laser particle size analyzer, the energy consumption of the mill was measured, and then the specific grinding energy was calculated. Additionally, the BET Surface Area (m^2/g), the mineral composition by XRD and the material structure by FT-IR analyses were performed for the characterization of the „total" specific surface area as well as the crystalline and material structure, respectively.

3.1 Particle size distribution

During the milling, the particle size distribution of the ground samples was determined by a Horiba LA-950V2 laser diffraction particle distribution analyzer in distilled water media. Before the measurement, ultrasonic treatment was applied for 1 min and 0.5 mL sodium pyrophosphate was used in a 50 gL^{-1} concentration for best dispersion. Samples were taken after 1, 3, 5, 10, 15, 30, 60,

90, 120, 150 and 180 min grinding. For these samples the calculated specific surface area was also measured by the Horiba LA-950V2 laser diffraction particle distribution analyzer.

3.2 Specific surface area by BET method

For the measurement of the total specific surface area Tristar 3000 from Micrometrics Corporation was used. Before each measurement the sample has to be degassed. From the adsorbed gas the surface area of the particle is determined. The BET (Brunauer, Emmett and Teller) method is used to determine the surface area of the particles. To prepare the samples for the BET measurement, the suspensions were dried overnight in a laboratory oven at 100°C. To degas the powder the samples were dried at 90°C for 30 minutes and additionally at 300°C for 180 minutes in flowing nitrogen gas in the Micrometrics Smartprep before the adsorption measurements were carried out. The complete adsorption-isotherms were recorded. The specific surface area was calculated by a 5 point Brunauer- Emmett-Teller (BET) method. The weight of each sample was approximately 1 g.

3.3 X-Ray powder diffraction (XRD)

XRD investigations were carried out on a Bruker D8 Advance diffractometer (Cu-Kα radiation, 40kV, 40mA) in parallel beam geometry obtained with Göbel mirror (0.6 mm exit slit, 2.5° axial Soller), equipped with Vantec-1 position sensitive detector (1° opening, 2.5° axial Soller). Measurements were recorded in 2-70°(2θ) range of angle with 0,007°(2θ)/28 secundum goniometer speed. Crystalline components were identified by Search/Match algorithm in Bruker DiffracPlus EVA software package using ICDD PDF-2 (2005) database. Quantitative results were obtained by Rietveld-refinement, in Bruker TOPAS4 software using ICSD database, unit cell parameters were refined and mean crystallite sizes determined.

3.4 FT-IR

To follow the structural changes due to grinding process Fourier Transformed Infrared Spectroscopic measurements of zeolite carried out by JASCO FT-IR 4200 type Fourier Transformed Infrared Spectrometer in Reflection mode, diamond ATR PRO470-H condition was used. 3 spectra was made from each samples. For making of one spectra of a sample, different parts of zeolite was measured. One spectra averaged of 64 FT-IR measurements with 4 cm^{-1} resolution number.

4. Experimental

The grinding experiments were carried out in a stirred ball mill (Netzsch MiniCer) in wet mode, using 180 min maximum grinding time. The first experimental serie is based on the influence of the circumferential velocity of the rotor, which was studied at 7.7, 9.3 and 10.8 m/s, and the second one is that of the grinding media size which was studied at 550 μm, 700 μm and 900 μm. All other grinding operating parameters were permanent. The suspension concentration was 10 w/w %.

5. Results

Figure 1 shows the particle size analysis results of the 180-min-ground final products. It shows the frequency curves. It can be observed that bimodal nature of the feed remained after grinding, but significantly shifted to the submicron size range. The finer mode of all of the ground products were finer than 200 nm, on the other hand the coarser mode was 2 μm approximately. Consequently, since the initial modes were 3 μm and 50 μm, not only the surface roughness of the particles was changed but breakage of the particles was occurred due to the grinding process.

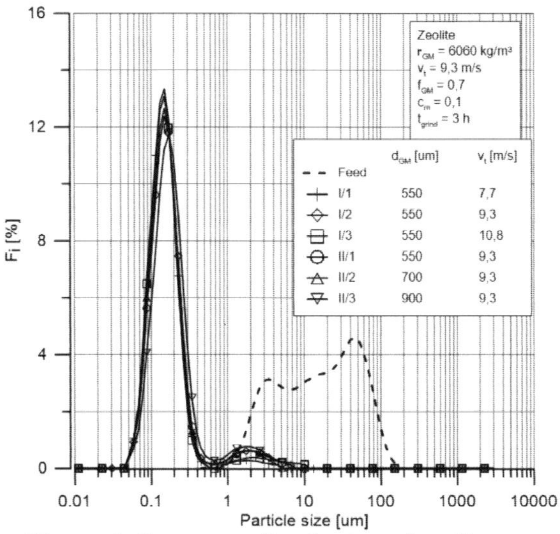

Figure 1. Frequency distribution of zeolite final product

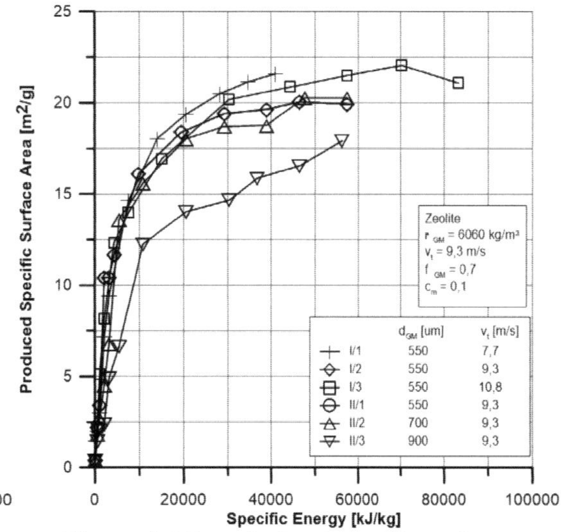

Figure 2. The calculated specific surface area as a function of the specific energy

Figure 2 shows the produced specific surface area of the ground samples during grinding to monitor the grinding kinetics. It shows the calculated specific surface area as a function of the specific energy input. From this point of view it can be seen that using the largest grinding media size resulted in the lowest specific surface area. To observe more significant difference in the effect of the operating parameters stress models were applied.

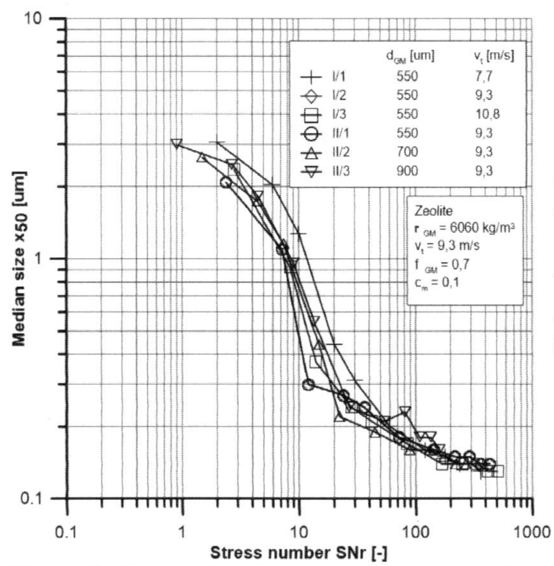

Figure 3. The product fineness as a function of the stress number

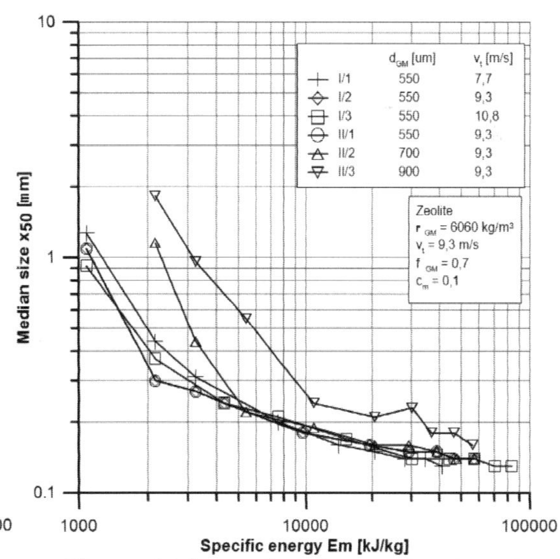

Figure 4. The median size as a function of the specific energy

Figure 3 shows the product fineness (median size) as function of the stress number for different grinding media size and circumferential velocity. Based on the relation of these parameters it can be established that grinding occurred with sufficient size of grinding media since the stress energy is sufficient to break the particles at the beginning of the grinding process.

Figure 4 shows the median size as function of the specific grinding energy. The main difference is given by the bigger grinding media size. In order to find the optimal stress intensity this figure can be used. It can be estimated in two ways. First way is to read the different values of the specific grinding energy which contains to an exact median size. In this case the 500 nm of median size was chosen.

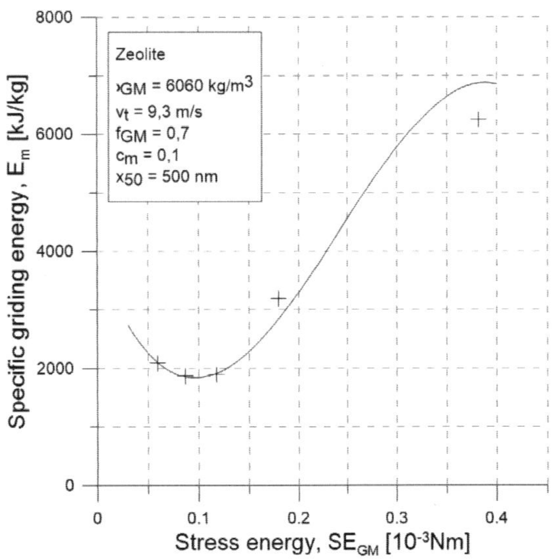

Figure 5. Specific grinding energy required to produce a median size of 500 nm as function of stress intensity

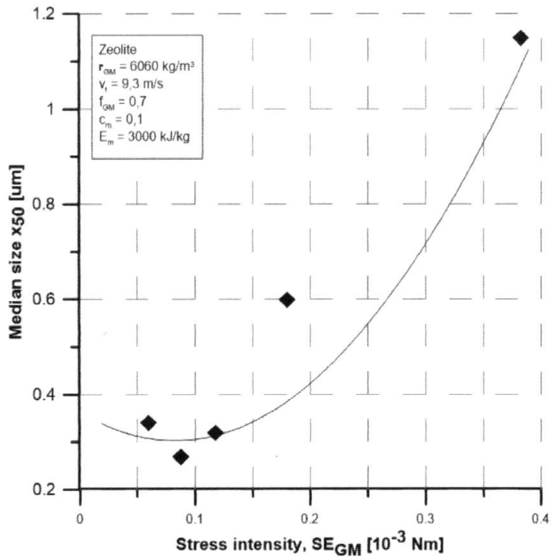

Figure 6. The product fineness at 3000 kJ/kg specific energy as a function of the stress energy

If the specific grinding energy required to produce a median size of x_{50} = 500 nm is constant as function of stress energy, the relation is shown in figure 5. At lower stress intensity high specific energy is necessary to produce a median size of 500 nm, because the stress intensity is too low for an effective grinding and therefore a large number of stress events are necessary. Therefore, in this range of stress energy the specific energy required for a certain product fineness decreases with increasing stress intensity. If the stress intensity is so high, that is average the feed particles can be broken, the specific grinding energy reaches a minimum value. If the stress energy is increased further, the energy utilization of each stress event decreases, so that with increasing stress intensity more specific grinding energy is required to produce a median size of 500 nm. The optimum stress intensity is $0.1 \cdot 10^{-3}$ Nm to produce a median size of 500 nm.

Besides the specific grinding energy at a constant product fineness, also the product fineness at a constant specific grinding energy input can be presented as a function of the stress intensity. In figure 6 the median size is produced at a specific energy input of 3000 kJ/kg are depicted as a function of the stress intensity. According to literature, for single particle stressing, the maximum energy utilization and thus, the optimum stress intensity can be found at fracture probabilities of 100 % [12]. The minimum median size is 300 nm.

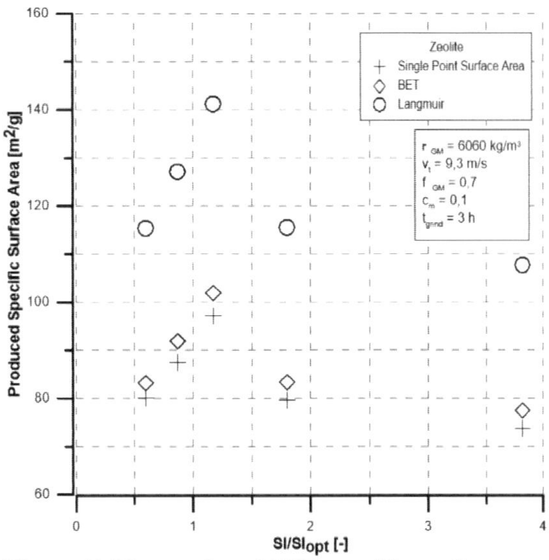

Figure 7. The produced BET specific surface area as a function of the SE/SE_{opt}

If the optimal stress energy is $0.1 \cdot 10^{-3}$ Nm and the grinding time is 180 min, the produced specific surface area (which was measured by BET method) as a function of the SE/SE optimum is presented in figure 7. It can be seen that the SSA increased until the ratio reached the optimum value, then it began to decrease.

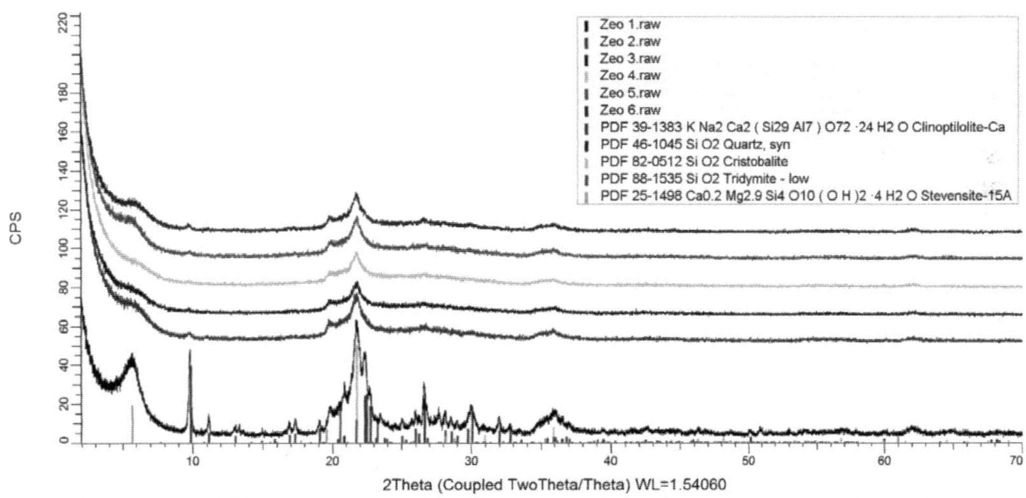

Figure 8. XRD measurement of the feed sample and the 180 min ground samples

The results of the XRD measurements of the feed sample and the 180 min ground samples show that amorphization occurred due to grinding (figure 8). Significant amorphization was detected since the content of amorphous phases in the feed material was only 7% which increased to 36% up to 50% in the ground products (depends on the circumstances). The ratio of clinoptilolite phase parallel decreased due to grinding.

In figure 9, the results of the 180-min-ground zeolite samples' FT-IR measurement in transmission mode can be seen. The feed (which is the first curve from the top) had the highest intensity of the 790 cm^{-1} peak related to the AlO_4 vibrations, and it was reduced significantly due to the effect of grinding. New peaks were appeared at 1472, 1098, 874 and 856 cm^{-1}. 1472 cm^{-1} is assigned to the stretching vibration (O–C–O), 1098 cm^{-1} is assigned to the asymmetric stretching vibration (T–O–Si, T=Si or Al), 874 and 856 cm^{-1} are assigned to the Si–O stretching, OH bending (Si–OH). Other peaks were disappeared due to grinding: at 1045 and 696 cm^{-1}. 1045 cm^{-1} is assigned to the asymmetric stretching vibration (T–O–Si, T=Si or Al), 696 cm^{-1} is assigned to the symmetric stretching vibration (Si–O–Si and Al–O–Si) [13]. These changes indicate the structural changes due to grinding.

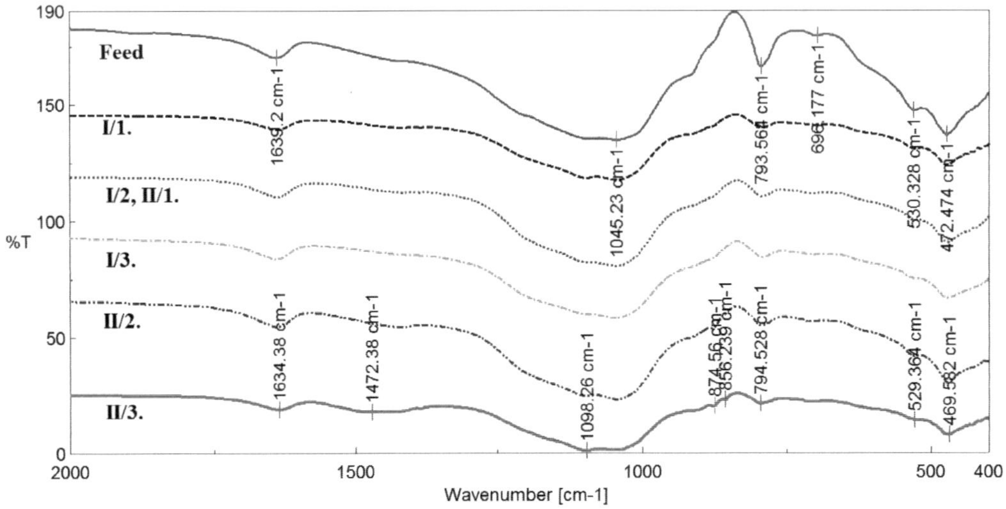

Figure 9. The results of the 180-min-ground zeolite samples' FT-IR measurement

6. Conclusions

Based on the results of the laboratory experiments it was found that the optimum stress intensity is $0.1 \cdot 10^{-3}$ Nm to produce a median size of 500 nm. The optimum stress intensity is $0.07 \cdot 10^{-3}$ Nm to produce minimum median size product at 3000 kJ/kg specific energy. It can be seen that the SSA - BET increased until the SE/SE_{opt} reached the optimum value than it began to decrease. The changing of intensity of FT-IR peaks indicates the structural changes due to grinding. Furthermore, the crystallinity of the samples also decreased.

Acknowledgements

The research work was performed in the framework of the Centre of Excellence in Sustainable Natural Resource Management at the Faculty of Earth Science and Engineering, University of Miskolc.

Furthermore authors appreciate for implementation and helping in evaluation the BET measurements for Tibor Ferenczi, in XRD investigations for Ferenc Kristály and in FT-IR investigations for Zoltán Molnár.

References

[1] Xie J, Kaliaguine S. 1997 *Appl Catal A Gen* **148** 415–23. doi:10.1016/S0926-860X(96)00234-7.

[2] Zielinski P a., Van Neste a., Akolekar DB, Kaliaguine S. 1995 *Microporous Mater* **5** 123 doi:10.1016/0927-6513(95)00050-J.

[3] Akçay K, Sirkecioğlu A, Tatlıer M, Savaşçı ÖT, Erdem-Şenatalar A. 2004 *Powder Technol* **142** 121. doi:10.1016/j.powtec.2004.03.012.

[4] Reháková M, Čuvanová S, Dzivák M, Rimár J, Gaval'ová Z. 2004 *Curr Opin Solid State Mater Sci* **8** 397–404. doi:10.1016/j.cossms.2005.04.004.

[5] Misaelides P. 2011 *Microporous Mesoporous Mater* **144** 15 doi:10.1016/j.micromeso.2011.03.024.

[6] Malekian R, Abedi-Koupai J, Eslamian S S 2011 *J Hazard Mater* **185** 970 doi:10.1016/j.jhazmat.2010.09.114.

[7] Ramesh K, Reddy DD. 2011 *Adv. Agron* **113** 219 doi:10.1016/B978-0-12-386473-4.00004-X.
[8] Rácz Á. Research and development of the grinding process for the production of ultrafine materials. PhD Thesis, University of Miskolc, 2014. http://www.doktori.hu/index.php?menuid=193&vid=12807
[9] Breitung-Faes S, Kwade A. 2008 *Chem Eng Res Des* **86** 390 doi:10.1016/j.cherd.2007.11.006
[10] Mucsi G, Rácz Á, Mádai V. 2013 *Powder Technol* **235** 163 doi:10.1016/j.powtec.2012.10.005
[11] Rácz Á. 2014 *Chem Eng Technol* **37** 865 doi:10.1002/ceat.201300671
[12] Kwade A. 2004 *Int J Miner Process* **74** S93 doi:10.1016/j.minpro.2004.07.027
[13] Panias D, Giannopoulou IP, Perraki T 2007 *Colloids Surfaces* A Physicochem Eng Asp **301** 246 doi:10.1016/j.colsurfa.2006.12.064

Selenium, tellurium and precious metal mineralogy in Uchalinsk copper-zinc-pyritic district, the Urals

I Vikentev

Institute of geology of ore deposits, petrography, mineralogy and geochemistry (IGEM), Russian Academy of Sciences, Moscow, Russia

E-mail: viken@igem.ru

Abstract. During processing the most of Au, Ag, Se, Te, Pb, Bi, Sb, Hg as well as notable part of Cu, Zn and Cd fail for tailings and became heavy metal pollutants. Modes of occurrence of Au, Ag, Te and Se covers two giant VMS deposits: Uchaly (intensively deformed) and Uzelginsk (altered by late hydrothermal processes) as well as middle-sized Molodezn and West Ozern deposits (nondeformed) have been studied. Mineral forms of these elements as well as their presence in disperse mode in common ore minerals (pyrite, chalcopyrite, sphalerite) have been studied using SEM, EPMA, INAA, ICP-MS and LA-ICP-MS.

1. Introduction

The Urals stretching longitudinally over 2000 km represents the largest ore belt containing about 2 billion tones of massive sulfide reserves with 70 MT of non-ferrous metals in volcanogenic massive sulfide deposits (VMS deposits). Major non-ferrous metal output in the Urals are cooper, zinc, gold and silver producing from ores of 18 small and 6 large VMS deposits [1]. The increasing amount of processed ores of VMS deposits (25 million tones per year) has aggravated a problem of gold recovery [2]. Loss of gold in the pyritic concentrate and tailings can amount from 13 to 15 tones annually (for example, by now reserves of gold at tailing dump of Uchaly plant range 80 t). Up to a half of primary silver contained in ore is also lost in the tailings.

Ores of VMS deposits of the Urals are characterized by multicomponent composition but only a small part of trace elements is extracted: 10-20% of primary quantities of Se, Te, Co, Ni; 15-40% Au, Tl, In, Ga, Hg; 30-60% Ag and 80% Cd contained in ore. Lead, antimony and arsenic are not extracted. During processing the most of these elements as well as notable part of Cu, Zn fails for tailings and become pollutants (heavy metals plus arsenic) [3]. Se and Te are very toxic elements and they usually occur as admixtures in sulfide ore in VMS deposits. Nevertheless modes of occurrence of these and other trace elements (the most of them are toxic elements) in VMS deposits are left unclear. There is only total agreement that they are dispersed over the common sulfides of ore (pyrite, chalcopyrite and sphalerite). There is rare exception – a few papers deal with telluride mineralogy of some VMS deposits of the Urals [2,4]. It is also important to know the ways of potential pollutants during processing and smelting [3,5]. In this connection it is especially vital to study the mode of occurrence of the toxic trace elements in ores and ore minerals.

Geological and geochemical overview. The most of VMS deposits are located on the east slope of the Urals within the Main greenstone belt of the Urals – Tagil-Magnitogorsk megazone [1,6] and there are the most of ore dressing plants and smelters in this megazone. VMS deposits of the Urals associate with transitional from tholeiitic to calc-alkaline Na-bimodal rhyolite-basalt series [1]. According to the modern genetic model their formation was related to shallow chamber of acidic magma resulted from differentiation of mantle-derived basalt [7]. Specific features of VMS deposits of the Urals are the following: (1) large dimension: nine deposits involve 80-480 Mt of ore and 3-10 Mt of (Cu+Zn) reserves – these nine deposits totally contain more than 65% of the bulk initial base metal reserves of the Urals; (2) seven deposits contain 120-500 t of associated gold; (3) ore bodies are usually lens-like or lenticular; (4) massive sulfide ores predominate with subordinate amount of disseminated ores; (5) grade of metamorphism varies from zeolite to amphibolite facies [1,2]. World-class deposits are located within Magnitogorsk megazone, they belong to ***Uralian*** (or Cu-Zn-pyritic) type (major one for the Urals), which can be divided on two subtypes: Cu>>Zn and Zn>>Cu. Gold and silver are unevenly distributed in massive sulfide ores [2,8]. Gold content in ores commonly ranges about 1 ppm, in some bodies it amounts 3-5 ppm and very rarely single samples with 70-90 ppm Au can been found. Common Ag contents are measured 10-30 ppm (up to 600 ppm).

2. Materials

We will focus on *Uchalinsk* copper-zinc-pyritic district [7,8,9], which contains significant metal reserves in three large and ten medium-sized VMS deposits (figure 1): 435 Mt non-ferrous metals ore, including 5.7 Mt of Cu, 12 Mt of Zn, 1.2 Mt of Pb, 635 t of Au and 11.1 Kt of Ag. VMS deposits of the district are belong to *Zn-dominated subtype* of Uralian type**.** Specific feature of the deposits of this region is the most "rich" geochemistry of trace elements with enrichment of Au, Ag, Pb, Se, Te, Sb, As, Sn and Cd being compared with other ore districts of the Urals [10]. Ore bodies comprise large bodies of massive sulfide ores. In some case (Uchaly, Uzelginsk) their thickness can reach 120-150 m.

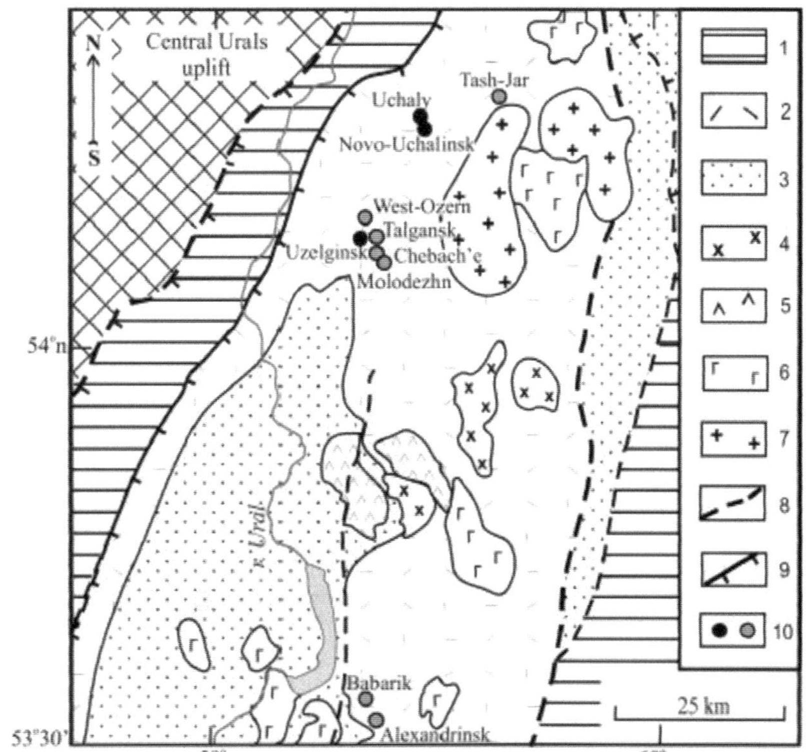

Figure 1. Scheme of Uchalinsk region (South Urals)
1 – ophiolites O–S1; (2,3) – island arc basalt- hyolite D2 (2) and andesite-basalt, basalt-trachyte complexes D3–C1, some limestone C1t (3); 4 – ranodiorite, plagiogranite (D2–D3fr); 5,6 – gabbro-granite series D3fm–C1t1 (5) and C1 (6); 7 – ranite P1; 8 – fault; 9 – overthrust; 10 – VMS deposit (black – large, grey – small and medium-sized)

3. Methods

The distribution of chemical elements has been studied for bulk samples of ores as well as for sulfide concentrates. The contents of Au, Ag, and base metals and some rare elements (Co, Ni, Se, Te, As, Sb, Cd, Hg, Ba) were examined by instrumental neutron activation analysis (INAA). PGE contents in 2-5g bulk samples, sulfide and ultra heavy concentrates were measured by wet-chemical analysis with preconcentration of sum of PGE and following ICP-MS determination. Laser ablation (LA) ICP-MS (IGEM RAS and LabMaTer at the Université du Québec à Chicoutimi) has been used for trace element analysis of sulfides from the Uchaly and Uzelginsk deposit. Microprobe analyses were carried out using an electron-probe microanalyser (EPMA) and a scanning electron microscope.

4. Results and Discussion

Geochemistry. There is broad variation of trace element contents in VMS ores (table 1, figure 2). Gold content in ores commonly ranges 1-1.5 ppm but in Au-rich zones of deposits can amount 5-20 ppm of Au. Silver content in ores ranges 5-100 ppm (up to 1000-3000 ppm) (figure 3). Cu-Zn ores with high Au content usually bear high concentrations of **Pb, Ba, As, Sb** and **Ag**, for some deposits (Uzelginsk) – Au-bearing ore is also enriched in **Hg, Se** and **Te** trace elements. Maximums of **ΣPGE** in VMS ore (0.5-0.6 ppm) are usually corresponding to high **Te** contents (250-1300 ppm) [11]. During processing the most of trace elements containing in ore of the deposits aren't extracted (Au, Ag, Pt, Pd, Pb, Se, Te, Sb, As, Sn, Co, Ni and Hg) and many of them (Pb, Se, Te, Sb, As, Co, Ni and Hg) become pollutants (heavy metals and As together with sulfur dioxide, Fe, Cu, Zn, Cd) (see also [3]).

Table 1. Common contents of trace elements (ppm) in ores of Uchalinsk ore district

Element	**Common contents**	**Maximum**	Element	**Common contents**	**Maximum**
Au	1-2	16.8	Hg	10-100	1500
Ag	5-50	2990	Sb	50-500	7250
Se	20-100	1324	As	1000-5000	23650
Te	20-150	3650	Pb	1000-5000	25564

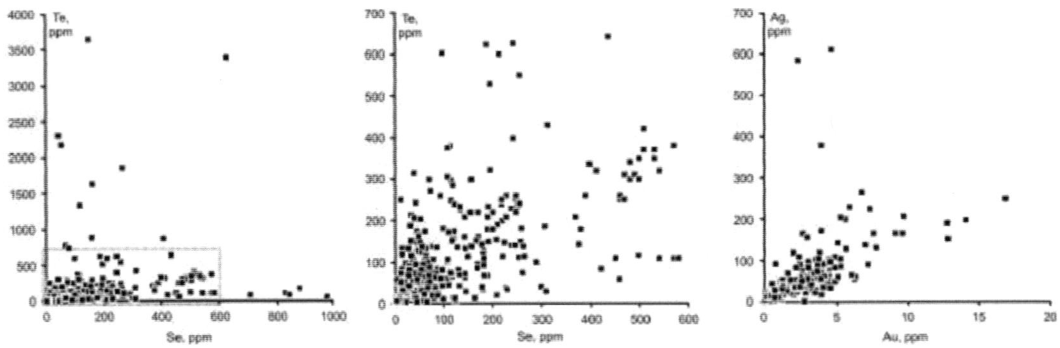

Figure 2. Te, Se, Ag and Au contents in VMS ores of Uchalinsk ore district

Selenium, tellurium and precious metal mineralogy. Pyrite, chalcopyrite and sphalerite are dominant ore minerals. Tennantite is a common mineral in most of the VMS deposits (0.1-1 vol%) while galena and bornite occur locally in smaller quantities. The ore samples which are anomalous in terms of gold concentration are mainly composed of *tennantite, tennantite-tetrahedrite, galena, bornite* with subordinated *tellurides, sulfotellurides* (tetradymite), *Ag-sulfursalts*, *Bi-sulfosalts* (aikinite, wittichenite), *Au-Ag-sulfides* (petrovskaite, uytenbogaardtite), *native elements* and *alloys* (native gold, native silver, electrum Au-Ag, native tellurium).

Gold was mostly concentrated in pyrite and chalcopyrite (1-20 ppm Au in the mode of "invisible" gold). Main trace toxic elements (Se, Te, As, Hg) occur in common sulfides (table 2) as well as Se is host by galena (up to 0.45 wt %, EPMA), pyrrhotite (up to 343 ppm, INAA), tellurobismuthite (up to 0.69-0.85 wt %, EPMA) and petzite (up to 0.02-0.49 wt %, EPMA). Selenium can be rarely found in pyrite and other common sulfides as tiny inclusions of clausthalite PbSe and galena-clausthalite Pb2SeS (Uchaly, West-Ozern deposits) and kawazulite-tetradymite Bi2Te1.9(Se0.57-0,65S0.3-0,26) (Molodezhn deposit). Tellurides, native tellurium, Au-Ag alloys, tennantite-tetrahedrite together with main sulfides are detected as microinclusions in pyrite [2]. A positive correlation between Au and Ag in ore and a high dispersion of the contents seems to reflect the occurrence of Au-Ag alloys (see figure 2). The similar tendency occurs for contents of Au and Ag in pyrite: pyrites with high Au content usually bear high concentrations of silver and tellurium trace elements. That can be related with tiny inclusions of Au-Ag alloys, Au-tellurides and Te-bearing (up to 8.9 wt % Te) tennattite-tetrahedrite in pyrite (figure 3).

Table.2. Trace element contents (ppm, INAA) in common sulfides of deposits of Uchalinsk district

Mineral	Au	Ag	Se	Te	As	Hg
Pyrite	0.21-17.11	4-326	2-511	<4-583	<1-5600 (up to 2.3 wt%)	0.1-1257
Chalcopyrite	0.4-0.15	1-6	7-313	1-314	<1-33 (up to 0.3 wt%)	0.2-792
Sphalerite	no data	no data	<6-471	28-1367	<1	40-581

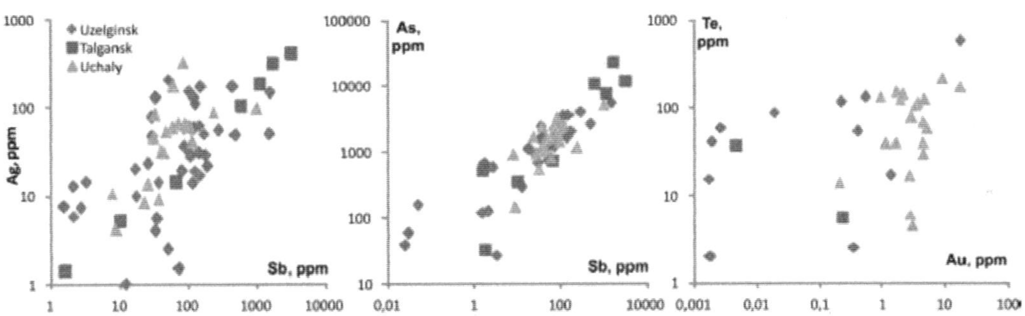

Figure 3. Trace elements in pyrites (INAA)

In the *Uchaly* deposit the bulk of Au and Ag occurs in a dispersed form in sulfides – sphalerite, pyrite and chalcopyrite. Important hosts of Au and Ag are native gold, sulfides (petrovskaite, argentite), tellurides (hessite, empressite, calaverite) and sulfosalts of silver (Ag-tetrahedrite, Ag-tennantite-tetrahedrite with 7-8 wt.% Ag and pirseit) (figure 4). Native gold is extremely rare, size from 5 to 30 microns (up to 150 μm), usually low-grade with the composition range from Au0.69Ag0.31 to Au0.34Ag0.59Hg0.1. Mercury-bearing native gold (11.3 wt.% Hg) was also found rarely.

In the *Uzelginsk* deposit neutron-activation analysis found Au enrichment up to 22 ppm as well as elevated Ag and As in reniform pyrite. The euhedral pyrite associated with this variety is characterized by an order of magnitude lower concentrations of Au, Ag and As [8]. Native gold (up to 200 μm) occurs in recrystallized ores of lower ore level (ore bodies 3 and 4) in association with Au-Ag and Ag tellurides (petzite, hessite, stützite). In some case grains of native gold is inhomogeneous and contains emulsion-like inclusions of altaite [8].

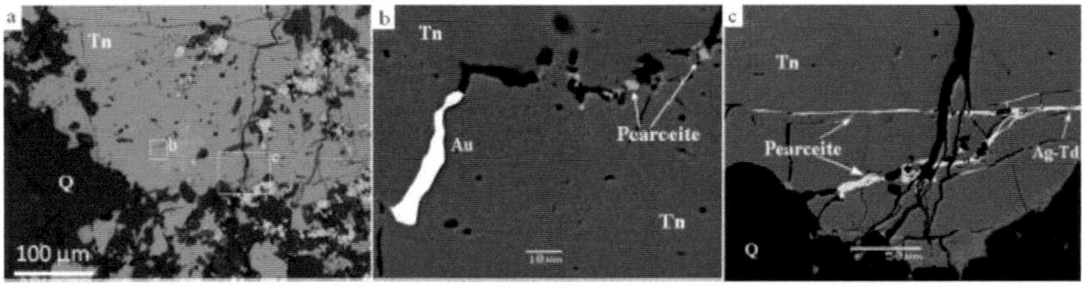

Figure 4. Electrum, Ag-tetraedrite and pearceite (Ag,Cu)16As2S11 occur in thin cracks in tennantite, Uchaly deposit

Silver mainly occurs as isomorphic component of tennattite-tetrahedrite. Low-iron tennantite (0.1-0.5 wt.% Ag) is dominant variety; Ag-tennantite (up to 8.4 wt % Ag), Ag-tennantite-tetrahedrite (7-8 wt % Ag) and Ag-tetrahedrite (8-11 wt % Ag) are scarce [2,8]. In addition, up to 0.14 Se (wt%, EPMA), 0.38 Pd, 0.05-0.3 Hg and 0.2-4.5 Te were found in tennattite-tetrahedrite from the Uzelginsk deposit. Positive correlation of Ag, Sb and As in pyrites is seems to related to tiny inclusions of Ag-tetrahedrite and tennantite-tetrahedrite occurring in pyrites (see figure 3). In some case tennattite-tetrahedrite is enriched in gold (up to 6 wt %).

Proper minerals of Au, Ag and Te routinely form fine submicron inclusions inside of common sulfides of the ores: sphalerite, chalcopyrite and pyrite. A lot of *tellurides* occurs in massive sulfide ores: these are altaite, coloradoite, tellurobismuthite and tellurides of gold and silver [2,8]. Tellurides of gold and silver include hessite, stützite, empressite, petzite, krennerite and sylvanite. Some tellurides bear gold admixture (as solid solution) especially altaite (up to 5.2 wt % Au), coloradoite (up to 4.2 wt % Au), hessite (up to 1.4 wt % Au) and tetradymite (up to 1.2 wt % Au), with gold content being much more than that in sulfides may be due to more metallic character of chemical bond in tellurides. Apparently, it is the presence of small inclusions of Au-Ag tellurides (and other tellurides with appreciable admixture of Au) in pyrite causes a significant correlation of gold and tellurium in the composition of the pyrite. In this way tellurides are a major carrier of gold and silver in VMS deposits of the Urals.

Native gold comprises grains and aggregates with size from first to 200 micrometers. In weakly transformed ore deposits of Uchalinsk ore district, native gold is observed as rare tiny grains (up to 20 µm). Larger grains (up to 200 µm) are detected in recrystallized ores and within late quartz+sulfide-barite-tennantite veinlets. Evidently, the appearance of gold minerals was mainly related to the release of isomorphous Au from sulfides as well as coarsening of tiny ("invisible") native gold grains during epigenetic hydrothermal alterations.

The sulfide precipitation occurred under conditions favourable for the incorporation of isomorphic gold into pyrite, chalcopyrite and sphalerite+chalcopyrite during high temperature stages of ore genesis [2]. Parageneses of native metals (Au-Ag alloys, native Te), tellurides and sulfosalts were mainly formed at the latest stage of hydrothermal process at below 200oC as well as during low grade metamorphism due to refining of the common sulfides and segregation of rare elements as well as coarsening of tiny gold grains [2,8].

PGE do not form proper minerals in VMS ores of the Urals. According to LA-ICP-MS data admixtures of Au, Ag and PGE in pyrite range between (ppm): 0.03-0.2 Pd, 0.04-0.26 Rh for Uchaly deposit [11] The LA-ICP-MS method has been used for trace element analysis of sulfides from the Uchaly and Uzelginsk deposits. The bulk of the PGE and gold occur invisibly in pyrite and chalcopyrite. Their contents range from ~0.01-42.8 ppm Au, 0.03-0.2 ppm Pd and 0.02-0.25 Rh ppm. The PGE enrichment (ppb, up to 1220 Pd, 375 Pt and 707 Rh for the Uzelginsk deposits) was established [8] in ultra heavy concentrates from Au rich ores.

5. Conclusion

There are broad variations of Se, Te, Au, Ag, Pt, Pd contents in VMS ores (but usually below 0.1 ppm for Pd and Pt). The bulk of the gold and PGE occur invisibly in pyrite and chalcopyrite with predomination of Au.

Tellurides appear to be major carriers of gold and silver (and probable Pt+Pd) in VMS deposits of the Urals – along with native gold. Tennantite-tetrahedrite is also important host of Au, Ag and Pt+Pd in VMS deposits.

Visible segregations of native gold (~1 μm or more) and tellurides appear in massive sulfide ores as a result of recrystallization during late (mainly low-temperature) hydrothermal processes and low-grade metamorphism.

During processing the most of pyrite, pyrrhotite, tennantite-tetrahedrite and galena bearing a major part of Pb, Se, Te, Co, Sb, As, Hg as well as notable part of Cu, Zn and Cd fails for tailings and became pollutants (heavy metals and As together with sulfur dioxide, Fe, Cu, Zn, Cd).

Mineralogical study of the VMS deposits may clarify the mode of occurrence of trace toxic elements in ore and in such way it can help to optimize ore processing with output of additional quantity of noble metals as well as to prevent the pollution of environment by toxic elements and make the process of mining and dressing of ore of VMS deposits more green to keep the nature for future generations.

This work was supported by Russian Science Foundation (Project 14-17-00693).

References

[1] Prokin V A and Buslaev F P 1999 *Ore Geol. Rev.* **14** 1
[2] Vikentyev I V 2006 *Mineral. Petrol.* **87** 305
[3] Shafigullina G T and Udachin V N 2009 *Prospect and Protection Miner. Res.* **1** 60
[4] Maslennikov V V *et al* 2013 *Mineral. Petrol.* **107** 67
[5] Udachin V *et al* 2003 *J. Sustainable Dev.* **11** 133
[6] Puchkov V N 2013 *Miner. Petrol.* **107** 3
[7] Karpukhina V S *et al* 2013 *Geol. Ore Dep.* **55** 125
[8] Vikentyev I V *et al* 2004 *Can. Mineral.* **42** 651
[9] Vikent'ev I V *et al* 2000 *Geol. Ore Dep.* **42** 221
[10] Vikentyev I V *et al* 2013 Volcanic-hosted massive sulfide deposits of the Urals, Russia: Evidence for a magmatic contribution of metals and fluid *Proc. 12th Bien. SGA Meet.* 1526
[11] Vikentyev I V *et al* 2014 PGE in minerals of volcanogenic massive sulfide deposits of the Urals: ore geochemistry and first LA-ICP-MS data *Abs.12th Int. Platinum Symp.* 326

Occurrence modes of As, Sb, Te, Bi, Ag in sulfide assemblages of gold deposits of the Urals

O Vikent'eva and I Vikentev

Institute of geology of ore deposits, petrography, mineralogy and geochemistry (IGEM), Russian Academy of Sciences, Moscow, 119017, Russia

E-mail: ovikenteva@rambler.ru

Abstract. Review of occurrence modes of trace toxic elements ("potential pollutants") in ores from large gold deposits (the Urals) of different genetic types is presented. Mineral forms of these elements as well as their presence in main minerals from gold-bearing sulfide assemblages according to SEM, EPMA, INAA, ICP-MS and LA-ICP-MS are demonstrated.

1. Introduction
At the present time, great attention is paid to the complex use of precious and non-ferrous metals ores, both in Russia and abroad. Besides the main components the gold deposits contain copper, arsenic, antimony, lead, zinc, tungsten, uranium, mercury, bismuth, thallium, selenium, tellurium, etc. Some of these elements have the commercial interest but other are "potential pollutants". During ore processing the "pollutants" fall into waste and represent the risk to the environment and health [1, 2]. The importance of the present study is due to the sharply increased in recent years demands for comprehensive utilization of raw and the environmental protection from the effects of toxic industrial waste. It is necessary to study the modes of these elements occurrence in the ore, and their variability in different physical states and the possibility of changes in the technological cycle of ore processing.

2. Materials
The Urals is the oldest (266 years) gold-mine province of Russia [3] where about 60 million tons of sulfide-bearing tailings (gravity separation and flotation) are accumulated. Objects of this study are large gold deposits of different genetic types – Berezovsk (Au), Svetlinsk (Au-Te) and Vorontsovsk (Au-As-Hg). The comparative characteristics of the studied deposits are given in table 1.

3. Methods
Determination of the occurrence modes of potentially harmful elements in the gold ores was studied using complex of methods. Proper mineral forms of these elements were identified by optical and scanning electron microscopy and were investigated using an electron-probe microanalysis (EPMA). Contents of the trace elements in monomineral probes of the main sulfides from gold assemblages were examined by instrumental neutron activation analysis (INAA) and by mass-spectrometry with inductively coupled plasma (ICP-MS). Contents of trace elements in gold-bearing pyrite were determined by laser ablation-inductively coupled plasma-mass spectrometry (LA-ICP-MS) firstly done for the gold deposits of the Urals. All investigations have been carried out in IGEM RAS, while the LA-ICP-MS was studied also in LabMaTer at the Université du Québec à Chicoutimi (Canada).

Table 1. Characteristics of studied gold deposits

Deposit, region	**Vorontsovsk,** North Urals	**Berezovsk,** Middle Urals	**Svetlinsk,** South Ural
Host rocks	Volcanic-sedimentary (S2-D1)	Volcanic-sedimentary (O-S), gabbro, serpentinite, granite	Metamorphosed volcano-sedimentary rocks (D-C)
Geochemical type	Au-As-Hg	Au	Au-Te
Ore bodies	Vein-disseminated zones; rare veinlets	Veins	Vein-disseminated zones; veins
Wall rock alteration	Propilite, quartz-seritsite, argillizite, jasperoid	Berezite, listvenite	Quartz-biotite (with amphi-bole), quartz±biotite-sericite
Stage of mineral formation	Arsenopyrite-pyrite→Pyrite-realgar→Sulfosalt-polymetalic→Polymetalic	Ankerite-quartz→Pyrite-quartz→Polymetalic→ Carbonate	Quartz-pyrite→Au-Te-polymetalic→ Quartz-carbonate-sulfide
Au reserves (CAu)	~60 t (2.8 g/t)	350 t (2.4 g/t)	~80 t (1.8-2.8 g/t)

4. Results and Discussion

Studied chemical elements occur as invisible form in common sulfides as proper minerals (table 2).

a. Arsenic

Mineral forms of As have been found in all studied deposits (table 2) but on the Vorontsovsk deposit its are widespread and the most typical. On the Vorontsovsk deposit arsenopyrite predominates and it contains 10-1018 ppm Au (INAA data), realgar and orpiment being also common. Minerals of tennantite-tetrahedrite series occur on all studied deposits (figure 1). Ratio As/(As+Sb+Bi) ranges from 0.1 to 1 (Vorontsovsk), from 0.04 to 0.07 (Svetlinsk) and from 0.52 to 1 (Berezovsk). According to data LA-ICP-MS pyrite of the Vorontsovsk deposit contains 5.6-12710 ppm of As (figure 2).

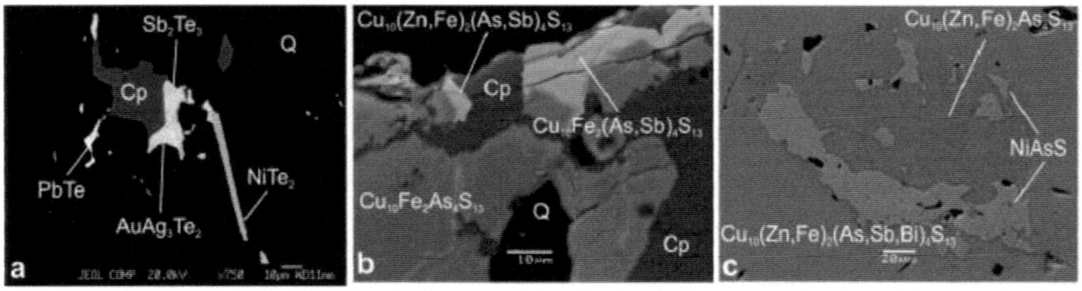

Figure 1. Minerals of As, Sb, Te, Ag in ores of Svetlinsk (a), Vorontzovsk (b) and Berezovsk (c) deposits. Q – quartz, Cp – chalcopyrite.

b. Antimony

The most common mineral form of Sb in all studied deposits is tennantite-tetrahedrite. On the Vorontsovsk deposit sulfides of Sb, sulfoantimonides of Pb, Cu, Mn, Fe and sulfosalts of Tl-(Hg)-Sb,As are present in the ores (table 2). Neutron activation analysis revealed antimony in pyrite (0.5-367 ppm), As-pyrite (19.5 ppm - 3.8 wt %), arsenopyrite (50.7-678 ppm) and sphalerite (1-63 ppm) from the Vorontsovsk deposit, as well as in pyrite (5-1147 ppm) and galena (3 ppm – 4 wt %) from the Berezovsk deposit. Pyrite from altered rocks of both deposits has the lower concentrations of antimony - 1.7-21.2 and 12.49.4 ppm for Vorontsovsk and Berezovsk deposits, respectively. Pyrite of the Vorontsovsk deposit contains 0.01-350 ppm of Sb (LA-ICP-MS) (figure 2).

c. Bismuth

Broad variety of Bi minerals occurs only at the Berezovsk deposit (table 2). Reserves of Bi in ores of this deposit amount to ~1.2 t. Bismuth minerals are presented by native form, sulfides, selenides, sulfotellurides, sulfobismuthites of Pb, Cu, Ag, all these minerals are rare except of aikinite. Furthermore isomorphic Bi is included into galena (0.3-5.36 wt %) and tennantite-tetrahedrite minerals (0.13-3.6 wt% for the Berezovsk, 0.12-1.18 wt% for the Svetlinsk, up to 2.25 wt% for the Vorontsovsk deposit). Native gold rarely contains up to 0.3 wt% of Bi (EPMA data). On the Svetlinsk deposit Bi tellurides and sulfotellurides are found whereas sulfobismuthites Pb and Cu occur in ores of the Vorontsovsk deposit.

Table 2. Mineral forms of trace elements for gold deposits of the Urals (data of authors and [4])

	Vorontsovsk	Berezovsk	Svetlinsk
As	Arsenopyrite FeAsS, löllingite FeAs2, orpiment As2S3, realgar AsS, tennantite Cu12As4S13, tennantite-tetrahedrite Cu12(As, Sb)4S13, routhierite TlHgAsS3, aktashite Cu6Hg3As4S12, native As, pierrotiteTl2(Sb,As)10S17, geocronite Pb14(Sb,As)6S23	Tennantite Cu12As4S13, tennantite-tetrahedrite Cu12(As, Sb)4S13, arsenopyrite FeAsS, gersdorffite NiAsS, löllingite FeAs2, nickeline NiAs	Löllingite FeAs2
Te	Coloradoite HgTe, hessite Ag2Te	Hessite Ag2Te, tetradymite Bi2Te2S	Melonite NiTe2, frohbergite FeTe, altaite PbTe, calaverite AuTe2, sylvanite AuAgTe4, krennerite (Au,Ag)Te2, petzite AuAg3Te2, hessite Ag2Te, tsumoite BiTe, tellurantimony Sb2Te3, volynskite AgBiTe2, tetradymite Bi2Te2S
Sb	Tetrahedrite-tennantite Cu12(Sb,As)4S13, klerite MnSb2S4, stibnite Sb2S3, pierrotite Tl2(Sb,As)10S17, zinkenite Pb9Sb22S42, chalcostibite CuSbS2, boulangerite Pb5Sb4S11, jamesonite Pb4FeSb6S14, bournonite PbCuSbS3, plagionite Pb5Sb8S17, geocronite Pb14(Sb,As)6S23	Tetrahedrite-tennantite Cu12(Sb,As)4S13, bournonite PbCuSbS3, livingstonite HgSb4S8	Tetrahedrite Cu12Sb4S13, tellurantimony Sb2Te3
Bi	Aikinite CuPbBiS3, wittichenite Cu3BiS3	Aikinite CuPbBiS3, cosalite Pb2Bi2S5, tetradymite Bi2Te2S, bismuthinite Bi2S3, wittichenite Cu3BiS3, matildite AgBiS2, emplectite CuBiS2, guanajuatite Bi2Se3, galenobismutite PbBi2S4	Tsumoite BiTe, volynskite AgBiTe2, tetradymite Bi2Te2S
Ag	Hessite Ag2Te, electrum AuAg, kustelite Ag3Au	Native silver Ag, electrum AuAg, acanthite Ag2S	Electrum AuAg, sylvanite AuAgTe4, krennerite (Au,Ag)Te2, petzite AuAg3Te2, hessite Ag2Te, volynskite AgBiTe2
Hg		Aktashite Cu6Hg3As4S12, cinnabar HgS, routhierite TlHgAsS3, coloradoite HgTe	Cinnabar HgS, livingstonite HgSb4S8

d. Tellurium

Tellurides are typical for the Svetlinsk deposit (table 2, figure 1); native gold (fineness 873-948) is closely associated with them in quartz-sulfide veins. Tellurides of Ni, Fe, Pb, Bi, Sb, Ag and Au were identified in the ore assemblages. Tellurium is present in galena (147-977 ppm) and pyrite (64.4 ppm) of the Berezovsk deposit according to INAA. Pyrite of the Vorontsovsk deposit contains 1.64-76.5 ppm of Te (LA-ICP-MS) (figure 2).

e. Silver

Main Ag minerals are tellurides of silver, whereas the Au-Ag alloys, native silver and silver sulfide are occasionally found (table 2). Silver is a constant impurity in tennantite-tetrahedrite minerals and galena. Native gold always contains silver (EPMA, wt %): 1.4-14 and 9.3-26.7 for early and late generation of native gold in the Berezovsk deposit, recpectively; 10.7-21.2 for the Svetlinsk deposit. On the Vorontsovsk deposit besides of native gold (7.1-30 wt% Ag) electrum (44.2-48 wt% Ag) and kustelite (69.6-77.5 wt% Ag) are found. According to EPMA tennantite-tetrahedrite minerals contain Ag up to 0.9 wt% (Berezovsk), 1.9-4.5 wt% (Svetlinsk) and up to 22.5 wt% (Vorontsovsk). Sulfides of the Berezovsk deposit are enriched of Ag: in pyrite 9.2-260.6 ppm (INAA) and 1.2-62.5 ppm (ICP-MS); in chalcopyrite 67.2-75 ppm (ICP-MS) and up to 0.18 wt% (EPMA); in galena 0.1-0.4 wt% (EPMA) and 639 ppm-2.5 wt% (INAA); in tennantite-tetrahedrite 0-0.89 wt% (EPMA) and 1062-3212 ppm (INAA).

f. Other elements

Mercury determines the geochemical type of Vorontsovsk deposit. In addition to its proper mineral forms (table 2) mercury is constantly present in native gold (up to 5 wt%) of this deposit as well as in sphalerite (up to 21 wt%) and tennantite-tetrahedrite (up to 0.21 wt% [4]).

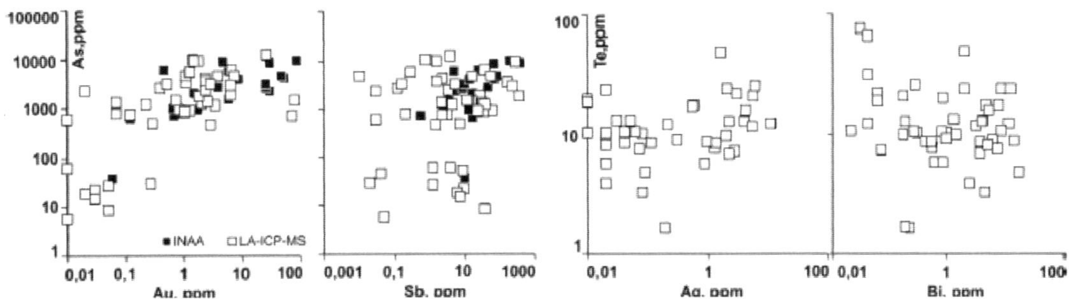

Figure 2. Distribution of Au, As, Sb, Bi, Te and Ag in pyrite from Vorontzovsk deposit

5. Conclusion

High concentrations of chemical elements - "potentially polluting substances" - were found in the ores of gold deposits of the Urals. The proper minerals of these elements and disseminated form in the main sulfides are demonstrated. The results of the present study should be considered when selecting technological scheme of gold ore processing for reducing the risk of potential environmental problems.

This work was supported by Russian Science Foundation (Project 14-17-00693).

References

[1] Kyle J H *et al Hydrometallurgy* **107** 91
[2] Kyle J H *et al* 2012 *Hydrometallurgy* **111-112** 10
[3] Sazonov V N *et al* 2001 *Economic Geology* **96** 685
[4] Sazonov V N *et al* 1998 *Geology of Ore Deposits* **40** 157

Investigation of the properties of Sb doping on tin oxide SNO_2 materials for technological applications

Z Hachoun[1,2] A Ouerdane[2], M Bouslama[3], M Ghaffour[3]; A Abdellaoui[3], Y Caudano[4] and A Ali benamara[1]

[1]Université Hassiba Benbouali de Chlef. Algeria
[2]Université de Khemis Miliana. Algeria
[3]LabMat Ecole Nationale Polytechnique d'Oran Mnaouer, Algeria
[4]Facultés Universitaires Notre-Dame de la Paix Rue de Bruxelles, Namur, Belgium.

E-mail: hachoun_zoubir@yahoo.fr

Abstract: The conductivities of the oxide SnO_2 is dependent on the nature of the surrounding gas. This property stems from the adsorption or desorption on the surface of oxide grains. These phenomena are usually accompanied by electronic transfer between the adsorbed molecule and the semiconductor material, changing its conductivity. Tin oxidation and Sb doping were realized without and with heating process. The XPS technique and the TEM microscopy showed the synthesized nanocrystals. Simulated Monte Carlo program Casino is used for a scanning its profile. The surface characteristics are highlighted in the aim to be used as spatial gas sensors.

1. Introduction

Tin oxide SnO_2 is a semiconductor with a band gap of 3.6 eV. The oxygen vacancies, formed by the transfer of an oxygen atom of a normal site in the tin gaseous state, are used to obtain an n-type semiconductor. In fact, the released electrons can bind to tin atoms Sn^{4+}, they become Sn^{2+} and behave as electron donors [1]. The adsorption is generally reversible for many gases at temperatures of the order of 400 to 500°C. For this, SnO_2 material is usually used in the field of monitoring air pollution and toxic gas detection. Furthermore, SnO_2 films stem others applications as solar cells, optoelectronic devices, thin film resistors, antireflection coatings, photochemical devices and electrically conductive glass. Tin oxide is also known for its catalytic properties, it facilitates the detailed breakdown of many hydrocarbons above 350°C.These properties are attracted the attention of scientists who have attempted to improve electrical performance by different experimental and simulated methods (microstructural stabilization, doping). [2,3]The doping of SnO_2 by antimony Sb is our goal in order to perform its conductivity to be better in detection of poisonous gas.

2. Experimental

The samples Sb: SnO_2 thin films were prepared using several techniques as reactive radio frequency (RF), sputtering process, spray pyrolysis technique, hydrothermal method from initial material $Sncl_{4,5}H_2O$, polymeric precursor method by polymerization of a metallic citrate with ethylene glycol. An attempt has been made in our study to prepare Sb:SnO_2 films by the simple method of alternating electro-deposition in the Sn and Sb coatings electrolytes on SiO_2 substrate. The reaction mixture of SnO2 and Sb powders was in 10% weight ratio.The performance of the growth sample film was largely influenced by the substrate material, its position and its temperature. In this case the electrical properties and the texture of the film are affected. We confirm that SnO_2 becomes Sb-doped using this technique and when the substrate is submitted to an heated effect. [4]

3. Results and discussion

3.1. XPS analysis and binding energies of the chemical elements in Tin oxide

The relative atomic concentrations of Sn and O were analyzed semi-quantitatively based on XPS spectra (Fig. 1), In the as-prepared sample, broad Sn3d5/2, Sn3d3/2, and O1s peaks exist around 486, 495 and 532.4 eV, and the concentrations of Sn and O are respectively about 38 and 52%, exhibiting evidence of non-stoichiometry. When the sample was annealed above 500°C for 8 h and more, Sn concentration decreases to 34% and O concentration increases to 66%, which corresponds to stoichiometric SnO_2. Fig.1a and fig.1b show respectively the binding energy in electron volt of the two elements oxygen with its O1s scan and the tin element with Sn d3/2 scan which corresponds to a good oxidation process giving rise to the SnO_2 compound. [5]

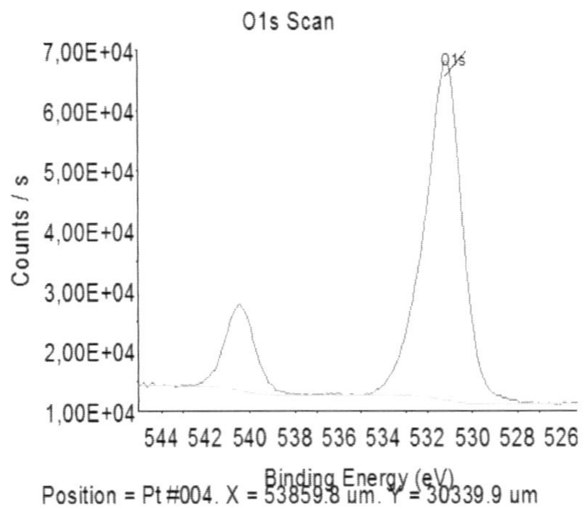

Figure 1a. Binding energy of the element Oxygen O1s scan

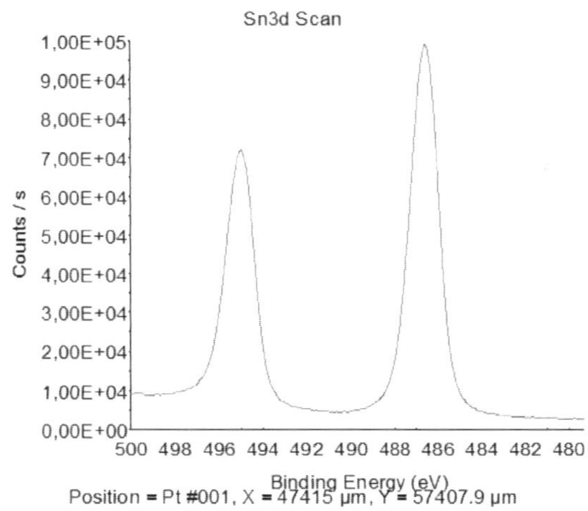

Figure 1b. Binding energy of the element tin Sn3d scan

3.2. Electron diffraction EDAX study and TEM spectroscopy

The EDX analysis shows the chemical composition of the Sb-SnO_2 sample. The concentration of Sb in the SnO_2 crystal is 10% as described as and confirmed by the same analysis. The corresponding EDAX spectrum shown in Fig.2 is in agreement with the electron diffraction micrography results. The peak broadening pattern clearly indicates that very small nanocrystals are present in the samples. The surface morphology of the SnO_2 sample is followed by Transmitted Electron Microscopy (TEM) apparatus which indicates a nanostructures grew onto the surface. In fig.3 a, b, c and d the electronic microscopy give us a good map of the surface of SnO_2-Sb doped and heated at the temperature above 500°C. The grains are so small, the morphology of the surface is uniform giving rise to crystallographic structure of Sb-SnO_2 with a good stoichiometric surface distribution Increase of annealing time leads to further change of microstructure closed-packed crystallites with sharp boundary [6]

IOP Conf. Series: Materials Science and Engineering **123** (2016) 012029 doi:10.1088/1757-899X/123/1/012029

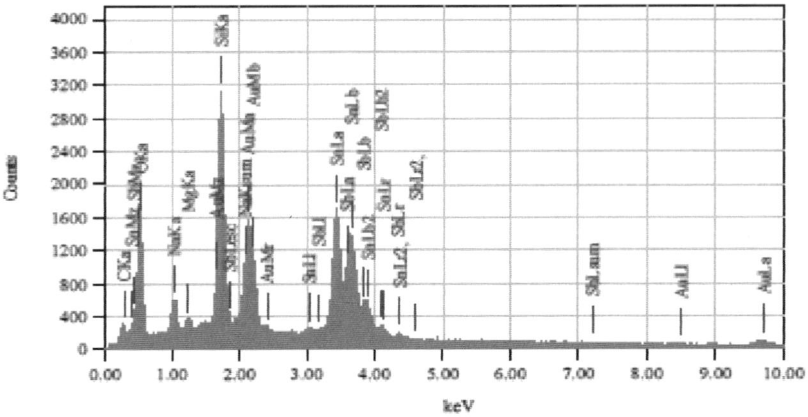

Figure 2. Energy Dispersive (EDX) showing the chemical elements inside the compounds with their energy position lines for Sb-SnO_2

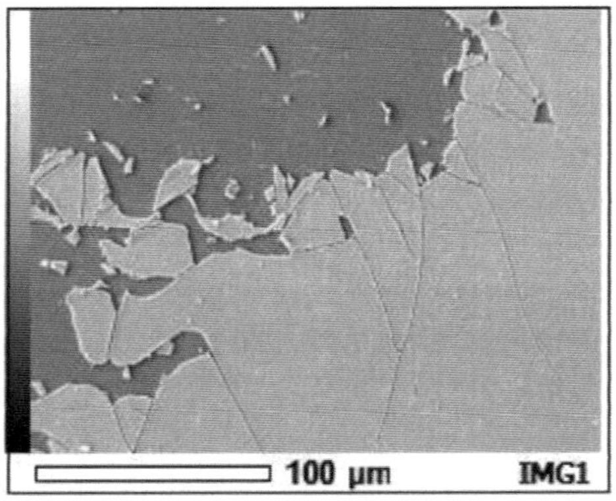

Figure 3a: Heated Sb doped SnO_2: TEM analysis showing the SnO_2 (grey color) and SnO island (white color)

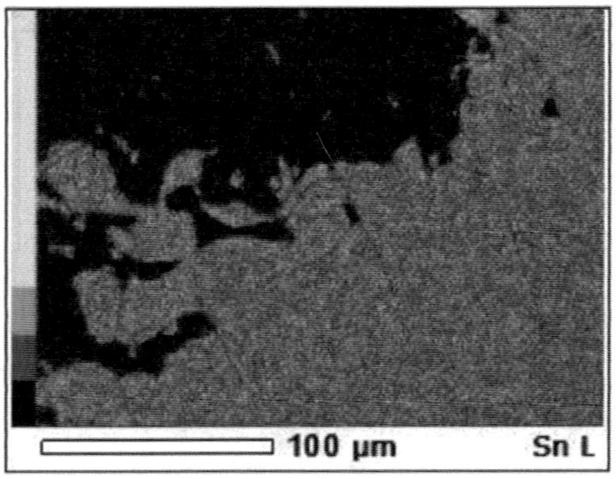

Figure 3b. Heated Sb doped SnO_2: TEM analysis showing tin element distribution on the surface of the sample

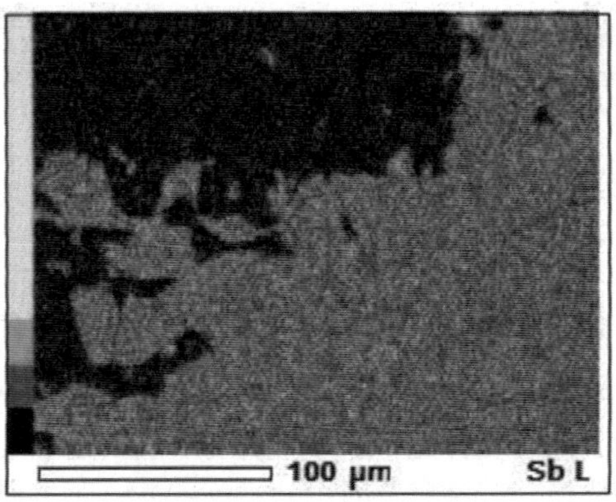

Figure 3c. Heated Sb doped SnO_2 :TEM analysis showing the antimony element distribution on the surface of the sample

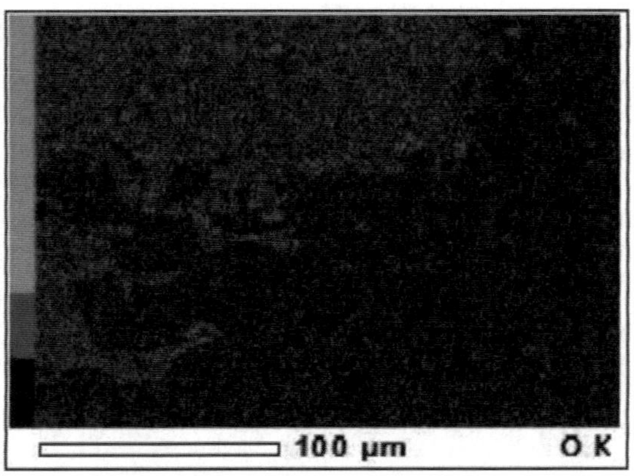

Figure 3d. Heated Sb doped SnO_2 :TEM analysis showing the oxygen element distribution on the surface of the sample

4. SEM Casino program

The simulated program Casino was used as scanning electron microscopy to analysis the compounds Sb-SnO_2 on the substrate SiO_2. We give in fig.5a the electron X Ray absorbed by the matrice SnO_2 red color and the X-Ray diffused into the compound (blue color). In fig.5b the absorbed X-ray by the oxygen element present in the chemical formula of Sb SnO_2 showing the disposition of the grains on surface compounds.

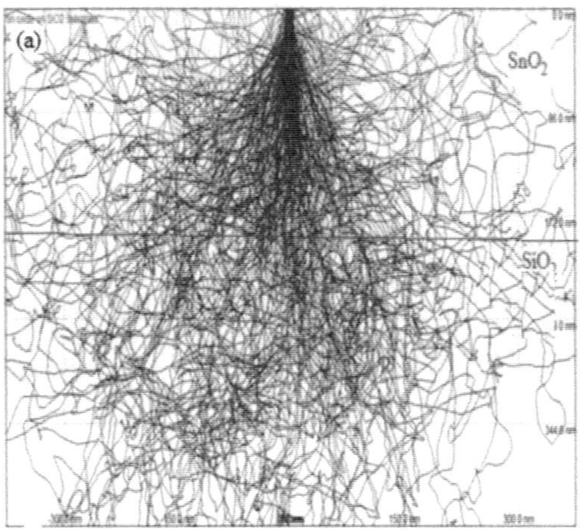

Figure 4a. Casino ray dispatching into the crystal SnO2 / SiO_2

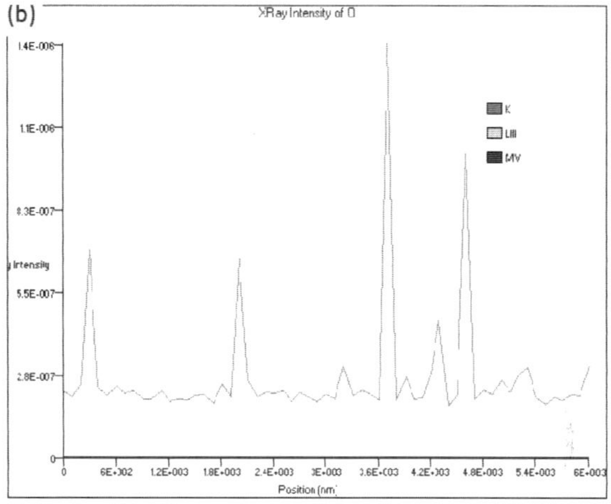

Figure 4b. Electron scanning microscopy SEM of SnO_2 on SiO_2 substrate by Casino

5. Conclusion

Our work was based on the characterization of the SnO_2 oxide using spectroscopic methods. The EDS technique recorded spectra of as grown SnO2 film showed that no metal phase was detected. The TEM microscopy images of SnO2 films give us a good ideas on the films microstructures as grown and after annealing at 500°C for more than two hours and for one day. Increase of annealing time leads to further change of microstructure closed-packed crystallites with sharp boundary formed agglomerates of considerably smaller size (80-150 nm) with orientation parallel to the substrate.

6. References

[1] H.Hui Huang, Chiew Keat Lim, Man Siu Tse,Jun Guo And Ooi Kiang 2012 *Tan Nanoscale* **4** 1491
[2] Chien-KunWang et al. 2010 *Journal of Marine Sciences and Technology* **18** 825
[3] Z.Chelahi Chikr, A.Mokadem, M Bouslama, F.Besahraoui, M.Ghaffour, A. Ouerdane, K.Boulenouar, N. Chauvin and B.Benrabah 2013 *Surface Review Letter* **20** 1350050
[4] J.X Zhou, M.S Zhang, J.M.Hong and Z.Yin *Solid State Communications* **138** 242
[5] Hui Huang , Chiew Keat Lim , Man Siu, Jun Guo and Ooi Kiang Tan 2012 *Nanoscale* **4** 1491
[6] A.Forleo,L.francesco,S.Capone,F.Casino,P.Siciliano,O.K.tan,H.Hui, 2011 *Sensors and Actuators B:Chemical* **154** 280

Effect of temperature and concentration of precursors on morphology and photocatalytic activity of zinc oxide thin films prepared by hydrothermal route

S Heinonen[1], J-P Nikkanen[1], H Hakola[2], E Huttunen-Saarivirta[3], M Kannisto[2], L Hyvärinen[1], M Järveläinen[1], E Levänen[1]

[1]Department of Materials Science, Tampere University of Technology Tampere, Finland
[2]Department of Chemistry and Bioengineering, Tampere University of Technology Tampere, Finland
[3]VTT Technical Research Centre of Finland, Materials Performance, Tampere, Finland

E-mail: saara.heinonen@tut.fi

Abstract Zinc oxide (ZnO) is an important semiconductive material due to its potential applications, such as conductive gas sensors, transparent conductive electrodes, solar cells, and photocatalysts. Photocatalytic activity can be exploited in the decomposition of hazardous pollutants from environment. In this study, we produced zinc oxide thin films on stainless steel plates by hydrothermal method varying the precursor concentration (from 0.029 M to 0.16 M) and the synthesis temperature (from 70°C to 90 °C). Morphology of the synthesized films was examined using field-emission scanning electron microscopy (FESEM) and photocatalytic activity of the films was characterized using methylene blue decomposition tests. It was found that the morphology of the nanostructures was strongly affected by the precursor concentration and the temperature of the synthesis. At lower concentrations zinc oxide grew as thin needlelike nanorods of uniform length and shape and aligned perpendicular to the stainless steel substrate surface. At higher concentrations the shape of the rods transformed towards hexagon shaped units and further on towards flaky platelets. Temperature changes caused variations in the coating thickness and the orientation of the crystal units. It was also observed, that the photocatalytic activity of the prepared films was clearly dependent on the morphology of the surfaces.

1. Introduction

The state of environment is one of the most important issues faced by modern society. Sufficiency and availability of energy and water as well as environmental pollution are today's main concerns. Industrial growth caused by increased consumption has resulted in severe environmental problems due to the release of harmful and toxic substances into water and air. This has led to development of cheap and effective technologies to remove and destroy the toxins and pollutants. [1–5]

Since 1972 when Fujishima and Fonda [6] discovered that titanium dioxide can split water into hydrogen and oxygen under ultraviolet illumination, TiO2 photocatalysis has been widely studied and its efficiency in mineralization of severe organic pollutants as well as bacteria and other organic contamination has been recognized. [7–14] Another wide band gap (3.37 eV) semiconductor material that has recently attained focus due to its photocatalytic ability is zinc oxide. In addition to photocatalytic properties, nanostructured ZnO has optical and electrical properties which make it suitable for several potential applications, e.g., solar cells, gas sensors, piezoelectric materials and ceramic devices. ZnO nanostructures, such as nanotubes, nanosheets, nanowires, mesoporous films, nanorods and nanoparticles have been synthetized using different methods and results have been widely reported. [1, 15–23] It has also been shown that structure directing agents, such as ethanol and urea, can be used to control the formation of ZnO structures. [23–24] Photocatalytic performance depends on the crystal structure and the composition of the material. Since photocatalysis is a surface controlled mechanism, increase in the surface area emphasizes the photoactivity of the material. [25–26]

In this study, we have used hydrothermal method to produce nine different ZnO coatings each with unique nanostructure by varying synthesis temperature and precursor concentration. The aim of the study was to investigate the effect of surface morphology on photocatalytic performance of the ZnO coatings.

2. Experimental

2.1. Preparation and characterization of zinc oxide films

Zinc oxide films were prepared by a two-step hydrothermal method. At first, a seed layer was formed on an AISI 304 stainless steel (SS) substrate (75mm x 25 mm) followed by the growth of ZnO nanorods. Before the process, substrates were cleaned by wiping with 2-propanol (C3H7OH, 99.5 %, VWR) and then rinsed with ethanol (C2H5OH, 99.5 %, Altia Oyj). A solution for the seed layer was prepared mixing 2-methoxyethanol (CH3OCH2CH2OH, ≥99,5%, Sigma-Aldrich) and ethanolamine (NH2CH2CH2OH, ≥ 99.0 %, Sigma-Aldrich) (96:4) and dissolving 0.23 M zinc acetate (C4H6O4Zn·2H2O, ≥ 98%, Sigma-Aldrich) into the mixture. Solution was then stirred at 60°C for 2 h. After cooling down, the solution was spin coated (1500 rpm, 20 s) on the substrates and then the samples were heat treated at 350 °C for 20 min. The solution for growth of ZnO layer with nanotopography was prepared mixing hexamethylenetetramine (C6H12N, 99%, Sigma-Aldrich) and zinc nitrate (Zn(NO3)2, 98%, Sigma-Aldrich) volume ratio being 1:1. The concentrations were 0.029 M, 0.058 M and 0.16 M. Coatings were fabricated at three temperatures: 70°C, 80°C and 90°C. The plates were placed in the solution the seed layer side upside down and the growth was carried out for 2 hours. After removing the samples from the solution they were rinsed with de-ionized water and dried in air. The samples were finally heat-treated at 300°C for 30 min.

The prepared zinc oxide films were characterized using field emission scanning electron microscopy (FESEM, Zeiss ULTRAplus) for surface morphology studies, energy-dispersive X-ray spectroscopy (EDS, INCA Energy 350) for elemental analysis and X-Ray diffraction (XRD, Panalytical Empyrean Multipurpose Diffractometer) to define the crystal structure of the samples.

2.2. Photocatalytic measurements

Photocatalytic measurements were performed in an aqueous solution of methylene blue (MB). The sample plates were placed in a decanter glass with 30 ml of 0.0085 mM MB solution. In order to reach adsorption equilibrium, the test solution was first kept in the dark for 60 min after which the ultra violet (UV)-lamp was turned on. Illumination was done from above with a 100 W/m2 power using Ledia NIS330U-M UV-Gun (peak maximum at 365 ± 5 nm). The concentration of MB in the solution was followed by UV-VIS spectroscopy (Shimadzu UV-2501PC Spectrophotometer). At one hour intervals a 3 ml MB solution was taken for absorbance measurement. The absorbance was measured at the wavelength of 665 nm. A dark experiment without any irradiation but under otherwise identical without sample plates was also tested to confirm that irradiation does not compose MB intrinsically.

Photocatalytic activity of the coatings was also tested by irradiating the samples with UV-light of $50W/m^2$ power with the same equipment described above. Coatings were irradiated for 30 min and the water contact angle (WCA) was measured after 10, 20 and 30 min of irradiation.

3. Results and discussion

3.1. Morphology and crystal structure of ZnO films grown in different precursor concentration solutions and temperatures

Nine types of ZnO coatings were prepared varying synthesis temperature and precursor concentration. Samples prepared at precursor concentration of 0.029 M at 70°C, 80°C and 90°C are denoted as $C_1$70, $C_1$80 and $C_1$90, respectively. Samples prepared at precursor concentration of 0.058 M at 70°C, 80°C and 90°C are denoted as $C_2$70, $C_2$80 and $C_2$90, respectively, and samples prepared at precursor concentration of 0.16 M at 70°C, 80°C and 90°C are denoted as $C_3$70, $C_3$80 and $C_3$90, respectively. Both synthesis temperature and precursor concentration impacted significantly the final nanotopography of the surfaces (Fig. 1–3). Sample $C_1$70 was only partly covered by small nano-sized nubs which were mostly concentrated on the grain boundaries of the stainless steel. This may be explained by the fact that the grain boundaries are often thermodynamically favourable nucleation sites [27] and the low temperature and precursor concentration were not sufficient for nucleation at more energetically demanding surface nucleation sites. The same effect could be seen with samples $C_1$80 and $C_1$90: the coating was thickest and the structure most uniform at the grain boundaries whereas some small uncoated areas were observed in the middle of the grains.

Sample $C_2$70 had flat hexagonal crystals whereas samples $C_2$80 and $C_2$90 consisted of hexagonally shaped rods. The structure of $C_2$80 was dense and the rods grew perpendicularly to the substrate contrary to the rods of $C_2$90 that had approximately the same rod diameter but the rods were pointing irregularly to different directions (Fig. 2c). This phenomenon was also distinguished with the sample $C_1$90. On samples $C_3$70, $C_3$80 and $C_3$90, ZnO formed a flaky structure flake size being highest at $C_3$90. The flaky morphology is probably due to the high precursor concentration which led to fast nucleation at several different sites resulting in less organized structure as compared to $C_2$80 and $C_2$90. Nucleation also occurred to a great extent in the solution which could be observed during the synthesis as ZnO powder precipitated on the bottom of the synthesis container.

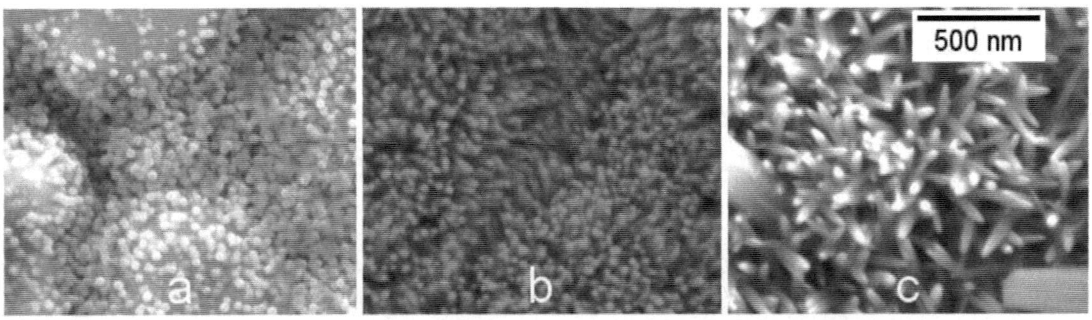

Figure 1 a–c. FESEM images of ZnO coatings grown at a) 70°C, b) 80°C and c) 90°C precursor concentration being 0.029 M (C1).

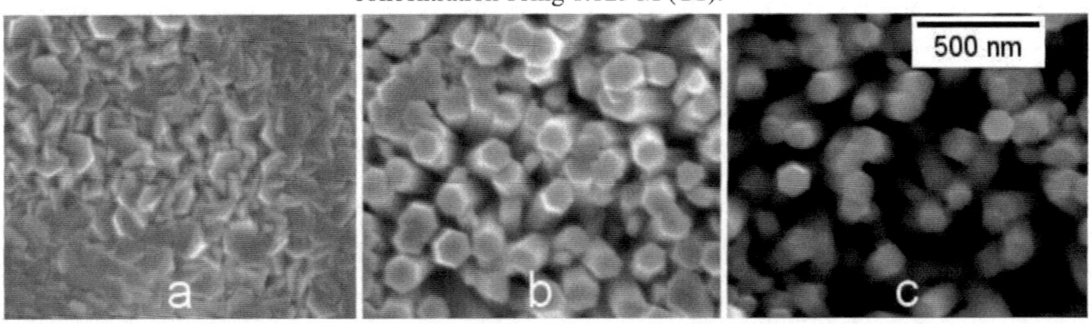

Figure 2 a–c. FESEM images of ZnO coatings grown at a) 70°C, b) 80°C and c) 90°C precursor concentration being 0.058 M (C2).

Figure 3 a–c. FESEM images of ZnO coatings grown at a) 70°C, b) 80°C and c) 90°C precursor concentration being 0.16 M (C3)

The thicknesses of the coatings were investigated by imaging the cross-sections of the coated plates (Table 1). The thicknesses were of the same magnitude (1.1–1.6 μm) for $C_1$80, $C_1$90, $C_2$80 and $C_2$90. Instead, for $C_1$70, $C_2$70, $C_3$70 and $C_3$80, the coating thicknesses were in the range of 180–300 nm, i.e., much smaller than in the other cases. For $C_3$90, the coating thickness was 630 nm. The lower thicknesses of $C_3$70, $C_3$80 and $C_3$90 can be explained with the nucleation occurring in the solution leading to less molecules precipitating on the surface and forming the coating. With $C_1$70, $C_2$70 and $C_3$70, the low thickness is probably due to the low temperature, which decelerates or hinders the reaction.

Table 1. Coating thicknesses of the prepared samples

c (mol/l)	70°C	80°C	90°C
0.29 M	180 nm	1380 nm	1100 nm
0.58 M	230 nm	1100 nm	1600 nm
0.16 M	300 nm	210 nm	630 nm

XRD analyses were conducted for all nine coatings and all of them showed clear peaks of hexagonal wurtzite structure. Apart from peaks of wurtzite and stainless steel substrate, no other peaks were found. EDS analyses were in of line with the received XRD patterns showing peaks of Zn, O and peaks resulting from the stainless steel substrate.

3.2. MB degradation results

The dark tests performed before the actual measurements showed that the concentration of the solution did not change during one hour in dark and thus the samples did not adsorb methylene blue. It is also notable that the irradiation of MB without the samples did not influence the concentration of MB, which means that irradiation did not intrinsically decompose MB. Based on MB degradation test results, the sample $C_2$90 had the highest photocatalytic activity although all the prepared coatings showed relatively good photocatalytic performance (Fig. 4). The difference between samples $C_2$80 and $C_2$90 could be due to the more randomly oriented structure of $C_2$90 leading $C_2$90 to have larger surface area being in contact with the MB-solution. The nanorods of the sample $C_2$80 seem to have fused together particularly near the substrate but also partly all the way along the rods. The same explanation could be applied to the difference in the photocatalytic activity between the coatings $C_1$80 and $C_1$90.

The results obtained from the WCA measurements of the irradiated coatings (Table 2) were consistent with the MB degradation results. Reference stainless steel substrate did not show any change in water contact angle during the irradiation test. In contrast, complete wetting of the surface was obtained after 30 min of irradiation with UV-light with samples $C_1$90, $C_2$90 and $C_3$80 which can be attributed to high photocatalytic activity of the films. Some questions still remain, since the sample $C_3$80 was clearly less efficient at MB degradation test (76 % of the MB degraded after 7 h of irradiation) compared to C290 (91 % of the MB degraded after 7 h of irradiation) but it showed omplete wetting of the surface after 30 min of irradiation along with $C_1$90 and $C_2$90. In our future studies, it is essential to explain these observations.

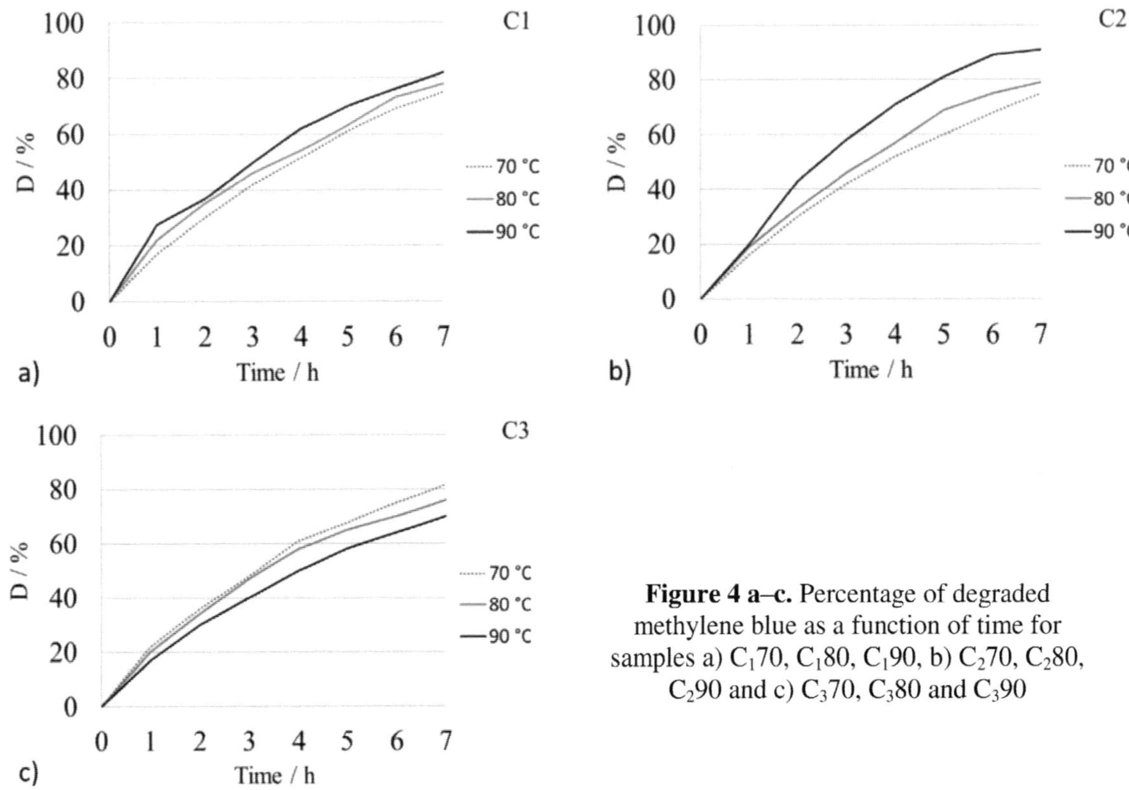

Figure 4 a–c. Percentage of degraded methylene blue as a function of time for samples a) $C_1$70, $C_1$80, $C_1$90, b) $C_2$70, $C_2$80, $C_2$90 and c) $C_3$70, $C_3$80 and $C_3$90

Table 2. Water contact angles of stainless steel and the ZnO coatings after 0, 10, 20 and 30 min irradiation with UV-light

Sample	0 min (°)	10 min (°)	20 min (°)	30 min (°)
$C_1$70	40	26	16	6
$C_1$80	20	4	10	7
$C_1$90	78	9	0	0
$C_2$70	83	32	25	22
$C_2$80	81	28	22	10
$C_2$90	61	12	0	0
$C_3$70	35	11	8	7
$C_3$80	71	15	12	0
$C_3$90	76	60	39	21
SS	77	78	70	78

The results presented in this paper could be concluded as follows: 1) both concentration of the precursor solution and synthesis temperature have a notable impact on the surface morphology and coating thickness. At lower concentrations the structure of the surface is needle-like and it changes towards hexagonal shaped units as the concentration increases. Finally, when the concentration is further increased, the surface changes from hexagonal units to flaky structure due to the rapid nucleation in the solution. 2) Although all the tested coatings showed relatively good photocatalytic performance, the activity is clearly related to nanostructure of the coatings. Indeed, the high photocatalytic activity of the two most active coatings can be explained by fragment position of ZnO nanorods which obviously enhances the surface area of the coatings. In addition to the impact of surface area, the polarity difference in different crystal orientations of the nano-sized crystals could also affect the photocatalytic performance due to different adsorption properties of different plains. This will be investigated more in forthcoming studies.

References

[1] Khan R, Hassan M S, Cho H-S, Polyakov A Y, Khil M-S and Lee I-H 2014 *Mater. Lett.* **133** 224
[2] Gürlük S 2009 *Ecol Econ* **68** 2327
[3] Loganathan N, Shahbaz M and Taha R 2014 *Renew. Sust. Energ. Rev.* **38** 1083
[4] Barakat M A 2011 *Arabian Journal of Chemistry* **4** 361
[5] Aarthi T and Madras G 2008 *Catal. Commun.* **9** 630
[6] Fujishima A and Honda K 1972 *Nature* **238** 37
[7] Wang Y, Zhu S, Chen X, Tan Y, Jiang Y, Peng Z and Wang H 2014 *Appl. Surf. Sci.* **307** 263
[8] Yu T, Tan X, Zhao L, Yin Y, Chen P and Wei J 2010 *Chem. Eng. J* **157** 86
[9] Yang J, Bai H, Tan X and Lian J 2006 *Appl. Surf. Sci.* **253** 1988
[10] Nikkanen J-P, Huttunen-Saarivirta E, Zhang X, Heinonen S, Kanerva T, Levänen E and Mäntylä T 2014 *Ceram. Int.* **40** 4429
[11] Goswami D Y, Trivedi D and Block S S 1997 *J. Solar. Energy Eng.* **119** 92
[12] Jacoby W A, Maness P, Wolfrum E J, Blake D M, Fennell J A 1998 *Environ. Sci. Technol.* **32** 2650
[13] Vohra A, Goswami D Y, Deshpande D A and Block S S 2006 *Appl. Catal. B-Environ.* **65** 57
[14] L Zhong, F Haghighat and C-S Lee 2013 *Build. Environ.* **62** 155
[15] Saarenpää H, Niemi T, Tukiainen A, Lemmetyinen H and Tkachenko N 2010 *Sol. Energ. Mat. Sol. C* **94** 1379
[16] Saarenpää H, Sariola-Leikas E, Pyymäki Perros A, Kontio J M, Efimov A, Hayashi H, Lipsanen H, Imahori H, Lemmetyinen H and Tkachenko N V 2012 *J. Phys. Chem. C* **116** 2336
[17] Polsongkram D, Chamninok P, Pukird S, Chow L, Lupan O, Chai G, Khallaf H, Park S and Schulte A 2008 *Physica B* **403** 3713
[18] Eswar K A, Rouhi J, Husairi F S, Dalvand R, Alrokayan S A H, Khan H A, Rusop Mahmood M and Abdullah S 2014 *J. Mol. Struct.* **1074** 140
[19] Zhou Y, Liu C, Zhong X, Wu H, Li M and Wang L *Ceram. Int.* **40** 10415
[20] Chen X, Zhai Y, Li J, Fang X, Fang F, Chu X, Wei Z and Wang X 2014 *Appl. Surf. Sci.* **319** 216
[21] Koao L F, Dejene F B and Swart H C 2014 *Mat. Sci. Semicon. Proc.* **27** 33
[22] J Gupta, Bhargava P and Bahadur D 2014 *Physica B* **448** 16
[23] Zhu Y F, Fan D H, Dong Y W and Zhou G H 2014 *Superlattice Microst.* **74** 261
[24] Foe K, Namkoong G, Abdel-Fattah T M, Baumgart H, Jeong M S and Lee D S 2013 *Thin solid films* **534** 76
[25] Nikkanen J-P, Heinonen S, Huttunen-Saarivirta E, Honkanen M and Levänen E 2013 *IOP Conf. Ser.: Mater. Sci. Eng.* **47** 012066
[26] Maury-Ramirez A, Nikkanen J-P, Honkanen M, Demeestere K, Levänen E and De Belie N 2014 *Mater. Charact.* **87** 74
[27] Ng H T, Li J, Smith M K, Nguyen P, Cassell A, Han J and Meyyappan M 2003 *Science* **300** 1249

Investigation of InGaAsN for laser diode application

A Ouerdane[1, 2], **Z Hachoun**[1]

[1]Université de Khemis Miliana route de Theniet El Had, Algeria
[2]Laboratoire Matériaux LabMat ENPO Mnaouer Oran, Algeria

E-mail: ouerdanea@yahoo.fr

Abstract. The knowledge of influence of nitrogen on optical transition energies in InGaAsN Quantum Wells (QWs) or Quantum Dots (QDs) structures is very important due to the possibility of adjustment of the optical transition energies to the telecommunication wavelength range when the system GaAsN is submitted to further indium elements giving rise to the quaternary alloys (In)GaAsN. We add a computational method in quantum mechanics to study the Schrödinger equation for one and many quantum wells configurations. The photoluminescence PL measurement obtained by simulated technique is used to predict the different lines and energy levels of our compound. The simulated results are followed by solving the coupled nonlinear rate equations describing the complex electric field and the carrier density in a simple model of the GaN_xAs_{1-x} or (In)GaAsN semiconductor laser.

1. Introduction

GaInNAs constitute actually a novel material system proposed and developed nowadays because of their physical properties and potential device especially in fiber devices applications. High performance and cheaper long wavelength (1300–1550 nm) emitting diode lasers with InGaAsN active region are essential components for optical fiber communication systems. Theoretical considerations and results on the effect of nitrogen incorporation on the oscillator strength of optical transitions in InGaNAs/GaAs quantum wells (QWs) are presented. The fundamental transition energy and its oscillator strength considering Quantum Wells configuration have been systematically investigated. Furthermore, the effect of the bandgap discontinuities and its variation with the concentration of nitrogen on the transitions intensity recorded in the Photoluminescence spectra has also been considered. Such structures are now realized and are emitting in the telecommunication wavelength range [1].

2. Theoretical model

The calculations of the related energy states in the QW region are obtained by solving the stationary Schrödinger equation assuming parabolic dispersion curves for the particles. The band gap is given by the band anticrossing (BAC) model which includes the influence of nitrogen and Indium on the band structure and estimates the effective band gap reduction [2]. The Schrödinger equation has a significant role in calculation of the parameters such as gain, effective barrier (SCH) height, energy levels of electron and hole in the well and consequently output wavelength which are directly or indirectly used in the rate equations.

$$\left[-\frac{h^2}{2_m}\frac{d^2}{dz^2}+V(z)\right]\psi(z)=E\psi(z) \quad (1)$$

$$V(z) = \begin{cases} \upsilon_0 & |z| = L_{QW}/2 \\ 0 & |z| < L_{QW}/2 \end{cases} \tag{2}$$

Where $\psi(z)$ is wave function of each particle (electron, light hole and heavy hole) in the well, E is energy level of particle, L_{QW} is well thickness and $V(z)$ is barrier height that is corresponding to ΔE_c for conduction band and ΔE_v for valence band. In addition m is particle mass that equals m_w in the well and m_b in the barrier regions. By assuming even solutions as the results, the wave functions are as following:

$$\psi(z) = \begin{cases} A_e^{-\alpha(|z|-L/2)} & |z| = L_{QW}/2 \\ B\cos kz & |z| < L_{QW}/2 \end{cases} \tag{3}$$

$$k = \sqrt{2m_w E/h} \text{ and } \alpha = \sqrt{2m_b(\upsilon_0 - E)/h}$$

The numerical solution of the Schrödinger equation for the configuration of one two and more quantum wells structures is given on figure 2.

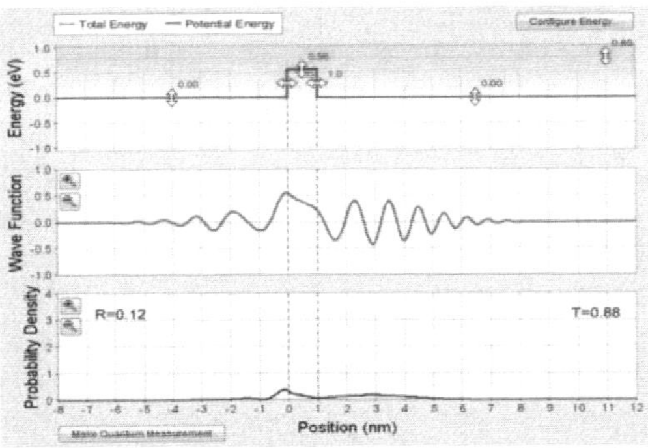

Figure 2a. Shrodinger equation solution for one quantum well structure with fixed reflected and transmitted coefficients

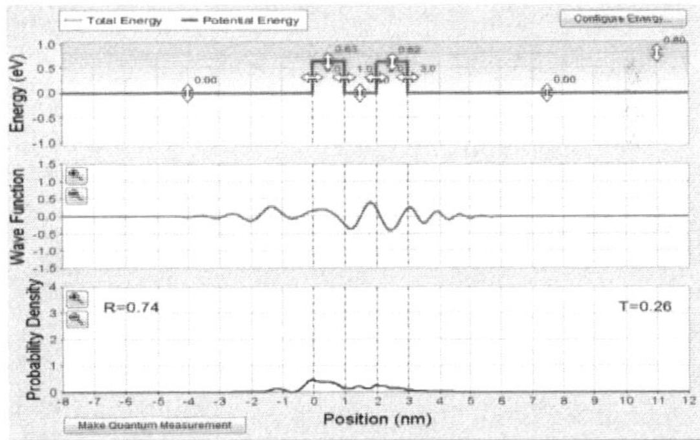

Figure 2b. Shrodinger equation solution for two quantum wells structure with fixed reflected and transmitted coefficients

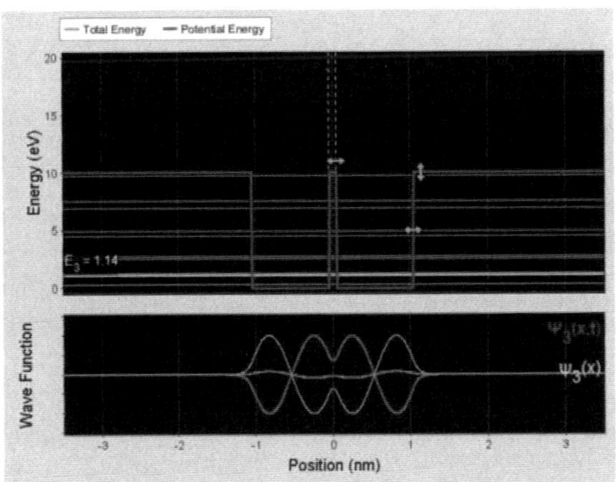

Figure 2c.:Shrodinger equation solution for two quantum wells structures with fixed energy at 1.14 eV

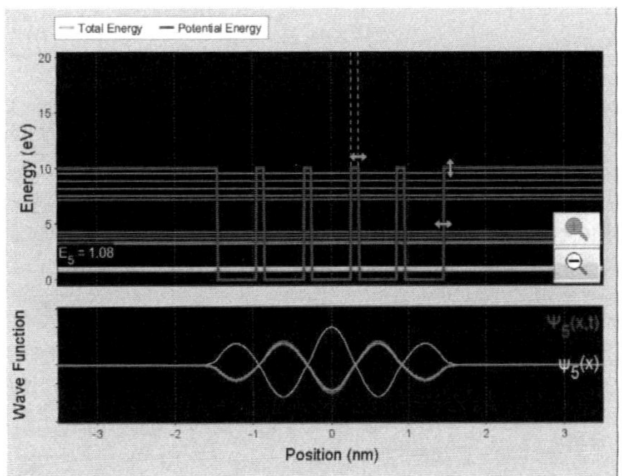

Figure 2d. Shrodinger equation solution for five quantum wells structure with fixed energy at 1.08eV

The photoluminescence peak due to band-to-band transitions can be calculated for each band gap E_g using the following formula:

$$I_{bb} = I_0(h\omega)^2(h\omega - E_g)\exp\left(-\left(\frac{h\omega - E_g}{K_bT}\right)^2\right) \quad (4)$$

We plot different curves corresponding to each concentration of nitrogen as follow:

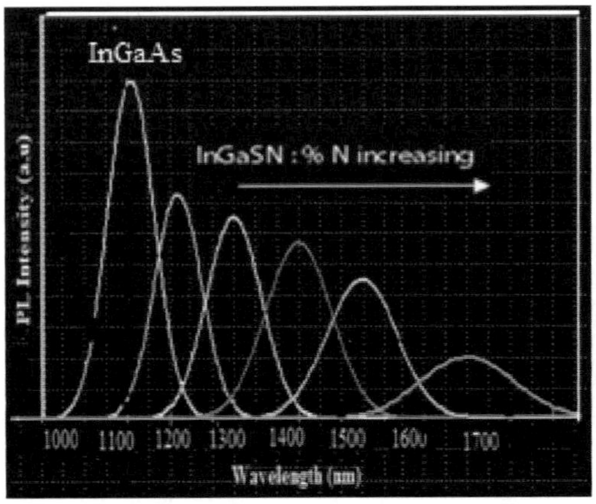

Figure 3a. Photoluminescence peak variation with nitrogen concentration (simulated curves)

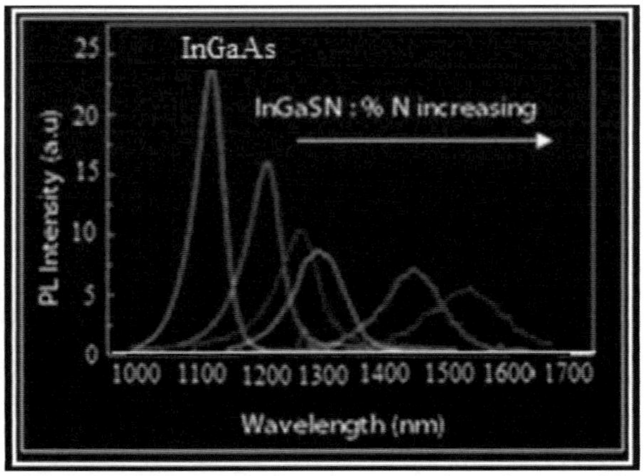

Figure 3b. Photoluminescence peak variation with nitrogen concentration (experimental curves)

Fig.3a shows that the photoluminescence intensity diminishes and the bandwidth of the curves increases when the nitrogen concentration increases.Fi3b is the similar curves variation with the nitrogen concentration obtained by experimental measurement [3,4].

3. Simulation of the diode laser InGaAsN.

The optical power emitted by the diode laser is proportional to the current supplied to the InGaAsN semiconductor heterojunction. The output of the solver is the time, injection current, and the transformed components of the state vector, computed at discrete time steps. The rate equations are Integrated using the fourth order Runge-Kutta computation[9].The laser dynamics can be modeled by coupled rate equations which describe the relation between the carrier number $N_p(t)$, the photon density $S_p(t)$ and the optical phase $\varphi(t)$ [1-6].

$$\frac{dN_p(t)}{dt} = \frac{I(t)}{q} - \frac{N_p(t)}{\tau_n} - g(N,T)\frac{N_p(t) - N_0}{1 + \varepsilon S_p(t)}S_p \quad (5)$$

$$\frac{dS_p(t)}{dt} = \Gamma g(N,T)\frac{N_p(t) - N_0}{1 + \varepsilon S_p(t)}S_p(t) - \frac{S_p(t)}{\tau_n} + \Gamma\frac{\beta N_p(t)}{\tau_n} \quad (6)$$

$$\frac{d\varphi(t)}{dt} = \frac{\alpha_0}{2}g(N,T)\left[N_p(t) - N_0\right] \quad (7)$$

N_0 is the carrier number at transparency, τ_p is the photon lifetime, τ_n is the carrier lifetime, Γ is the optical confinement factor describing the confinement mode in the active region, β is the spontaneous emission factor, ε is the gain compression factor, $g(N,T)$ is the optical gain coefficient dependent on the carrier density and the temperature function, $I(t)$ is the injected current, α_o is the linewidth enhancement factor and q is the electron charge, η is the quantum efficiency, h the Planck constant and υ the radiation frequency. The steady-state solution to the rate equations is obtained by setting all the time derivatives $\frac{dN_p(t)}{dt}$ and to $\frac{dS_p(t)}{dt}$ zero. The carrier concentration that satisfies a given steady-state injected current is obtained by iterative self-consistent solutions of the two coupled equations. The output power the threshold current and the carriers density are given respectively by the following expressions [1, 10-13].

$$P(t) = \frac{\eta h v}{q}(I - I_{th}) \quad (8)$$

$$I_{th} = \frac{q}{\tau_n}\left\{N_0 + \frac{1}{g_0\tau_p}\right\} \tag{9}$$

$$N = \frac{\frac{I(t)}{qV} - \frac{\tau_n S_p}{\tau_p \Gamma}}{\beta - 1} \tag{10}$$

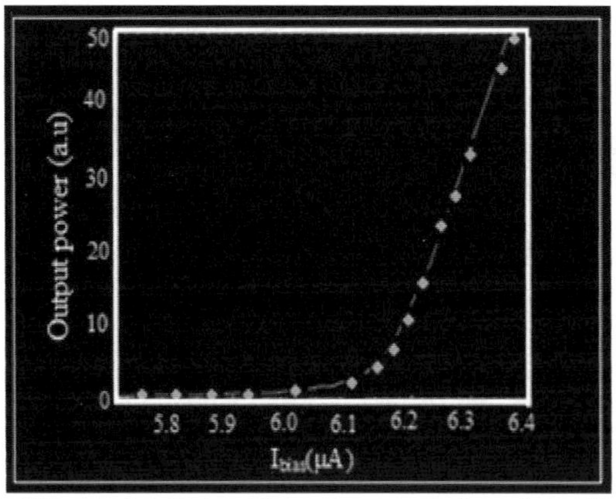

Figure 4a. Output laser power as function of the bias current at room temperature

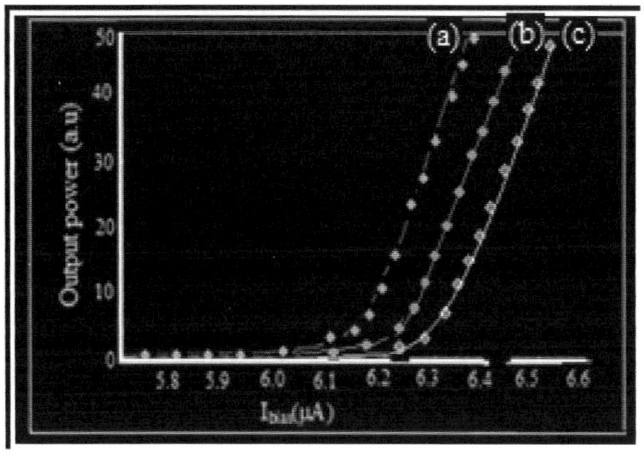

Figure 4b. laser threshold current temperature dependence (a) 20°C; (b) 40°C and (c) 60°C

4. Numerical Simulation and Results.

Fig.5a records the variation of carrier and the photons generated by the injected current. Fig.5b gives a single photon signal following the driver electrical signal when fig.5c is a modulated optical signal used in telecommunication. We record in Fig.5d the frequency signal modulation of the laser power centered on the wavelength λ=1550 nm. In practice, optical modulation is a way to minimize the effects of electrical parasitics. Within the model, this is equivalent to adding an optical perturbation to optical signal generated by the carriers.

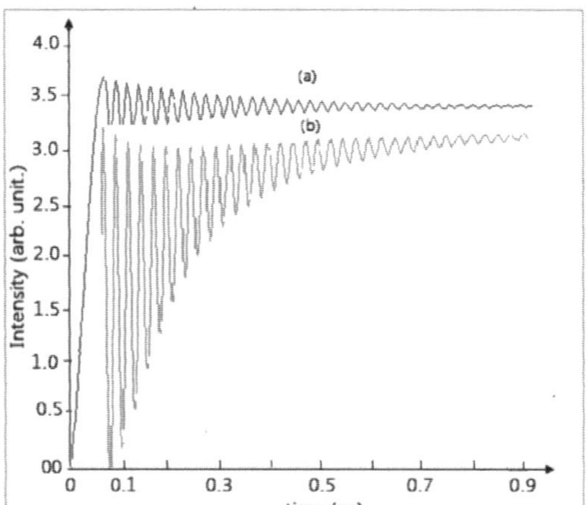

Figure 5a. Transient dynamics of carrier density (a) and photon number (b)

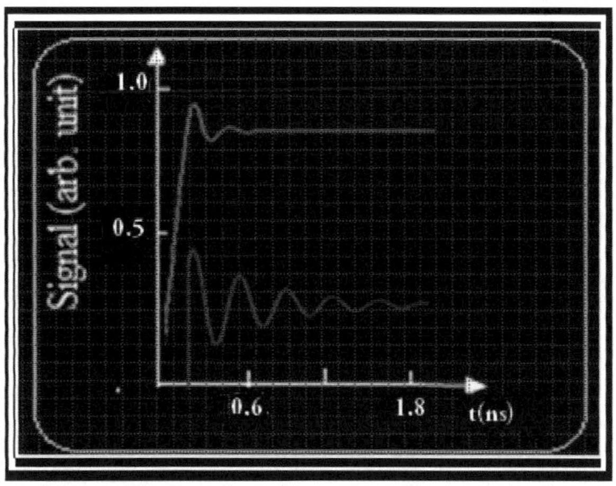

Figure 5b. Transient dynamics of carrier density and photon number for a deterministic single mode

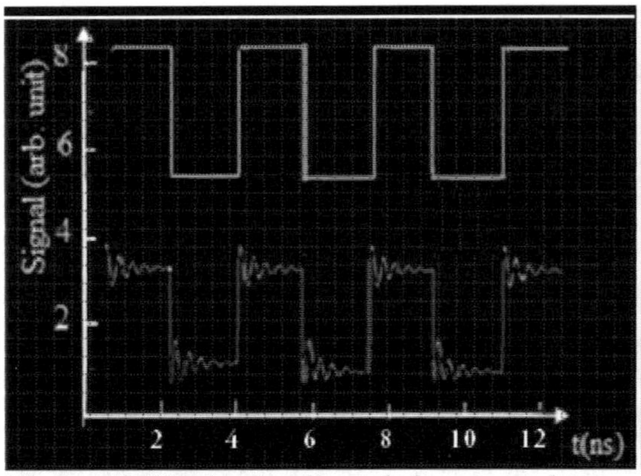

Figure 5c. Carrier density and photon number signals for high-speed light wave telecommunication systems

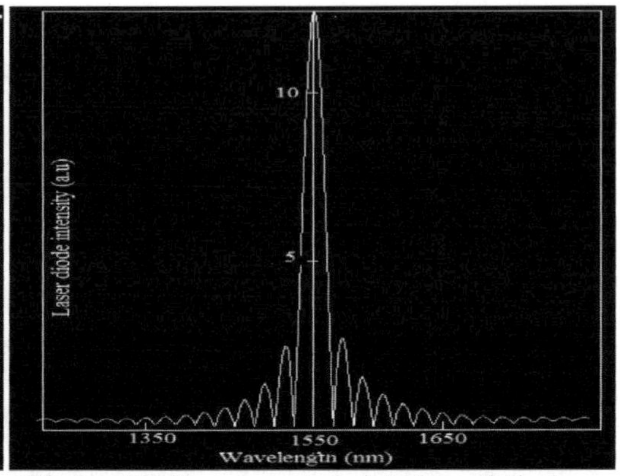

Figure 5d. Frequency modulation of diode laser with emission of the wavelength λ=1550 nm

The damping due to the injected current variation is simulated and recorded in fig.5c, the latter being significantly influenced by parasitics and adiabatic phenomenon. Hence as cited above, the introduction of weak amount of nitrogen in InGaAs compound allows great flexibility in terms of band gap energy and stress state [1, 14].

5. Conclusion

Our material system InGaNxAs1-x was in its configuration as quantum wells. The numerical solution of the Schrödinger equation enables us to show the oscillation of the carriers in one, two and many quantum wells as well. Considering InGaNxAs1-x as active region of a diode laser, we determine the photoluminescence spectrum which shifts towards the red color when the nitrogen concentration increases. An efficiency computational program has been used to study the modulation response of laser diode emitting in different wavelengths following the concentration of the Nitrogen. The simulation program developed can be used to illustrate the performance of a waveguide laser diode as function of device parameters for all semiconductors electrically pumped laser diode. Adding of weak amount of nitrogen in InGaAs compound allows great flexibility in terms of band gap energy, appropriated wavelengths for telecommunication systems and stress state visibility.

References

[1] A.Ouerdane, M.Bouslama, A.Abdellaoui, M.Ghaffour and Y. Al Douri, 2013 Computational Simulation of Ultraviolet ZnO Diode Laser, Journal of Physics: Conference Series 435, 012028
[2] Arkadiusz Mika et al., 2013 *Optica Applicata*, XLIII **1** 53
[3] Haeyeon Yang et al. 2014 *Applied Surface Science* **296** 8
[4] Makato Inagaki et al. 2011 *Japanese Journal of Applied Physics* **5004**, DP14 5004DP14
[5] H. Baaziz, Z. Charifi, Ali Hussain Reshak, B. Hamad,Y. Al-Douri, 2012 *appl. Phys.* A **106** 687
[6] F. Maskuriy, M. S. Alias, S.M. Manaf, 2011 *Journal of Applied Optoelectronics and Advanced Materials* **13** 1213
[7] R.A. Abdullah, K. Ibrahim 2010 *Optoelectronics and Advanced Materials–Rapid.Com* **4** 568
[8] Jeffrey O. White et al. 2011 *ARL-TN* **1** 0451
[9] A.Nouri, Z.Lounis, A.Ouerdane, M.Ghaffour, M.Bouadi, H.Dumont, L. Auvray, M.Bouslama, 2007 *Vacuum* **7** 979
[10] Babar Bachir, 2009 Electronics/Telecommunications University of Gävle Supelec 592,35
[11] Safwat W.Z. Mahmoud, 2007 *Egypt Journal Solids* **30** 277
[12] H.K. Liang, S. F. Yu and H.Y. Yang 2010 *Applied Physics Letters* **97** 241107
[13] Vitezslav Jerabek, Ivan Huttel, 2011 *Radioengineering* **20** 486
[14] Krishna Myneni, 2011 *Semiconductor Laser Rate Equation Solver* **44** 1, 37

Initiation of the Adiabatic Wave of Combustion for Obtaining the Substances with the Free Valence

A Baideldonova, G Ksandopulo, L Mukhina

Institute of Combustion Problems, Almaty, Kazakhstan

E-mail: abaid@inbox.ru

Abstract. According to the task of obtaining substances with the free valence for the linkage of the nano-powders, the procedure of the synthesis of materials under the extreme nonequilibrium conditions is presented. The combustion of multilayer aluminothermic systems in the revolving reactor was investigated. Experiments were carried out in the reactor of high-temperature centrifuge. The initiation of process realizes by electric pulse in the effective layer. Further the wave of combustion was propagated along the axis of the reactor. The particles of the restored metal penetrated the underlayers of fresh mixture under the action of centrifugal acceleration and created the additional centers of ignition. The higher the density of metal, the higher speed and depth of penetration. An increase in the centrifugal acceleration strengthens the activity of process also. The speed of the motion of metallic particles grows. According the theoretical calculations it reaches 90 m/s in the case of tungsten.

1. Introduction

As is known one of the type of combustion is SHS - Self-propagating high-temperature synthesis, which based on the strong exothermic reaction (combustion reaction) in which the heat localize in the layer and spread from one layer to another by conduction. The main purpose of SHS - the synthesis of substances and materials. SHS has a great technological potential, since it includes such features as a high rate of combustion (0.01-0.2 m/s) and temperatures (2300-3800 K) the relative simplicity of the equipment, low energy (heat energy of the process of SHS), the ability to use raw materials of different origin, obtaining a wide range of materials with different properties etc. The above-mentioned advantages makes the SHS process is attractive for implementation in the production of a variety of materials [1].

Undertaken studies of the combustion process performed by the Self-propagating High Temperature Synthesis (SHS) based on the nickel, iron, and molybdenum oxides and affected by a centrifugal acceleration have revealed specific properties and high power capacity of a wave propagating in the reactor. High velocity of the thermal combustion front advancement cause by contribution of the rotation kinetic energy in the total energy of the metal clusters motion in the SHS front. A purposeful use of the clusters generated energy find its practical application. The main objective is to develop technology based on the energy of an adiabatic combustion wave to obtain the non-trivial material.

2. Formulation of the problem and method of solution

These studies have been carried out using the unit specially designed to model the combustion process under the centrifugal force effect [3] (see figure 1). This unit includes a shaft-installed engine, and a crosspiece with three cylindrical reactors fixed on it in a balanced manner at 120o angle. Inside the reactor the combustion process is initiated by an electric impulse when set rotation frequency is achieved. The number of the centrifuge revolutions was regulated by changing the electrical motor. The rotation working frequency varies from 500 rpm up to 5000 rpm in accordance with the centrifugal acceleration in the ignition point varying between 25 and 2000 g, correspondingly. There is a narrow cut has been made both in one of the three rotating reactors and in the unit cover to observe advance of the combustion wave front. High-speed video camera is used for supervision [2].

Usually solid-phase combustion in the regime SHS has features of combustion in the gas phase, but without the participation of mass transfer. The thermal wave propagated by means of the thermal diffusivity is supported by high-temperature and high-speed heat emission.

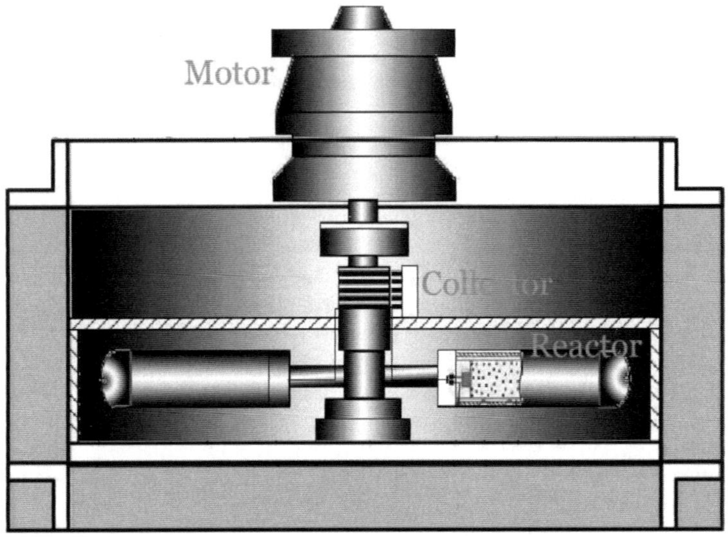

Figure 1. High-temperature centrifuge

With the combustion of oxide systems, the temperature in the front is 2500 - 3700 K. The melt of products, which the traveling wave leaves after itself, divided (separate) into parts in accordance with the density in the gravitational field, until low viscosity has continued. This effect becomes more intensively under rotation conditions.

The size of the particles of the metal at the time of their formation within the limits of the front of SHS-wave averages $2\text{-}3\cdot10^{-6}$ m. If reaction occurs under the conditions of rest(the natural gravity) then these particles coalesce without difficulty, forming at first amalgamated aggregates, and then ingot. Under rotation, conditions the particles of metal do not remain in the wave of combustion (see figure 2). Their trajectory is directed to the side of fresh mixture with respect to the directivity of the resulting vector of centrifugal force F_N and Coriolis effect F_K It is obvious that the depth ε of the penetration of hot (T_c = 2800 - 3500 K) particles into fresh reaction mixture is increasing with the increase of F_N, other conditions being equal. After achievement of a certain critical value ε_{cr} the process passes into the explosive regime. As opposed to the steady conception, in our case the super adiabatic combustion progresses with the broadening (expansion), of the combustion front, but not with the contraction (narrowness) (see figure 3). It is important that the mass transfer participate in SHS-process under conditions of rotation as opposed to the SHS-process under the conditions of normal gravitation.

Figure 2. Metal clusters distributed in the final product of burning

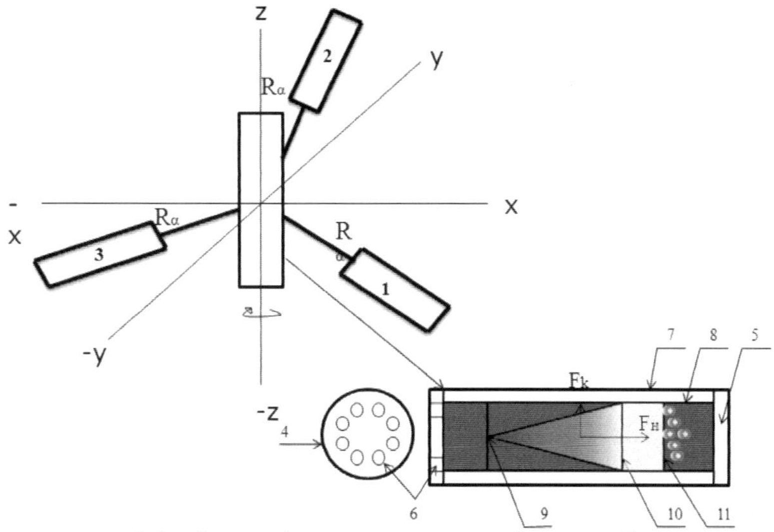

1,2,3 - rotating reactors; 4,5 – front and rear reactor covers; 6 - gas outflow apertures in the front cover; 7 - steel casing; 8 - quartz tube; 9 - ignition point; 10 – SHS front;11 – adiabatic wave

Figure 3. Scheme of installation for receiving an adiabatic wave

Stoichiometric mixtures of tungsten, cobalt, and nickel oxides with aluminum was chosen as active layers. These mixtures have been diluted with Al_2O_3. A low calorie layer of the stoichiometric mixture of boron oxide with aluminium [4].

$$3NiO + 2Al = Al_2O_3 + 3Ni + 2944\ kJ/kg; \quad 3FeO + 2Al = Al_2O_3 + 3Fe + 2810\ kJ/kg;$$
$$Fe_2O_3 + 2Al = Al_2O_3 + 2\ Fe + 3455\ kJ/kg; \quad WO_3 + 2Al = Al_2O_3 + W + 2890\ kJ/kg; \qquad (1)$$
$$B_2O_3 + Al = Al_2O_3 + 2B$$

Total thermal and kinetic energy of reduce metal flux is enough to overcome the Arrhenius barrier of any exothermic reaction or to break the chemical bonds in the oxides of metals.

A high-speed video filming was used for record of process of combustion in real time. Expanding the video into separate frames, we have identified the combustion wave transition in the adiabatic regime-sharp increase speed of displacement of the front. Here we put less active layer- a mixture of boron oxide with aluminum or pure boron oxide.

Based on this video record we have managed to build up a diagram of the co-ordinate dependent front advance along the reactor axis. Noteworthy that combustion process in the second less active layer develops similarly to the first one. In addition, you can put another layer on the way the second adiabatic waves.

Slow velocity increase is then replaced by its sharp increase. Afterwards the wave slows downat the layers separation boundary. Then the whole process is repeated but with less energy. Therefore, we have placed the third, i.e. the most active layer, in the wave action field (see figure 4).

Actually, the burning process in the three-layer system lasts 0.8 seconds. This time is sufficient for the formation of gradient material. The material structure changes from metal (tungsten) to ceramics (aluminum oxide, corundum). The basic phases take 91.9 %. Other involved substances are boron-containing solid tungsten and boron compounds. Similar results have been obtained when used samples active layers are based on cobalt and nickel oxides [5].

If the energy of the metal particle flux is designed to break the chemical bonds in the molecule, the chemical compound with a free valence is formed in the product of synthesis (see figure 5).

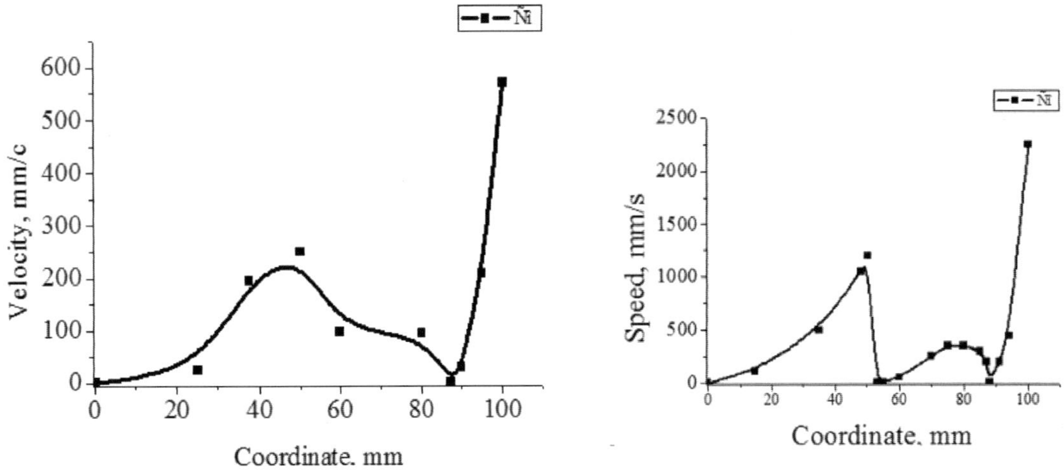

Figure 4. The combustion front advance velocity within the (Co_3O_4+Al+Al_2O_3) (B_2O_3+Al) system (a – in two-, b – in three-layer system)

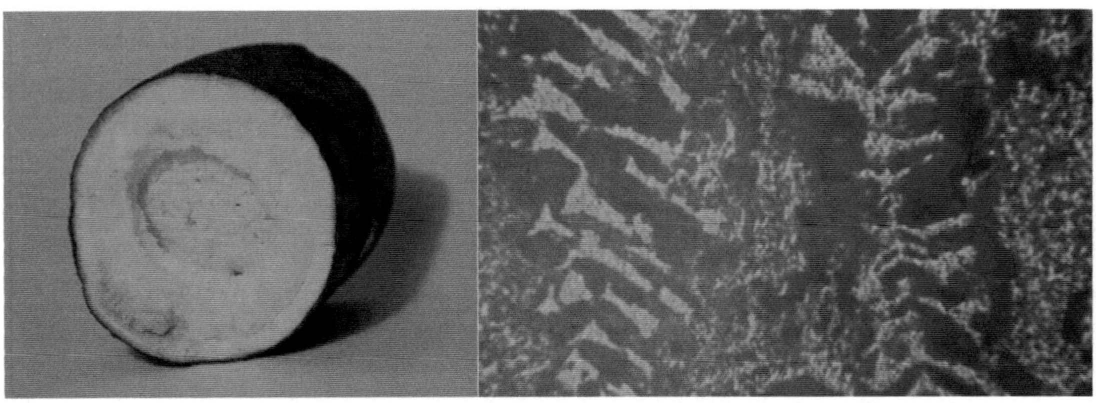

Figure 5. Formation of aluminum oxide macrostructure in boron presence

3. Results and Discussion

The combustion wave velocity is not equal to that of the reduced metal particles. According to the experimental data, the flame front moves with 20 m/s velocity. Theoretically, calculated velocity of the particles movement through a loose compacted material reaches 90 m/s at 3000 r/m rotation frequency. In such conditions, viscosity of the synthesized intermediate and final products is not already of any importance for the front advance along the reactor axis. However, viscosity remains a determining factor of the synthesis products structure formation in the cross-section direction. Its importance sharply increases in the direction from the reactor axis to its rather cold walls. Fast hardening of occurs near the internal surface of the reactor quartz lining, this process resulting in formation smelt of a ceramic tube. Depending on the initial components composition it can have a corundum, spinel, and mullite structure. Consequently, experimental study of the adiabatic combustion wave propagating under the centrifugal acceleration affect in the layer system has led to creation of new materials and products.

An interesting material has been produced at combustion of the three-layer system based on tungsten and boron oxides –an extremely porous aluminum oxide crystallized in its three modifications such as α, δ and γ (see figure 6).

Figure7 illustrates a sample electron paramagnetic resonance (EPR), data testifying to emergence of a free valence in $Al_{\alpha}O_{\beta}B_{\gamma}$ system.

Performed studies allowed for production of materials with various structures and compositions, namely spinels and mullites (see figure 8 and tabte 1). There have been synthesized materials with both cubic and dendrite-shaped structures [6].

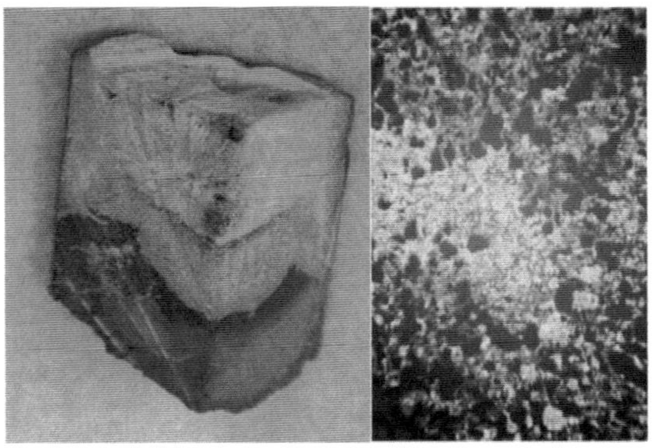

Figure 6. Results of three-layer system combustion (WO_3-Al-Al_2O_3)-(B_2O_3)-(WO_3-Al-Al_2O_3)

Tabte 1. Phase structure of the products

Crystal phase of the product	Mass contents, %
Al2O3	50.6
W	41.3
Al	4.8
Al20B4O36	1.5
WB2.O	0.7
W2B	0.6
WB	0.4

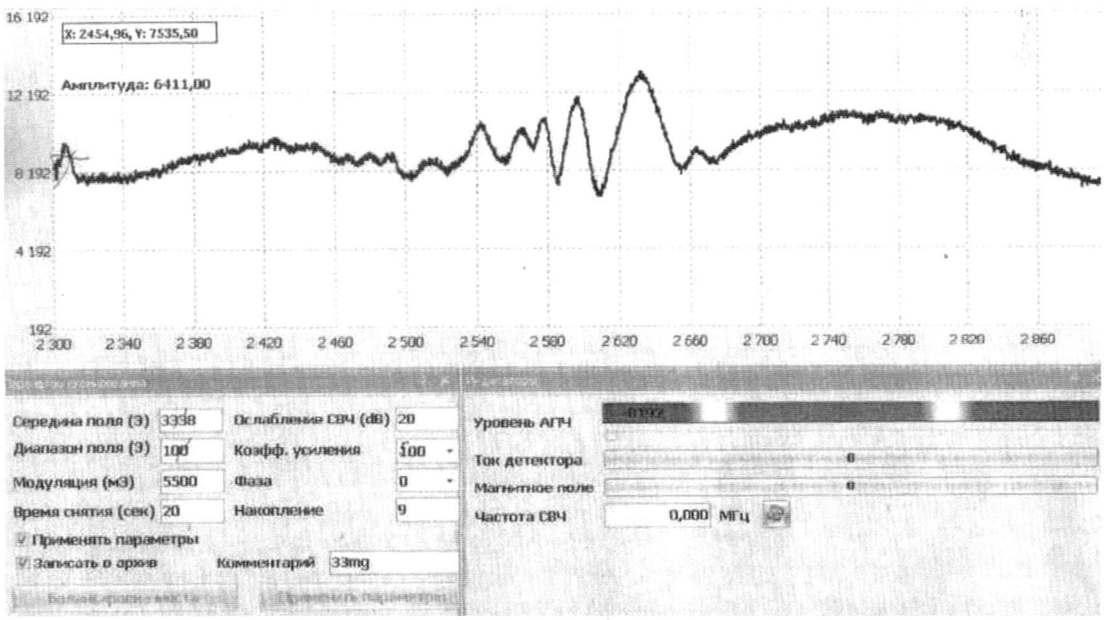

Figure 7. Electron paramagnetic resonance (EPR) of $Al_{\alpha}O_{\beta}B_{\gamma}$ system

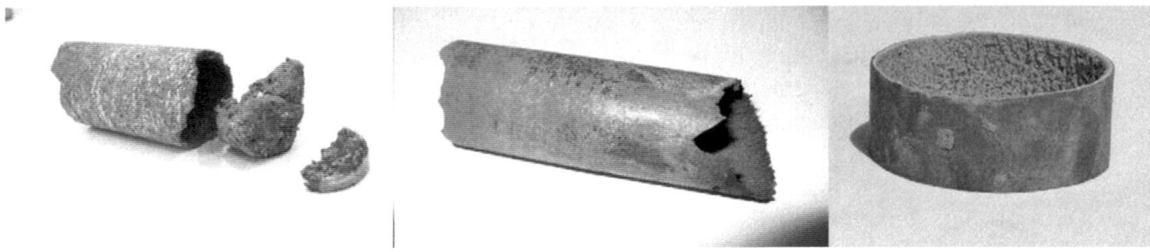

Figure 8. Refractory Tubes

Thus, sufficient centrifugal acceleration allows initiating of the adiabatic combustion wave and application of its thermal and kinetic properties to start chemical processes or to bond breakage and to formed gradient materials or substance with free valence.

4. Conclusion

The arrangement of the layer of inert substance on the way of the adiabatic wave of combustion leads to a change in its structure. Energy of the adiabatic wave of combustion was used also for the initiation of reactions with the high energy barrier. As a result the complex oxide compound Al20B4O36 was obtained. The presence of free radicals in it is confirmed by the method EPR.

References

[1] Ksandopulo G.I, Baideldinova A.N. 2011*Combustion and Plasma Chemistry* **9**, No 4. 241

[2] Ksandopulo G.I. 2011*Int. J. Self-Propag. High-Temp. Synth* **20** 220

[3] G.I. Ksandopulo, V.N. Shevchenko, A.N. Baideldinova. High-temperature centrifuge / RoK Patent No 68317 as of 26.05.2010

[4] Ksandopulo G.I. 2011 *SHS Journal* **20**, No. 4. 220.

[5] Ksandopulo G., Baydeldinova A., Ainabayev A., Arkhipov M., Omarova K. 2011 *Eurasian ChemicoTechnological Journal* **V.13**, №3 4. 155

[6] Ksandopulo G., Aynabaev A., Arkhipov M., Baydeldinova F., Omarova R. 2013 *IOP Conference Series: Materials Science and Engineering (MSE)* **44**

Wear mechanisms maps of Si_3N_4/carbon nanotube nanocomposites

Zs Koncsik, MB Maros, L Kuzsella, A Kovacs

University of Miskolc, Hungary

E-mail: metkzs@uni-miskolc.hu

Abstract. Silicon nitride-based nanocomposites with different amount (0, 1, 2 wt%) of multiwall carbon nanotubes (MWCNT) have been prepared by hot isostatic pressing. Tribological tests were carried out at room temperature, in dry sliding condition using pin-on-disc test apparatus, applying different normal loads and sliding speeds. Wear phenomena was analysed by scanning electron microscopic investigations of the wear tracks. Based on the controlling wear mechanisms and wear types two-dimensional wear mechanisms maps were created for all composition of the investigated ceramic nanocomposites.

1. Introduction

Silicon nitride ceramics are known as a group of one of the strongest structural ceramics. Their monolithic and even reinforced types possess an advantageous combination of mechanical and other physical properties. They have low density, low thermal expansion, thermal shock resistance, corrosion resistance, and appropriate hardness and strength even at high temperatures. Their application as tribological material, especially at high temperatures is widely known. Nowadays Si_3N_4 nanocomposites with different types of carbon additives have been created with the aim of producing electrically conductive materials with reasonable mechanical and electrical properties [1, 2]. The microstructural and electric properties of these carbon added nanocomposites have already been investigated by many authors [3, 4, 5, 6]. However details relating to their tribological performance– e.g. processes controlling the friction and wear behavior of these novel materials in various wear systems – have not been exactly clarified [7, 8 9].

The basic aim of the current research work is to investigate the wear mechanisms of Si_3N_4 ceramic composites containing different amount of MWCNT applying pin-on-disc wear test methods.

2. Experimental work

2.1. Test and evaluation methods

Pin-on-disc wear tests were carried out using a CETR-UNMT1 high-precision micro-nano surface tester, according to the standard ISO 20808 [10], applying different normal loads: 10 N, 40 N, 80 N, different sliding velocity: 10 mm/s, 100 mm/s, 200 mm/s, at constant sliding distance: 100 m and sliding radius: 3 mm. SiC ball was used as counterpart with r = 5,545 mm radius. Wear tests were carried out in dry conditions at ambient atmosphere.

The main objectives of the research work were to describe and identify the controlling wear mechanisms and wear types occurred in the investigated different tribo-systems,based on which wear mechanisms maps for the investigated materials could be created. These 2D maps are able to visualize the controlling wear mechanism regions of the given tribo-system as a function of the applied load, *F* and the sliding velocity, *v*.

The first step of creating these type of maps is the adequate systematization of wear mechanism and wear types, a question that is rather confusing and incomplete in the current technical literature. Therefore a new classification system of tribological processes has been worked out (Fig.1.) and published [11, 12]. Beside our aim for providing a suggestion for classification the definitions of the wear mechanisms and wear types are also given and well differentiated.

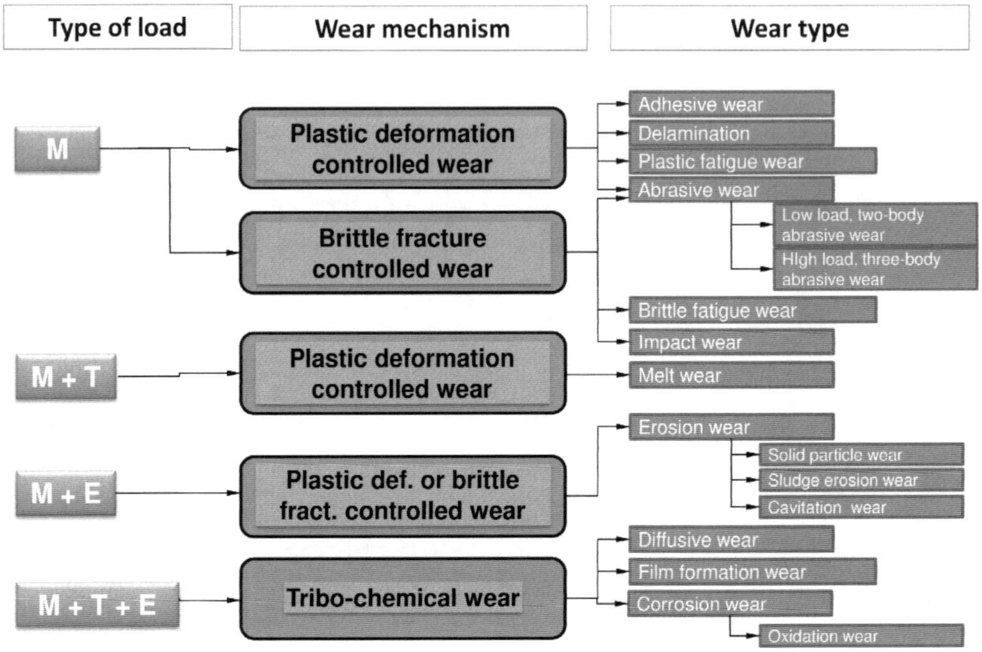

Figure 1. Classification of tribological processes
(M: mechanical, T: thermal, E: environmental)

To create a two-dimensional wear mechanisms map the following steps have been taken:

1. assigning the numerical value of the corresponding test parameters, i.e. the normal load and the sliding velocity pairs;
2. scanning electron microscopic investigation of the wear track produced in a tribo-system with the given test parameter combination;
3. morphological analyses of the wear track and identification of related wear mechanisms and wear types;
4. choosing the controlling wear mechanism(s) and wear type(s) for the given tribo-system. The wear mechanisms and wear types may change during the tribological process, therefore we usually able to identify only the main appearing tribological phenomena from which the controlling wear mechanism and wear type should be chosen;
5. creating two-dimensional diagrams of wear mechanisms, and wear types as a function of the applied load and sliding velocity, and determining their validity ranges.

2.2. Material characterisation

Test specimens were produced by the Research Institute for Technical Physics and Materials Science, Department of Ceramics and Nanocomposites Budapest, Hungary. The chemical compositions of the investigated samples are listed in Table 1.The initial powder mixtures were milled in water in a high-speed HDDM type attritor mill using zirconia balls for 5 hours, with 3000 rpm. After drying the dust, screening was applied with 100 μm sieve. Green samples were obtained by cold pressing at 200 MPa. Sintering process was carried out at 1700 °C, for 3 hours, with 20 MPa pressure, in high purity N_2-gas.

Table 1. Chemical composition and measured density of the samples

Samples	Chemical composition, [wt %]				Density, [g/cm^3]
	Si_3N_4	Al_2O_3	Y_2O_3	MWCNT	
0% MWCNT	90	4	6	0	3.434
1% MWCNT	89	4	6	1	3.386
2% MWCNT	88	4	6	2	3.293

The experimental work was executed on these 3 types of specimens on 3-3 samples for each composition. The initial disc shape samples with the diameter of ϕ 27.3 … 27.6 mm and height of 2.2 … 2.3 mm are shown in Fig.2.

XRD analyses proved that beside the presence of the initial compounds and phases of the ceramic materials formation of new phases during sintering have occurred. Relating to the basic Si_3N_4matrix material a SiAlON complex phase has formed, while relating to the Y_2O_3sintering aid, it reacted with the ZrO_2(a residuum from the zirconia milling balls) and formed a complex ZrYO phase.

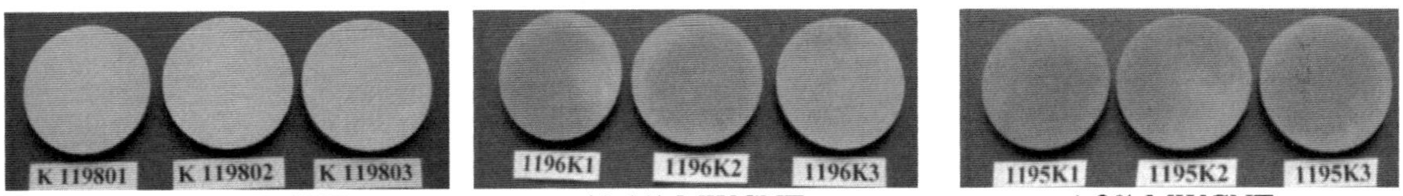

a) 0% MWCNT b) 1% MWCNT c) 2% MWCNT

Figure 2. The initial samples used for the wear tests

3. Test results, wear mechanism mapping

High (magnification: 300×) and low (magnification: 50×) resolution pictures were taken by scanning electron microscope to describe the wear process of the samples and of the tribo-systems.

Analyzing the wear tracks evolved in the different tribo-systems the controlling wear phenomena were identified. In each cases, i.e. for the monolithic and even for the composite samples three different dominant wear mechanisms could be identified, in addition four different wear types occurred in case of the monolithic samples, while five different ones were observed in case of the composite samples, which were identified according to the own classification system presented in Fig. 1.

Table 2.contains the regions of controlling wear mechanisms and wear types indicating the measured ranges of the related steady state friction coefficients, while Fig. 3. illustrates the two dimensional wear mechanisms/wear types maps of the investigated materials. The main characteristics of the different regions of the wear behavior are as follows:

Region I.: Delamination of thin silicon-hydroxidetribofilm occurs by plastic deformation, due to unidirectional shearing of surface. In this case a new, plastically deformed hydroxide layer is forming, while under the surface pores can be overlapped, forming cracks. Due to repeated loading of surface the cracks will propagate and smaller-larger platelets will erupt from the surface by fatigue wear. The steady state friction coefficients ranged from 0,43 to 0,46.

Region II.: Due to plastic shearing of a more massive hydroxide film the newly developed sheets can overlap and stack on each other. In addition, due to the SiC counterpart traces of grooving in the plastically deformed surface layer can be observed. In general the best wear performance accompanied with the lowest frictional coefficient falling into the range of 0,36-0,43.

Region III.: At higher normal loads or sliding speeds an intensive tribofilm formation of surface complexes is possible due to thermochemical reaction between sliding surfaces. This new tribofilm forms a continuous coating on the surface, which is mostly an oxide film and its physical properties will basically influence the wear process. The higher normal load and frictional heat results in amore expressed plastic deformation, and abrasive wear increasing the friction coefficient up to 0,42-0,58.

Region IV.: Melt wear occurs, as very high temperature between the sliding surfaces occur, so some μm^2 of the surface can melt and operate as lubricant, with friction coefficient of 0,49-0,54.

Region V.: Massive tribo-chemical film formation is followed by fatigue wear due to repeated load, by applying high normal loads and low sliding velocity. The pores being present due to the MWCNT particles enhances the crack initiation and propagation, with a friction coefficient range of 0,43-0,46.

Table 2. The controlling wear mechanisms, wear types and the steady state friction coefficient values

Region	Controlling wear mechanism	Main wear types	$\mu_{steady\ state}$
I.	plastic deformation	weak hydroxide film formation, fatigue wear, delamination	0,43-0,46
II.	plastic deformation	hydroxide film form. layer stacking, growing	0,36-,043
III.	tribo-chemical wear, plastic deformation,	strong, oxide film formation, abrasive wear	0,42-0,58
IV.	tribo-chemical wear, plastic deformation,	strong, oxide film formation, melt wear	0,49-0,54
V	tribo-chemical wear, brittle fracture	massive tribofilm formation followed by fatigue wear and delamination	0,43-,046

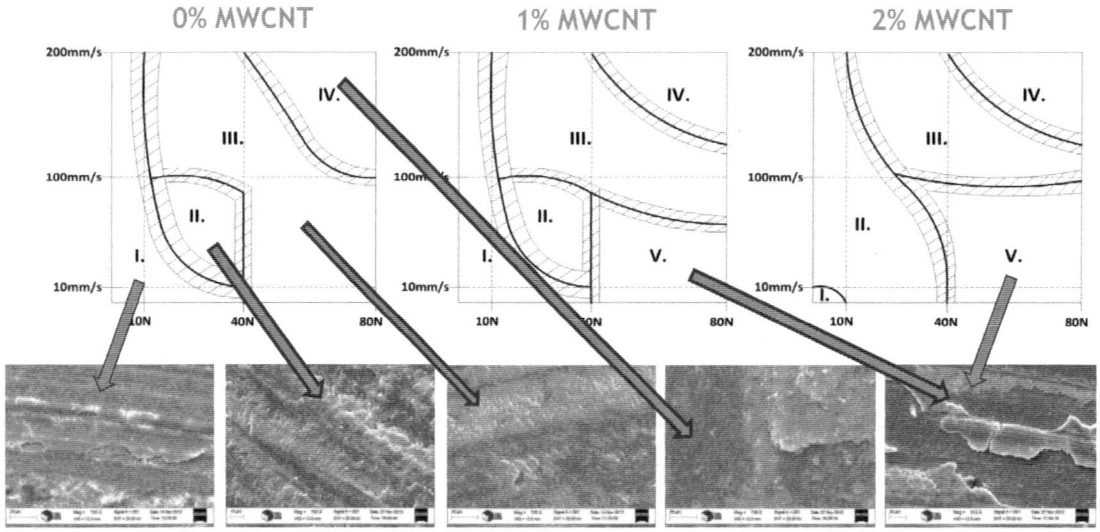

Figure 3. Wear mechanisms/wear types maps of the investigated materials

In case of monolithic samples the ranges I. – IV. appeared, while in case of composite samples the range V. additionally occurred. In general the plastic deformation as a controlling wear mechanism turned up in the I-IV.regions, while in region V. the massive tribofilm formation was accompanied by a brittle fracture. The wear maps of samples containing 0 and 1 wt% MWCNT were very similar. The only difference was observed in the loading range of 40-80 N at low (v=10 mm/s) sliding velocity where a new wear regime, i.e. the Region V. appeared for the samples containing 1 wt% MWCNT.

This happened at the expense of Regions III., as well as of the ability for plastic deformation. In case of samples containing 2 wt% MWCNT the area of Region II. is enlarged, demonstrating that this amount of MWCNT can serve as a solid lubricant lowering the friction coefficient and the wear damage. At the same time with an increasing MWCNT content the Region I. is changed by region V. at the low sliding speeds. It can be explained by the higher sensitivity to surface cracking due to the higher porosity developing during sintering caused by the insufficient dispersion of the MWCNT particles in the matrix. Increasing the sliding velocity at higher loads the temperature of the contact area is increasing and a more massive tribofilm formation can be experienced. The fatigue wear is changed by a more intensive plastic deformation resulting in abrasive wear. At the highest loads (F= 80 N and v = 200 mm/s) the oxide film formation process is competing with a melt wear process.

4. Summary

A novel proposal for classification and description of wear processes – such as wear mechanisms and wear types – has been worked out. Using this classification a systematic analysis of wear mechanisms controlling the wear damage process of monolithic and MWCNT reinforced Si_3N_4 ceramics was completed investigated in different pin-on-disc test conditions. The observations have been summarized in two-dimensional wear maps as a function of applied load and sliding velocity.

For the monolithic and even for the composite samples three main wear mechanisms – plastic deformation, tribochemical wear and brittle fracture could be identified –while five different wear types – tribofilm formation, fatigue wear, delamination, abrasive wear and melt wear – appeared. Each of the investigated materials showed more or less plastic deformation, but MWCNT composites due to the more porous matrix tended to brittle fracturing at low sliding speeds and high loads.

The most advantageous wear behavior is represented by the Region II. wear tribofilm of good plastic deformation capability and low friction coefficient provides to realize the lowest wear damage.

Si_3N_4 ceramics containing 0 and 1 wt% MWCNT possess a very similar wear mechanism maps, while in case of 2 wt% MWCNT composites the carbon additive could serve as a solid lubricant.

The composed 2D wear maps can contribute to a more easy and purposeful tribological application of these new composites, and can help a more reliable prediction of their wear behavior.

References

[1] Cs. Balázsi, Z. Kónya, F. Wéber, L. P. Biró, and P. Arató, 2003.*Mat. Sci. Eng.* C **23**/6-8 1133
[2] J. Tatami, T. Katashima, K. Komeya, T. Meguro, and T. Wakihara 2005.*J. Am. Ceram. Soc.* **88** 2889
[3] B. Fényi, N. Hegman, F. Wéber, P. Arató, and C. Balázsi, 2006 *Materials and technologies of the future* **139**. 6. sz. 41 (in Hungarian)
[4] Cs. Balázsi, et al., 2006 *Composites* Part B **37** 418
[5] Cs. Balázsi, Z. Kónya, F. Wéber, L. P. Biró, P. Arató 2003 *Materials Sceince and Engineering* C **23** 1133
[6] M. B. Maros, 2010 Mechaical properties of Si_3N_4 technical ceramics ISBN 978-963-7064-24-1 pp. 233-241. (in Hungarian)
[7] Cs. Balázsi, 2010 *Materials Science Forum* **659**. ISSN 0255-5476 121
[8] J. Pfeifer, et al. 2010.*Materials Sceince Forum* ISSN 0255-5476 Vol. **659**. 235
[9] C.P.Dogan, J. A. Hawk 2001, *Wear* **250**, 256
[10] ISO 20808: Fine ceramics (advanced ceramics, advanced technical ceramics), Determination of friction and wear characteristics of monolithic ceramics by ball-on-disc method, 2004.
[11] Zs. Koncsik, Systematization of tribological properties of technical properties. Technical Letters 2010., HU ISSN 2060-7954, pp 251-257. (in Hungarian)
[12] Zs. Koncsik, Examination of tribological and mechanical properties of Si_3N_4 technical ceramics, PhD thesis, Miskolc, 2013. pp 1-240. (in Hungarian)

Combustion of layers inorganic systems under rotation to produce composite and gradient materials

B Nurakhmetov, A Sergaziyev, N Sabirov, A Baideldonova, L Mukhina, G Ksandopulo

Institute of Combustion Problems, Almaty, Kazakhstan

E-mail: milabrega@yandex.ru

Abstract. The new way of production of gradient materials is presented. Course of self-propagating high-temperature synthesis reactions in layered systems under the influence of centrifugal force and some methods of analysis of the process and products of reaction is described.

1. Introduction

This method is based on carrying out SHS in a centrifugal installation – a high-temperature centrifuge [1]. Three identical cylindrical reactors rotate at up to three thousand revolutions per minute (see figure 1). The horizontal orientation was chosen to avoid the non-uniformity of the mass forses field [2]. As usual, the ignition is made from the center and then the SHS reaction wave moves to the edge of the circle, but the reverse option is also possible. A quartz tube is placed inside the reactor, and the sample, the mixture of the base metal oxide powders and aluminum, is placed in the tube. The reaction of restoring metal with aluminum occurs. The process proceeds very rapidly and usually requires the retardation by adding a dilutant to the initial mixture – the final reaction product, or one of the components in excess.

2. Formulation of the problem and method of solution

Under steady-state conditions due to the effect of gravity, we obtain a layered product. In rotating, the picture changes and it is possible to obtain gradient materials. In the rotating system, the clusters of combustion products have a large initial acceleration [3]. The less inert ones fly to the walls of the reactor, and the heavy ones rush along the radius. They hit the surface of the compressed mixture of powders. As the mixture has porosity, the clusters go inside the sample, through a surface layer. The penetration depth depends on the size and temperature of clusters, i.e. their viscosity. The kinetic energy of these particles at the time of blow is converted into the thermal energy and along with a simple heat and mass transfer, it increases the temperature of the pre-reaction mixture. Getting on the boundary between two grains of different components, the glowing clusters originate the point fire hazards (see figure 2). All this leads to the continued expansion of the combustion boundaries, the acceleration of its progress along the axis of the sample, the continuous increase in temperature (the accelerated super adiabatic regime – a non-chain distribution of the SHS reaction) [4].

Due to the high-energy efficiency of the reaction, we have the opportunity to make the experiments with such pairs of components that cannot be ignited under normal conditions. Placing the two different reaction mixtures consistently, we initiate combustion in the combustible pair of the components, and the clusters of this reaction's products ignite the second flash-resistant layer. There can be a few such layers there. In the high-energy active layers, the mixture of tungsten oxide, nickel, and iron with powder aluminum is used, and in the low-energy ones, the mixture of boron oxide or silicon aluminum is used. The sample of the four-layer system is shown in figure 3. The final product is a gradient material with a gradual transition between metal and ceramic. The composition is selected due to the relative value of the heat effects of the reaction [5].

Samples of the gradient materials received at different compositions of initial mix and different frequencies of rotation of installation are given in figure 4.

After the active reactive layer, the attacked layer, for example nanographite, can be considered. To estimate the kinetic energy of the metal clusters, we have sought to catch, get, and examine the particles. As the screen the compacted graphite, boron oxide and a plexiglass tablet were used. The experiment with plexiglass gave the interesting results, the metal clusters of different sizes went to the different depths, and are easily visible to the naked eye. After sawing the sample, we got the pictures of the particles (see figure 5), it gave an impression of their average size, let to calculate the estimated kinetic energy component, which can then be used for the theoretical calculations of the SHS wave temperature.

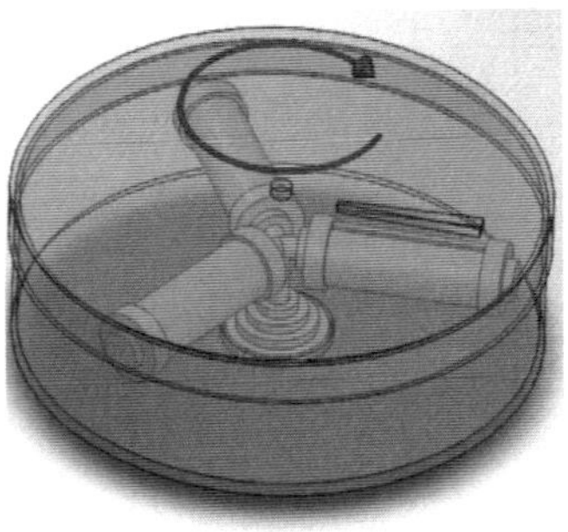

Figure 1. High-temperature centrifuge

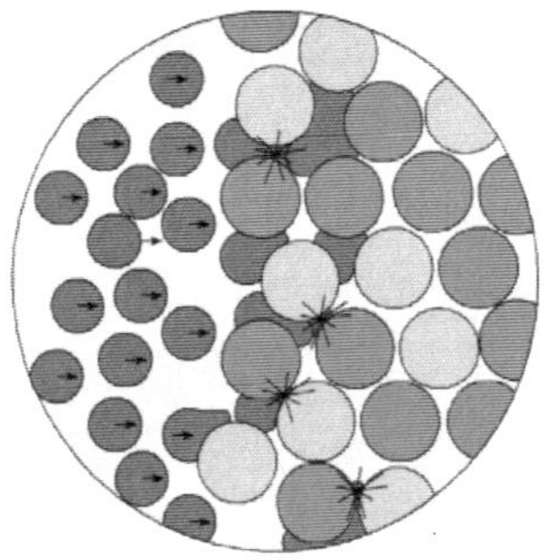

Figure 2. Penetration of clusters

structure of a layer	relative height
WO_3 +Al + 30% Al_2O	¼
B_2O_3 + Al	¼
WO_3 +Al + 50% Al_2O_3	¼
B_2O_3 + Al	¼

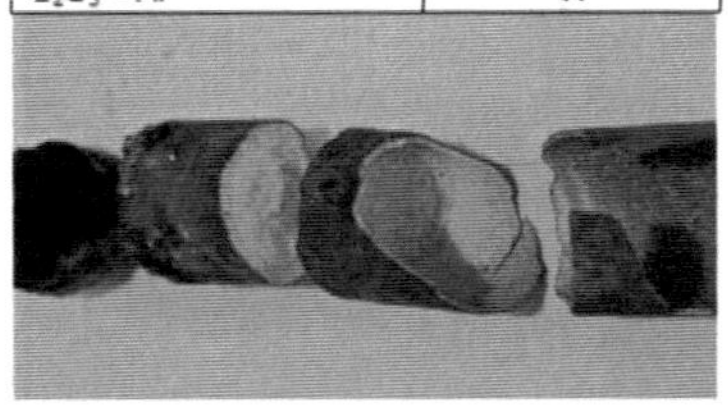

Figure 3. Example of layered system

Figure 4. Examples of the received gradient structures

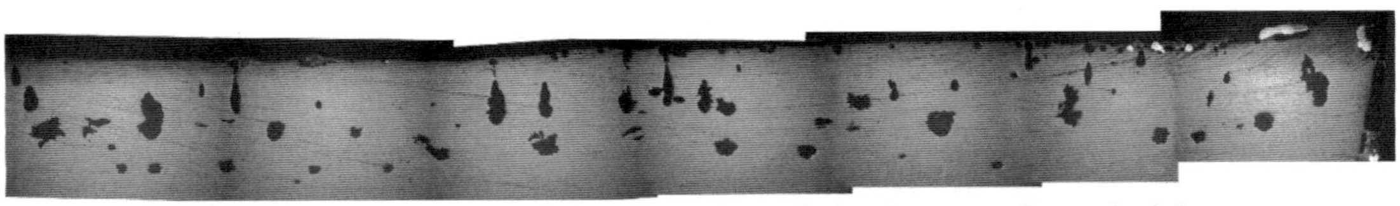

Figure 5. Photos of the particles of metal caught by the screen from plexiglass

After the reaction the various analyses of the products were carried out, that is, the X-ray, the method of electronic optical fluoroscopy etc. However, the particular interest is in getting the information in a dynamic mode, in that short period, in which the process takes place. First, we are interested in such parameters as the velocity of the SHS wave and the temperature as a function of the coordinates at the time. Using the high-speed video recording allowed us to solve the first problem, and receive the additional data on the frequency of the reactors' rotation. To shoot in one of the three reactors there is one slot and another slot is made in the cover of the installation (see fig. 1), in the course, we obtain the images at the moment of the geometrical coincidence of these two "windows". At the speed of 400 shots per second, every 14th or 22th shot is informative depending on the rotational speed. In some pictures, the flame boundary is distinctly seen. Graphic way of measurement of speed of the front of burnings given in figure 6. The task of measuring the temperature was more difficult. We cannot use the thermocouples due to the inertia of their reaction and too high values of the measured temperature. The pyrometric method also takes time, and the reaction is too fast for it. We attempted to use pictures of the process in real time, to treat them in software and thus to measure the temperature.

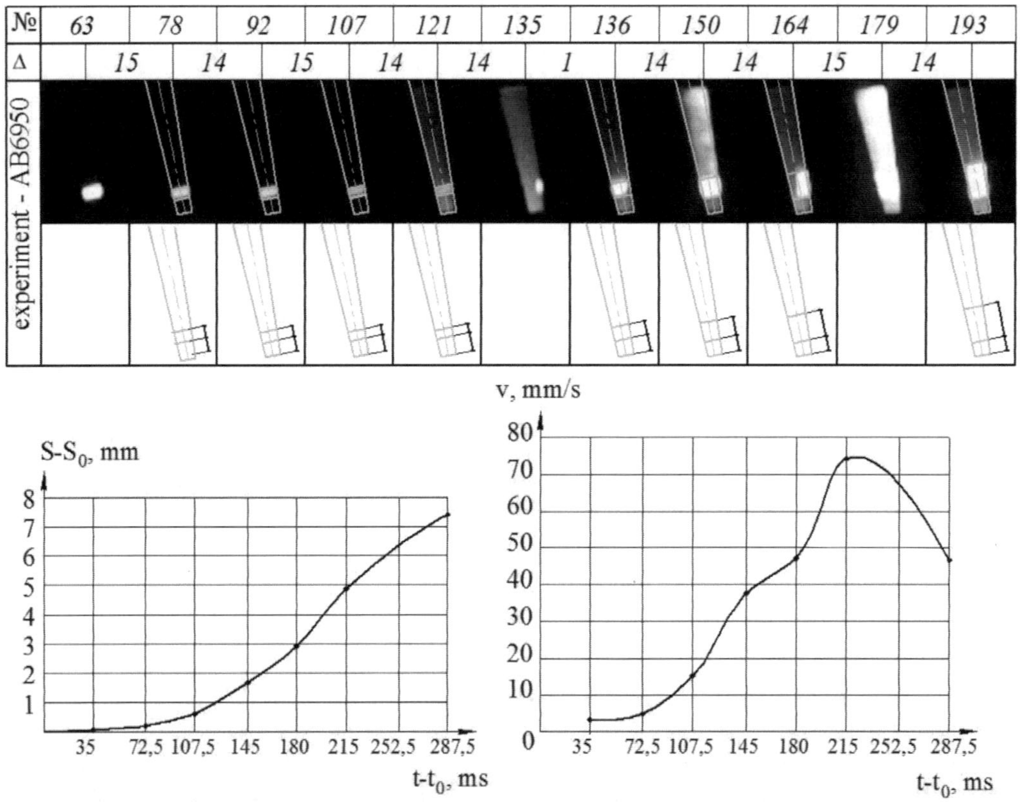

Figure 7. Graphs of dependence of movement and speed of the front from time (Δ - quantity of empty frames between two informative ones, as number of frames per second is 400 that each frame- 1/400 seconds; in this example time interval is equal 14/400 or 15/400 s)

3. Conclusion

For better visual representation of the process, we give some numbers. The increasing of the speed from 500 to 5000 revolutions per minute gives an advantage over the force of gravity settling from 80 to 1000 times. The normal front propagation velocity at rest is up to 10 centimeters per second. At the boundary width of 2 millimeters the characteristic time of the process is 20 milliseconds, and the time of contact of the metal particles on the boundary is 20-60 nanoseconds. The maximum temperature of the aluminothermy process reaches 2500-3500 Kelvin. The tungsten particle radius of 1 millimeter has a mass of 80 milligrams, the kinetic energy 718 mill joules per second, speed of 90 meters per second. If the content of tungsten metal in the first part of the reactor is 45 grams, the total capacity of the stream of particles passing through the attacked layer of the thickness of 2 millimeters per 20 milliseconds is 2 mega joule per second.

References

[1] Ksandopulo G.I. 2011 *Int. J. Self-Propag. High-Temp. Synth.* **20**. 220

[2] Ksandopulo G.I, Baideldinova A.N. 2011 *Combustion and Plasma Chemistry*. **9** No 4. 241

[3] G.I. Ksandopulo, V.N. Shevchenko, A.N. Baideldinova. High-temperature centrifuge / RoK Patent No 68317 as of 26.05.2010

[4] Ksandopulo G., Baydeldinova A., Ainabayev A., Arkhipov M., Omarova K 2011 *Eurasian ChemicoTechnological Jornal*. **13**, №3-4 155

[5] Kcandopulo G., Aynabaev A., Arkhipov M., Baydeldinova F., Omarova R. 2013.Out-of-furnace synthesis of high-temperature ceramic materials in the revolving reactor, IOP Conference Series: Materials Science and Engineering (MSE) **47**

IOP Conf. Series: Materials Science and Engineering **123** (2016) 012035 doi:10.1088/1757-899X/123/1/012035

Bifurcation and "self-organization" of a system

M Aldabergenov[1], G Balakaeva[2]

[1]Al-Farabi Kazakh National University, Almaty, Kazakhstan
[2]K.I.Satpaev Kazakh National Technical University, Almaty, Kazakhstan

E-mail: aldabergenov_m@mail.ru

Absrtact. Triangulation of a multicomponent system was shown on example of the CaO-SiO_2-H_2O system. "The Gibbs function normalized to the total number of electrons" was applied in order to reflect all the possible transformations of components of the system at non-equilibrium as well as at equilibrium conditions. The bifurcation points of the system are located at the each intersection of the line connected compositions of the interacting components and the stable secant line. It was shown the possibility of the "self-organization" processes which is based on the exchange as well as that reactions.

Introduction

Natural and technological objects are represented by multicomponent systems. Chemical and physico-chemical interactions of these systems are defined by the substance microstates. Geometric images of the multicomponent systems can be represented by the phase diagram which is based on the continuity and consistency principles [1] which reflects the transformation processes between components of a system. The "structure-property" diagram is the precise geometric model of the complex functions which are establishing the relationship between temperature, volume, concentration and the other physical and chemical factors which are determining the state of the open system. In other words, the "structure-property" diagram merges together chemical transformations of matter and space geometric transformations into one whole. The phase diagram allows to define phase boundaries of the different phases existing in the system as well as to reveal not pronounced chemical processes. Additionally it allows to high light the weak interparticle interactions that do not lead to the formation of new compounds or to the decomposition of existing phases [2]. The phase diagrams enables widely consider problems of physical chemistry and chemical thermodynamics, reflecting modern ideas about the properties of individual substances as well as their mixtures, phase and chemical equilibria, and surface phenomena jointly. They are widely used in search of new intermetallic compounds [3-5], the glass-forming systems in the crystalline, glassy and molten states [6-8], solid solutions [9,10], predicting of the properties of metastable states [11-13] , analysis of phase transitions and critical phenomena [14,15], the fabrication of composite materials [16,17], the processes of separation and purification of substances [18,19], the development of algorithms for constructing phase diagrams [20-21], and others [22-25] presently.

Study of the phase diagrams implies partitioning of the multidimensional figures on individual cells - media of the invariant points which are called triangulation. The topological [1,2], geometric [26,27] and thermodynamic [28,29] triangulation techniques are available at this moment. The main feature of the systems triangulation is the determination of the constituents forming only eutectic composition (single phase units). The constituents of a single phase unit do not form new compounds and do not mutually interact, but those of different phase units interact with each other. Therefore the system triangulation gives possibility to determine the phase transitions occurring during, for example, exchange reactions, reactions of hydration and dehydration, reactions of polycondensation, conversion and complex formation. It allows predicting all the way of bifurcation of components and predict the existence of new compounds additionally.

The triangulation of the multicomponent open CaO-SiO_2-H_2O system was applied, as a collection of equilibrium and nonequilibrium states of the system components. A method of determination of the bifurcation points and "self-organization" of systems is developing in the present paper.

1. Method, results and discussion

Triangulation of the CaO-SiO_2-H_2O system (see Fig.1) is based on the Kurnakov theory [1] using the "the Gibbs function normalized to the total number of electrons" [30]. In the CaO-SiO_2-H_2O system the formation of 86 compounds was considered (Table I). The corresponding thermodynamic parameters were calculated using the ion increments method by Aldabergenov's increments [30]. Our calculations show good consistency with the reference data presented by Glushko VP [31], the relative error does not exceed 0.9%. The Gibbs free energy of formation of many compounds presented in Table 1 were calculated for the first time.

A new concept of "The Gibbs function normalized to the total number of electrons" ($\Delta\bar{G}^{o}_{298}$) [32] was proposed. Its physical meaning consists in determination of the chemical bonding as a collective effect of the electron-nuclear interaction. It is calculated by dividing of the Gibbs free energy of formation value of the compound to the total number of electrons in this compounds. In other words this function determines the density of the formation energy of the compound per electron. "The Gibbs function normalized to the total number of electrons" appropriately responds on the all changes occurring in the crystal structure of a compound. Additionally the function allows to assess compounds' reactivity and stability [32]. It also can determine the value of the chemical potential and affinity, calculate entropy's production and flow, calculate: - thermodynamic force and flux, - thermodynamic "probability" of interaction, - degree of conversion of components in a chemical reaction. Finally using the "The normalized Gibbs function" allows to perform the triangulation of multicomponent systems and establish the reaction mechanisms.

The triangulation is performed by partitioning of the large basic triangle into several small "subtriangles", which are named by Kurnakov as simplex [1]. In Ref. [28] A.S.Trunin calls single phase unit (SPU). SPU's are characterized by the formation of eutectic composition. Compounds belonged to one single SPU do not interact with each other, however those belonged to different SPU engage in interaction. In similar way it is possible to consider any interaction between components of the system.

Let us consider the interaction between $Ca_2SiO_4.4H_2O$ (compound #42) and $Ca_3Si_2O_6.(OH)_2.2H_2O$ (compound #48) (see Table 1) in the exchange reactions. For this we draw a straight line 42-48 connecting the corresponding compositions, which intersects the following stability secants: 18-69; 39-69; 39-78; 38-78; 78-17; 78-37; 78-79; 69-79; 69-17; 69-82 (Fig. 1). The points of the intersections are called Radishchev's points of conversion [33]. At these points the system becomes unstable and new solutions of evolution appear due to bifurcations. Therefore these points are also named as bifurcation points. The line 42-48 indicate that in the presence of a small displacement of the initial components of the system is abruptly switching between the stable states. These points can be named as symmetric bifurcations [34].

At Radishchevs' points [33] of conversion (or bifurcation points) the following reaction occur (we present only two reactions). The corresponding numbers are presented in Fig.1 for convenience.

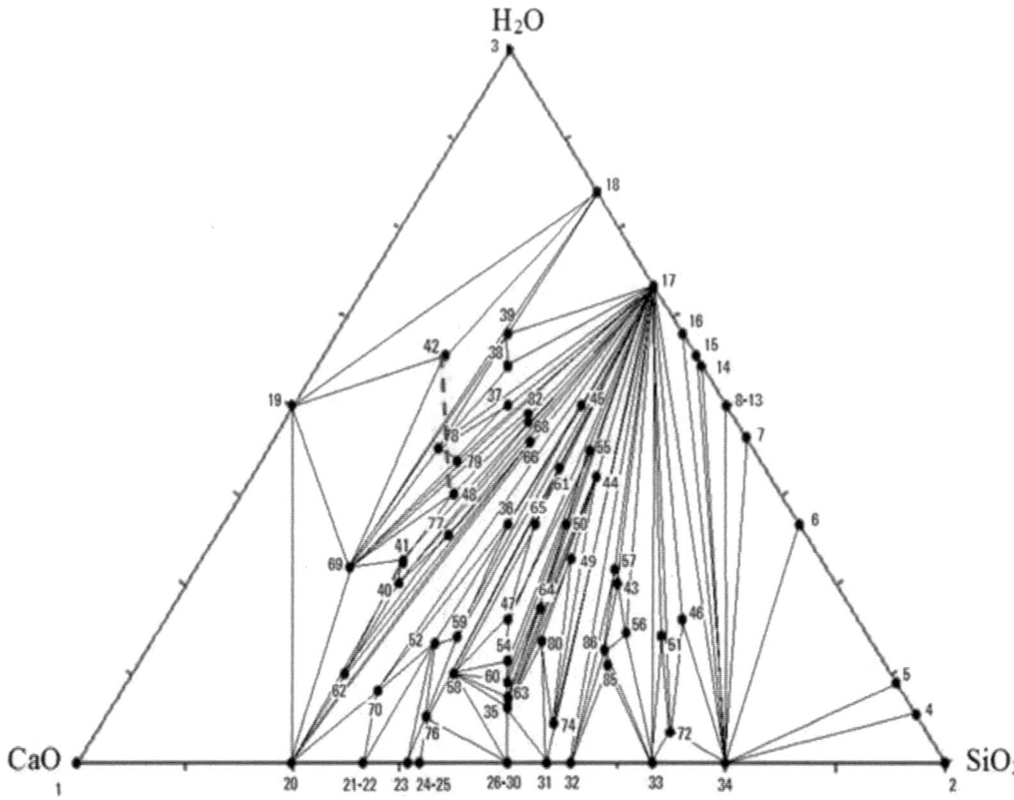

Figure 1. Triangulation system CaO-SiO2-H2O
(Compounds #48 and #53 have identical compositions. Similar coincides are existing for other compounds #66 and #67; #59 and #73; #49 and #75; #65 and #81; #35 and #71, #83; #41 and #84; v8 and #9, #10, #11, #12, #13; #21 and #22; #24 and #25; #26 and #27, #28, #29, #30. Therefore only one number is shown for the coincided compounds)

Table 1. The values of the Gibbs free energy of formation ($-\Delta_f G_{298}$) and "The Gibbs function normalized to the total number of electrons" ($-(\Delta\overline{G}^{o}_{298})$ of the compounds existing in the CaO-SiO_2-H_2O system

Compounds	$-\Delta_f G_{298}$, kJ/mol		% av.	$-(\Delta\overline{G}^{o}_{298})$, kJ/mol.el
	by increments	[31]		
1	2	3	4	5
1. CaO		603.5		21.55
2. SiO_2		892.6		29.75
3. H_2O		237.2		23.72
4. $H_2Si_{14}O_{29}$	12437.23			28.92
5. $H_2S_{i8}O_{17}$	7274.17			29.10
6. $H_2Si_2O_5$	2111.11			30.16
7. $H_{10}Si_6O_{17}$	7113.49			30.93
8. H_2SiO_3	1250.60			31.26
9. $H_4Si_2O_6$	2501.19			31.26
10. $H_6Si_3O_9$	3751.79			31.26
11. $H_8Si_4O_{12}$	5002.38			31.26

1	2	3	4	5
12. $H_{10}Si_5O_{15}$	6252.98			31.26
13. $H_{12}Si_6O_{18}$	7503.58			31.26
14. $H_{10}Si_4O_{13}$	5392.47			31.72
15. $H_8Si_3O_{10}$	4141.87			31.86
16. $H_6Si_2O_7$	2891.28			32.13
17. H_4SiO_4	1640.68			32.81
18. H_8SiO_6	2099.28			29.99
19. $Ca(OH)_2$		897.5		23.62
20. Ca_3SiO_5	2788.84	2785.1	0.13	24.46
21. Ca_2SiO_4	2185.34	2191.3	0.27	25.41
22. $Ca_6Si_3O_{12}$	6460.24			25.04
23. $Ca_8Si_5O_{18}$	9548.14			25.53
24. $Ca_3Si_2O_7$	3724.29	3756.7	0.87	25.86
25. $Ca_6Si_4O_{14}$	7362.80			25.57
26. $CaSiO_3$	1538,95	1543.9	0.32	26.53
27. $Ca_2Si_2O_6$	3077.90			26.53
28. $Ca_3Si_3O_9$	4616.85			26.53
29. $Ca_6Si_6O_{18}$	9233.70			26.53
30. $Ca_7Si_7O_{21}$	10772.65			26.53
31. $Ca_5Si_6O_{17}$	8587.31			26.84
32. $Ca_3Si_4O_{11}$	5509.41			27.01
33. $CaSi2O_5$	2431.51			27.63
34. $CaSi_3O_7$	3324.07			28.17
35. $CaSiO_3$.0,16H_2O	1574.15			26.41
36. $CaSiO_3.H_2O$	1758.95			25.87
37. $CaSiO_3.2H_2O$	1978.95			25.37
38. $CaSiO_3$.2,5H_2O	2088.95			25.17
39. $CaSiO_3.3H_2O$	2198.95			24.99
40. $Ca_2SiO_4.H_2O$	2405.34			25.06
41. Ca_2SiO_4.1,17H_2O	2442.74			25.00
42. $Ca_2SiO_4.4H_2O$	3065.34			24.33
43. $CaSi_2O_5.H_2O$	2651.51			27.06
44. $CaSi_2O_5.2H_2O$	2871.51			26.59
45. $CaSi2O_5.3H_2O$	3091.51			26.20
46. $Ca[Si_3O_6](OH)_2$	3553.37			27.76
47. $Ca_2Si_2O_6.H_2O$	3297.90			26.17
48. $Ca_3Si_2O_6(OH)_2.2H_2O$	4393.59			25.25
49. $Ca_2Si_3O_8.2H_2O$	4367.57			26.31
50. $Ca_2Si_3O_8$.2,5H_2O	4477.57			26.18
51. $Ca_2Si_5O_{12}$.1,5H_2O	6085.58			27.54
52. $Ca_3Si_2O_7.H_2O$	3944.29			25.61
53. $Ca_3Si_2O_7$. $3H_2O$	4384.29			25.20
54. $Ca_3Si_3O_9.H_2O$	4836.85			26.29
55. $Ca_3Si_6O_{15}.7H_2O$	8834.53			26.45
56. $Ca_3Si_6O_{13}(OH)_4$	7753.13			27.30
57. $3Ca_3[Si_6O_{13}](OH)_4.4H_2O$	24139.39			27.06
58. $Ca_4Si_3O_{10}.H_2O$	5473.24			25.82
59. $Ca_4Si_3O_{10}$.1,5H_2O	5583.24			25.73

1	2	3	4	5
60. $Ca_4Si_4O_{12}.H_2O$	6375.80			26.35
61. $Ca_4[Si_6O_{13}](OH)_6.4H_2O$	9508.82			26.27
62. $Ca_5Si_2O_9.H_2O$	5108.40			24.33
63. $Ca_5Si_5O_{15}.H_2O$	7914.75			26.38
64. $Ca_5Si_6O_{17}.3H_2O$	9247.31			26.42
65. $Ca_5Si_6O_{17}.5,5H_2O$	9797.31			26.13
66. $Ca_5(Si_6O_{18}H_2).8H_2O$	10576.61			25.80
67. $Ca_5Si_6O_{17}.9H_2O$	10567.31			25.77
68. $Ca_5Si_6O_{17}.10H_2O$	10787.31			25.68
69. $Ca_6[Si_2O_7](OH)_6$	6222.69			24.12
70. $Ca_6Si_3O_{12}.H_2O$	6680.24			24.93
71. $Ca_6Si_6O_{18}.H_2O$	9453.70			26.41
72. $Ca_7Si_{16}O_{39}.H_2O$	18768.35			27.36
73. $Ca_8[Si_6O_{17}](OH)_6$	11085.71			25.54
74. $Ca_8Si_{10}O_{28}.H_2O$	14316.72			26.81
75. $Ca_8Si_{12}O_{30}(OH)4.6H_2O$	17960.44			27.05
76. $Ca_9Si_6O_{21}.H_2O$	11221.31			25.39
77. $Ca_9Si_6O_{21}.7H_2O$	12541.31			24.98
78.$Ca_9(Si_3O_9H)(Si_2O_7H)(OH)_8.\ 6H_2O$	12651.03			24.71
79. $Ca_9(Si_6O_{18}H_2)(OH)_8.\ 6H_2O$	13467.81			24.85
80. $2Ca_{10}[Si_{12}O_{31}](OH)_6.3H_2O$	36385.04			26.56
81.$Ca_{10}[Si_{12}O_{31}](OH)_6.8H_2O$	19622.52			26.16
82.$Ca_{10}[Si_{12}O_{31}](OH)_6.18H_2O$	21822.52			25.67
83. $Ca_6[Si_6O_{17}](OH)_2$.	9463.00			26.43
84. $Ca_{12}[Si_6O_{17}]_2(OH)_4.12Ca(OH)_2$	29696.00			25.34
85. $Ca_{14}Si_{24}O_{58}(OH)_8.2H_2O$	31828.10			27.16
86. $Ca_{14}Si_{24}O_{61}(OH)_2.6H_2O$	32020.20			27.09

<u>42 + 48 = 18 + 69:</u>

$$3[Ca_2SiO_4.4H_2O] + 2[Ca_3Si_2O_6(OH)_2.2H_2O] = 3H_8SiO_6 + 2[Ca_6Si_2O_7(OH)6] \quad (1)$$

ΔrG^o = -760.02 kJ (-253.34 kJ/mol per $Ca_2SiO_4.4H_2O$) (according Gess law)

Each chemical reaction is characterized by affinity (A), which can be found using the theory of de Donder [35]: A = 4.53 kJ / mol.el.

<u>42 + 48 = 39 + 69:</u>

$$9[Ca_2SiO_4.4H_2O] + 7[Ca_3Si_2O_6(OH)_2.2H_2O] = 15[CaSiO_3.3H_2O] + 4[Ca_6Si_2O_7(OH)_6] \quad (2)$$

ΔrG^o = 468.18 (52.02 kJ/mol per $Ca_2SiO_4.4H_2O$). A = -0.47 kJ/mol.el

The first reaction occurs with a large gain in energy.

Let us consider the dehydration process of the $Ca_3Si_2O_6(OH)_2.2H_2O$ compound (see #48 in Fig. 1). The dehydration process obeys the following postulates [36]:

1. The water removal takes place only along the phase equilibria line which is connecting the compound with the water corner (see Fig.1);
2. If the compound under going dehydration is not directly related to the pole of water, then at the reaction of conversion occurs or bifurcation take place for compounds related to the pole of this compound.

The $Ca_3Si_2O_6(OH)_2.2H_2O$ compound (#48) is not directly connected with the water corner and it is not affected by the conversion reaction then. However, in accordance with a second postulate, the dehydration process begins from the decomposition (bifurcation) of $Ca_3Si_2O_6(OH)_2.2H_2O$ to compounds #41, #69, #17 and #82 which are located on the stability diagonals connected with the corner of the parent silicate. Additionally so-called asymmetric bifurcation also occurs, which is associated with the conversion of the original salt into the four following compounds of the system:

48 = 41 + 69 + 17 + 82:

$$126[Ca_3Si_2O_6(OH)_2.2H_2O] = 100[Ca_2SiO_4.1.17H_2O] + 13[Ca_6Si_2O_7(OH)_6] + 6H_4SiO_4 + 10[Ca_{10}Si_{12}O_{31}(OH)_6.18H_2O] \quad (3)$$

$\Delta rG^o = 354.09$ kJ (2.81 kJ/mol per original salt). A = 82.35 kJ/mol.el.

In this reaction H4SiO4 is connected with the water corner only and subjected to the dehydration process with the water elimination by the reaction (4) and (5):

17 = 18 + 16:

$$5H_4SiO_4 = H_8SiO_6 + 2H_6Si_2O_7 \quad (4)$$

$\Delta rG^o = 321.56$ kJ (64.31 kJ/mol on the initial salt). A = 29.31 kJ/mol.el.

The evolving of water from H4SiO4 occurs only by the conversion reaction:

18 = 3 + 17:

$$H_8SiO_6 = 2H_2O + H_4SiO_4 \quad (5)$$

ΔrG^o =-15.8 kJ/mol on the initial salt A = 26.54 kJ/mol.el.

The conversion reaction may be continue until silica is produced:

16 = 17 + 15:

$$2H_6Si_2O_7 = H_4SiO_4 + H_8Si_3O_{10} \quad (6)$$

ΔrG^o= 0 kJ/mol on the initial salt. A = 32.54 kJ/mol.el.

15 = 16 + 14:

$$2H_8Si_3O_{10} = H_6Si_2O_7 + H_{10}Si_4O_{13} \quad (7)$$

ΔrG^o= 0 kJ/mol on the initial salt. A = 32.54 kJ/mol.el. rG^o= 0 kJ/mol on the initial salt. A = 31.99 kJ/mol el. and so on until SiO_2 is produced.

As it is shown in (6) and (7), the reactions are characterized by high degree of affinity and the conversion process does not actually require for external influence. The conversion reactions along the SiO_2-H_2O (2-3) line leads to the formation of various silicic acids and SiO_2.

The compounds formed by the reaction (3), also undergo the bifurcation transformations further in accordance with the second postulate, for example:

69 = 20 + 19 + 42 + 18 + 39 + 78 + 79 + 17 + 82 + 48 + 41:

$$143[Ca_6Si_2O_7(OH)_6] = 137Ca_3SiO_5 + 211Ca(OH)_2 + 2[Ca_2SiO_4.4H_2O] + H_8SiO_6 + CaSiO_3.3H_2O + [Ca_9Si_3O_9HSi_2O_7H(OH)_8.6H_2O] + [Ca_9Si_6O_{18}H_2(OH)_8.6H_2O] + 20H_4SiO_4 + [Ca_{10}Si_{12}O_{31}(OH)_6.18H_2O] + 100[Ca_2SiO_4.1.17H_2O] + [Ca_3Si_2O_6(OH)_2.2H_2O] \quad (8)$$

ΔrG^o= 0 kJ/mol on the initial salt. A = 32.54 kJ/mol.el. rG^o = -21450.37 kJ (-143.0 kJ/mol to original salt). A = 261.56 kJ/mol.el.

In this reaction the products H_4SiO_4 and $Ca(OH)_2$ are connected with the water corner. The dehydration process of H_4SiO_4 proceeds by reactions (4) and (5). The dehydration process of $Ca(OH)_2$ proceeds according to the following conversion mechanism with the formation of water and calcium oxide:

19 = 1 + 3:

$$Ca(OH)_2 = CaO + H_2O \quad (9)$$

ΔrG^o= 56.8 kJ/mol on initial salt. A = 21.65 kJ/mol.el.

After wards the dehydration processes of $Ca_{10}Si_{12}O_{31}(OH)_6.18H_2O$ (#82), $Ca_9Si_6O_{18}H_2(OH)_8.6H_2O$ (#79), $Ca_9Si_3O_9HSi_2O_7H(OH)_8.6H_2O$ (#78), $Ca_2SiO_4.1.17H_2O$ (#41) and $CaSiO_3.3H_2O$ (#39) continue in accordance with the second postulate.

Analysis of the dehydration reaction of $Ca_3Si_2O_6(OH)_2.2H_2O$ shows that water is able to be released from the system only along the lines 1-3 and 2-3. All the compounds formed by reactions (3), (8) [except for compounds H_4SiO_4 (#17), H_8SiO_6 (#18) and $Ca(OH)_2$ (#9)] undergo bifurcations in accordance with the second postulate.

Let us consider the case where the dehydration process of compound #48 was stopped under certain conditions, e.g. partial dehydration of $Ca_3Si_2O_6(OH)_2.2H_2O$. Under these conditions certain portions of water is removed from the system and the system goes out of equilibrium. The following compound remain in the system (e.g.): CaO (#1), $H_{10}Si_4O_{13}$ (#14), $H_8Si_3O_{10}$ (#15), $H_6Si_2O_7$ (#16), H_4SiO_4 (#17), H_8SiO_6 (#18), $Ca(OH)_2$ (#19), Ca_3SiO_5 (#20), $Ca_5Si_6O_{17}$ (#31), $CaSiO_3.2H_2O$ (#37), $CaSiO_3.3H_2O$ (#39), $Ca_2SiO_4.H_2O$ (#40), $Ca_2SiO_4.1,17H_2O$ (#41), $Ca_2SiO_4.4H_2O$ (#42), $Ca_3Si_2O_6(OH)_2.2H_2O$ (#48), $Ca_4Si_3O_{10}.H_2O$ (#58), $Ca_5Si_2O_9.H_2O$ (#62), $Ca_6[Si_2O_7](OH)_6$ (#69), $Ca_9(Si_3O_9H)(Si_2O_7H)(OH)_8.6H_2O$ (#78),$Ca_9(Si_6O_{18}H_2)(OH)_8.6H_2O$ (#79), $Ca_{10}[Si_{12}O_{31}](OH)_6.18H_2O$ (#82). All of these compounds are tending to the natural structure or the equilibrium state, i.e. to the "self-organization" through the exchange reactions:

$$3CaO + Ca_6Si_2O_7(OH)_6 = 3\ Ca(OH)2 + 2Ca_3SiO_5 \quad A = 2.41 \text{ kJ/mol.el} \quad (10)$$

And so on.

It should be noted that "self-organization" of the order and the structuring of systems are based on the exchange and conversion reactions. The "self-organization" process initiates structuring and, simultaneously, it gets stochastic. It is the result of a sequence of nonequilibrium phase transitions. A property of the "self-organizing" systems is the self-similarity of the different hierarchical levels [37]. In principle all the possible reactions between the components should be considered using triangulation of systems. It is necessary due to the thousands of reactions which can be represented by a group of compounds which can be formed and structured in a system.

Conclusions

The singular triangulation is a rational splitting of the primary simplex (triangle) on secondary triangles using singular secants. It helps to determine direction of reactions, to establish the nature and character of the individual phases in any part of systems and make conclusions about the character of the system under investigation. The most easiest way of triangulation can be carried out using so called "The Gibbs function normalized to the total number of electrons".

The dehydration reaction occurring in the CaO-SiO2-H2O system showed a possibility to become the natural structure as well as to reach the equilibrium state through "self-organization". It was established that "self-organization" occurs in the result of the exchange and conversion reactions.

Reference

[1] Kurnakov N.S. Selected labors. Moscow: 1960 *Academy of Science of USSR*, **565**
[2] Anosov V.J., Ozerov M.I., Fialkov U.Y. 1976 *Nauka* **504**

[3] Chen M., Hallstedt B., Gauckler L.J. 2005 *Solid State Ionics* **176** 1457
[4] Lutsyk V.I., Vorobyova V.P. 2011 *Advanced Materials* **13**. 191
[5] Kosjakov V.I. 2008 *Journal of inorganic chemistry* T.**53**. 6 1020
[6] Pavlyuchkov D., Savinukh G., Fabrichnaya O. 2013 *Advanced Engineering Materials* **15**, N7, 618
[7] Lutsyk V.I. 2012 *Bulletin* BSC SB RAS. **1** (5) 78

[8] Ushakov V.M., Borisova N.V. Vitrification and alkali silicate melts alkali borates / Materials XV Inter. Conf. on Chemical Thermodynamics in Russia. Moscow, 2005. - P.35.
[9] Trahtenberg B.F., Kenis M.S., Krestyanov V.I., Troshina L.V. Diagnostics of the main mechanisms of technological heredity // Foundry. 1999. №1. p.13-15.
[10] Kosjakov V.I., Shestakov V.A., Grachev E.V. Enumeration fusion diagrams of ternary systems with compounds of constant composition // Journal. Inorganic Chem. 2010. T.55. №4.- P.662-670.
[11] Nazmutdinov A.G., Nesterov T.N., Nesterov I.A., Vodenkova N.N. Investigation and prediction of properties of binary mixtures of organic substances / Materials XV Inter. Conf. on Chemical Thermodynamics in Russia. Moscow. 2005. - P.113.
[12] Anisimov M.P., Fominih E.G., Akimov S.V., Trilis A.V. Elements of topology region of existence of metastable states of the two-component vapor // Journal. Fiz. chem. 2010, T.84, №3, P.427-433.
[13] Ray A.K., Chalam M., Peters L.K. Homogeneous nucleation of binary vapours partially misciblein liguid state //J.Chem. Phys. 1986. V.85. №4. P.2161.
[14] Aliev A.M., Stepanov G.V. Effect of extractables on the critical parameters of supercritical fluids / Materials of XV Inter. Conf. on Chemical Thermodyn.in Russia. Moscow. 2005. - P.137.
[15] Anisimov M.P., Hopke P.K. Nucleation Rate Surface Topologies far Binary Systems //J.Phys. Chem. B. 2001, 105. P. 11817-11822.
[16] Wu P., Eriksson G., Pelton A.D. Critical Evaluation and Optimization of the Thermodynamic Properties and Phase Diagrams of the CaO-FeO, CaO-MgO, CaO-MnO, FeO-MgO, FeO-MnO, and MgO-MnO Systems//J.Am.Caram. Soc. 1993, 76(8). P.2065-2075.
[17] Aldabergenov M.K, Balakayeva G.T, Massimov K.K, Tusupbekov K.I. Composite borosilicate - conteynery for radioactive waste storage /Thesis.pres. VII All-Russian Scientific. Conf. "Ceramics and composite materials", Syktyvkar. - 2010.- P.7-8.
[18] Lutsyk V. Phase diagram as a tool of metallurgist and materials scientist //roceedings of 46th International October Conf. on Mining and Metallurgy. Bor, Serbia, 2014. P. 148-151.
[19] Klimova M.V. Trunin A.S. Modelling and identification of trees of four-phase reciprocal salt systems with different types of chemical interaction. Samara: Samara State Techn.Univ. 2005-120p.
[20] Lutsyk V.I., Vorob'eva V.P. Computer models of eutectic type Txy diagrams with allotropy. Two inner liquidus fields of two low-temperature modifications of the same component // Journal of Thermal Analysis and Calorimetry. 2010. V. 101. No 1. P. 25-31.
[21] Connell R.G. Tutorial on Flow Diagrams: A Tool for Developing the Structure of Multicomponent Phase Diagrams //J.Phase Equilibria and Diffusion. 1994. V.5. No 1.–P.6-19.
[22] Cheynet B., Bonnet C., Stankov M. GEMINI-DiagPlot: 2D & 3D ternary phase diagrams. CALPHAD. 2009; 33:312-6.
[23] Miura S. Geometrical approach to reaction schemes of multicomponent phase diagrams // J.Phase Equilib. Diffus. 2006. – P.34-36.
[24] Chuvakov A.V., Trunin A.S., Morgunova O.E. Theory and Methodology of differentiation of multicomponent systems. Samara: Samara State Techn.Univer. 2007.–117p.
[25] Lutsyk V.I. Heterogeneous design: phase diagram - microstructure - materials genotype // Phase Diagrams in Materials Science. Stuttgart, Germany: Materials Science International Services, GmbH. 2004. - P.304-306.
[26] Posypayko V.I. Methods of investigation of multicomponent systems.M.:Nauka,1978.- 255 p.
[27] Trunin A.S. A comprehensive research methodology of chemical interaction and heterogeneous equilibria in multicomponent salt systems // Journal of Applied Chem.-1983. T.28, V.1.- P.174-179.

IOP Conf. Series: Materials Science and Engineering **123** (2016) 012035 doi:10.1088/1757-899X/123/1/012035

[28] Aldabergenov M., Balakaeva G., Kokibasova G. Triangulation of the MgO-Al2O3-H2O system as applied to the hydration of magnesium aluminates and the dehydration of their hydrates //J.of Physical Chemistry. 2002, Vol.76, No. 5. P.791-795.
[29] Aldabergenov M.K, Balakayeva G.T Triangulation of multicomponent systems based on thermodynamic functions / "Actual problems of modern science", Proceedings of 3 Intern..forum, Samara, 2007. - Part 9., Issue 1. - P.39-43. http://ROSMAN.NET.RU.
[30] Aldabergenov M., Balakaeva G., Kassenov B. Applied Chemical Thermodynamics. Berlin .: LAP Lambert, 2012. - 350p.
[31] Glushko V.P. (Ed.), Thermal constants of compounds. Moscow, VINITI, 1970, Vol.4, 320p.
[32] Balakaeva G., Aldabergenov M. The Gibbs function normalized to the total number of electrons // Journal of Materials Science and Engineering B. 2 (6), 2012. P.394-403.
[33] Radishchev V.P. Multicomponent sistems.- M .: ANSSSR, 1964.- T. I.- 540 p.
[34] Prigogine I., Kondepudi D. Modern Thermodynamics from heat engines to dissipative structures. Moscow: Mir, 2002. - 462p.
[35] De Donde T. L'Affinite. - Paris: Gauthiers-Villars, 1927.
[36] Aldabergenov M.K., Balakaeva G.T. Reactions of dehydration and polycondensation the system K2O-P2O5-H2O // Doklady NAS RK - 1994, №2.- P.54-58.
[37] Haken H. Information and self-organization. Moscow: Mir, 1991. - 240 p

4D space models of quaternary systems for the phase diagrams graphics correction

V I Lutsyk[1,2]**, A E Zelenaya**[1] **and E R Nasrulin**[1]

[1]Institute of Physical Materials Science (Siberian Branch of Russian Academy of Sciences), Ulan-Ude, Russian Federation
[2]Buryat State University, Ulan-Ude, Russian Federation

E-mail: vluts@ipms.bscnet.ru

Abstract. The methods of T-x-y-z diagrams invariant point determination by the "geometrical calculations" are analyzed, and the appearance of ambiguous and multiple solutions are shown. The inaccuracies of experimentally constructed isopleths and invariant point detection are analyzed.

1. Introduction

The space model of phase diagram contains information about each geometric element. The computer 3D model of T-x-y-z diagrams can be visualized both in the concentration projections and in the projections taking into account the temperatures [1]. Using of T-x-y-z diagrams computer models gives the possibility not only to analyze the structure of studied systems, but also to choose the arrangement of vertical and horizontal sections, since already on the arrangement of section it is possible to assume the section view and to avoid the errors with its interpretation [2].

The determination of invariant points compositions within the T-x-y-z diagrams is possible both by the selection of the hyper surfaces models and by the use of tie-lines method. In the first case the invariant points are calculated as the intersection of hyper surfaces (in the simplified variant their additive approximation by hyper planes is used) [3-5].

In the second case the coordinates of invariant points is determined from the series of vertical sections [1-2]. Let's consider the calculation methods for the invariant points coordinates and the detection of errors at the interpretation of experimental data by the use of 4D space model for T-x-y-z diagram of eutectic type.

2. Calculation of invariant points by the solution of equations of liquidus (hyper)surfaces of T-x-y-z diagrams

The linear approximation of liquidus surfaces for T-x-y diagrams with the calculation of invariant points coordinates can give both the good agreement with the experimental data and the negative values of coordinates. For example, the linear models of ternary systems $(NaNO_3)_2$-$(LiNO_3)_2$-$Ba(NO_3)_2$, AgCl-$ZnCl_2$-$CdCl_2$, $(KCl)_2$-K_2WO_4-K_2SO_4, CsJ-NaJ-KJ, $FeCl_3$-$PbCl_2$-AgCl, $FeCl_3$-$PbCl_2$-$PdCl_2$, $SnCl_2$-NaCl-$CeCl_3$, $SnCl_2$-$FeCl_2$-$CeCl_3$, $SnCl_2$-$FeCl_2$-$LaCl_3$, RbF-LiF-NaF, NaOH-NaJ-NaBr, $(CsNO_3)_2$-$(NaNO_3)_2$-$Ba(NO_3)_2$, $(NaNO_3)_2$-$(KNO_3)_2$-$Ba(NO_3)_2$, NaCl-$SrCl_2$-$BaCl_2$, CsCl-TlCl-KCl, $(NaCl)_3$-$(NaF)_3$-Na_3AlF_6, $(KCl)_3$-$(KF)_3$-K_3AlF_6, $CaCl_2$-$BaCl_2$-$CaMoO_4$, $(KF)_3$-$(LiF)_3$-K_3AlF_6, $(KCl)_3$-$(LiCl)_3$-K_3AlF_6, LiCl-NaCl-$LiBO_2$, LiCl-RbCl-LiF, $NaBO_2$-$LiBO_2$-NaCl, Na_2SO_4-$PbSO_4$-$PbMoO_4$, $K_2Cr_2O_7$-$(NaNO_3)_2$-$Na_2Cr_2O_7$, $(NaCl)_2$-Na_2SO_4-$BaSO_4$, $BaCl_2$-$(NaCl)_2$-$BaSO_4$, $(AgJ)_2$-$(AgCl)_2$-$CdCl_2$

contradict their physico-chemical sense, because the coordinates of three planes intersection is situated outside the concentration triangle [6]. Analogous problems should appear at the construction of linear models of T-x-y-z diagrams. Furthermore, the ambiguous solutions are possible, and they depend of the methods of hyper plane determination.

2.1. Calculation of invariant points by the solving of three equations system for the surfaces $e_{IJ}E_{IJK}E_{IJL}$ on the liquidus hyper surfaces intersection

In [3] authors suggest the calculation method of the point of quaternary eutectic point by intersection of three planes passing through one binary and two ternary eutectic points $e_{IJ}E_{IJK}E_{IJL}$ (figure 1). Since T-x-y-z diagram includes six such surfaces, then four different systems of three equations can be combined, which gives the ambiguous solution. Using as an example a T-x-y-z diagram model we produced the different variants of calculation of quaternary eutectic point ε by intersection of three planes adjoining to different vertexes of tetrahedron in two coordinate systems: T-Σz_i and Σz_i. As a result 8 variants of coordinates of quaternary eutectic point were obtained (figure 1, table 1). The intersection of planes adjoining to vertex D gives the negative values of coordinates at the calculation in coordinates Σz_i. The calculation in the coordinates T-Σz_i give two cases with negative coordinates for the planes adjoining to vertexes B and D.

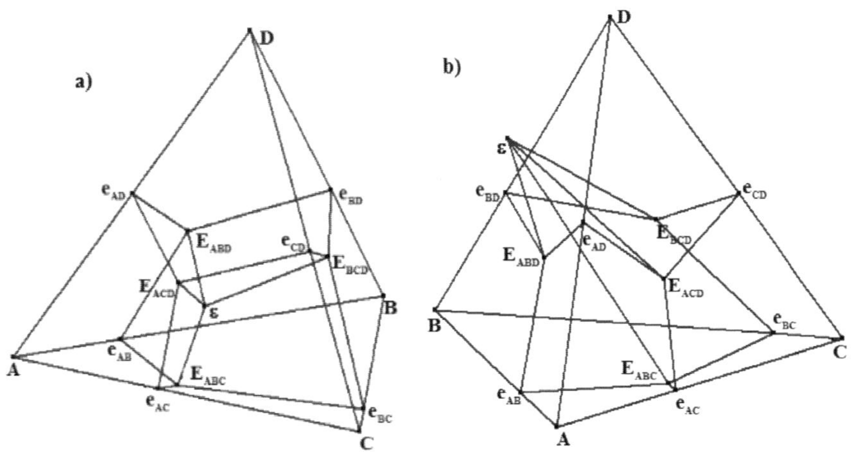

Figure 1. Calculation of point ε by the intersections of three planes adjoining to vertex A ($e_{AB}E_{ABC}E_{ABD}$, $e_{AC}E_{ABC}E_{ACD}$, $e_{AD}E_{ABD}E_{ACD}$) in coordinates Σz_i (a) and T-Σz_i (b)

Table 1. Calculation of coordinates of quaternary eutectic point by the intersection of three planes $e_{IJ}E_{IJK}E_{IJL}$ at component I

	in coordinates z1-z2-z3-z4				in coordinates T-z1-z2-z3-z4			
z1	A (a)	B	C	D	A (б)	B	C	D
z2	0.387	0.356	0.263	0.435	0.253	-0.115	-0.294	0.420
z3	0.093	0.180	0.187	-0.017	0.261	0.493	0.121	-0.017
z4	0.311	0.134	0.270	0.284	-0.184	-0.248	0.245	0.275
	0.209	0.330	0.280	0.298	0.670	0.870	0.928	0.322

2.2. Calculation of invariant points by the solving of system of four equations $Ie_{IJ}E_{IJK}E_{IJL}$ [4]

Calculation method of quaternary eutectic point of system Cd-Pb-Sn-Bi with binary incongruently melting compound and two quaternary points (eutectic and quasiperitectic) is based on the solving of the system of four hyper planes. As the authors said, "because the presence of the peritectic compound Pb_3Bi has essentially no influence on the liquidus surface, Bi, Pb, Sn and Cd were used as components of the system" [4, P. 78]. Each their hyper plane passes through the vertex with initial component (I), the binary eutectic point (e_{IJ}) and two ternary eutectic (E_{IJK}, E_{IJL}): $Ie_{IJ}E_{IJK}E_{IJL}$. The authors obtained a good agreement between the calculated and experimental data. But this method does not have the unique solution because each liquidus can be defined by three equations and therefore there are different variants of equations system.

2.3. Calculation of invariant points by the solving of system of four equations $Ie_{IJ}e_{IK}e_{IL}$

The description of the liquidus by hyper planes $Ie_{IJ}e_{IK}e_{IL}$ [5] has a unique solution, and the coordinates of ternary and quaternary eutectics are calculated. The coordinates of initial component and binary points were used as initial data. But the positive and negative coordinates of calculated points can be obtained in depending of the structure of T-x-y-z diagrams. E.g., the present model of T-x-y-z diagram (table 2) gives the negative values of coordinates of ternary eutectic points E_{ABC} and quaternary eutectic ε (table 3a, figure 2a). To solve this problem, we propose to use a nonlinear model of T-x-y-z diagram (table 3b, figure 2b). Using of the curvature of liquidus hyper surfaces increases the equations degree, but it makes possible to avoid the intersection outside of the concentration tetrahedron.

Table **2.** Initial data for linear model

	A	B	C	D	eAB	eAC	eAD	eBC	eBD	eCD
z1	1	0	0	0	0.71	0.58	0.50	0	0	0
z2	0	1	0	0	0.29	0	0	0.17	0.60	0
z3	0	0	1	0	0	0.42	0	0.83	0	0.55
z4	0	0	0	1	0	0	0.50	0	0.40	0.45
T	600	850	710	900	500	310	400	680	550	520

Table 3. Calculated coordinates of ternary and quaternary eutectic points

	a) linear model					b) non-liner model				
	EABC	EABD	EACD	EBCD	ε	EABC	EABD	EACD	EBCD	ε
z1	0.857	0.397	0.422	0	0.494	0.528	0.405	0.430	0	0.381
z2	-0.732	0.226	0	0.291	-0.239	0.056	0.254	0	0.261	0.096
z3	0.875	0	0.272	0.333	0.403	0.416	0	0.277	0.392	0.280
z4	0	0.377	0.306	0.376	0.342	0	0.341	0.293	0.347	0.243
T	248.14	371.29	289.8	500	267.14	301.69	346.66	278.36	477.15	269.55

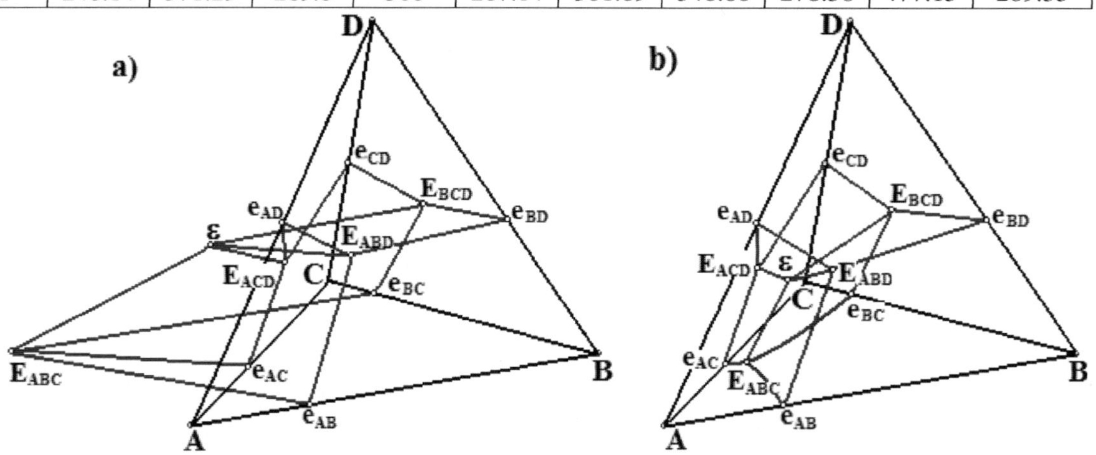

Figure 2. Linear model of T-x-y-z diagram with negative value of coordinates of points E_{ABC} and ε (a), non-liner model (b)

2.4. Modernized tie-lines method

The alternative method of determination of invariant points coordinates is the tie-lines method. Earlier we proposed the modernization of this method, based on the construction of three nonplanar two-dimensional vertical sections [1-2].). In the first stage the section for determination of point belonging to the general two-dimensional simplex $I_\varepsilon J_\varepsilon \varepsilon$ of two linear hyper surfaces Q^r_{IJK} and Q^r_{IJM} at the temperature of quaternary eutectic ε is constructed. It is necessary to take into account the arrangement of section relative to the forming two-dimensional simplexes of different liner hyper surfaces. Let's consider the possible variants of section arrangement near the edge AB (figure 3). The section s_1s_2 is located so that the point $s_1 \in ABC$ is inside simplex $AE_{ABC}B$ and the point $s_2 \in ABD$ is inside simplex $AE_{ABD}B$ (figure 3a). Such section (figure 3b) intersects the hyper surface of liquidus Q_A (line 1-2), linear hyper surface Q_{rAB} with forming segment Ae_{AB} (line 3-4), two linear hyper surfaces with forming two-dimensional simplexes Q_{rABC} (5-6), Q_{rABD} (6-7) and hyperplane H_ε at T_ε (8-6-9). The point s_4 of section s_3s_4 is located in simplex $AE_{ABD}D$ (after the line AE_{ABD}). Such section additionally intersects two linear hyper surfaces Q_{rAD} and Q_{rADB} with one-dimensional (4-7) and two-dimensional (4-10) forming simplexes, but lines of hyper surfaces Q_{rAB} and Q_{rABD} is decreased: (3-4) and (4-6) (figure 3c). Since the ending points of section s_5s_6 are arranged in simplexes $AE_{ABC}C$ and $AE_{ABD}D$, then else two linear hypersurfaces (Q_{rAC} and Q_{rACB}) are intersected: Q_{rAB} (4-5), Q_{rAC} (3-4), Q_{rAD} (5-8), Q_{rABC} (4-6), Q_{rACB} (4-7), Q_{rABD} (5-6), Q_{rADB} (5-9) (figure 3d).

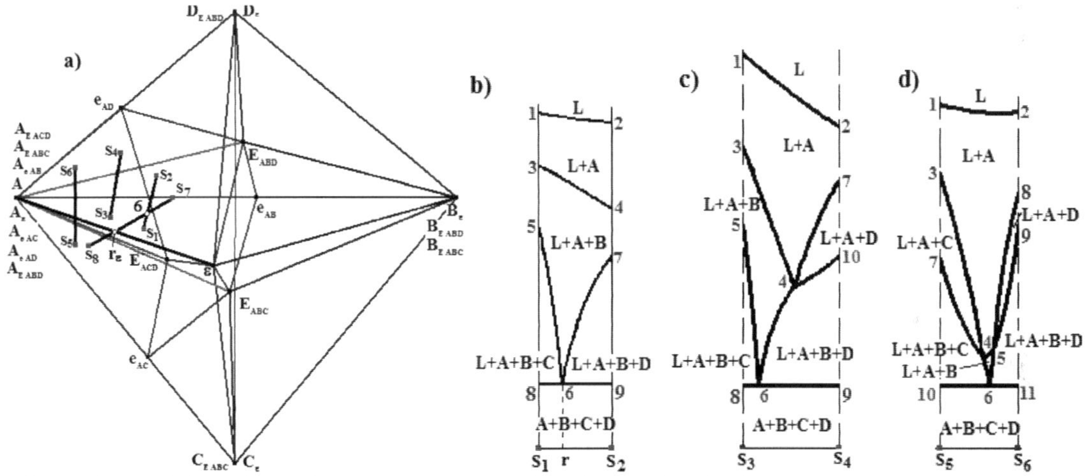

Figure 3. Possible variants (a) of arrangement of section s_is_j for the determination of point (6) on the forming two-dimensional simplex $A_\varepsilon B_\varepsilon \varepsilon$ of two linear hyper surfaces Q_{rABC} and Q_{rABD} at the temperature of quaternary eutectic: s_1s_2 (b), s_3s_4 (c), s_5s_6 (d)

The point 6 of sections s_1s_2, s_3s_4 and s_5s_6 is located on the forming simplex $A_\varepsilon B_\varepsilon \varepsilon$ at T_ε, therefore any section (s_7s_8) constricted through the arbitrary point of edge AB and given point (6) intersects the one-dimensional forming section $A_\varepsilon \varepsilon$ of linear hyper surfaces Q^r_{AB} (in point r_ε). Then third section is constructed along a ray Ar_ε and includes the point of quaternary eutectic. The matrix conversions is used for the calculation of points coordinates, which determine the arrangement of vertical sections [7].

3. Graphic correction of T-x-y-z diagrams

The view of section can be assumed based on its arrangement into tetrahedron and avoid the mistakes in interpreting of experimental data. Authors of [7] are located the two dimensional vertical section into tetrahedron so that its ends are arranged in the different simplexes (as the section s_3s_4 in figure 3a). Whereas the authors give the section with the intersection of liquidus hyper surface and three linear hyper surfaces similar the section s_1s_2 (figure 3b). Whereas the section with such arrangement really intersect five liner hyper surfaces as the section s_3s_4 (figure 3c).

The using of computer models also allows to correctly interprets the experimental data according with determination of eutectic point. Authors of [8] at the study of system Li,Ba,Mg,Zr//F suggest that the section passing through vertex ZrF_4 and opposite ternary eutectic point Li,Ba,Mg//F contain the quaternary eutectic point.

While the tie-line constructed from the vertex and containing ternary and quaternaru eutectic points may corresponds to very particular case of phase diagram structure. Figure 4 shows the possibility sections constructed from tetrahedron vertexes and the ternary eutectic points for the model of T-x-y-z diagram.

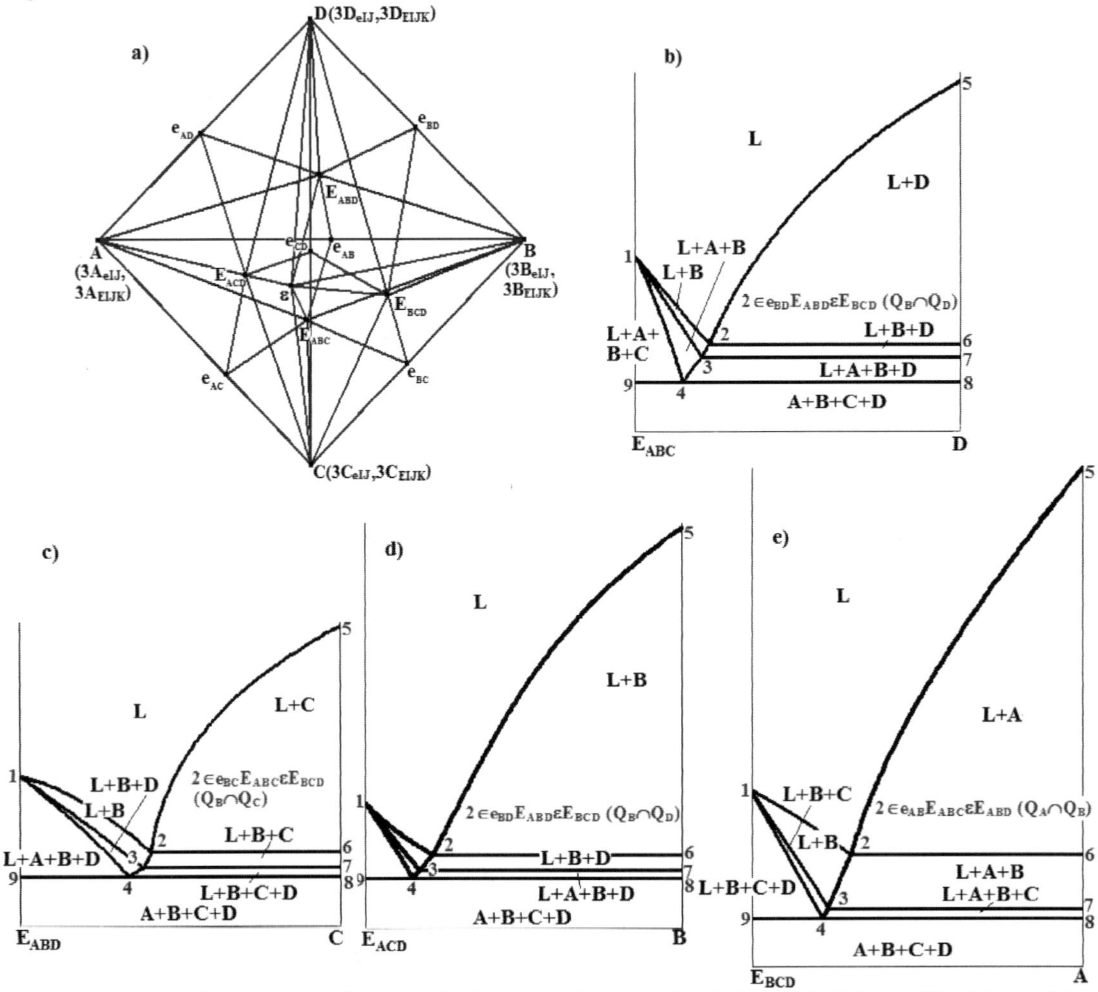

Figure 4. T-x-y-z diagram model (a), vertical sections D-E_{ABC} (b), C-E_{ABD} (c), B-E_{ACD} (d), A-E_{BCD} (e)

The modernized tie-line method are described in [10]. In the first step the three dimensional vertical section mnk is given. Then two-dimensional section fg for determination of point r is constructed in plane mnk (figure 5a). In the next step the section sv (not belonging to the plane mnk) passing through point r (figure 5b) is calculated by simultaneous reducing of the concentrations of components A and D in the ratio given by the point s and the increase at the same value of the concentration of components B and C. Such method provides the moving on the plane $A_\varepsilon B_\varepsilon \varepsilon$ and the finding of tie-line $A_\varepsilon \varepsilon$ in point r'ε. In the last step section Ap is constructed along the ray $A_\varepsilon r'_\varepsilon$, which contains the quaternary eutectic ε (figure 5c). Authors combine second and third sections (figure 5b and 5c) on one figure as $rr'_\varepsilon\varepsilon$ and give the smooth curve of intersection of liquidus hyper surfaces. But the sections rr'_ε (figure 5b) and $r_\varepsilon\varepsilon$ (figure 5c) are constructed on the axes located with different angles. Therefore the combination of these two sections should give the inflection on the liquidus isopleth 1-2-3 in point 2 (figure 5d).). The liquidus curve can be smooth only when the sections sv and $A_\varepsilon r'_\varepsilon$ belong to one line.

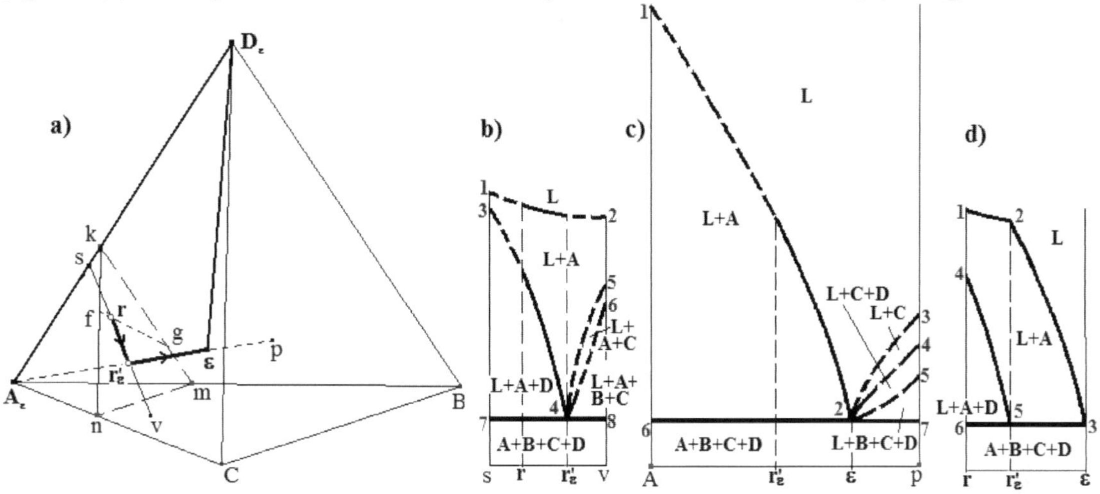

Figure 5. Schema of sections arrangement (a), sections sv (b), Ap (c), $rr'_\varepsilon\varepsilon$ (d)

4. Conclusions

The computer model of T-x-y- z of diagram is not only the example for 3D visualizations of studied system, but also it contains the information about all geometric elements. Using these models allows to understand the structure of the phase diagram and predict the view of horizontal or vertical sections on its arrangement relative to the elements of phase diagram in the concentration tetrahedron. And then - to avoid the errors at its interpretation. The model of phase diagram also can be used to predict the experimental work. It can be constructed using partial experimental data and then rebuilt at the new data appearance.

Acknowledgements

The work was supported by the Russian Foundation for Basic Research (project number 14-08-00453-a and 14-08-31468-mol-a).

References

[1] Lutsyk V and Zelenaya A 2012 *Solid State Sci.* **14** 1604
[2] Lutsyk V I, Sumkina O G, Savinov V V and Zelenaya A E 2011 *IOP Conf. Ser.: Mater. Sci.* **18** 162021
[3] Gasanaliev A M, Gamataeva B Yu, Rasulov A I, Umarova Yu A and Mamedova A K 2009 *Russ. J. Inorg. Chem.* **54** 1494
[4] Mamedov A N and Mekhdiev I G 1995 *Thermochim. Acta* **269/270** 73
[5] Lutsyk V I and Vorob'eva V P 1989 *Russ. J. Inorg. Chem.* **43** 1253
[6] Lutsyk V I 1987 *Analysis of the Ternary Systems Liquidus Surface* (Moscow: Nauka Publ. House) (in Russian)
[7] Lutsyk V I and Vorob'eva V P 2008 *Z. Naturforsch A.* **63a** 513
[8] Gubanova T V, Kondratyuk I M and Garkushin I K 2004 *Russ. J. Inorg. Chem* **49** 1087
[9] Li G, Takagi R and Kawamura K 1991 *Denki Kagaku* **59** 800
[10] Verdiev N N 2012 Simulation of chemical equilibria in multicomponent salt systems: Abs. dis. doctor. chem. sciences (Moscow) (in Russian)

Topological Correction of Multicomponent Systems Polyhedration

V I Lutsyk[1] and V P Vorob'eva

Institute of Physical Materials Science (Siberian Branch of Russian Academy of Sciences), Ulan-Ude, Russian Federation

E-mail: vluts@ipms.bscnet.ru

Abstract. An algorithm (Topological Correction of Lists of Simplexes of Different Dimensions) for polyhedration of quaternary reciprocal systems is presented. It can control all polyhedration stages, accelerates the search of internal diagonals and takes into account their possible competition.

1. Introduction

Melting Salt Reactor is one of the concepts of the IV generation nuclear reactors with the molten salt fuel compositions. According to the rules of selection, the preferable salt compositions are the metal fluorides. Nevertheless, despite the fact that the chloride's systems have a high vapor pressure and low thermodynamic stability at high temperatures than the fluoride's ones, they are less aggressive in relation to the structure of the material and have lower melting points. However, they can be used only in the fast reactors, but not in the thermal ones. Therefore, in order to ensure reliable operation of new generation reactors the comprehensive information on equilibria and chemical processes in the reciprocal fluoride-chloride systems are to be provided. One of the necessary conditions for the implementation of a nuclear power new generation is to detect the regularities, which permit to predict the change of physical and chemical properties of fluoride-chloride systems with the accumulation of products of nuclear transformations. Such regularities are preferable to extract from the information on the phase equilibria accumulated in the multidimensional phase diagrams of these systems [1].

Phase diagrams of reciprocal systems are a special class among others, because the concentration base is not a simplex, but a complex. Known calculate methods and phase diagrams simulation programs are developed mainly for the simplexes (triangles, tetrahedra, ...). Therefore, the main questions are the correct concentration coordinate system [2], and the polyhedration or dividing the concentration complex into the energetically stable subsystems.

2. Polyhedration of multicomponent systems

Polyhedration of a multicomponent system is very important stage of its investigation because it can minimize the experimental data. In this case, it is generally unknown (or only partially known) what phases are formed inside the system, as well as the information on the boundary systems may be incomplete or wrong. If the system is studied completely, the meaning of polyhedration is to check the consistency of triangulation of its faces and also to pass to the prediction of new variants of polyhedration at other temperatures and pressures.

2.1. Polyhedration algorithms

Usually the reciprocal system concentration complex has been shown as a graph and it is described by the adjacency matrix. Known algorithms are based on graph decomposition by or the adjacency matrix zero elements (the missing links between vertices), or, on the contrary, by the unit elements, that correspond to the connected graph vertexes. Such algorithms are widely used in practice and often cited (see, e.g., [3,4]). The software for multicomponent systems polyhedration was created [5]. It is based on the graph decomposition by the multiplying of adjacency matrix zero elements on the law of absorption and inversion of its result.

The problem occurs if the adjacency matrix contains unknown elements, when we do not know, whether there are links between the graph vertices or not. As all links between the graph vertices in boundary systems are known before the polyhedration, internal diagonals remain mainly unknown. In some cases they may be found without great problems. So, you can use the decomposition algorithm for the graph, simply sorting out the possible links between the vertices of the graph, which could become the internal diagonals. There are also the methods, using "matrixes of vertices indexes" and "metathesis reactions analysis" to identify internal diagonals [3,4].

The newly developed algorithm is based on the topological correction of the lists for the simplexes of different dimension (TCLSDD) [6,7]. It is based on the scanning of all the relationships between the vertices of the graph that describes the concentration complex and selecting those variants of polyhedration, which correspond to the ratios between the quantities of different geometrical elements within the investigated complex (vertices of the graph, the links between them, two- and three-dimensional simplexes). And only those internal elements (diagonals and planes) are selected, that can implement the possible variant of polyhedration. If several versions of the polyhedration are possible as a result of the internal diagonals competition, the final decision should be confirmed by the experiment.

2.2. Formulas, connected the number of 0-, 1-, 2-, 3-dimension simplexes within the initial complex

The TCLSDD algorithm uses formulas, connected the number of the graph vertices (0D) and links between them (1D) with the number of 2D and 3D simplexes. If to represent the triangular prism of the quaternary reciprocal system A,B,C||X,Y with its stoichiometric compounds as a graph, then V (the number of graph Vertexes) and L (the number of Links between them) in the adjacency matrix are related by expressions [8]:

$$V=V_0+V_E+V_F+VI \text{ and } L=L_E+L_F+L_I, \qquad (1)$$

where $V_0=6$ is the number of the prism vertices, V_E, V_F, V_I are the number of binary, ternary, quaternary compounds, that are the points on Edges, Faces and Inside the polyhedron; L_E, L_F, L_I are the number of Edges (simple binary subsystems without compounds) and diagonals (quasi-binary sections), on Faces and Internal diagonals, respectively.

L is equal to the number of unit elements of the adjacency matrix, V•V is the number of all elements of the square adjacency matrix, and $(V^2-V)/2$ is the number of all elements above (or below) of its main diagonal; therefore, L can be calculated as

$$L=(V^2-V)/2-L_0-L_?,$$

where L_0 is the number of all zero elements of the adjacency matrix, and $L_?$ is the number of all links, unknown from the boundary elements. Each unknown element in the adjacency matrix can be labeled with the subscript "?". Further, all of these $L_?$ elements will take the value zero or unity, depending on whether or not the link corresponding to such element becomes a stable internal diagonal.

Because the triangular prism of the quaternary reciprocal system A,B,C||X,Y has $L_E=9$ edges and $S_0=5$ faces (two triangular and three quadrangular ones), the edges are divided by V_E points into $L_E=9+V_E$ segments. The numbers L_F of diagonals, S_F of 2D Simplexes on Faces, SI of Internal secant planes, and T of 3D Tetrahedrons depend on the numbers V_E, V_F, V_I of binary, ternary, quaternary compounds, respectively, and the number LI of internal diagonals [8]:

$$L_F=3+2V_E+3V_F,\ S_F=8+2V_E+2V_F,\ S_I=2+V_E+V_F-2V_I+2L_I,\ T=3+V_E+V_F-V_I+L_I \qquad (2)$$

There are known the formula of Euler V - L + S = 2 and Schlefli-Stringem V - L + S - T = 1, which links the number of simplexes of different dimensions: vertices and edges of the graph, 2D and 3D simplexes. Surely they are useful to verify the polyhedration. However, they are not sufficient, when you search the internal secants of the polyhedron. Because it is required to distinguish among the L and the S simplexes those ones, that belong to the faces of the original complex, and those, that are within the polyhedron. Similar, the simplexes S and T can be created either using the internal diagonals, or without them.

2.3. Polyhedration of systems with competition of internal diagonals

If to form the planes $x_ix_jx_k$ from unit elements of the adjacency matrix or segments x_ix_j and from all its elements labeled with the subscript "?", these planes are divided into the following groups: S_F planes on faces; S_I* internal planes formed without using internal diagonals; S_I** internal planes in which at least one of the sides is a possible internal diagonal and its corresponding element in the adjacency matrix is subscripted "?".

If the equality S_I=S_I* is valid, i.e., if the number of internal planes S_I*, formed without internal diagonals, satisfy formulas (2) at L_I=0, then there can be the only variant of polyhedration without internal diagonals. Then all elements, which labeled as "?", are assigned zero.

If S_I*<$2+V_E+V_F-2V_I$, then L_I>0, and missing internal planes should be taken from the sets of S_I** by selecting those satisfying formulas (2).

Table 1. Prediction of internal planes (S_I) and tetrahedra (T) in the system with two binary compounds (V_E=2, V_F=V_I=0) (Matrix R_1), which belong to different ternary systems and to the same reciprocal ternary system, for two variants of polyhedration (Figure 1)

a) $S_I=2+V_E+V_F-2V_I+2L_I=2+2+0-0+2=6$ (x_3x_8 or x_6x_7)

elements with "?"	with element x_3x_8	with element x_6x_7
$x_1x_6x_8$, $x_3x_5x_7$	$x_1x_3x_8$, $x_3x_5x_8$, $x_3x_6x_8$, $x_3x_7x_8$	$x_1x_6x_7$, x_3x_6x, $x_5x_6x_7$, $x_6x_7x_8$
2	4	4

b) $T=3+V_E+V_F-V_I+L_I=3+2+0-0+1=6$ (x_3x_8 or x_6x_7)

without elements with "?"	with element x_3x_8	with element x_6x_7
$x_1x_4x_6x_8$, $x_2x_3x_5x_7$	$x_1x_3x_6x_8$, $x_1x_3x_7x_8$, $x_3x_5x_6x_8$, $x_3x_5x_7x_8$	$x_1x_3x_6x_7$, $x_1x_6x_7x_8$, $x_3x_5x_6x_7$, $x_5x_6x_7x_8$
2	4	4

Thus, combinations of nonzero elements of the adjacency matrix, taken three at a time, give lists of all 2D simplexes. From them according to formulas (1) and (2), groups of simplexes are formed.

Suppose, for example, two binary compounds x_7 and x_8 in the quaternary reciprocal system belong to the same ternary reciprocal system and to the different ternary systems (Figure 1, Table 1). The correspondent adjacency matrix

$$R_1 = \begin{array}{c} \\ x_1 \\ x_2 \\ x_3 \\ x_4 \\ x_5 \\ x_6 \\ x_7 \\ x_8 \end{array} \begin{array}{cccccccc} x_1 & x_2 & x_3 & x_4 & x_5 & x_6 & x_7 & x_8 \\ * & 0 & 1 & 1 & 0 & 1 & 1 & 1 \\ & * & 1 & 0 & 1 & 0 & 1 & 0 \\ & & * & 0 & 1 & 1 & 1 & ? \\ & & & * & 0 & 1 & 0 & 1 \\ & & & & * & 1 & 1 & 1 \\ & & & & & * & ? & 1 \\ & & & & & & * & 1 \\ & & & & & & & * \end{array}$$

has two unknown elements x_3x_8 and x_6x_7. They are labeled as "?". If to write from R_1 all nonzero element combinations, taken by three, then we have got the list of triangles:

$x_1x_3x_4$, $x_1x_3x_6$, $x_1x_3x_7$, $x_1x_3x_8$, $x_1x_4x_6$, $x_1x_4x_7$, $x_1x_4x_8$, $x_1x_6x_7$, $x_1x_6x_8$, $x_1x_7x_8$, $x_2x_3x_5$, $x_2x_3x_7$, $x_2x_5x_7$, $x_3x_5x_6$, $x_3x_5x_7$, $x_3x_5x_8$, $x_3x_6x_7$, $x_3x_6x_8$, $x_3x_7x_8$, $x_4x_6x_8$, $x_5x_6x_7$, $x_5x_6x_8$, $x_5x_7x_8$, $x_6x_7x_8$.

Planes $x_1x_3x_4$ and $x_1x_4x_7$ mast be deleted because links x_3x_4 and x_4x_7 are absent. Further $S_F=8+2V_E+2V_F=8+4+0=12$ simplexes are deleted from the list, because they are belong to faces:

$x_1x_3x_7$, $x_2x_3x_7$, $x_4x_6x_8$, $x_5x_6x_8$, $x_1x_3x_6$, $x_1x_4x_6$, $x_1x_4x_8$, $x_1x_7x_8$, $x_2x_5x_7$, $x_5x_7x_8$, $x_2x_3x_5$, $x_3x_5x_6$, and the rest internal planes are grouped (Table 1a): S_I*=2 – $x_1x_6x_8$, $x_3x_5x_7$ - are formed by unit elements of the R_1 matrix, S_I**=4 – one side of triangles $x_1x_3x_8$, $x_3x_5x_8$, $x_3x_6x_8$, $x_3x_7x_8$ is a possible internal diagonal $x_1x_3x_8$. There is another group (S_I**=4) of four possible internal planes - $x_1x_6x_7$, $x_3x_6x_7$, $x_5x_6x_7$, $x_6x_7x_8$. They are formed by other unknown element of the R_1 matrix - x_6x_7.

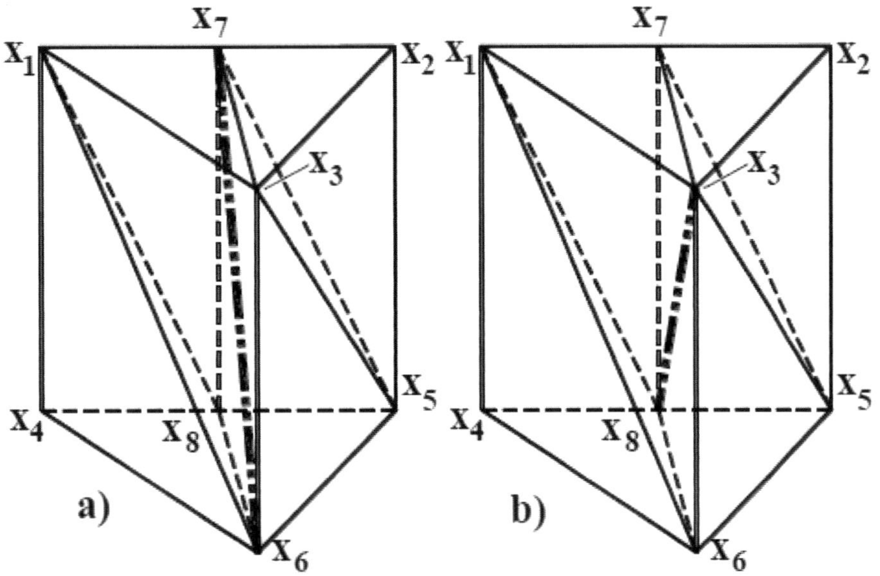

Figure 1. Polyhedration of the quaternary reciprocal system with two binary compounds (x_7, x_8 belong to different ternary systems and to the same reciprocal ternary systems) and the only one internal diagonal: x_6x_7 (a) or x_3x_8 (b)

Table 2. Prediction of internal planes (S_I) and tetrahedra (T) in the system with two binary compounds (V_E=2, V_F=V_I=0) (Matrix R_2), which belong to different ternary and reciprocal ternary systems, for two variants of polyhedration (Figure 2)

a) $S_I=2+V_E+V_F-2V_I+2L_I=2+2+0-0+4=8$ ($x_2x_8+x_7x_8$ or $x_6x_7+x_7x_8$)

without elements with "?"	with element x2x8	with element x6x7	with element x_7x_8	with elements $x_2x_8+x_7x_8$	with elements $x_6x_7+x_7x_8$
x1x5x8	*x2x3x8, x2x5x8, x2x6x8*	*x2x6x7, x3x6x7, x5x6x7*	*x1x7x8*, $x_3x_7x_8$, $x_5x_7x_8$	*x2x7x8*	*x6x7x8*
1	3	3	3	1	1

b) $T=3+V_E+V_F-V_I+L_I=3+2+0-0+2=7$ ($x_2x_8+x_7x_8$ or $x_6x_7+x_7x_8$)

without elements with "?"	with element x_2x_8	with element x6x7	with element x_7x_8	with elements $x_2x_8+x_7x_8$	with elements $x_6x_7+x_7x_8$
x1x4x5x8	*x2x3x6x8, x2x5x6x8*	*x2x3x6x7, x2x5x6x7*	*x1x3x7x8, x1x5x7x8*	*x2x3x7x8, x2x5x7x8*	*x3x6x7x8, x5x6x7x8*
1	2	2	2	2	2

Expression $S_I=2+V_E+V_F-2V_I+2L_I=2+2+0+0-0+2L_I=4+2L_I$ satisfies to formulas (2), if $S_I=S_I*+S_I**=2+4=6$, or if to take two planes S_I*=2 and one of two groups S_I**=4. As a result, there are two competitive elements of the R_1 matrix - x_3x_8 (Figure 1a) and x_6x_7 (Figure 1b). The final answer, which element receives the designation "1" - and the corresponding link become the internal diagonal, - and which receives the designation "0", will give the experiment. The list of tetrahedra is formed after analysis of the planes that were selected as possible internal ones (Table 1b).

If two binary compounds x_7 and x_8 belong to differ ternary and ternary reciprocal systems (Figure 2, Table 2), and the R_2 matrix has three unknown elements

$$R_2 = \begin{array}{c} \\ x_1 \\ x_2 \\ x_3 \\ x_4 \\ x_5 \\ x_6 \\ x_7 \\ x_8 \end{array} \begin{array}{cccccccc} x_1 & x_2 & x_3 & x_4 & x_5 & x_6 & x_7 & x_8 \\ \hline * & 0 & 1 & 1 & 1 & 0 & 1 & 1 \\ & * & 1 & 0 & 1 & 1 & 1 & ? \\ & & * & 0 & 0 & 1 & 1 & 1 \\ & & & * & 1 & 0 & 0 & 1 \\ & & & & * & 1 & 1 & 1 \\ & & & & & * & ? & 1 \\ & & & & & & * & ? \\ & & & & & & & * \end{array},$$

then selection of 21 nonzero elements (18 "1" and 3 "?") by three gives $S_F=8+2V_E+2V_F=8+6+0=14$ triangles on faces and 12 inside the prism (Table 2a). The only plane is formed without elements, labeled as "?". Three planes are formed with a help of one of unknown elements. And there are planes, formed by a pair of elements "?": $x_2x_8+x_7x_8$ or $x_6x_7+x_7x_8$.

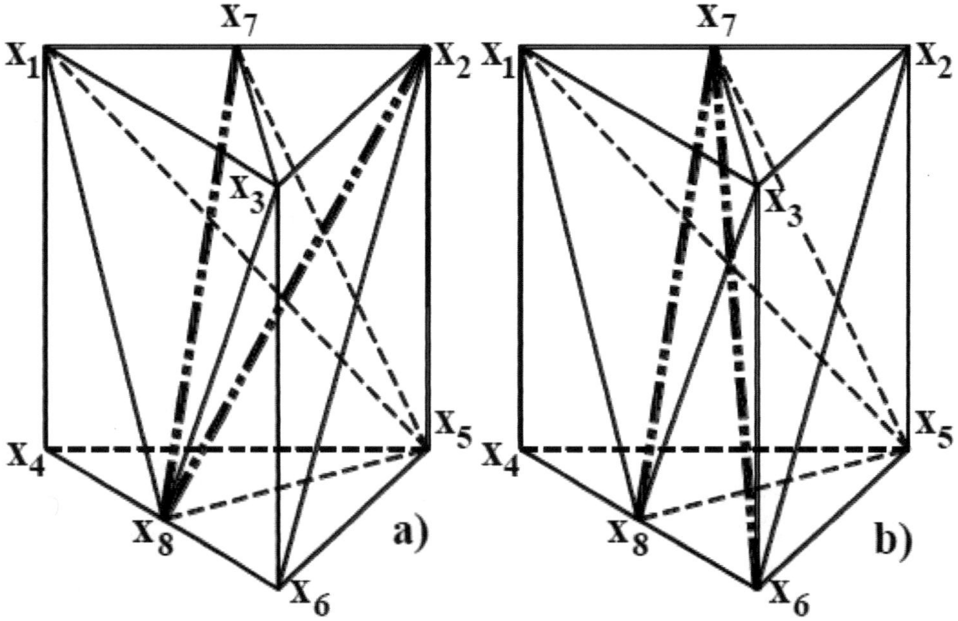

Figure 2. Polyhedration of quaternary reciprocal system with two binary compounds (x_7, x_8 belong to the different ternary and reciprocal ternary systems) and with the two internal diagonals: $x_2x_8+x_7x_8$ (a) or $x_6x_7+x_7x_8$ (b)

Selection of planes in accordance to formulas (2) gives two alternative groups; every of them has 8 planes. The element x_7x_8 becomes the internal diagonal in any case, and the experiment is necessary only for one of two other diagonals.

As a result, we have the polyhedration method, which gives a possibility to control every stage of the process of dividing of the initial complex into simplexes and simultaneously to understand, which of possible internal secant elements (diagonals ans planes) are able to participate in polyhedration. The method helps not only to verify the polyhedration correctness, but to find all possible variants of polyhedration at competition of internal secant elements.

3. Results and Discussion

By the TCLSDD algorithm [6,7]:

1. in the Ba,Ca,K||F,WO_4 system with three binary compounds, the only variant of polyhedration was determined: with one internal diagonal CaF_2-K_2WO_4•$BaWO_4$;
2. in the Na,K,Ba||WO_4,F system with four binary compounds, two variants of polyhedration were found: with one of the competing internal diagonals, $(NaF)_2$-K_2WO_4•$BaWO_4$ or BaF_2-K_2WO_4•Na_2WO_4;
3. in the K,Na,Ca||Cl,NO_3 system with two binary compounds and one ternary compound, two variants of polyhedration were determined, with the stable internal diagonal $(NaCl)_2$-$(KCl)_2$•$2Ca(NO_3)_2$, and with one of the two competing internal diagonals, $(NaCl)_2$-$2(KNO_3)_2$•$Ca(NO_3)_2$ or $(NaNO_3)_2$-$(KCl)_2$•$2Ca(NO_3)_2$;
4. among eight quaternary reciprocal systems (Li,Sr,Na||Cl,NO_3, Ca,Na,K||F,MoO_4, K,Na,Li||Cl,NO_3, Ba,Ca,K||F,WO_4, Ba,Ca,K||F,MoO_4, Ba,Na,K||F,MoO_4, Na,K||Cl,NO_3,NO_2, Li,K,Ba||F,WO_4), the polyhedration of only three systems - Li,Sr,Na||Cl,NO_3, Ba,Ca,K||F,WO_4, Ba,Ca,K||F,MoO_4 - with participation of one internal diagonal was performed correctly [3,4]. In the polyhedration results for the other five systems the contradictions were detected.

4. Conclusion

The first variant of the TCLSDD algorithm, presented in [6], is possible to simplify, if to select according to formulas (2) the internal planes only (Table 1a and Table 2a). In this case the final result of polyhedration will be got automatically in the tetrahedral form (Table 1b and Table 2b). TCLSDD algorithm is a powerful tool to divide the multidimensional complex into the simplexes or more simple microcomplexes, and to enquire the contradictions in the published results of the multicomponent systems polyhedration.

Acknowledgements

The work was supported by the Russian Foundation for Basic Research (project number 14-08-00453-a).

References

[1] Lutsyk V and Vorob'eva V (2014) Topological Correction on Multicomponent Systems Polyhedration In L.A. Goemze (Editor) 3rd International Conf. on Competitive Materials and Technology Processes, Miskolc-Lillafured, Hungary, pp. 193
[2] Lutsyk V and Vorob'eva V 2008 *Z Naturforsch A*. **63a** (7-8) 513
[3] Gasanaliev A, Akhmedova P and Gamataeva B 2010 *Rus J Inorg Chem.* **55** (12) 1970
[4] Gasanaliev A, Akhmedova P and Gamataeva B 2012 *Rus J Inorg Chem.* **57** (2) 274
[5] Lutsyk V and Vorob'eva V 2011 *IOP Conf. Ser.: Mater. Sci.* **18** (22) http://iopscience.iop.org/1757-899X/18/22/222005/pdf/1757-899X_18_22_222005.pdf
[6] Lutsyk V and Vorob'eva V 2014 *Rus J Inorg Chem.* **59** (9) 956
[7] Lutsyk V and Vorob'eva V 2014 *Rus J Inorg Chem.* **59** (10) 1123
[8] Lutsyk V, Vorob'eva V and Sumkina O 2012 *Crystalorg Repts.* **57** (7) 115

IOP Conf. Series: Materials Science and Engineering **123** (2016) 012038 doi:10.1088/1757-899X/123/1/012038

Prediction of precipitate evolution and martensite transformation in Ti-Ni-Cu shape memory alloys by computational thermodynamics

A Povoden-Karadeniz[1], D C Cirstea[2], E Kozeschnik[1]

[1]Vienna University of Technology – Institute of materials science and technology, Vienna, Austria

[2]National Institute for Research and Development in Electrical Engineering INCDIE ICPE-CA, Advanced Materials Department, Bucharest, Romania

E-mail: erwin.povoden-karadeniz@tuwien.ac.at

Abstract. Ti-50Ni to Ti-55Ni (at.%) can be termed as the pioneer of shape memory alloys (SMA). Intermetallic precipitates play an important role for strengthening. Their influence on the start temperature of the martensitic transformation is a crucial property for the shape memory effect. Efforts for increasing the martensite start temperature include replacement of a part of Ni atoms by Cu. The influence of Cu-addition to Ti-Ni SMA on *T0*- temperatures and the character of the austenite-martensite transformation is evaluated using a new thermodynamic database for the Ti-Ni-system extended by Cu. Trends of precipitation of intermetallic phases are simulated by combining the assessed thermodynamics of the Ti-Ni-Cu system with assessed diffusion mobility data and kinetic models, as implemented in the solid-state transformation software MatCalc and are presented in the form of time-temperature-precipitation diagrams. Thermodynamic equilibrium considerations, complemented by predictive thermo-kinetic precipitation simulation, facilitates SMA alloy design and definition of optimized aging conditions.

1. Introduction

Cold-forged Ti-Ni alloys close to Ti50Ni50 (at.%) have a martensitic B19' structure and transform thermo-elastically to their original austenitic B2-phase during re-heating [1,2]. Precipitation of intermetallic phases during special aging strengthens the shape memory alloy (SMA) [2-5]. Addition of Cu to Ti50Ni50 can stabilize the additional B19-phase at low temperatures, associated with a two-step transformation from austenite to B19′martensite via B19-phase [6-8]. Direct transformation of B2-phase to B19 martensite at higher Cu-contents showed significantly narrowed hysteresis [7].

However, the reported critical Cu-content limiting the single-stage from the two-stage the transformation characteristics is quite scattering. Heat-treated Ti50Ni50-*x*Cu*x* showed fine, nano-sized distributions of Cu-containing intermetallic phase of the type Ti(Cu,Ni)2 inside the grains [8,9].

However, the understanding of the effect of this phase on the transformation behaviour and the knowledge of required Cu-alloying to provoke its precipitation is scarce. Thermodynamic CALPHAD analysis [10,11] constrains the important parameters for transformation and precipitation, namely the critical Cu-content for two-step transformation and the thermodynamic stability of intermetallic precipitate phases theoretically. Povoden-Karadeniz et al. [12] established a thermodynamic Ti-Ni SMA database, containing the metastable precipitate phases Ti3Ni4 and Ti2Ni3 additionally to the thermodynamic equilibrium phases of the Ti-Ni system. Two thermodynamic assessments of the Ti-Ni-Cu system are available [13,14], with two different descriptions of the austenitic B2-phase. Tang´s [13] assessment focused on the SMA-relevant martensite phases and does not contain model parameters of the ternary Ti(Cu,Ni)2 solid solution. On the other hand, Zhang´s assessment [14] lacks the low-temperature martensite phases, but comprises a comprehensive descriptions of all observed high-temperature phases [15] of the Ti-Ni-Cu system. In our present reassessment of thermodynamic model parameters, we aimed on merging these "partial" assessment to one consistent set of optimized model descriptions which allows computational thermodynamics calculations for all phases of the Ti-Ni-Cu system. In a computational thermo-kinetic evaluation of temperature- and aging time-dependent precipitation in Cu-alloyed Ti-Ni SMA, the assessed thermodynamics are combined with kinetic models for diffusion, nucleation and growth of precipitates, as implemented in the solid-state transformation software Matcalc [16].

2. Computational methods

2.1. Thermodynamic re-optimisation of model parameters

The Ti-Ni-Cu system contains ternary extensions of binary alloy and intermetallic phases, listed in Table 1. Their proposed structural CALPHAD model descriptions [14] were maintained. The use of SMA Ti-Ni database [12] in the Ti-Ni-Cu extension however required some re-adjustments of binary Ti-Cu parameters. For re-optimized Gibbs energies of their ternary extensions, also revised interaction parameters were added to the constitutive Ti-Ni [14], Ti-Cu [17] and Cu-Ni [18] end-member parameters. Model parameters of Tau1 to 6 [14] are ternary phases that are not constituted of stable binary end-member compounds were also re-optimized. These re-adjustments, listed in Table 1, were necessary to merge previous partial assessments to a consistent database of the whole Ti-Ni-Cu system.

Table 1. Re-assessed model parameters of phases in the Ti-Ni-Cu system

Phase	Model type	Model formula	Parameter	Opimised values
fcc	Solution[a)]	(Cu,Ni,Ti)Va	0L(Cu,Ni,Ti:Va)	-200000+25T
B2	Ordering[b)]	(Cu,Ni,Ti)0.5(Cu,Ni,Ti)0.5(Va)3	Cu:Ti:Va	-7000
			Ti:Cu:Va	-7000
			0L(Cu,Ni:*[d)]:Va)	-3000-3T
			0L(*:Cu,Ni:Va)	-3000-3T
			0L(*:Cu,Ti:Va)	+7000
			0L(Cu,Ti:*:Va)	+7000
TiCu	CEF	(Cu,Ni,Ti)(Cu,Ni,Ti)	Cu:Ti	*G*(Cu)[e)]+*G*(Ti)-19000+3.6T
			0L(Cu,Ni:Ti)	-83000+19.8T
Ti2Cu	CEF	(Cu,Ni)(Ti)2	0L(Cu,Ni:Ti)	-15000
TiCu2	CEF	(Cu,Ni)2(Ti)	0L(Cu,Ni:Ti)	-30000
Ti2Cu3	CEF	(Cu,Ni)3(Ti)2	0L(Cu,Ni:Ti)	-65000
Ti3Cu4	CEF	(Cu,Ni)4(Ti)3	0L(Cu,Ni:Ti)	-120000-6.7T
TiNi3	CEF	(Cu,Ni,Ti)(Cu,Ni,Ti)3	0L(Cu,Ni:Ti)	-18000
Ti2Ni	CEF	(Cu,Ni,Ti)2(Cu,Ni,Ti)1	0L(Cu,Ni:Ti)	-18000
			0L(Ti:Cu,Ni)	-15000
			1L(Ti:Cu,Ni)	+15000
			2L(Ti:Cu,Ni)	-15000
Tau1	CEF	(Cu,Ni)2(Ti)	Cu:Ti	2*G*(Cu)+*G*(Ti)-25000+11T
			Ni:Ti	2G(Ni)+G(Ti)-95820-T
			0L(Cu,Ni:Ti)	-54500
Tau2	CEF	Cu0.175Ni2.825Ti2	Cu:Ni:Ti	0.175G(Cu)+2.825G(Ni)+2G(Ti)-180250+7.2T
Tau4	CEF	Cu0.05Ni0.7Ti0.25	Cu:Ni:Ti	0.05G(Cu)+0.7G(Ni)+0.25G(Ti)-35500
Tau6	CEF	Cu0.25Ni0.5Ti0.25	Cu:Ni:Ti	0.25G(Cu)+0.5G(Ni)+0.25G(Ti)-25000-3.5T
B19'	CEF	(Cu,Ni,Ti)0.5(Cu,Ni,Va)0.5	Cu:Ni	0.5G(Cu-bcc)+0.5G(Ni)+20000
			Ti:Cu	0.5G(Cu-bcc)+0.5G(Ti-bcc)-3900+15T
B19	CEF	(Cu,Ni,Ti)0.5(Cu,Ni,Va)0.5	Ti:Cu	0.5G(Cu-bcc)+0.5G(Ti-bcc)- 7400+T
Liquid	Solution	(Cu,Ni,Ti)	Cu,Ti	-19330+7.3T

[a)]Regular solution model, [b)]Split model of ordering [19], [c)]* denotes Cu,Ti or Ni, [d)]Gibbs energies of elements adopted from Dinsdale [20]

2.2 Thermo-kinetic precipitation simulation

A thermo-kinetic simulation of the precipitate evolution in Ti-Ni SMA and Ti-Ni SMA alloyed by Cu was performed with MatCalc [16]. The physical modeling base implemented in MatCalc is schematized in Figure 1. It comprises the formalisms of nucleation theory [21], the SFFK model for the growth of precipitates [22] and takes into account CALPHAD-assessed [10, 11] thermodynamic and diffusion mobility data stored in the assessed Ti-Ni-Cu Matcalc databases mc-sma.tdb and mc.sma.ddb supplemented to this paper.

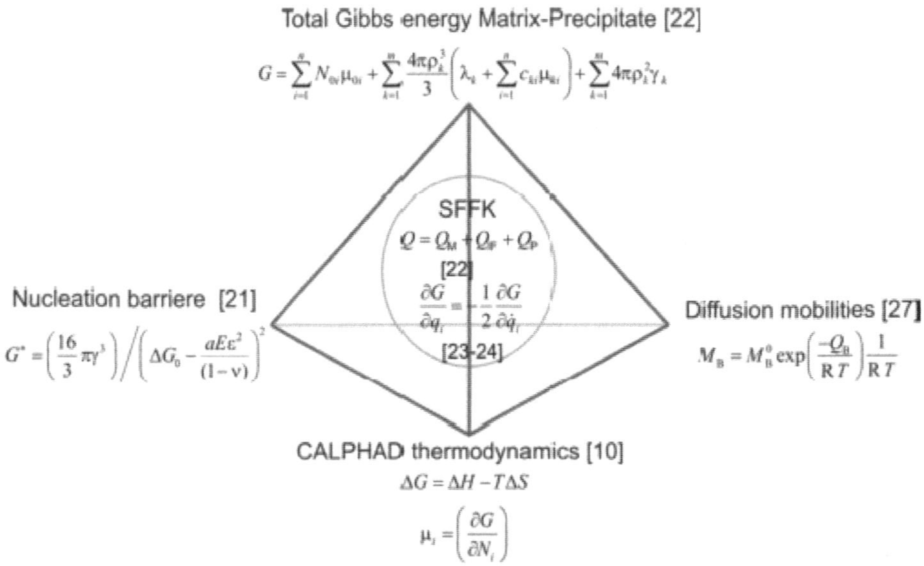

Figure 1. Schematized physical fundament for the precipitation simulation with MatCalc

The rate of radius change and chemical composition change of each precipitate is evaluated by application of the thermodynamic extremum principle [23-25]. Real matrix-precipitate systems are typically in a highly non-equilibrated state and they have thus a high driving force to evolve towards minimum energy state as a function of time and temperature. In the SSFK model three parts contribute to the energy dissipation during the system evolution. These are dissipation due to diffusion inside of the precipitation matrix, dissipation due to interface migration and dissipation due to diffusion inside of the precipitates. In order to evaluate the energy dissipation diffusivity data are thus required. Diffusivities in the matrix are stored in the form of a diffusion mobilities database. Diffusion mobilities contain activities for the diffusion and a pre-factor which basically represents the jump frequency of a solute atom. The diffusivities inside of precipitates were roughly related to the diffusivity in the matrix by a factor of 1/100. The numerical of the evolution equations were presented by Svoboda et al. [22] and Kozeschnik [26].

2.3. Assessment of diffusion mobilities in B2-ordered austenite matrix

For the definition of diffusion mobilities in the B2-ordered austenite matrix phase the modeling of activation energy Q for the diffusivity as suggested by Helander and Agren [27] was followed

$$Q = Q^{dis} + \Delta Q^{ord} \tag{1}$$

where dis stands for disordered bcc-phase, i.e. the "parent" reference phase from which the thermodynamics of B2-phase is derived by CALPHAD model parameters describing chemical ordering [19] and ΔQ^{ord} thus denotes a contribution to the activation energy from chemical ordering with

$$\Delta Q^{ord} = \sum_i \sum_{j \neq i} \Delta Q_{ij}^{ord} \left[y_i^a y_j^b - x_i x_j \right] \tag{2}$$

x_i is the mole fraction of component I and y_i^a is the atomic fraction of component I on the crystallographic a sublattice sites of the B2-phase.

3. Result and discussion

3.1. Equilibrium thermodynamics of the TI-Ni-Cu system

As shown in the isothermal section of Figure 2 (a) satisfactory reproduction of experimental high-temperature phase equilibria is obtained with our re-assessed database.

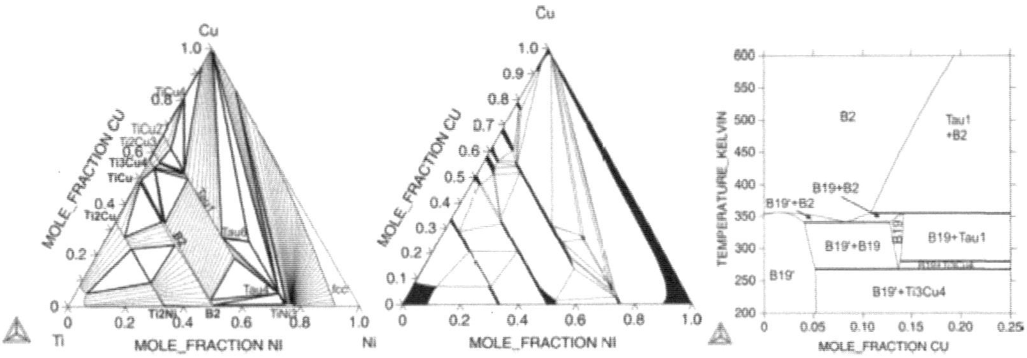

Figure 2. Calculated Ti-Ni-Cu equilibrium phase diagram (a) at 1143 K compared to experiments [15] (b) and (c) calculated Ti-Ni SMA equilibrium phase diagram with varying Cu-content, a small amount of Ti_2Ni is present in all phase regions of Fig.2 (c)

Figure 2 (c) shows the calculated equilibrium phase transformations among martensitic phases at low temperatures. The sequence of stable phase from low to high Cu-content goes along with the observed transformation sequence B19' – B19' + B19 – B19. Our calculation suggests x(Cu)>0.14 for the transition from two-step martensite formation to single B19 martensite.

3.2. Diffusivities in austenitic SMA matrix

In Figure 3 optimized diffusivities of Ti and Ni in B2-austenite are compared with experimental data. Note that for Cu-alloyed Ti-Ni-B2-phase no experimental data are available. As a first approximation we set the Cu-related ordering contributions to diffusivities of (metastable) binary Cu-Ti and Cu-Ni compounds of this phase to the same values as for the Ti-Ni compound. Assessed diffusivities of the ordered austenitic SMA matrix phase are used for the numerical solution of energy dissipation during precipitation, a pre-requisite of predictive simulation of the precipitate evolution discussed in the following section.

Figure 3. Calculated interdiffusion int he B2-ordered austenitic TiNi phase. Experimental data (symbols) from Divinski et al. [28] are included

3.3. Precipitate evolution in TI-Ni and Cu-alloyed Ti-Ni SMA

Thermodynamic calculations of individual composition-temperature states, obtained by using the assessed Ti-Ni-Cu database allow pre-definitions of proper heat-treatments for controlled precipitation already prior to kinetic simulation. Computational thermodynamics suggest negligible formation of intermetallic precipitates in Ti50Ni50-χCuχ, χ<0.2. At χ>0.2 a significant equilibrium fraction of Ti_2Ni and tau1 was calculated at temperatures below 280°C. As a consequence in Ti50-Ni25-Cu25 (TI50CU25) the temperature of full solid solution treatment to form single B2-phase was strongly increased (680°C) relative to TI50CU20 (485°C). In both alloys intermetallic Ti-Cu phases and Ti_2Ni occurred. Ti-overstoichiometric Ti50.5CU25 revealed increased fractions of potential precipitate phases relative to TI50CU25 and allowd full solution treatment at 800°C. Second phase formation was neglilgible at a low temperature of 310°C for TI50CU20. Towards lower temperatures tau1 and Ti_2Ni became more stable. Testing composition deviations from equal atom fractions of Ti and (Ni+Cu) again, Ti-overstoichiometrix TI50.5CU20 represented an interesting composition in regard of solution treatment and precipitation. While being almost fully solution-treatable at 870°C (0.01 mol.% Ti_2Ni) the Ti_2Ni-stability was strongly increased relative to Ti50Ni30 (at 500°C 3 mol.% Ti_2Ni). On the other hand in TI49CU20 tau2-phase formation was strongly increased, but this alloy would not be solution-treatable towards high temperatures. After these initial equilibrium consideration predictive (all kinetic calibration parameters in MatCalc kept default) isothermal thermo-kinetic precipitation simulations for potentially precipitation-strengthened TI50Cu25, TI50.5CU25, TI50CU20 and TI50.5CU20 were performed stepwise from high to low temperatures. From our equilibrium pre-investigation we identified intermetallic phases occurring in those compositions and considered Cu_4Ti_3, CuTi, Ti_2Cu, Ti_2Ni as potential precipitates. The resulting isolines of 10% precipitate fractions relative to maximum fraction are presented in the time-temperature-precipitation (TTP)-diagrams of Fig. 4. Isothermal aging steps with relevant precipitation were inserted to show absolute precipitate fractions in mol.%.

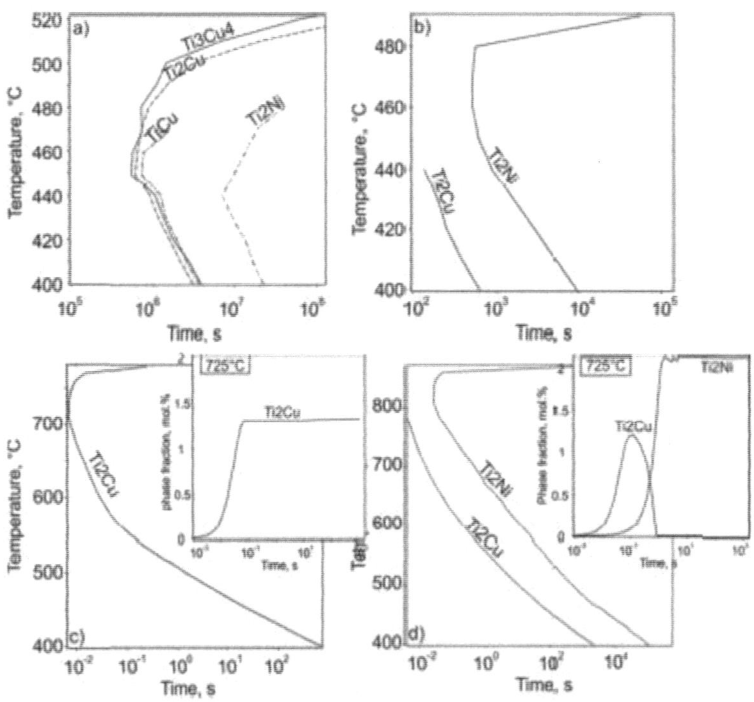

Figure 4. Simulated time-temperature-precipitation in TI50CU25 (a), TI5CU20 (b), TI50.5CU25 (c) and TI50.5CU20 (d)

The TTP-diagrams confirmed that the chosen Ti-overstoichiometry (c, d) shifted the start of precipitation towards higher temperatures and shorter aging times. The inserts in Fig. 5 (c, d) reveal a significant fraction of precipitates at the TTP-nose in TI50.5CU20 and TI50.5CU25, whereas precipitation in the TI50-variants (a, b) took too long to be of technological strengthening interest (TI50CU25), or was too weak (TI50CU20). In TI50.5CU25, Ti2Cu dominated, with precipitation starting around 780°C. Nucleation and growth at this temperature was rapid. At 400°C, the 10 % isoconcentration line was crossed after approx. 10 minutes of aging. Ti2Ni was the first precipitate to form in TI50.5CU20 at a high temperature with its TTP-nose around 850°C. Towards lower temperatures, Ti2Cu was precipitating prior to Ti2Ni. As in TI50.5CU25, kinetics became slower with decreasing temperatures, likely allowing better control of nucleation and growth during technological heat treatments: At 500°C, the 10 % iso-concentration line of Ti2Cu was crossed after approx. 10 minutes of aging.

For high-performance SMA, there is demand of combing single-step martensitic B2 to B19 transformation at relatively high temperatures with precipitation strengthening. This, in turn requires the examination of B2-matrix composition modulated by precipitation. Interestingly, calculated *T0*-temperatures (the highest theoretic temperature for the start of martensitic transformation, where *G*m(B2-austenite) equals *G*m(martensite) with the same composition) of TI50.5CU25 and TI50.5CU20 altered by precipitation were increased by 40°C and 20°C, respectively, relative to the TI50 variants.

Our simulation results thus indicate that an alloy composition between T50.5CU20 and T50.5CU25 may be a promising choice for both optimal martensitic transformation behavior and precipitation strengthening.

4. Conclusion

Based on CALPHAD-assessed thermodynamic and diffusion mobility databases, the kinetics of precipitation in Ti-Ni-Cu SMA was evaluated. Ti2Ni and metastable Ti2Cu were identified as most important precipitation phases. Computational thermodynamics confirmed that the martensitic transformations strongly depend on the amount of Cu-alloying. For single-step transformation associated with minimal transformation hysteresis, and thus optimal shape memory effect our calculation suggests alloying with Cu above approx. 15 at.%. Precipitation took place under strong control of alloying. Precipitation could change the mean matrix composition and associated martensite transformation temperature significantly. Our simulations suggest alloying of Cu from 20 to 25 at.% for precipitation strengthening in balance with high relatively martensite transformation temperatures. Slight Ti-overstoichiometric nominal composition (Ti around 50.5 at.%) increases precipitate fraction of Ti_2Ni and/or Ti_2Cu during aging of solution-treated SMA. Computational simulations with our compiled thermodynamic and diffusion mobility databases, combined with precipitate nucleation and growth models are suitable for defining ingenious heat treatments for advanced SMA performance.

References

[1] Buehler, W J; Gilfrich, J W; Wiley, R C 1963 *J. Appl. Phys.* **34** 1475
[2] Otsuka, K.; Ren, X. 2005 *Prog. Mater. Sci.* **50** 511
[3] Khalil-Allafi, J, Ren, Xy; Eggeler, G 2002 *Acta Mater.* **50** 793
[4] Khalil-Allafi, J; Eggeler, G; Schmahl, W, Sheptyakov, D 2006 *Mater. Sci. Eng. A-Struct Mater. Prop. Microstruct. Process.* **438-440** 593
[5] Akin, E. 2010, Thesis, Texas A&M University.
[6] Moberly, W J; Proft, J L; Duerig, T W; Sinclair, R. 1990 *Mater. Sci. Forum* **56-58** 605
[7] Nam, T H; Saburi, T; Shimizu, K 1990 *JIM* **31** 959
[8] Fukuda, T; Kitayama, M; Kakeshita, T; Saburi, T 1996 *Mater. Trans. JIM* **37** 1540

[9] Cesari, E; Pons, J; Santamarta, R; Segui, C; Stroz, D; Morawiec, H In: Morawiec H, Stroz D, editors. Proc XVIII Conference on Applied Crystallography. Singapore: World Scientific; 2001. p. 171

[10] Kaufmann, L; Bernstein, H 1970 *Academic Press*, New York

[11] Saunders, N; Miodownik, A P 1998 Calphad Calculation of Phase Diagrams. *Pergamon Materials Series* **1** Elsevier Science Ltd.

[12] Povoden-Karadeniz, E; Cirstea, D C; Lang, P; Wojcik, T; Kozeschnik, E 2013 *CALPHAD* **41** 128

[13] Tang, W; Sandstrom, R; Myazaki, S 2000 *J. Phase Equilib.* **21** 227

[14] Zhang, H; He, Y; Yang, F; Liu, H; Jin, Z 2013 *Thermochim. Acta* **574** 121

[15] Van Loo, V J J; Bastin, G F; Leenen, A.J H 1978 *J. Less-Common Metals* **57** 111

[16] http://matcalc.tuwien.ac.at/. Author: Kozeschnik, E. Last date of access: 2014-12-01; current MatCalc version is 5.61.057.

[17] Wang, J; Liu, C.; Leinenbach, C.; Klotz, U. E.; Uggowitzer, P. J.; Löffler, J. F. 2011 *CALPHAD* **35** 82

[18] an Mey, S. 1992 *CALPHAD* **16** 255.

[19] Ansara, I; Dupin, N; Lukas, H.L; Sundman, B. 1997 *J. Alloys Cmpd.* **247** 20

[20] Dinsdale, A T 1991 *CALPHAD* **15** 317

[21] Volmer, M; Weber, A. 1926 *Z. Phys. Cemie* **119** 277

[22] Svoboda, J; Fischer, F D; Fratzl, P, Kozeschnik, E 2004 *Mater. Sci. Eng. A-Struct. Mater. Prop. Microstruct. Process.* **385** 166

[23] Onsager, L 1931 *Phys. Rev.* **37** 405

[24] Onsager, L 1931 *Phys. Rev.* **38** 2265

[25] Svoboda, J; Turek, I; Fischer, F D 2005 *Philos. Mag.* **85** 3699

[26] Kozeschnik, E; Janssens, K; Bataille, C 2012 *Modeling Solid-State Precipitation. Momentum Press*, New York.

[27] Helander, T; Agren, J 1999 *Acta Metal.* **47** 3291

[28] Divinski, S V; Stloukal, I; Kral, L; Herzig, C 2009 *Defect Diffusion Forum* **289-292** 377

Cesari E, Pons J, Santamarta R, Segu
ı C, Str
oz D, Morawiec H.

Investigation of alumina samples cast from concentrated polymer suspensions

L Angyal[1], Zs Mátyás-Karácsony[2], A Kállay-Menyhárd[1] and Gy Bánhegyi[2]

[1]Department of Physical Chemistry and Materials Science, Budapest University of Technology and Economics, Hungary
[2]Bay Zoltán Nonprofit Ltd. for Applied Research, Hungary

E-mail: angyal_lilla@hotmail.com

Abstract. In this research we investigated a new ceramic forming process. The technology is based on a concentrated ceramic suspension with ethylene acrylic acid copolymer binder system. After the forming procedure to get the final products, the samples were dried, debinded and sintered on high temperature. The samples were studied by different ways during the process: rheological studies, surface properties (optical and scanning electron microscopy), mercury porosimetry.

1. Introduction

Various ceramic processing technologies utilize concentrated ceramic suspensions to produce shaped ceramic final products (e.g. slip casting, tape casting, gel casting, low pressure injection molding etc.). In these technologies the "green body" is obtained by drying the suspension followed by debinding (thermal decomposition of the polymer component) and finally by sintering of the "brown body". The advantage of these techniques over powder injection molding (PIM) is that they usually require less organic binder, thus they are cheaper and more environmentally friendly.

The aim of the research was to invent a new technology for making various shape of alumina products. The most important requirements for the technique are the simplicity, the cost-effectiveness, the size limitation and the good mechanical properties of the products. To reach these requirements the proper composition of the suspension is needed.

2. Materials and methods

In this research we investigated a special system containing alumina filler and ethylene-ethyl acrylate (EEA) solutions as liquid phase. As EEA dissolves in water only under high pH conditions, we have tried to use both NaOH and concentrated aqueous ammonia solution to produce the polymer solution. It has to be noted, that not all EEA grades dissolve in water even under such conditions, therefore first we had to select the proper polymer grade. When preparing the samples we varied the following parameters: NaOH or NH_4OH for dissolving the polymer, the concentration of the polymer solution (the relative amount of the filler and of the liquid phase was kept constant), and the composition of the filler: in some samples we replaced a part of Al_2O_3 by $Al(OH)_3$. It is expected that in the case of $Al(OH)_3$ the greater concentration of surface Al-OH groups as compared to alumina (where such groups are formed only by partial hydration) produces stronger physical network. The potential advantage of using ammonia instead of NaOH is that the ammonium salts decompose and evaporate at high temperature, not causing further contamination in the pure Al_2O_3 product to be prepared. It is more important in the case of high purity alumina samples [1-6].

The details of the examined samples are shown in the table 1. There were two outstanding samples which are containing only alumina powder with low polymer concentration and NaOH base.

During the research different measurement methods were used to study the samples. These were the reological studies, optical and scanning electron micrography, and mercury porosimetry.

3. Rheological properties of the EEA suspensions

The rheological properties of the EEA solutions and the suspensions made of them were studied by rotation viscometry as a function of frequency and shear rate respectively. Nonlinearities as function of shear rate were observed which is called dilatancy [7].

The following table contents the data of the measured samples:

Table 1. The details of the examined samples

Samples:	**1**	**4**	**6**	**9**	**11**
EEA solution concentration (with NaOH basis)	5%	5%	10%	10%	15%
Powder	Al_2O_3	$Al_2O_3 + Al(OH)_3$	Al_2O_3	$Al_2O_3 + Al(OH)_3$	Al_2O_3

The diagram on the figure 1. shows that the top of the processability window is at 10% polymer content. If the polymer solution is more concentrated the viscosity is too high for easy handling.

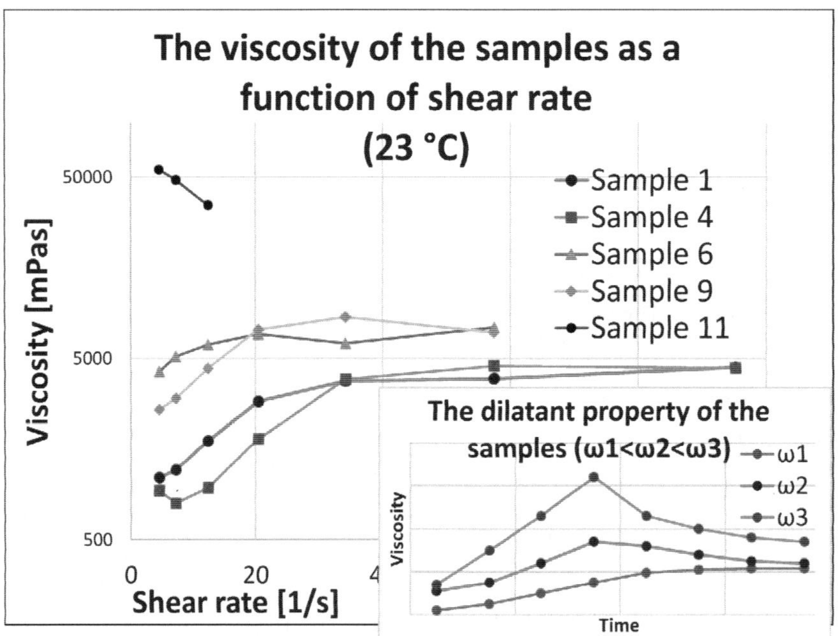

Figure 1. The reological studies

4. Surface study

For the optical studies a Keyence VHX-2000 digital optical microscope was used. During the test different surface defects were detected (Figure 2., 3.). The samples with higher $Al(OH)_3$ content had this kind of defect and these samples also had surface deformation after sintering e.g. bubbles, meniscus, deflection of the sample.

A few samples showed outstanding surface properties. The scanning electron micrograph (Hitachi TM-1000) of sample 1 is shown on figure 4. Both sample 1 (containing 98,83% Al_2O_3 powder and 1,17% EEA with equimolar NaOH) and sample 6 (containing 97,64% Al2O3 and 2,36% EEA with equimolar NaOH) has almost the same surface properties.

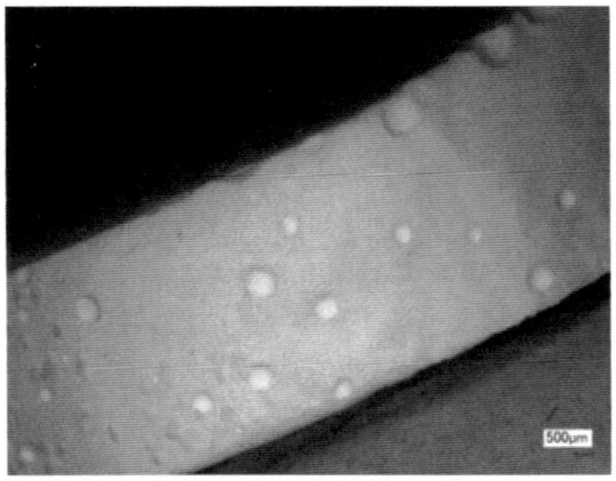

Figure 2. Surface defects – bubbles – on optical microscope

Figure 3. Surface defects – cracks – on optical microscope

Figure 4. The surface of an outstanding sample (SEM micrograph)

5. Mercury porosimetry

3 chosen samples were studied by mercury porosimetry. From the results of this measurement the distribution of the pores can be calculated.

During the measurement, the height of the mercury level in the capillary was recorded. While the pressure is increasing the mercury penetrates more and more into the pores of the sample and this causes the decrease of the mercury level in the capillary. [8]

The following table (table 2.) shows that the sample 1 has the smallest distribution of pore size. Moreover we can see that the increase of the sintering temperature causes a quite big decrease in the pore volume.

Table 2. The results of the porosimetry studies

Measurement data	Sample 1	Sample 6	Sample 6 (sintered at 1650°C)
Range of pore diameter [μm]	0,088-26	0,035-37,5	0,035-48
Most frequent pore size range [μm]	0,65-2,5	0,6-3,4	0,7-3,4
Cumulative pore volume [cm^3/g]	0,096	0,119	0,078
Pore volume [%]	19,4	22,6	16,3

6. Conclusions

Negative results: The green bodies which were made with NH_3 solution were too fragile to place into the furnace and these samples got relevant deformity during the drying. The density of the sintered samples were significantly lower than the theoretical density of the Al_2O_3 which points to major closed porosity.

The samples which contain both Al_2O_3 and $Al(OH)_3$ powders also became fragile after the drying or the debinding process. Moreover several surface defects were observed after the heat treatment.

Positive results: The sample 1 and sample 6 showed the best properties in every study. These contain only Al_2O_3 powder. The difference between them is in the concentration of the EEA solution. Both of them had outstanding surface properties and the SEM micrograph proves the homogenous structure. There wasn't any surface deformation after the drying or the sintering and both of them had proper stability after the debinding.

Overall the sample 1 showed better results in almost every test e.g. viscosity of the suspension, density, porosity, and mechanical properties. This sample has similar mechanical properties as in other researches.

Acknowledgements

The authors gratefully acknowledge the TÁMOP 4.2.2.A11/1/KONV2012 0075 project supported by the European Union and co-financed by the European Social Fund.

References

[1] F. S. Rana, B. B., A. Sultan, F. Shaheen, 2009 *J. Chem. Soc. Pakistan* **31** 1
[2] Y. K. N. Omura, Y. Hotta, K. Sato, 2005 *J. Ceram. Soc. Japan*,
[3] R. Moreno and M. I. Nieto, 2001 *Mater. Lett.*, no. **February** 324
[4] M. S. Centre, 2002 *Bull. Mater. Sci.* **25**, no. 6. 565
[5] A. Bouvy, 2007 High Performance EAA Copolymer Dispersions for Applications in Coatings **23** 9 44.
[6] T. Csányi, 2009 *Építőanyag* **61** 6
[7] E. Ewais, A. Zaman, and W. Sigmund, 2002 *J. Eur. Ceram. Soc.*, **22** 16, 40
[8] Z. Wagner, 1981 *Építőanyag* **33** 150

Structure and properties of ZrO_2–MgO powders

S P Buyakova[1,2,3], E S Kalatur[2], A S Buyakov[1,2], and S S Kulkov[1,2]

[1]Tomsk State University
[2]Institute of Strength Physics and Materials Science SB RAS
[3]Tomsk Polytechnic University

E-mail: sbuyakova@ispms.tsc.ru

Abstract. It have been studied magnesia stabilized zirconia (ZrO_2–MgO) powders produced by high-frequency decomposition of water salt solutions. It have been shown that due to precursor dissociation powders consist of spherical particles with smooth surface and irregular hollow aggregates resembling foam. An increase in the magnesium concentration in the precursor has an effect on the powder properties, namely, the specific surface decreases significantly and the bulk density increases. Zirconia in the nanostructured state is present only in the powders with 8.6 and 13.9 mole% MgO, with the size of the c-ZrO_2 crystallites in them no more than 100 nm. In powder with 43.3 mole% of MgO the magnesia content in c-ZrO_2 solid solution corresponds to its solubility limit of 22.2 mole% at temperature 2100°C.

1. Introduction

Plasma spray synthesis is one of the most efficient methods of synthesizing ultrafine powder systems of various chemical compounds [1]. A high dissociation temperature of precursors in plasma and a high cooling rate of reaction products provide a high nucleation rate and low nucleus growth rate, which makes possible the formation of the nanostructured state in the system.

A highly nonequilibrium phase formation in plasma spray synthesis makes this method attractive for the production of solid solution powders with limited solubility of components under normal conditions. This group of solid solutions includes the ZrO_2–MgO system, as magnesia is not dissolved in zirconia under normal conditions [2]. Partial substitution of zirconium atoms by magnesium atoms in the lattice ensures the stability of high-temperature β or γ modifications of ZrO_2 under normal conditions. The phase transition of high-temperature to low-temperature ZrO_2 modifications is accompanied by an increase in the unit cell volume, leading to the formation of a defect structure when the ceramics is cooled down to the sintering temperature of the powder mixture. Owing to this fact, only zirconium ceramics on the basis of β or γ modifications have found practical application.

A characteristic feature of the ZrO_2–MgO ceramics is a unique combination of high strength and fracture toughness. The type of the ZrO_2 lattice and, consequently, properties of the ceramics depend on the MgO content in solid solution determined by the ratio of components in the precursor. It should be expected that the precursor composition affects not only the phase composition of the powder but also its particle size distribution, which is important for the structure and properties of the ceramics as a material produced by powder technology.

The aim of this paper is to study how the magnesia content in the ZrO_2–MgO system influences the phase composition, crystal structure and morphology of powder particles produced by plasma chemical synthesis.

2. Materials and experimental procedure

Magnesia stabilized zirconia (ZrO_2–MgO) powders were produced by denitration of aqueous solutions of zirconium and magnesium nitrate salts in high-frequency plasma. The ratio of zirconium and magnesium salts in the precursor was calculated from the magnesia content in the synthesized ZrO_2–MgO powders: 8.6, 13.9, 25.4, 35 and 43.3 mole%.

The powder particle morphology was analyzed using scanning and transmission electron microscopy. The specific surface of the powders was determined by the BET method. The phase composition and crystal structure were analyzed by the results of X-ray diffraction with CuK_{α} radiation. The ZrO_2 phase content was calculated from the ratio of the integral intensities of diffraction lines in X-ray diffraction patterns.

Powder compaction was analyzed by diagrams in the coordinates $lgV=f(lgP)$, where V is the powder volume and P is the compaction pressure, under quasi-static compression in a steel mould.

3. Results and discussion

Electron microscopic analysis of the particle morphology has shown that the ZrO_2–MgO powders of all studied compositions consist of spherical particles and irregular aggregates that resemble solidified foam (Fig. 1). The spherical particles had a smooth surface, appearing on the fracture surface as spheres with a wall thickness of more than 0.25 μm. Most of the aggregates were transparent to the electron beam even at low 40 kV accelerating voltage of the electron beam in the microscope.

a) b) c)

Figure 1. Powders of ZrO_2-MgO with a) 8,6 mol. %; b) 25,35 mol. %; c) 43,3 mol. % MgO

Variation in the ratio of the precursor components had an effect on the powder particle size distribution. The proportion of spherical and irregular particles in the studied powders was different. The fraction of spherical particles in the powders increased with a growing magnesia content in the precursor. For example, in the ZrO_2 – 8.6 mole% MgO powder the amount of spherical particles did not exceed 20%, and in the powder with the largest MgO content 43.3 mole% the amount of spherical particles exceeded 70% (Fig.2).

In powders synthesized by pyrolysis of aqueous salt solutions in low-temperature plasma some particles inherit the shape of aerosol droplets sprayed into the plasma (spherical particles), while other particles are fragments of spheres broken by the release of overheated intradroplet gas. The observed increase in the amount of spherical particles in the ZrO_2–MgO powders can be related to a decrease in the surface tension of aerosol droplets at a growing concentration of magnesia in the precursor, due to which the release of reaction gases was facilitated and the amount of unbroken spherical particles increased.

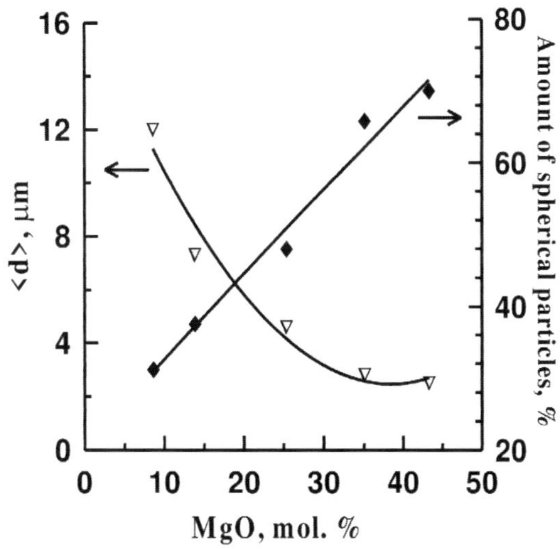

Figure 2. Mean size of grains and amount of spherical particles in ZrO_2 –MgO

Changes in the precursor composition influenced the average powder particle size <d> that was significantly reduced with the growing concentration of magnesia. The average particle size in the powder with the smallest MgO content 8.6 mole% was 12 μm, and in the powder with 43.3 mole% MgO it was 25 μm (Fig. 2). In this case, the average size of spherical particles did not depend on the precursor composition, being the same in all studied powders, while the size of aggregates was reduced.

Naturally, the powder particle geometry variation with increasing MgO content exerted an effect on the powder properties, [3, 4] (Fig. 3). An increase in the amount of spherical particles with smooth surface led to a decrease in the specific surface S_a of the ZrO_2–MgO powders. The powder with the smallest MgO content 8.6 mole% had the largest value S_a=15 m^2/g. The specific surface of the powder with 43.3 mole% MgO was 5.5 m^2/g. The growing amount of spherical particles and, correspondingly, the reducing amount of hollow aggregates led to an increase in the bulk density of the studied powders (Fig. 3).

Variation in the particle size distribution of the ZrO_2–MgO powders also had an effect on their compaction. The growing MgO content in the powders led to an increase in the parabolicity coefficient of the compaction equation (Fig. 3) determined as the slope of the compaction diagram in the coordinates $lgV=f(lgP)$. Indirectly, the parabolicity coefficient of the compaction equation of a dispersed system reflects the reduction rate of its volume in the mould under increasing pressure. The observed compaction rate growth for the studied powders is related to a decrease in interparticle friction due to an increase in the amount of spherical particles with smooth surface in them.

According to X-ray diffraction results, zirconia in the powders with the magnesia content less than 35 mole% was in the cubic (c-ZrO_2), tetragonal (t-ZrO_2) and monoclinic (m-ZrO_2) modifications. In this case, the c-ZrO_2 fraction in these powders was more than 90%. In the powders with 35 and 43.3 mole% MgO zirconia was only in the cubic modification. Along with the ZrO_2 reflections in the diffraction patterns of the powders with the MgO content above 13.9 mole% there were intensive reflections corresponding to magnesium zirconate $Mg_2Zr_5O_{12}$ and less intensive reflections corresponding to MgO. The integral intensity of the diffraction lines of $Mg_2Zr_5O_{12}$ increased with the growing MgO content in the ZrO_2–MgO system.

The calculation of average crystallites by X-ray reflection broadening for each phase showed that zirconia in the nanostructured state was present only in the powders with 8.6 and 13.9 mole% MgO. The size of the c-ZrO_2 crystallites in them did not exceed 100 nm. The average size of $Mg_2Zr_5O_{12}$ crystallites in all powders was less than 25 nm.

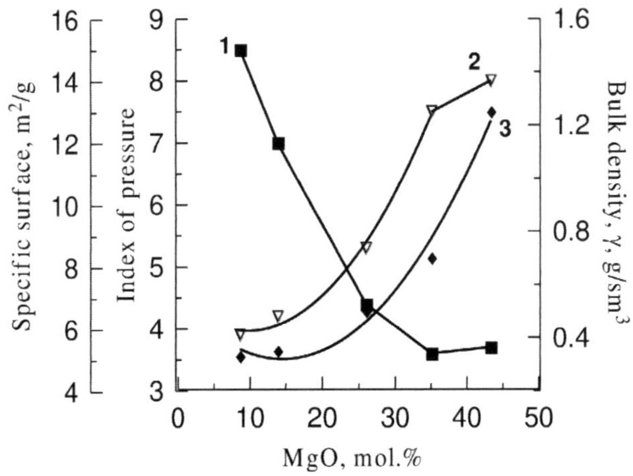

Figure 3. The raw powder density, specific surface and mould ability vs. magnesia content in system

The magnesia content growth in the studied ZrO_2–MgO powders was accompanied by a decrease in the cell parameter a_c of the cubic ZrO_2 modification, which can be, first of all, a result of substitution of Zr^{4+}(r=0.8) cations by Mg^{2+}(r=0.72) cations of smaller radius. Comparison of cell parameters a_c in the studied powders with the data provided in [5] for known MgO concentrations in a solid solution on the basis of cubic zirconia showed that the amount of magnesium cations substituting zirconium cations is less than it could be at complete dissolution of magnesia present in the studied powders. For example, for the powder with 8.6 mole% MgO the magnesia content in solid solution was 7.9 mole%, and for the powder with 35 mole% MgO its content in solid solution did not exceed 16 mole% (Fig. 4). However, in the ZrO_2 – 43.3 mole% MgO powder the magnesia content in c-ZrO_2 solid solution corresponded to its solubility limit of 22.2 mole% at 2100°C, according to the phase diagram given in [2].

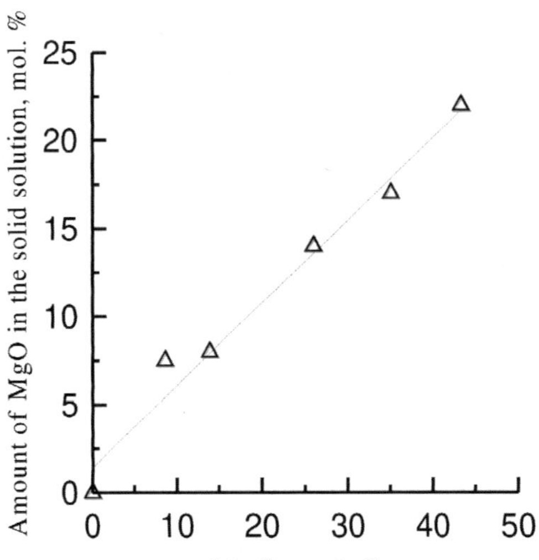

Figure 4. MgO content in a powder solid solution vs. magnesia content in system

Calculation of unit cell parameters a and c for $Mg_2Zr_5O_{12}$ showed that their value increases with the growing MgO content in the ZrO_2–MgO powders, which is related to the closeness of $Mg_2Zr_5O_{12}$ to the stoichiometric composition corresponding to the 28 mole% magnesia content in the system. In the powder with 43.3 mole% MgO the parameter values for $Mg_2Zr_5O_{12}$ correspond to those given in the ASTM cards (Nos. 41–185): a=0.94445 nm and c = 0.87450 nm.

4. Conclusion

- The magnesia stabilized zirconia (ZrO_2–MgO) powders produced by precursor dissociation in a high-frequency plasma consist of spherical particles with smooth surface and irregular hollow aggregates resembling foam. As a result of increasing magnesium salt nitrate concentration in the precursor, the amount of aggregates decreases and the amount of spherical particles grows.
- An increase in the magnesia concentration in the precursor has an effect on the powder properties, namely, the specific surface decreases significantly and the bulk density increases.
- Zirconia in the nanostructured state is present only in the powders with 8.6 and 13.9 mole% MgO, with the size of the c-ZrO_2 crystallites in them no more than 100 nm.
- The amount of magnesium cations substituting zirconia cations is smaller than it could be at complete dissolution of magnesia present in the studied powders. However, in the ZrO_2 – 43.3 mole% MgO powder the magnesia content in c-ZrO_2 solid solution corresponds to its solubility limit of 22.2 mole% at temperature 2100 °C.

5. Acknowledgments

This work has been partial financial supported by Tomsk State University Competitiveness Improvement Program and RFBR Project No. 14-08-31087/14

References

[1] Gurav A., Kodas T., Pluym T., Yun Xiong. 1993 *Aerosol Science and technology* **19** 411
[2] Garvie, R., Nicholson, H. 1972 *J. Am. Ceram. Soc.* **55** (6) 303
[3] Buyakova, S., Kulkov, S.: 2010 *Inorganic Materials* **46** (10) 1155
[4] Buyakova, S., Promakhov, V., Kulkov, S.: 2012 *Powder Metallurgy and Metal Ceramic* 1
[5] B.C. Muddle, R.H.J. Hannink 1986 *J. Am. Ceram. Soc.* **69** (7) 547
[6] Seung Kun Lee, Rajan Tandon, and Michael J. Readey 2000 *J. Am. Ceram. Soc.* **83** (6) 1428.
[7] Kalatur E, Buyakova S, Kulkov NS, Gotman I, Kocserha I 2014 *Építőanyag JSBCM* **66** 2 31
[8] Gömze AL, Gömze LN 2013 *IOP Conf. Ser.: Mat. Sci. Eng.* 47012033
[9] Orosz V, Gömze A.L 2012 *Int. J. Microstructure and Material Properties* **7** (2/3) 205
[10] Kulkov NS, Buyakova PS, Gömze AL 2014 *Építőanyag JSBCM* **66** (1) 2

Production of refractory chamotte particle-reinforced geopolymer composite

T Kovářík, L Kullová, D Rieger

New Technologies - Research Centre, University of West Bohemia, Plzeň, Czech Republic

E-mail: toko@ntc.zcu.cz

Abstract. Geopolymer resins are obtained by alkaline activation of aluminosilicate sources where raw calcined clays are one of the suitable potentialities. Besides the fact that chemical composition has an essential effect on final properties of the geopolymer binder, the type of filler strongly affected resulting properties of such granular composite. However, very few comparative studies have been done on detail description of composite systems: binder – granular filler, in relation to aggregate gradation design and rheology properties of the mixture. The aim of this work is to develop and describe granular composite concerning workability of the mixture and kinetics of geopolymerization/polycondensation through flow behaviour. The rheological measurements indicated that initial viscosities of the mixtures and their evolution are different for various proportions of the filler. Moreover, it was demonstrated that increase in complex viscosity responds to the creation of chemical bonds and the formation of structural network. Finally, a correlation of the mechanism of geopolymer formation was carried out by differential scanning calorimetry (DSC).

1. Introduction

Geopolymer binders can be considered as a class of inorganic polymers having predominantly amorphous structure consisting of tetrahedrons SiO4 and AlO4 mutually bonded by oxygen bridges 1]. Thus, the complex structure of polymer network is formed based on Si-O-Al chains. The negative charge in the aluminosilicate network is balanced by alkali cations (mostly Na+, K+, Ca2+) in the cavities of framework [2,3]. Geopolymers are generally prepared via alkaline solutions by mixing of an aluminosilicate raw material in form of powder derived from natural minerals (calcined clays, natural pozzolans) or industrial products (fly ash, blast-furnace slag, silica rich waste sludge) [4-6]. The mechanism for geopolymer formation involves a reaction between aluminosilicates and alkali metal hydroxides/silicates [7]. The first disaggregation process entails the severance of the Me–O (where Me is Na+ or K+), Si–O–Si, Al–O–Al and Al–O–Si bonds on the surface of solid particles. Dissolution of the solid aluminosilicate source by alkaline hydrolysis produces aluminate and silicate species [8]. Also, the effect of hydroxy groups on the Al-O-Si and Si-O-Si bonds results in disaggregation/dissolution of aluminates-silicates and formation of hydroxylated complexes such as Si(OH)4 −, Al(OH)4 − or Al(OH)3−6 [9]. Once, the species and complexes are released and incorporated in aqueous phase by dissolution, the accumulation enhances contact among the disaggregated products and coagulated structure is formed [10]. This results, in concentrated solutions, into the gel precipitation through oligomers networking in aqueous phase by condensation process.

The geopolymer binders with a cured rigid structure are represented by good resistance to high temperatures, acid resistance and mechanical properties [11,12]. Hence, it is an attractive alternative to standard Portland cement where advanced technical properties of binders are requested. The physical properties of geopolymer mixtures are strongly dependent on the type and chemical composition of raw materials, Si/Al molar ratio of prepared slurry and cured conditions [13-15]. The processing properties of geopolymer binders also depend on the rate of water add to the mixture, proportion of raw materials and stirring conditions. Also, plate/lamellar-like particle shape of clay minerals affects rheological properties of the fresh paste, such as thixotropic behaviour with a relatively high yield stress [16]. The rheology of fresh geopolymer pastes is highly complex process because of the mutual interactions such as surface forces during dissolution including chemical and mineralogical factors, contact forces between particles, ratio of water, mixing conditions as well as proportion and type of filler [17].

The aim of the present study was to investigate the effect of granular filler of ceramic grog on the rheological properties of variously filled geopolymer paste based on potassium silicate solution. Differential scanning calorimetry (DSC) was used to study the curing process at elevated temperature through reactions of heat change.

2. Experimental and methods

The geopolymer paste was synthesized as a two component system consists of alkaline potassium silicate solution with silica modulus 1.6 and reactive schistous clay residues as a main source of solid silica and alumina as presented in Table 1 with particle size distribution details as follows: d50 at 3.98 μm, d90 at 6.87 μm and BET surface area 12.99 m2 g-1. The mixture was formulated by molar ratios: SiO_2/Al_2O_3 = 2.5, SiO_2/K_2O = 6.80, Al_2O_3/K_2O = 2.80, H_2O/K_2O = 19.00. The ceramic grog as granular filler was used with maximum grain size 1 mm and specific aggregates gradation design as presented in Figure 1.

Table 1. Main chemical composition of solid and liquid components (mass %)

	SiO_2	Al_2O_3	CaO	MgO	TiO_2	K_2O	Fe_2O_3
Clay residues	50.20	45.10	0.12	0.15	1.45	0.62	1.00
Silicate solution	17.57	-	-	-	-	17.10	-

The geopolymer paste was prepared in the mass ratio 1/0.8 (solid/liquid). The materials were proportioned by weight ratio for optimum workability and the mixing was conducted in a laboratory automatic blender for 6 min (1000 rpm). After that the batch of A111 filler was added and the final mixture was blended for an additional 5 min.

The filler gradation design was applied with various proportion of ceramic grog to clay residues expressed as mass% in the range 260 – 320 mass%.

Dynamic rheological measurements were performed using a rotary rheometer (Ares G2, TA Instruments) operated in oscillation mode with small amplitude. Time sweep experiments were performed at an angular frequency 10 rad/sec (1.59 Hz) and strain rate 0.01% in plain-plain geometry as presented in Figure 1, detail (a, b). The curing kinetic at 30, 60 and 80°C was evaluated by differential scanning calorimetry (DSC Q200, TA Instruments) using Tzero hermetic alodined auminum pans/lids. The purge gas was nitrogen at flow rate 50 mL min-1.

3. Results and Discussion

3.1. Rheology

As shown in Figure 2, the rheological time sweep measurements indicate the increasing of viscosity in time for all filled mixtures. The time zero is taken as a + 15 min in virtue of preparation of the mixture for each measurement at 30 °C ± 0.1°C. The higher proportion of granular filler increased viscosity rapidly in comparison with the other states, as presented in the Figure 2 (4). This evolution corresponds to the creation of chemical bonds within the reactive mixture [18] and evidently to the rapid desiccation of the gel infilled by fine particles of ceramic grog. As presented in Figure 2 (0), the unfilled paste indicates minimal viscosity change in setting time (8000 s). Whereas, the filled mixtures started notable increasing their viscosities after 2000 s. The geopolymer mixture filled of 260 mass% reached its viscosity value to 5 GPa.s at 8000 s. Trend indicates that the increasing of proportion of the filler led to increasing of viscosity value and shorter setting time. As presented in Figure 2 (4), the highly filled paste increased its ultimate viscosity value to 7.5 GPa.s during 5000 s. The mixture was spontaneously detached plain-plain geometry after viscosity value reached its maximum.

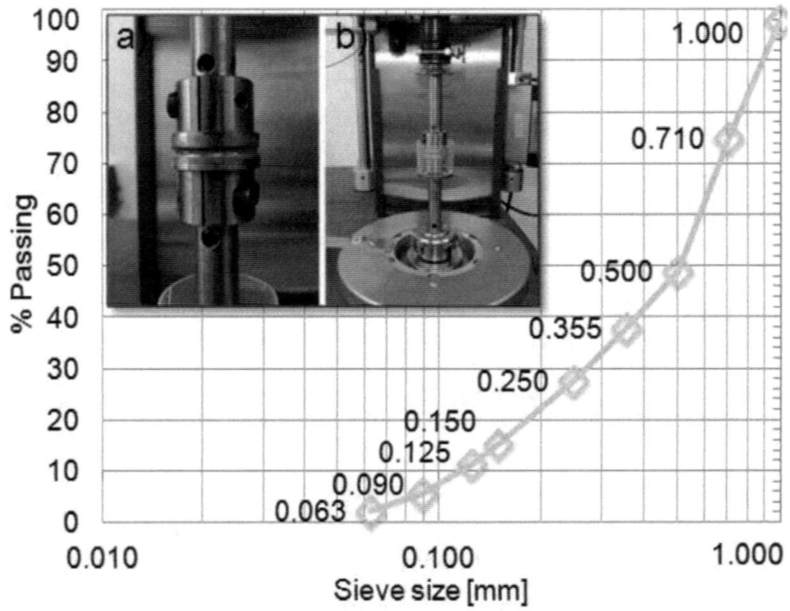

Figure 1. Gradation curve for ceramic grog with a detail of plain-plain geometry (a, b) for dynamic rheological measurements

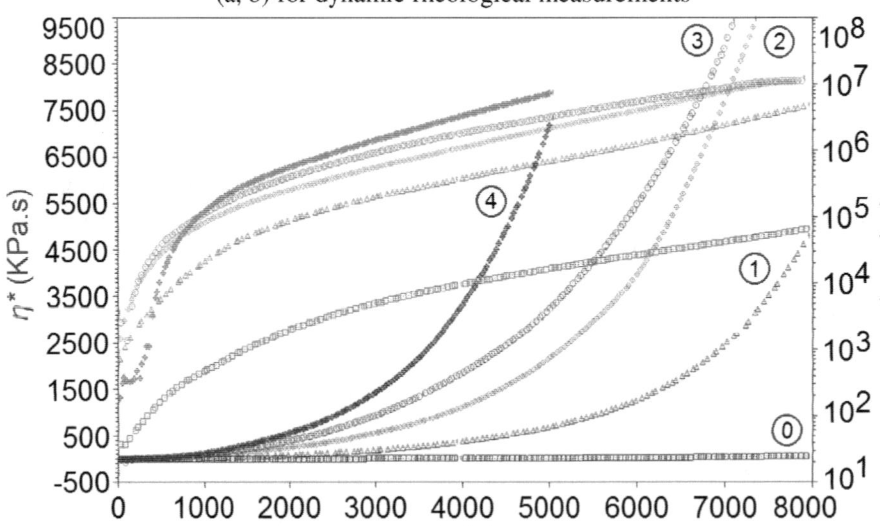

Figure 2. Viscosity curves in function of time of filled mixtures (1) 260 mass %, (2) 280 mass %, (3) 300 mass %, (4) 320 mass % and (0) unfilled paste

3.2. Curing kinetics

The geopolymer paste based on potassium silicate solution was measured by differential scanning calorimetry and started at a time + 10 min in virtue of stirring and moulding/setting the DSC run. The obtained thermographs indicate curing reactions as a heat evolution under isothermal conditions at 30, 60 and 80°C. The exothermic peaks appeared immediately after starting the measurements indicating the process of partial dissolution/disaggregation of solid particles in the strong alkaline aqueous solution as presented in Figure 3 (A1, B1, C1) [19]. Further, the formation of alumina/silica-hydroxy species and consolidation of larger network started as showed in Figure 3 (B2, C2) [20].

It is evident that the elevated temperature affects the mixture to release more heat within the increasing temperature. Also, the reaction rate in Figure 3 (C1) at 80°C is higher than that of A1 at 30°C implying that dissolution of raw material is accelerated by elevated temperature. This indicates that elevated temperature 60-80°C speed up curing time to ~ 20 min and both mechanism disaggregation and reorganization proceeds in short time period.

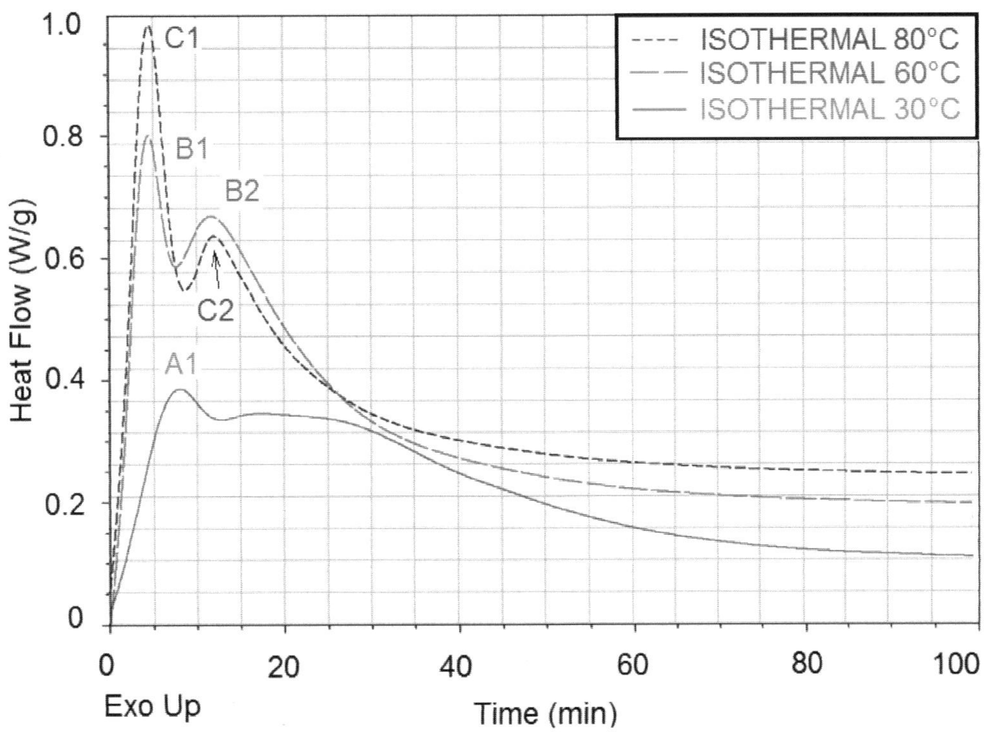

Figure 3. The effects of heat evolution under isothermal conditions at (A) 30 °C, (B) 60 °C, (C) 80°C

4. Conclusion

In this investigation, the influences of the proportion of granular filler to geopolymer paste were studied. Therefore, the ceramic grog with specific aggregate gradation design was applied in various proportions. It was showed, that increasing of proportion of the filler led to increasing of viscosity value and shorter setting time. Moreover, it was demonstrated that increasing of complex viscosity indicates the formation of structural network and gradual desiccation of the gel.

The curing kinetic of experimental mixture was described by effects of heat evolution at normal and elevated temperatures under isothermal conditions. The exothermic effects indicating acceleration of structural changes by elevated temperature in the range 60-80°C.

Acknowledgements
The result was developed within the CENTEM project, reg. no. CZ.1.05/2.1.00/03.0088, cofunded by the ERDF as part of the Ministry of Education, Youth and Sports OP RDI programme and, in the followup sustainability stage, supported through CENTEM PLUS (LO1402) by financial means from the Ministry of Education, Youth and Sports under the National Sustainability Programme I.

References

[1] Davidovits, J. 1989 *J. Thermal Anal.* **35** 429
[2] Šesták, J.; Foller, B. 2012 *J. Therm. Anal. Calorim.* **109** 1
[3] Alonso, S.; Palomo, A. 2001 *Mater. Lett.* **47** 55
[4] Palomo, A.; Alonso, S.; Fernández-Jiménez, A.; Sobrados, I.; Sanz, J. 2004 *J. Am. Ceram. Soc.* **87** 1141
[5] Fernández-Jiménez, A.; Palomo, A.; Sobrados, I; Sanz, J. 2006 *Micro. Meso. Mat.* **91** 111
[6] García, R.; Vigil de la Villa, R.; Vegas, I.; Frías, M.; Sánchez de Rojas, M. I. 2008 *Con. Build. Mat.* **22** 1484
[7] Duxson, P.; Fernández-Jiménez, A.; Provis, J. L.; Lukey, G. C.; Palomo, A.; Van Deventer, J. S. J; 2007 *J. Mater. Sci.* **42** 2917
[8] Swaddle, T. W.; Salerno, J.; Tregloan, P. A. 1994 *Chem. Soc. Rev.* **23** 319
[9] Shi, C.; Feznández-Jiménez, A.; Palomo, A. 2011 *Cem. Con. Res.* **41** 750
[10] Duxson, P.; Lukey, G. C.; Separovic, F.; Van Deventer, J. S. J. 2005 *Ind. Eng. Chem. Res.* **44** 832
[11] Duxson, P.; Provis, J. L.; Lukey, G. C.; Mallicoat, S. W.; Kriven, W. M.; Van Deventer, J. S. J. 2005 *Colloid. Surf. A: Physicochem. Eng. Asp.* **269** 47
[12] Palomo, A.; Blanco-Varela, M. T.; Granizo, M. L.; Puertas, F.; Vazquez, T.; Grutzeck, M. W. 1999 *Cem. Con. Res.* **29** 997
[13] Burciaga-Diaz, O.; Escalante-Garcia, J. I.; Gorokhovsky, A. 2012 *Cem. Con. Comp.* **34** 18
[14] Duxson, P.; Mallicoat, S. W.; Lukey, G. C.; Kriven, W. M.; Van Deventer, J. S. J. 2007 *Colloid. Surf. A: Physiochem. Eng. Asp.* **292** 8
[15] Rovnaník, P;. 2010 *Con. Build. Mat.* **24** 1176
[16] Provis, J. L.; Duxson, P.; Van Deventer, J. S. J. 2010 *Adv. Pow. Tech.* **21** 2
[17] Romagnoli, M.; Leonelli, C.; Kamse, E.; Gualtieri, M. L. 2012 *Con Build. Mat.* **36** 251
[18] Poulesquen, A.; Frizon, F.; Lambertin, D. 2011 *J. Non-Crys. Sol.* **357** 3565
[19] Zhang, Z.; Wang, H.; Provis, J. L.; Bullen, F.; Reid, A.; Zhu, Y. 2012 *Thermochim. Acta.* **539** 23
[20] De Silva, P.; Sagoe-Crenstil, K.; Sirivivatnanon, V. 2007 *Cem. Con. Res.* **37** 512

Porous and Dense Cordierite Ceramic from Illite Clay

M Rundans[1], I Sperberga[1] and G Sedmale[1]

[1]Institute of Silicate Materials, Faculty of Material Science and Applied Chemistry, Riga Technical University, Riga, Latvia

E-mail: maris.rundans@rtu.lv

Abstract. Ternary system $MgO-Al_2O_3-SiO_2$ contains many technically valuable crystalline phases such as enstatite, forsterite, spinel and cordierite. For most of these phases coefficient of thermal expansion (CTE) is extremely low (as low as $1.5 \cdot 10^{-6} K^{-1}$) which allows for them to be used in conjunction with other thermally demanding materials like quartz, for example. Also, other important properties like their high mechanical strength and good dielectric permeability is of value for potential use in practical applications. Formation of cordierite from oxide powders takes place above 1350 °C. Rational preparation of such material requires modifications in synthesis rout, i.e., addition of flux forming agents or presence of volatile compounds. In this work the use of Latvian illite clay as partial raw material for preparation of both dense and porous cordierite ceramics was studied. No additional artificial flux and/or volatiles were used. Obtained dense and porous ceramic samples were tested for their phase composition, mechanical strength and porosity. It was determined that the use of illite clay of no less than 1/3 of total mass was enough to form both extremely dense and tough (compressive strength of about 400 MPa) and extremely porous (about 59% apparent porosity) materials by just adjusting thermal treatment regime. The X-Ray diffraction of the samples showed that formation of single-phase crystalline cordierite can also be achieved in relatively lower temperatures, e.g., as low as 1300°C.

1. Introduction

Modern ceramic materials created in the system $MgO-Al_2O_3-SiO_2$ has a broad range of applications in various fields, as this system can be found such refractory phases as enstatite ($MgO \cdot SiO_2$), forsterite ($2MgO \cdot SiO_2$), spinel ($MgO \cdot Al_2O_3$), cordierite ($2MgO \cdot 2Al_2O_3 \cdot 5SiO_2$), mullite ($3Al_2O_3 \cdot 2SiO_2$), etc. [1]. These crystalline phases have low or even negative linear thermal expansion coefficient (CTE), high mechanical, chemical and thermal durability. For example, cordierite ceramics with its low CTE ($\alpha = 1.5 - 4.0 \cdot 10^{-6} K^{-1}$) safely withstand sudden changes in temperature and is used as a refractory material [1,2]. These materials also have good dielectric properties and therefore are used for such purposes as the integrated circuits, electronic packing and others.

Naturally occurring magnesium aluminum silicate, such as cordierite and sapphirine are rareoccurring natural minerals, and therefore these crystalline materials must be obtained synthetically. The most widely used method is the high-temperature solid-state synthesis, rarely – sol-gel synthesis and glass melt re-crystallization. Various magnesium, aluminum and silicon oxides containing raw materials are used for the synthesis, of which the most common are talc and kaolin [3]. Dense cordierite ceramic material bulk density reaches 2.60 g/cm^3, and compressive strength approaches 450 MPa [4]. Such materials may be acquired only from pure oxides, synthesizing them at temperatures above 1350 °C for several hours [5]. To reduce the required synthesis temperature, fluxing agents such as Bi_2O_3 are added to the oxide mixtures [6].

In this report obtaining and properties of high-temperature porous and dense ceramic materials using raw local Latvian Quaternary clay and quartz sand resources have been described.

IOP Conf. Series: Materials Science and Engineering **123** (2016) 012042 doi:10.1088/1757-899X/123/1/012042

2. Materials and methods

Mineral raw materials of Latvian origin - ground quartz sand (≈ 98% pur.) and carbonates-containing illite clay were used in pre-ceramic powder preparation; oxide composition of clay used is given in Table 1. By adjusting required stoichiometry with the help of $MgCO_3$ and $Al(OH)_3$ (> 99.99% pur., Sigma Aldrich) green powders' compositions were matched to that of a cordierite ($2MgO \cdot 2Al_2O_3 \cdot 5SiO_2$). These compositions are given in Table 2.

Table 1. Composition of illite clay in wt. %

SiO_2	Al_2O_3	MgO	CaO	Fe_2O_3	Na_2O+K_2O	Loss of ignition
51.0	16.5	2.4	6.8	6.3	5.0	11.4

Starting mixtures were carefully weighed and placed in an agate grinding containers and afterwards milled in water medium (1:1.6 ratio) for 2 hours in planetary ball mill. The sand was ground beforehand in similar way (3 hours in water medium) to the powdered state. After grinding all slurries were dried in oven at 60 °C for 24 hours and crushed in a mortar into powders afterwards. Powderstate mixtures were dosed and pressed into cylindrical samples with varying height: diameter ratio (from 1:1 for use in compression testing to 1:5 for use in CTE testing) with the uniaxial hydraulic press (pressing pressure of 15 MPa). The samples were annealed in a furnace in normal atmosphere at maximum temperatures of 1200, 1250 and 1300 °C with heating rate of 6 °C/min and dwell time of 3 hours.

Table 2. Compositions of prepared powder samples in wt. %

Sample	Illite clay	Sand	$MgCO_3$	$Al(OH)_3$
C0	-	44.1	21.5	34.4
C25	18.7	32.0	19.0	30.3
C50	33.2	22.5	17.0	27.3

Crystalline phase composition was determined with powder XRD using Rigaku Ultima, Japan apparatus (Cu K_α radiation; scanning rate of 4 °/min and 2θ range of 5-50°). Apparent porosity and volume density was determined using Archimedes method. Compressive strength (in MPa) was tested using ToniNorm by Zwick (Germany) test plant. Samples M25 (1300 °C) and M50 (1250 and 1300°C) were taken polished surface. Sample CTE was determined in temperature range of 20 – 1000°C and heating rate of 10°C/min. using horizontal dilatometer (Linseis L76 Platinum Series, Germany).

3. Results and discussion

3.1. X-ray diffraction

All samples containing the illite clay (C25 and C50) have cordierite as main crystalline phase. This phase had already formed at 1200 °C, in contrast to the sample C0 – composition without illite clay content (Figure 1). It indicates that the required cordierite synthesis temperature has been reduced when using illite clay. Samples without clay content, by contrast, begins to form cordierite only above 1200°C (Figure 2). Spinel and anorthite are formed as secondary phases. Anorthite crystalline phase above 1200 °C begins to melt and forms the glassy state, thus it is not detected at higher temperatures. Samples without clay additive retain both modifications of SiO_2 (quartz and cristobalite) as well as sapphirine throughout all of the temperature range of synthesis. On the contrary – samples of composition C50 can be characterized as consisting of crystalline cordierite as it can be seen in XRD (Figure 2) + glassy phase only.

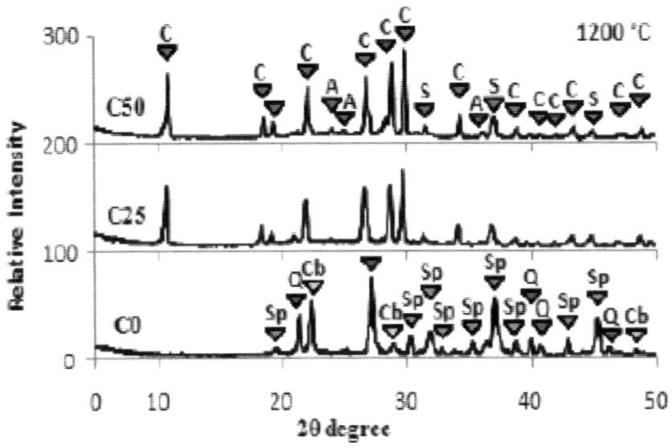

Figure 1. XRD of samples C0, C25, C50 at 1200°C. C – cordierite, A – anorthite, S – spinel, Q –quartz, Cb – cristobalite, Sp – sapphirine

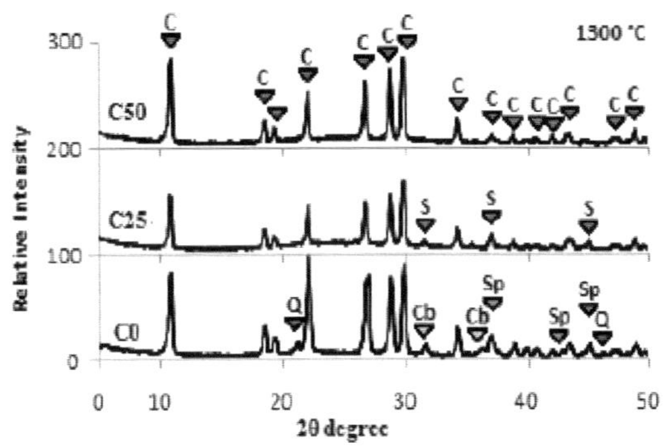

Figure 2. XRD of samples C0, C25, C50 at 1300°C. C – cordierite, S – spinel, Q – quartz, Cb –cristobalite, Sp – sapphirine

3.2. Sample porosity and compressive strength

Results of hydrostatic weighting as well as compressive strength testing are summarized in Table 3. As can be seen, the sample C50 which is obtained at highest temperature (1300°C) are extremely porous and sponge-like as it is evident from the high water absorption (58.80%) and high porosity (96.71%) with sample pore walls being of less than 1 mm thickness. However, samples obtained from the same composition at 1250 °C are also porous, but the sample pore size is much smaller, measured in microns range as well as wall thickness being greater. In both cases, the pore size and the layout is relatively homogeneous, but the pore size and the quantity significantly affects both the sample compressive strength, i.e., highly porous, sponge-like sample C50/1300 compressive strength is below 5 MPa, but the sample C50/1250 compressive strength is sufficient for practical uses.

Meanwhile, samples annealed at 1200°C are extremely dense as it can be seen from their compressive strength values reaching as high as nearly 400 MPa for composition C25. This value is close to the theoretical value of dense cordierite; it also far outclasses strength of pure cordierite composition samples C0 at this temperature as the presence of illite clay has improved clay-containing sample densification rate.

Table 3. Water absorption, density and compressive strength of samples

Sample	Apparent porosity, %	Density, g/cm^3	Compressive strength, MPa
C0/1200	25.2	1.70	43
C0/1250	10.4	2.06	148
C0/1300	1.1	2.35	154
C25/1200	1.1	2.55	117
C25/1250	1.2	2.38	399
C25/1300	4.8	2.23	110
C50/1200	0.9	2.46	390
C50/1250	29.2	1.46	21
C50/1300	58.8	1.02	≈ 5

3.3. Coefficient of thermal expansion

Results of CTE testing are visualized in Figure 3. Sample which does not contain cordierite crystalline phase has very high coefficient that ranges at about 7.5 – 9.0·10^{-6} °C^{-1} (not displayed). On the contrary, each of the other samples which contain cordierite shows a satisfactory small value of CTE with largest being for samples obtained at 1200°C. Generally, CTE value lowers with both increasing amount of clay and temperature of synthesis reaching as low as 1.2 – 2.4·10^{-6} °C^{-1} for sample C50 synthesized at 1300. Slight increase is detected for samples without clay C0 at around 600 °C which could be contributed to α-β quartz transition impact as this phase is present only in these samples as shown in XRD. For samples with clay content rapid increase in CTE is observed at 800°C followed by a decline afterwards and could be explained with the presence of anorthite rich glassy phase.

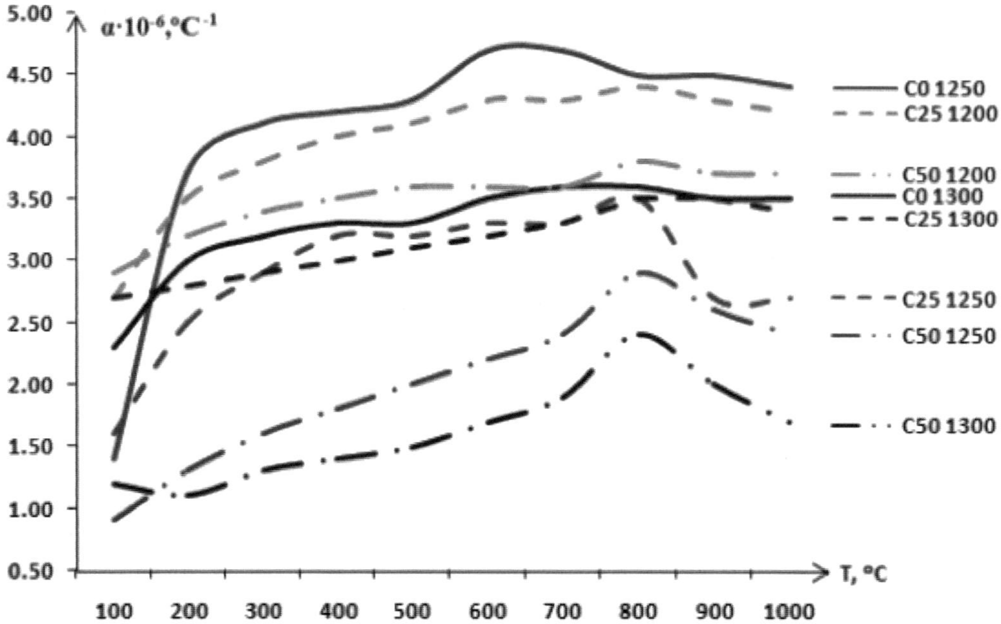

Figure 3. CTE for samples C0, C25 and C50 synthesized at temperatures 1200, 1250, 1300°C.

4. Conclusions

In this work both dense and highly porous cordierite containing ceramic materials were obtained using mineral raw materials of Latvia – illite clay and quartz sand. Dense cordierite ceramic samples were obtained at 1200°C reaching compressive strength of 400 MPa while highly porous material was obtained at 1300°C. Coefficient of thermal expansion is satisfactory small for every ceramic sample with composition of 1/3 clay content (C50) being the lowest and thus most beneficial for practical use.

References

[1] Stolyarova V A, Lopatin S I and Fabrichnaya O B 2011 *Russ. J. Gen. Chem.* **81** 1597
[2] Goren R, Gocmez H and Ozgur C 2006 *Ceram. Int.* **32** 407
[3] Jankovic-Castvan I, Lazarevic S, Jordovic B, Petrovic R, Tanaskovic D and Janackovic D 2007 *J. Eur. Ceram.Soc.* **27** 3659
[4] Dittmer M, Muller M and Russel C 2010 *Mater. Chem. Phys.* **124** 1083
[5] Costa-Oliveira F A and Cruz-Fernandes J *Ceram. Int.* **28** 79
[6] Obradovic N, Dordevic N, Filipovic S, Nikolic N, Kosanovic D, Mitric M, Markovic S and Pavlovic V 2013 *Powder Technol.* **218** 157

The influence of clay fineness upon sludge recycling in a ceramic matrix

A M Szőke, M Muntean, M Sándor and L Brotea

Babeş-Bolyai University, Faculty of Environmental Science, Extension Sf. Gheorghe, Romania

E-mail: mariszoke@yahoo.com

Abstract. The feasibility of sludge recycling in the ceramic manufacture was evaluated through laboratory testing. Such residues have similar chemical and mineralogical composition with the raw mixture of the green ceramic body used in construction. Several ceramic masses with clay and various proportion of sludge have been synthesized and then characterized by their physical-mechanical properties. The fineness of the clay, the main component of the green ceramic body, has been considered for every raw mixture. The proportion of the sludge waste addition depends on the clay fineness and the sintering capacity also, increases with the clay fineness. The ceramic properties, particularly, the open porosity, and mechanical properties, in presence of small sludge proportion (7, 20 %) shows small modification. The introduction of such waste into building ceramic matrix (bricks, tiles, and plates) has a very good perspective.

1. Introduction

Nowadays society is not only concerned in finding new sources of raw materials, but also in recycling waste as new materials [1].

According to Directive 2008/98/EC on waste, Member States shall determine waste prevention programs by the end of 2013, which will be integrated into waste management plans or will operate as separate programs.

Among the main types of waste existing in our country which are suitable for using in construction, which volume and characteristics justifying being studied, the research focused on recovery from power plant ash, phosphogypsum, blast furnace slag and to a lesser extent on urban sewage sludge.

A possible use of sewage sludge is to be used in producing construction ceramic. When we speak about gross construction ceramic we refer to ceramic matters in which composition falls the alumino-silicates, alkali, alkaline earth iron respectively fusible, vitrifiable clays and less refractory clays [2,3]. The variety of chemical and mineralogical composition of the raw materials, processing methods and burning conditions provide the products the most varied characteristics.

The reuse of the sludge from the wastewaters of Saint George city is so reduced that practically, it can't be mentioned. The sludge from the Wastewater Plant of Saint George is going through the drying process, by passing the concentrator of humid sludge and biogas production, used as a source of thermal energy. The waste coming from the wastewater treatment process is thickened, fermented and dried, then transported to the landfill. The humidity of the dried sludge is 70-75%.

Partial incorporation of sewage sludge in crude construction ceramic is a way to use them [4,5].

The present work assesses the incorporation of sewage sludge up to 20% into ceramic materials. This reuse of sludge obtained from water treatment plant provides an alternative recycling and disposal solutions. The objective of the study is to characterize the raw materials and the products by physical and chemical methods and to evaluate the effect of incorporated sludge amounts and the influence of fineness of the clay upon sludge recycling.

2. The experimental part

Through specific operations performed, ceramics with a certain composition were obtained, having the granulation and homogenizing which matched the adopted molding process and the texture to be after heat treatment.

In this paper we studied, by sieving, influence of particle size of raw materials: clay of Bodoc and sludge of waste water plant of Saint-George city, on the properties of ceramics.

The principle is simple and consists of passing the clay and sludge through a set of meshes corresponding to a particle size scales stacked from up to down, in decreasing order. We used sieves the following sizes: 0.2, 0.09 and 0.063 mm.

The following aspects were tracked:

- Specific methods of obtaining different compositions
- Characterization of ceramics (compactness, resistance to bending)

3. Results and Discussion

The chemical analysis showed the compatibility of the materials on Table 1.

Table 1. The oxide composition of the raw materials

Oxide	**Standard**		**Ceramic mass/Calcinated ceramic mass %**									
	0		**5**		**7**		**10**		**15**		**20**	
SiO2	67.97	70.49	65.27	69.83	64.19	69.54	62.57	69.09	59.87	68.34	57.18	67.55
Al2O3	15.41	15.99	14.82	15.85	14.60	15.81	14.23	15.74	13.68	15.61	13.12	15.48
Fe2O3	4.88	5.06	4.70	5.03	4.63	5.02	4.53	5.00	4.35	4.96	4.18	4.93
CaO	1.66	1.73	2.24	2.40	2.47	2.67	2.82	3.16	3.40	3.87	3.98	4.69
MgO	1.59	1.64	1.51	1.63	1.52	1.62	1.50	1.60	1.45	1.58	1.27	1.49
K2O	2.43	2.52	2.30	2.47	2.26	2.44	2.18	2.40	2.06	2.35	1.94	2.29
Na2O	1.54	1.62	1.46	1.61	1.50	1.60	1.46	1.57	1.43	1.53	0.61	0.72
Other	0.94	0.95	1.16	1.20	1.16	1.28	1.27	1.44	1.39	1.58	2.42	2.85
LOI*	3.58	-	6.51	-	7.67	-	9.44	-	12.37		15.30	-

***LOI at 1000°C**

The introduction of small amounts of sludge, up to 10%, does not lead to significant changes in the oxide composition of the mixtures. Only in the presence of large amounts of sewage sludge (15-20%) a more strongly increase of calcium oxide content and a decrease of silica content and merely a small drop of the content of aluminium oxide and iron oxide can be observed.

However, even in the presence of large amounts of sludge (20%) in clay mixture, its chemical and oxidical composition is within the limits of the composition corresponding to construction ceramic masses: ceramic blocks, bricks, tiles, roofing tiles.

The analysis of granulometric distribution of the clay particles has been done with a MALVERN Mastersizer 2000 laser granulometry analyzer in liquid suspension with a measurement scale between 0.02-600 microns. The granulometric bend of Bodoc clay is shown in Figure 1. The size of the particles ranges between 0.45 to 600 microns. There is a significant amount of small particles of 10-12 μm, in a single-mode distribution. The specific surface area has a value of 0.781 m2/g [7,8].

The analysis of granulometric distribution of the sludge particles has been done on the base of a FRITSCH Analisette 22 laser granulometry analyzer, in liquid suspension 0.1-600 microns.

The granulometric bend of sludge is shown in **Figure 5** and it highlights particle size with values between 1-108μm. The line is continuous, and it shows a cumulative shift. The specific surface area has a value of 0.59 m2/g [7,8].

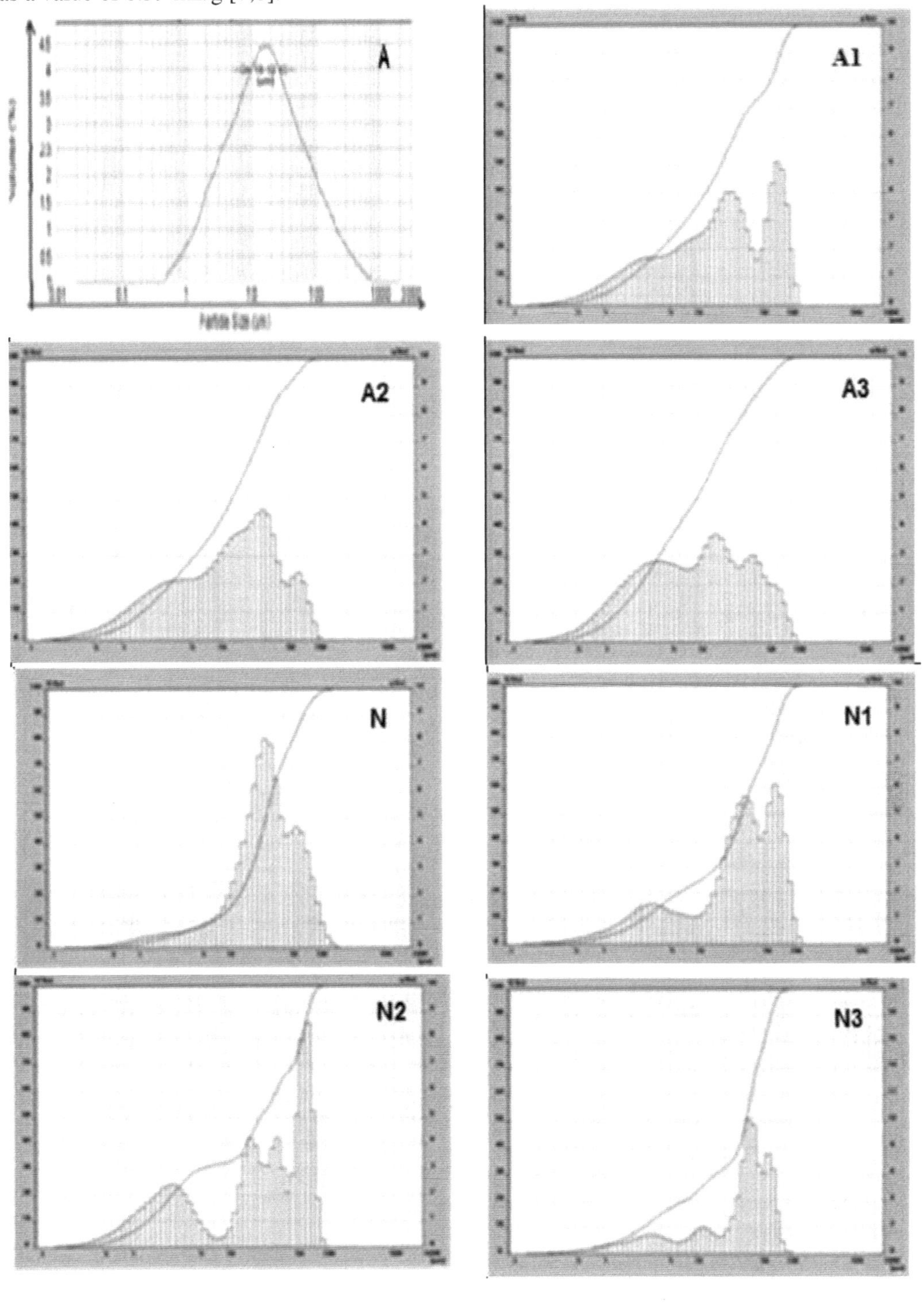

Figure 1-8. Grain size distribution of clay **(A)** and sewage sludge **(N)** before and after passing through **0.2 mm (1)**, **0.09 mm (2)** and **0.063 mm (3)** sieves

The advanced fineness of sewage sludge ensures a good condition for conducting reactions in solid phase in mixture with clay of Bodoc. On the other hand, in case of small amount of sludge in the clay mass, it is necessary a very good mixing to get a homogeneous mass.
Granulometric analysis was used for the determination of grain size by passing the raw materials through various sizes of mesh sieves, ranging from 0.063 mm to 0.2 mm, as indicated **in Figures 1-8.**
The prepared mixtures are presented in **Table 2,** but we focus more on the addition of 7 and 20 %.

Table 2. Prepared mixtures

Ceramic mass	Clay,%	Sludge,%
reference mass	100	-
1.	95	5
2.	93	**7**
3.	90	10
4.	85	15
5.	80	**20**

We determined the compactness characteristics of the ceramic masses, at the same time the resistance to bending of ceramics have been determined and they are presented in **Table 3**. These were burnt at 960 °C temperature.

Table 3. Compactness characteristics at 960°C (**D** – density, **A** – adsorption, **P** – porosity) and variation resistance to bending according to temperature (**Ri**)

Sieve dimension (mm)	Clay				Ceramic mass							
					7 % sewage sludge				20 % sewage sludge			
	D	A	P	R_i	D	A	P	R_i	D	A	P	R_i
0.2	1.90	15.07	29.54	3.55	1.81	17.03	30.74	3.25	1.54	25.19	38.39	3.17
0.09	1.88	15.70	28.76	4.12	1.80	18.11	31.04	4.02	1.61	24.59	38.77	2.71
0.063	1.80	18.20	32.79	10.54	1.79	19.06	33.70	8.40	1.63	23.55	39.66	7.13

As we expected, the increase of the heat treatment temperature positively influences the densification of ceramic masses with or without adding sewage sludge - Table 3. The apparent density slightly increases along with the increase of the burning temperature from 960 °C to 1100 °C, even in the case of sample containing 20% of sludge in the mass.

However, increasing the amount of sludge from 7% to 20% in the ceramic mass, there was an increase in the apparent porosity and consequently an increase in the values of water absorption in all cases.

The mechanical properties of ceramic masses, expressed by mechanical resistance to compression and bending were strictly dependent on the compactness of ceramic masses, so influenced by apparent density. The mechanical resistance was affected, firstly, by the temperature of the heat treatment - when these increased it increased too, and secondly by the amount of added sludge in ceramic mass – this decreased when the amount of sludge increases in ceramic mass. Reduction of mechanical strength is visible to the masses with 20% sludge - the values are given in Table 3.

4. Conclusion

Based on the study, we can conclude that increasing the fineness of particles in ceramic masses has its influence on the following properties:

- The increase of resistance to bending of the ceramic mixture is mainly due to the increase of the fineness of the particles.
- When you increase the fineness of the particles, the porosity values increase slightly.

- It is clear that the sewage sludge samples, used in the experiments, from Saint George (Covasna) are proper for gross construction ceramics (bricks).

The successful incorporation of 20% sludge amount into ceramic masses is a unique result, which cannot be found in the literature.

References

[1] Szőke A.M.; Muntean M.; Fazakas E.; Bartalis I. 2010 *Natura-Econ* **2** 122
[2] Szőke A.M.; Muntean M. 2009 *E. E. M. J.* **4** 907
[3] Szőke A.M.; Muntean M.; Dumitrescu O.; Mészáros S. 2014, *E.E.M.J.* **1** 15
[4] Teixeira, S.R.; Santosa, G.T.A.; Souzaa, A.E.; Alessioa, P.; Souzaa, S.A.; Souza, N.R. 2011 *Applied Clay Science*, **53** 4
[5] Leandro C.; Morais, R. S. C.; Vianna, Campos, V.; Fraceto, L. F.; Rosa A. H.; Büchler, P. M. 2010 *Materials Science, Advanced Powder Technology* **660-661** 7
[6] Manoliu, I.; Radulescu, N. 2010 *Geotechnics I,* Technical University of Construction Bucharest
[7] Szőke, A.M,; Muntean, M.,; Bartalis, I. 2011 *INCD Ecoind Bucharest* **I**
[8] Bercu, I. 2011 *Universitatea Tehnică Bucuresti*

IOP Conf. Series: Materials Science and Engineering **123** (2016) 012044 doi:10.1088/1757-899X/123/1/012044

Structure and machinability of thin-walled parts made of titanium alloy powder using electron beam melting technology

A Isaev, V Grechishnikov, M Kozochkin, M Volosova, Yu Petuhov and P Peretyagin

Moscow State University of Technology "STANKIN" Moscow, Russia

E-mail: a.isaev@stankin.ru

Abstract. The present study aims at study of structure and properties of raw and thermal treated titanium-based powder material used to produce the thin-walled components by electron beam melting technology. Producing the end product means also studying the finishing cutting process. Examining the quality of end product in terms of geometric tolerance and thin walls thickness stability consists of control the surface roughness is also included in this study.

1. Introduction

Nowadays, the additive manufacturing (AM) processes are extensively used to produce various kinds of machine components, including metal ones. One of the most widespread AM technologies is electron beam melting (EBM) which produces parts by melting metal powder layer by layer with an electron beam in a high vacuum. Unlike other techniques, e. g. direct metal laser sintering (DMLS), selective laser sintering (SLS) etc, parts made by the EBM are fully dense, void-free, and very strong.

Thin-walled parts are widely used in various products of aerospace, power engineering, fine mechanics and others, where light weight, high usability and ergonomics are required. In some cases, thin-walled components having complex geometry have to be produced using the AM technologies and further be machined either by "traditional" methods (turning, milling, grinding etc.), or by "non-traditional" ones, like electrical discharge machining or water jet cutting.

Milling is one of the most common machining method used in thin-walled part processing. Cutting and clamping forces acting during removal of chip produce the deflections of walls that are comparable with the machining stock thickness [1]. The specific microstructure of the parts made using the AM technologies causes additional instability of the machining process. Those deflections and process instability result in machining error that depends on wear of cutting tool, cutting conditions, parameters of the machine-workpiece-tool system, and other factors [2]. In practice, the type of milling (conventional or climb) dramatically influences the processing accuracy and machined surface roughness. Therefore, it is important to figure out how the machining errors vary with the change of operation conditions and material characteristics.

2. Material characterization

The raw (as-built) sample parts made of the VT-6 titanium alloy powder (equivalent to Ti6AI4V). with the average powder grain size ~70 μm using the Arcam A2 EBM system (Sweden) have the specific microstructure caused by the non-uniform thermal conditions of crystallizing and cooling of titanium alloy in vacuum chamber [3]. The X-ray diffraction phase analysis showed that the sample parts consist of the A2 type structure (β-Ti, up to 10% by volume) and the A3 type structure (α-Ti, up to 90% by volume). The typical microstructures of the raw parts in different cross-sections are shown in Figure 1. This microstructures are characterized with specific stroke pattern. When melting the powdered metal, the grains of β-Ti stretch along the direction of the metal flow. During cooling the parts in the vacuum chamber, when the temperature goes down to the α-Ti precipitation, the rate of β-Ti growth slows down. Regardless the cooling rate, the α-Ti phase precipitates on the β-Ti defects and along the metal flow. In Figure 1a) the microstructure of the layer parallel to the horizontal XY plane (base plane) where the melting is performed, is shown. In Figure 1b), long horizontal lines correspond the part's layers, while vertical lines in Figure 1c) coincide with the tracks columns.

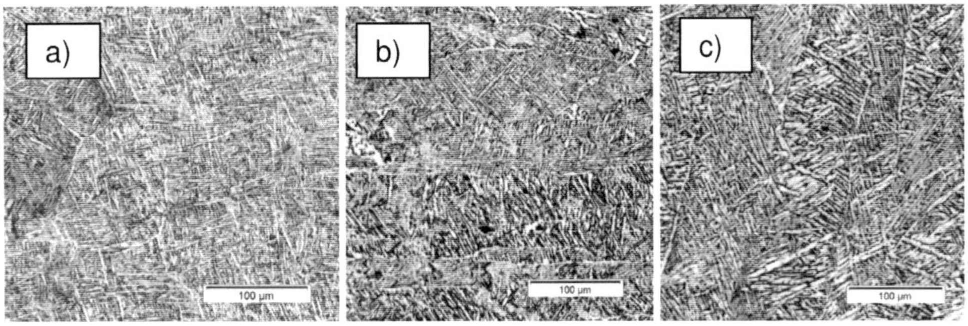

Figure 1. Microstructure of parts produced by the Arcam A2 EBM system in cross-section planes: a) XY; b) YZ; c) XZ

The raw material characterization in different cross-section planes is shown in Table 1. As it is shown there, the content of α-Ti and β-Ti depends on the cross-section plane orientation.

Table 1. VT-6 titanium alloy characterization

Phase	Lattice parameter	Volume ratio in XY plane	Weight content in XY plane	Volume ratio in ZY plane	Weight content in ZY plane	Volume ratio in ZX plane	Weight content in ZX plane
β-Ti	a=2.92 Å, c=4.683 Å	7.0 ± 0.2	7.3 ± 0.2	4.9 ± 0.1	5.1 ± 0.1	10.0 ± 0.1	10.4 ± 0.1
α-Ti	a=2.912 Å	93.0 ± 0.2	92.7 ± 0.2	95.1 ± 0.1	94.9 ± 0.1	90.0 ± 0.1	89.6 ± 0.1

At the same time, regardless the cross-section plane orientation, the same structure can be seen in the microsection. On the former β-Ti grain boundaries the spherical α-Ti grains start to grow (Figure 2a). With the cooling rate increasing (possibly when the electron beam moves away from the melting region), the new α-Ti grains becomes platy (Widmanstätten patterns). Between the plates of α-Ti some retained β-Ti grains can be observed, whilst the α-Ti grains contains a few number of micro-inclusions, that could be either needle-like precipitates of α-Ti oriented normally to the surface of microsection, or some unknown third phase.

The precipitating α-phase has the larger grain size. This is the reason of residual tensile stresses in the end product. Thus, the recrystallization annealing to lower these stresses is recommended. In the present work, the hot isostatic pressing (HIP) was applied to the sample part (hot isostatic press AIP8-30H (USA), 920°C, 1000 Bar, 120 min). The microstructure after the HIP is shown in the Figure 2.b. It is obvious that the structure became much more homogenous in all cross-section planes. Consequently, the mechanical properties of the part are more isotropic than in the raw material. On the other side, the effect of HIP is just about 3-4%. Actually, this small difference in properties between HIP and raw is an indication of a defect-free product.

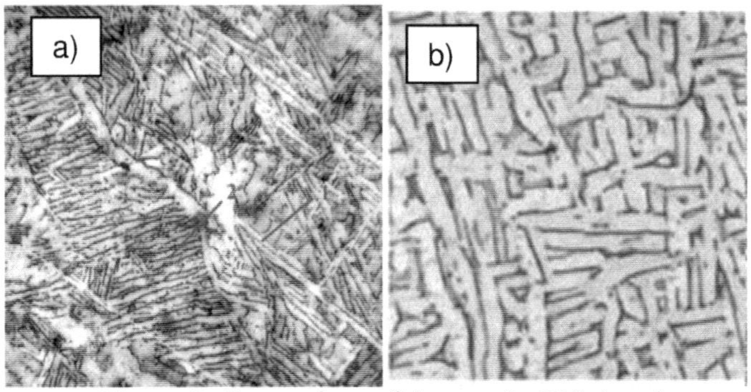

Figure 2. Comparison of the raw (a, 200x) and heat treated (b, 500x) microstructures

3. Experimental procedure

The geometric parameters of the EBM-parts don't match the initial 3D model. In addition, the side faces of the EBM-parts have stochastic irregular surface. Thus, to provide the end product quality, it is necessary to machine the parts using some 'traditional' methods like milling or grinding [4].

A sample part to be machined during the experiment is shown in Figure 3a. Four series of experiments were conducted with the wall thickness t of 1,5 mm, 1.0 mm, 0.5 mm and 0.3 mm. Finishing cutting speed: 188 m/min; feed rate: 0.03 mm/tooth; depth of cut: 0.5 mm; width of cut: 3.6 mm. The part was machined downwards in 5 passes. Walls thickness measurement scheme is shown in Figure 3b. To measure vibrations generated during the machining process, a measuring information system has been used [5].

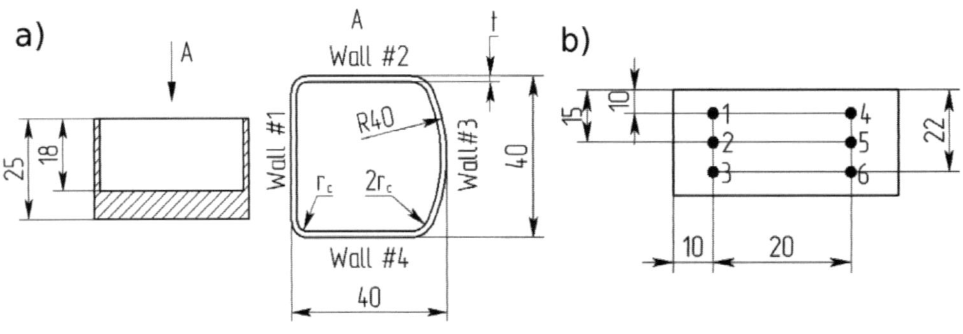

Figure 3. Sample part (a) and walls thickness measurement scheme (b)

The line graph of walls thickness measured in corresponding points from the top of the workpiece appears in Figure 4. Figure 4a shows that relatively thick walls (t=1.5 mm) gets thicker towards the bottom of the workpiece, especially for conventional milling, because in this type of milling a normal component of cutting force acts towards the axis of the cutter making the cutter's body to slightly increase its radial runout and to deviate from the relatively stiff wall being machined. While the point of contact between the cutting edge and the part's surface moves downwards during the machining process, the wall's stiffness increases and the cutter's deviations and runout become smaller, that leads to increasing wall's thickness towards the bottom. In addition, the chips disposed before the cutter can also affect the milling cutter and increasing the thickness of the walls. In climb milling, the width of the cut starts at the maximum producing less friction forces [6]. The normal component of the cutting force acts in a direction of the relatively stiff wall. The chips are disposed behind the cutter. This provides less cutter's deviations and hence less walls thickness variation.

This situation changes as the walls are getting thinner and their stiffness decreases (Figure 4b). In this case, the walls deform more than the cutter does. In climb milling, the top part of the walls deviates from the cutter's body and gets thicker than their lower part. Conversely, in conventional milling the walls become thicker to the bottom part of the workpiece, and the wall's mean thickness is closer to its nominal value.

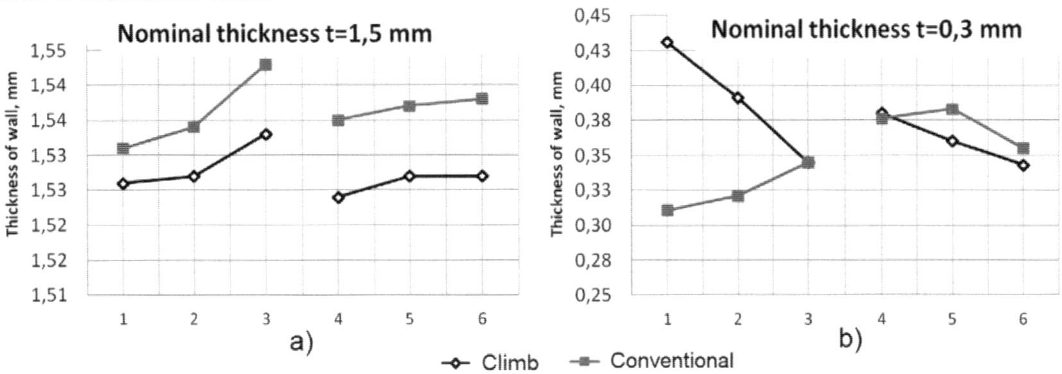

Figure 4. Walls thickness variations depending on the measuring point and milling type

The roughnesses of the wall #1 and the smaller inner corner surface was performed using the contact profilometer. The climb milling provides slightly better roughness (Ra=0.2—0.4 μm) than the conventional one (Ra=0.3—0.7 μm) when machining relatively thin walls (0.5 mm or thinner). Both conventional and climb milling produces very rough surface in the corners of small radii in comparison with the rest of surfaces being machined (Ra=2.0—7.8 μm). Though the roughness of the curved surface of relatively big radius (the wall #3) couldn't be measured directly with a contact profilometer, it has obviously a better roughness than other surfaces of the sample part. It can be explained by the higher stiffness on the curved wall that produces less chattering of the elastic system "machine-tool-workpiece".

Surfaces of the walls with t=0.3 mm machined under different cutting conditions are shown in Figure 5. In Figure 5a, the periodic milling marks left after climb milling are visible. In Figure 5b, the wall machined by conventional milling where there is a lot of spots left by chip, is shown. Figure 5c represents the curved wall #3 which has visibly lower roughness than other surfaces. Figure 5d shows a very rough surface in the corner of small radius which is equal to the radius of the milling cutter.

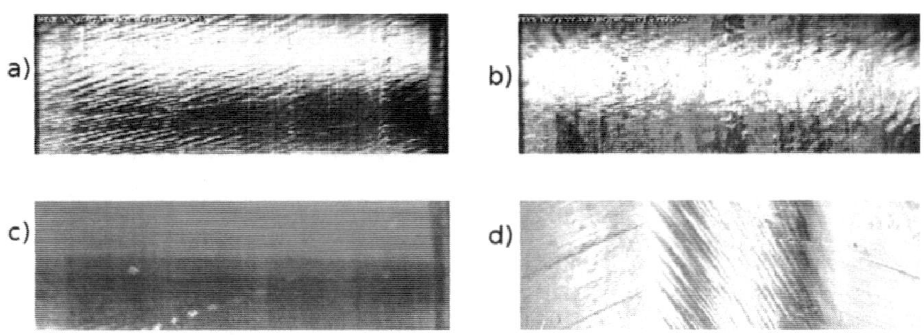

Figure 5. Surfaces of the walls with t=0.3 mm: a) climb milling; b) conventional milling; c) climb milling of the curved wall; d) climb milling of the corner with a small radius.

4. Vibroacoustic signal analysis

During the milling process a vibroacoustic signal was recorded using the accelerometer installed on the machine's table [7]. The samples of the vibroacoustic signal for conventional and climb milling are shown in Figure 6. In conventional milling, non-repetitive accidental peaks of up to 0.3 ms width have been observed. These peaks are caused by chip disposed before the milling cutter, as it was shown above, and result in a poor appearance of the wall's surface (spots and strokes), although a roughness parameter Ra has not been significantly changed on these parts of surface. The appearance of the waveform both for raw and heat treated materials is highly similar.

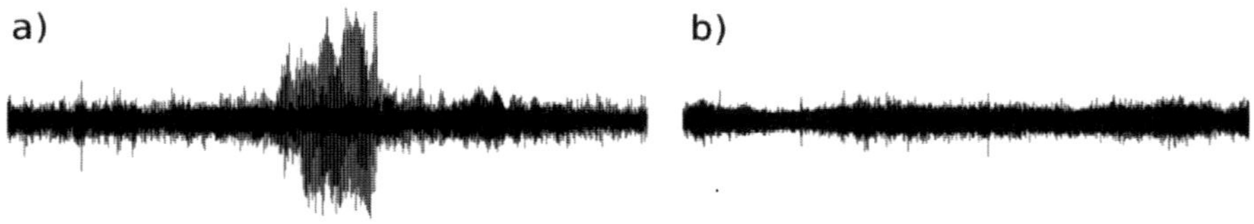

Figure 6. Vibroacoustic signal for conventional (a) and climb (b) milling processes

The vibration acceleration spectra for climb and conventional milling are represented in Figure 7a and 7b, respectively. These waveforms were obtained for the last pass of the cutter that have been moved along the bottom surface of the workpiece, where a large amount of chip affects the machining process.

The spectra presented in Figure 7 have their maximum amplitudes at double tooth frequency ($2f_{tooth}$) and the local maximum amplitudes at frequencies multiple to the cutter's rotation one (f_{rot}). However, for the conventional milling the amplitude of vibroacoustic signal at high frequencies is significantly lower than the maximum at $2f_{tooth}$. It means decrease of impacts when the teeth enter and exit the workpiece which is caused by the chip-disposal effect described above.

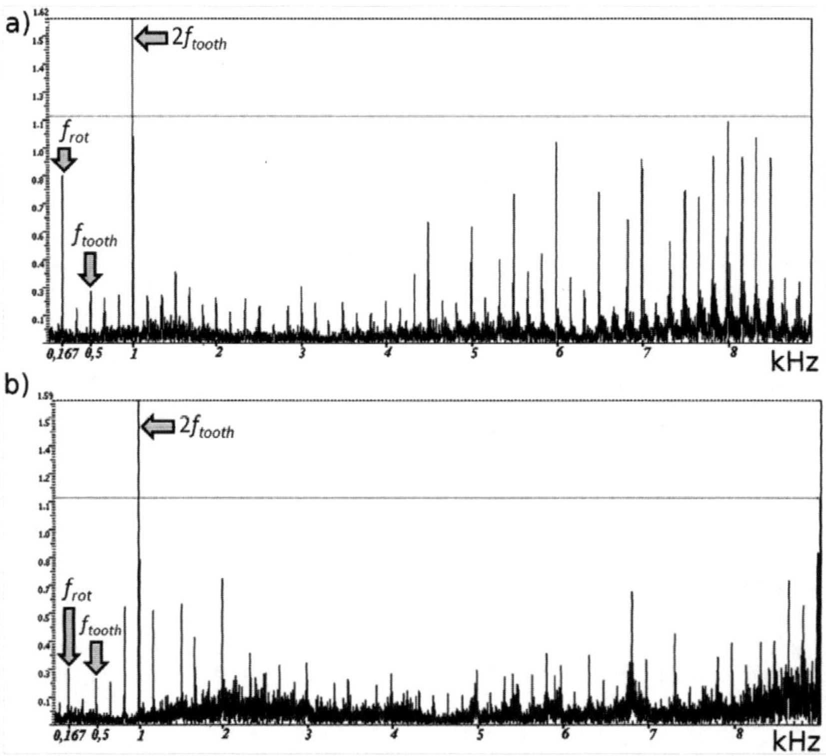

Figure 7. Vibration acceleration spectra of climb (a) and conventional (b) milling

5. Summary

As the studies results show, the microstructure and material properties of EBM parts in the base horisontal plane and in vertical ones are different, i. e. the raw EBM samples structure and properties are anisotropic. Annealing of the parts produced by EBM have no significant affect to machining process and the resulted surface properties;

When milling EBM thin-walled parts, the form accuracy and the surface roughness are the functions of variable wall stiffness which is estimated in a local cutting zone. This fact must be taken into account when selecting the cutting conditions.

The milling of surfaces with a low stiffness is a dynamically unstable process which produces vibrations (chattering) and relatively big displacements of the cutter and the workpiece. When using conventional milling these vibrations are lower, but the surface roughness is significantly higher because of the chip disposed to the cutting zone.

When designing thin-walled parts, it is desirable to replace flat surfaces with the curved ones, if possible, as the curved surface provides a higher stiffness and hence a better surface roughness and a less form deviations;

Use of vibroacoustic analysis when development of technological process of thin-walled parts machining seems to be feasible as it helps to efficiently detect and estimate the undesirable effects;

References

[1] Grechishnikov V A, Isaev A V and Markos S 2007 *Rus. Eng. Res.* **27** 291

[2] Grechishnikov V A and Isaev A V 2010 *Rus. Eng. Res.* **30** 413

[3] Luis A Díaz, Miguel A Montes-Morán, Pavel Y Peretyagin, Yuriy G Vladimirov, Anna Okunkova, José S Moya and Ramón Torrecillas 2014 *J. of Nanopart. Res.* **16** DOI 10.1007/s11051-014-2257-x

[4] Shaw M C 2005 *Metal Cutting Principles* (New York: Oxford University P.)

[5] Kozochkin M P, Kochinev N A and Sabirov F S 2006 *Meas. Techniques*, **49** 672

[6] Volosova M A and Gurin V D 2013 *J. of Frict. and Wear*. **34** 183

[7] Isaev A V and Kozochkin M P 2014 *Meas. Techniques*, **56** 1155

Physical mechanism of ultrasonic machining

A Isaev, V Grechishnikov, M Kozochkin, P Pivkin, Y Petuhov and V Romanov

Moscow State University of Technology "STANKIN" Moscow, Russia

E-mail: a.isaev@stankin.ru

Abstract. In this paper, the main aspects of ultrasonic machining of constructional materials are considered. Influence of coolant on surface parameters is studied. Results of experiments on ultrasonic lathe cutting with application of tangential vibrations and with use of coolant are considered.

1. Introduction

Ultrasonic machining has been known and studied for several decades [1], but so far it is not a wide spread technology because of some complexities of the implementation and certain cutting speed limits. Advantages of ultrasonic machining and its influence on machined surface finish, cutting tool durability and forces arising during machining, are controversial. At the same time, new constructional materials, up-to-date machine tool design and modern machining technologies allow to improve efficiency of ultrasonic machining. In particular, new ultrasonic units are capable of automatically adjusting frequency of ultrasonic generator on resonant frequency thus drastically reducing power consumption. There are also some new specific tasks where machining to precise (nanometric) tolerances producing ultra-high quality (e. g. machining of glass or other optical materials) which makes manufacturers to use ultrasonic machining technologies neglecting its disadvantages [2].

In present paper, the general theory of ultrasonic machining is considered. In addition, the common results of experiments on ultrasonic lathe cutting of stainless steel with application of tangential vibrations and with use of coolant are given.

2. Theoretical background

The main advantage of the ultrasonic cutting can be explained by size reduction of plastic deformation zone within the material being machined. Figure 1 shows the plastic deformation zones during conventional (a) and ultrasonic (b) lathe turning. The shapes of these zones were revealed by analysis of frames of high-speed shooting [3]. It is noticeable that shear angle becomes significantly larger when applying ultrasonic vibration to the cutting tool. It is evidence that chip shrinkage and plastic deformation rate of the cut off layer drastically fall. Since there is no cutting edge build-up, the ultrasonic lathe turning provides a better surface roughness even at low cutting speed.

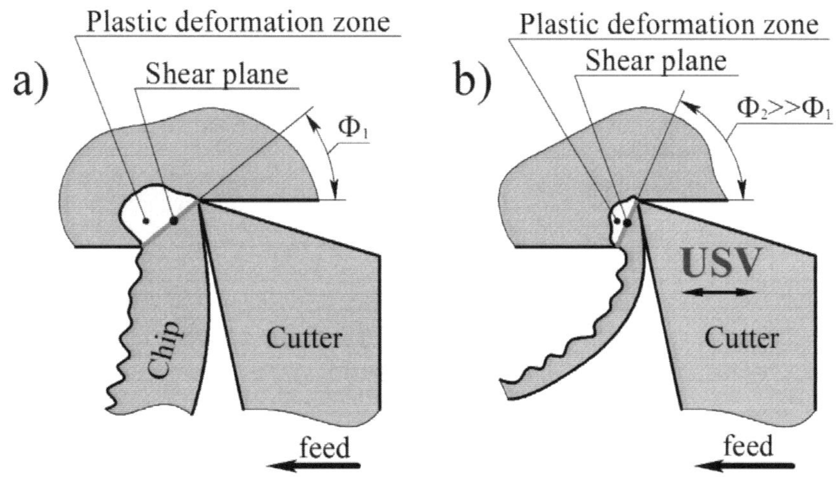

Figure 1. Geometry of plastic deformation zones when cutting without (a) and with (b) ultrasonic vibration

In Figure 2, the surfaces of workpieces made of stainless steel X5CrTi17 machined without (a) and with (b) application of ultrasonic vibration, are shown. It can be seen that in conventional cutting, the particles of the edge build-up remain on the machined surface. In addition, the cutting tool point trace in this case is not so distinct in comparison with ultrasonic cutting.

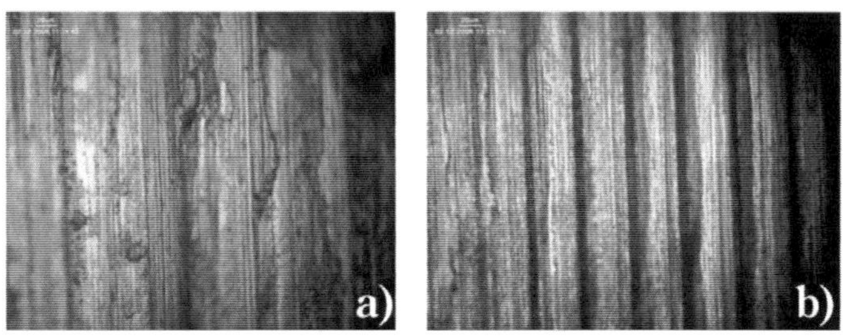

Figure 2. Machined surface of stainless steel X5CrTi17 without (a) and with (b) application of ultrasonic vibration

The main reasons of beneficial effect of ultrasonic cutting technologies are as follows: frictional force reduction in the cutting zone; cutting edge build-up elimination; initialization of dislocation movements in the machined material; better coolant access into the cutting zone; higher machining process stability.

In Figure 3, the forces acting to the chip element from the direction of rake surface of cutting tool are schematically shown. The vector of resulting force R formed by frictional force F and normal pressure force N can be decomposed to the components P_{shear} and P_{compr}, that are parallel and perpendicular to the shear line, respectively. The P_{shear} component generates a shearing force and a bending moment M arising in the chip and increases with distance between cutting edge and point of resulting force R on the rake surface. On the other hand, the compressing component P_{compr} prevents tension tending to arise under action of the moment M in the chip's layer located close to the rake surface, and inhibits crack-opening. These forces can be described as follows [4]:

$$M(t) = P_{\text{shear}} \cdot L = \frac{P_{\text{shear}} V}{\xi} \cdot \sin(\pi - \delta - \Phi) \cdot \kappa \quad (1)$$

$$P_{\text{shear}} = R \cdot cos(\rho + \Phi - \gamma) \quad (2)$$

$$P_{\text{compr}} = R \cdot \sin(\rho + \Phi - \gamma) \tag{3}$$

where L is the bending moment arm, mm; V — cutting speed, m/min; ξ — chip shrinkage coefficient; t — time, min; δ — cutting angle which is equal to $\pi/2-\gamma$; ρ — friction angle; γ — rake angle; Φ — shear angle; κ — coefficient that takes into account a speed decrease of point of resulting force R in comparison with the chip movement speed, $\kappa < 0.5$.

As equation (1) shows, at a first approximation, the bending moment grows proportionally to the time taken by the chip's element generation. The P_{shear} component is also time-dependent, but in a more complex way that M does, because of irregular distribution of pressure over the rake surface.

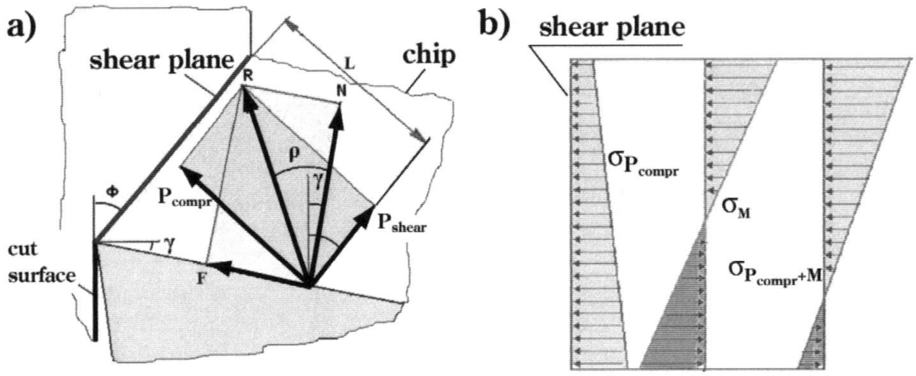

Figure 3. Forces acting to the element of chip from the direction of the rake surface (a) and stress distribution diagram on the shear plane (b)

The bending moment tends to turn the element of chip counterclockwise, thus generating the tensile stresses in the chip near the cutter's rake surface and compressing stresses on the opposite side of the chip. The maximum tensile stresses arise at the cutting tool point. Consequently, under action of the bending moment, the chip's element tends to come off the bulk workpiece near the tool point. The compressing stresses of P_{compr} partly compensate the tensile stresses tending to stop the crack-opening on the chip's surface oriented to the rake surface of the cutter. This compensation rate depends on the friction ratio between the chip and rake surface [5].

Applying of the ultrasonic vibration changes the force balance shown in Figure 3a. This happens because of significant decrease of friction angle ρ, increase shear angle and size of chip's element, changing friction forces and normal pressure. These overall changes don't allow the precise calculating of new value of P_{compr}, though the experiments [6—8] bring out clearly that applying of the ultrasonic vibration, under otherwise equal conditions, the radial and normal components of cutting force drastically decrease making it possible to machine non-rigid workpieces. Decrease of the chip's deformation rate also shows that when using the ultrasonic vibration the compressing force's contribution drastically decreases. Keeping the P_{compr} constant with increasing the M results in expanding the area of tensile stress on the shear plane and makes the chip to move easier. During the conventional cutting, the plastic deformation causes the concentration of dislocations and microcracks in the machined material before the cutting edge, which also form the macrocracks (including the advancing crack) in the chip under the action of tensile stress from the direction of the cutting edge. The surfaces of cracks near the tool's point remain rudimentary because there is no contact between them and atmospheric oxygen or coolant. As the chip moves to the area of compressing stresses, the microcracks can be healed by the molecular forces [9]. These forces also inhibit the advancing crack growth which makes the chip to remove easier.

Thus, the tangential ultrasonic will act efficiently if the forced vibration velocity exceeds the cutting speed value. In this case, the gap between the tool's rake surface and machined material arises periodically. If coolant is not used, the atmospheric oxygen is delivered into this gap, thus oxidizing the surfaces of microcracks and somewhat preventing the healing of the cracks. When the ultrasonic vibration applied, the chip becomes continuous (flow chip) decreasing the cutting and friction forces as well as chip strain hardening.

3. Experimental procedure

3.1. Microhardness of chip surface

The experiments show that using coolant with ultrasonic vibration applied helps to improve the machined surface roughness, increase the tool durability, change the chip hardness in a more efficient way than even when machining with ultrasonic vibration without coolant. In conventional cutting, the elements of chip adjacent to the rake surface have the highest hardness [10]. In Figure 4, the results of the chip microhardness measurement after the ultrasonic machining, are shown. The line 1 means the ultrasonic lathe turning without coolant, while the line 2 illustrates the ultrasonic machining using aqueous soap solution.

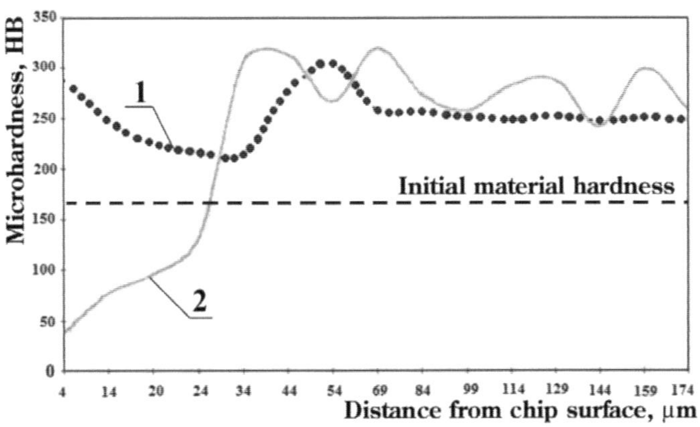

Figure 4. Microhardness of stainless steel X5CrTi17 depending on the distance from the rake surface: 1 — ultrasonic lathe turning without coolant; 2 — ultrasonic lathe turning with coolant

The graph analysis shows that when machining without coolant, the chip's hardness measured at the depth of up to 170 μm exceeds the initial hardness of machined material (166 HB) by a mean of 50% (line 1). When using coolant, the chip's hardness at the depth from the surface of 25 μm is much lower than the initial hardness (line 2). At the same time, the surface roughness when using coolant is at least 2 times lower. So it can believed that during ultrasonic lathe turning, the coolant application increases the microcracks generation rate on the chip's surface near the point of the cutter.

The experiments also show that when ultrasonic machining with coolant, the influence of secondary plastic deformation on the texture of chip becomes minimal. In Figure 5, the microslices of chips produced by the ultrasonic cutting are shown. In Figure 5a, a slight impact of the secondary deformation zone to the chip's texture can be observed, i. e. the chip's texture at the rake surface is slightly deformed in the direction opposite to the one of chip movement. In Figure 5b such deformations are not present. Moreover, the chip edge destruction is clearly visible, confirming the presence of microcracks located near the cutting edge. This leads to the chip microhardness reduction, as it was shown in Figure 4.

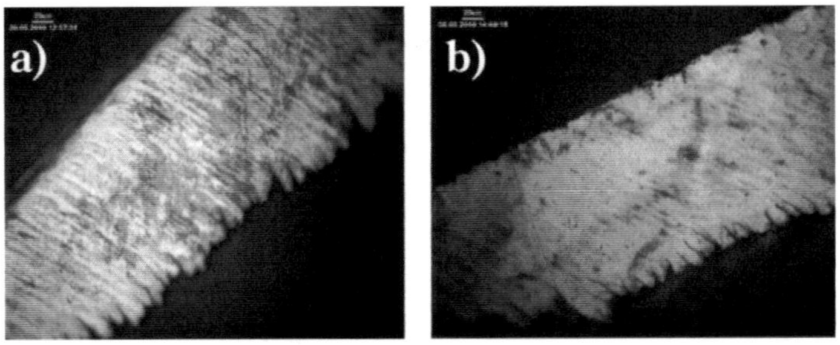

Figure 5. Longitudinal microslice of chip after lathe turning with tangential ultrasonic vibration: a) without coolant; b) with coolant (Note: top side of chip in the figure is oriented to the rake surface of the cutter)

3.2. Influence of coolant on ultrasonic machining by measuring vibrations

To study the influence of coolant on ultrasonic cutting process, the vibration acceleration was measured. In this experiment, an accelerometer was installed to the back center of lathe machine. The vibration acceleration was recorded within the range of tangential ultrasonic vibrations which is equal to 23.5 kHz.

In Figure 6, a sample waveform of amplitude spectrum of the vibration acceleration when ultrasonic machining with coolant, is shown. The spectral maximum at the frequency of 23.5 kHz matches the frequency of the forced vibration being applied to the lathe cutting tool.

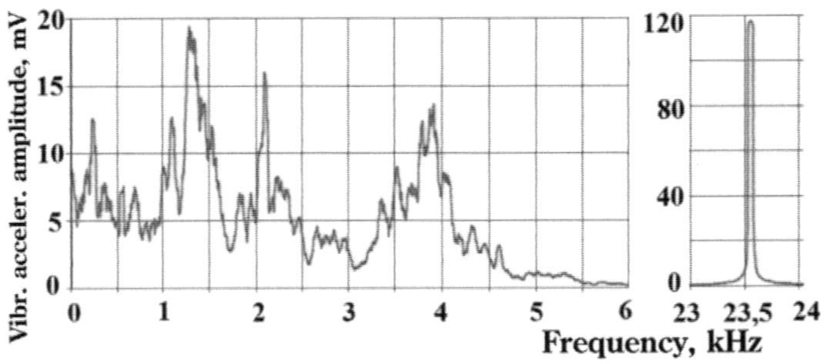

Figure 6. Vibration acceleration spectrum recorded on the back center lathe machine when turning with tangential ultrasonic vibration and with coolant applied to the rake surface

Despite dissipation of ultrasonic energy in the elastic system of the machine tool, it could be fixed on the back center [11]. In Figure 6, the low-frequency region depicts the components produced by cutting and friction processes. The vibration acceleration amplitudes of these processes are 6 times lower than one generated by the ultrasonic vibrations. In another experiment (without coolant), the signal amplitude at the frequency of 23.5 kHz was 4—5 times lower because of worsening contact between the cutter and the workpiece.

In addition, machine oil have been used as a coolant. In this case, the amplitude of the ultrasonic component is 2 times lower than when aqueous soap solution was used. Thus, the coolant composition, its viscosity and wetting capability strongly influence the machining process. The coolant viscosity effects the speed of filling of microcracks that open when the rake surface goes away from the bulk material, while wetting capability provides penetration of coolant into the microcracks by the capillary effect. The liquid penetrating into the microcracks creates the wedging effect (the wedging pressure in microcracks of about 1 nm may reach 100 N/mm^2), but it doesn't have a dominant influence on the chip hardness reduction during ultrasonic machining. The experimental studies also showed that even long-term application of ultrasonic vibration to the surface of part being completely dipped in the coolant, doesn't significantly influence the surface microhardness.

4. Summary

The chip surface microhardness reduction is caused by three conditions: cutting, ultrasonic vibration and coolant application. The ultrasonic vibration not only prevents chip adhesion, but also provides coolant supply into microcracks, including advancing crack. This causes the generation of liquid wedge allowing the growth of advancing crack and its movement by the cutting surface. The same mechanism leads to the chip material dispergation, finally resulting in the chip microhardness reduction.

The advancing crack allows chip to come off the bulk material not only by shear but also by its pressing out of the surface being machined.

The liquid wedge can drastically increase actual rake angle, although its value depends on geometry of advancing crack, so the final workpiece surface is produced by real cutting edge of the cutter. Under these condition, the dynamic rigidity of elastic system where the cutting process is performed, becomes crucial.

A big disadvantage of ultrasonic technologies in machining processes is cutting speed limits. Although the experiments have shown that ultrasonic machining with coolant is very promising and must be extensively studied.

References

[1] Li H and Cao Z Y 2010 *Appl. Mech. and Mat.* **43** 499

[2] Astashev V K, Andrianov N A, Kozochkin M P, Kolik L V, Krupenin V L and Solis-Pinargote N V 2009 *J. Mach. Manuf. and Rel.* **38** 566

[3] Kozochkin M P, Kochinev N A and Sabirov F S 2006 *Meas. Techn.* **49** 672

[4] Kainth G S and Gupta R C 1974 *J. Eng. for Ind.* **96** 1272

[5] Heard W B 2008 *Rigid Body Mechanics: Mathematics, Physics and Applications* (WILEY-VCH Verlag GmbH & Co. KGaA) 249

[6] Yu Altintas 2009 *Machine tool Vibrations, and CNC Design* (Cambridge: Cambridge University Press) chapter 2 4

[7] Astashev V K, Krupenin V L, Perevezentsev V N, Kolik L V and Andrianov N A 2011 *J. Mach. Manuf. and Rel.* **40** 463

[8] Suzuki N, Nakamura A, Shamoto E, Harada K, Matsuo M and Osada M 2003 *Proc. Int. Symp. on Micromechatronics and Human Science* (Japan: Nagoya) 221

[9] Rozenberg L D 1973 *Physical Principles of Ultrasonic Technology* chapter 3 ed L Balamuth (USA: Springer) 38

[10] Grechishnikov V A and Isaev A V 2010 *Rus. Eng. Res.* **30** 413

[11] Isaev A V and Kozochkin M P *Meas. Techn.* **56** 1155

IOP Conf. Series: Materials Science and Engineering **123** (2016) 012046 doi:10.1088/1757-899X/123/1/012046

An efficient method for solving the MAS stiff system of nonlinearly coupled equations: Application to the pseudoelastic response of shape memory alloys (SMA)

A Mourad and Z Kamel

Université de Batna, Département de Mécanique, Laboratoire d'Innovation en Éco-Conception des Structures et Mécanique Sismique (LIECS_MS), Batna, Algérie

E-mail: kzidani05@gmail.com

Abstract. A Shape memory alloy (SMA) actuators have great potential in advanced technology applications where space, weight and cost are crucial design factors. They are known for the shape memory effect which is the ability to recover an initial configuration by simple heating after deformation. SMAs also exhibit a behavior called "pseudoelasticity" also known as "superelasticity" which is the shape recovery associated with mechanical loading and unloading at temperatures above specific values. The key characteristic of SMAs is the martensitic phase transformation, brought about by temperature change and / or by application of stress. Martensitic transformation is usually accompanied by significant changes in mechanical, electrical and thermal properties that render them as prime candidates for the development of smart structures and devices. This work is a contribution to the study of the influence of parameters such as operating temperature and the mode of heat transfer (natural or forced convection) on the pseudoelastic response of SMA used as actuators. Based on the mathematical formalism of the Müller, Achenbach and Seelecke model known as the "MAS model", we develop through a new mathematical formalism a new iterative resolution methodology of the coupled equations. The new approach provides very significant results and allows a significant gain in computation time.

1. Introduction

Shape memory alloys (SMAs) are a unique class of metal alloys that exhibit unusual properties such as shape memory effects (one-way and two-way shape memory effects), superelasticity (pseudoelasticity), damping effect, etc. Due to their specific characteristics such as the ability to cover large apparent permanent deformations (strains), lightweight and high work output, SMAs have been implemented in high-tech industries mainly as actuators or sensors. More recently, they have become candidates for microactuators when fabricated as thin films. These unique properties are related to a so called martensitic transformation (MT) which is brought about by temperature change and / or by application of stress. Martensitic transformation is between a high-symmetric cubic austenitic phase (A) and a low symmetric martensitic phase (M). MT is usually accompanied by significant changes in mechanical, electrical and thermal properties that render them as prime candidates for the development of smart structures and devices. External thermodynamic variables that affect the martensitic transformation are temperature and / or mechanical stress. MT can be activated by a thermal mechanical or thermomechanical cycle (loading / unloading) [1]. SMA actuators reliability and functionality depend on a good knowledge of these specific behaviors which must be properly established mathematically and interpretable physically. SMAs exhibit a hysteretic nonlinear behavior with strong thermo-mechanical coupling. Temperature changes caused by latent phase transformation heats and Joule heating control explain the thermo-mechanical coupling. Thermo-mechanical behavior mathematical modeling and the martensitic transformation are very complex and still interesting scientific topics with innumerable scientific investigations [2]. Implementation in computer codes is difficult because of the representative elementary volume (REV) need [3]. In opposition, the so-called "phenomenological" models using macroscopic scale internal variables are best suited for computer codes implementation [4].

Generally, thermodynamic models are adapted for macroscopic responses capturing and allow explaining underlying physical phenomena. One of the most powerful and truly able to take into account both the strong thermo-mechanical coupling and the hysteretic nonlinearity models, is the model attributed to the authors (Müller - Achenbach and Seelecke) and known as "MAS model" [5,6]. Microscopic observations revealed that a SMA sample under a uniaxial load transforms into a twinned structure. Each twin is either in the austenitic phase denoted "A" or in a variant of the martensitic phase denoted "M_+" or "M_-". During the modeling, in addition to treating the nonlinearity problem, energy functions are introduced and associated to the different phases. Energies minima correspond to phases stability zones [7]. Temperature changes in the loaded SMA sample depend on latent heat emissions, phase changes absorptions, convection heat transfer on external surfaces and Joule heating. The mathematical formalism is based on statistical thermodynamics, probability of phase transition and latent heat. Thermodynamic laws application allows obtaining a coupled system of governing equations that describes the rate of change evolution of the present phases. Numerical solution of these coupled equations is very delicate and need very large machine time [8,9].The present work is devoted to the development of a new efficient resolution methodology which will be used to study the influence of operating conditions on the pseudo-elastic response of a SMA sample used as activator. Conditions of use include the temperature and cooling mode.

2. The martensitic transformation mechanisms

The basic mechanism is a phase transition in the crystal lattice structure [10]. During a unidirectional loading, the SMA sample exhibits a twinned crystallographic structure (Figures 1 and 2). Both variants of the martensitic phase “M_-” and “M_+” can be mechanically interpreted as the results of the shear deformations of the crystal lattice of the parent austenite phase “A”. In the absence of external load, the martensitic phase is stable at low temperature while the austenite phase is stable at high temperature. Observation of a specimen during the phase transition reveals a structure of alternating layers of austenite and martensite. (Figure 2) illustrates the behavior of these layers in a tensile experiment. Initially, at low temperature, the body is in a martensitic state, half of the layers “M_+”, the other half “M_-”. At a critical load level, the “M_-”-layers flip into the “M_+”-phase, thus contributing to a considerable length change. Upon removal of the load, the “M_-”-layers do not flip back into their original phase, but when the specimen is heated, all the layers transform into the austenitic phase. This causes the body to shorten again and thus gives rise to the well-known shape memory effect. Subsequent cooling finally completes the cycle by having the martensitic twins occur again.

3. MAS time-dependent thermomechanical behavior model

In order to describe the above behavior, the original MAS model takes the metallic layers as basic elements. For a representative elementary volume (REV), the microstructure of a sample consists of twins forming phases (martensite variants) “immersed” in a parent austenite phase.

Figure 1. Phases present in the martensitic transformation

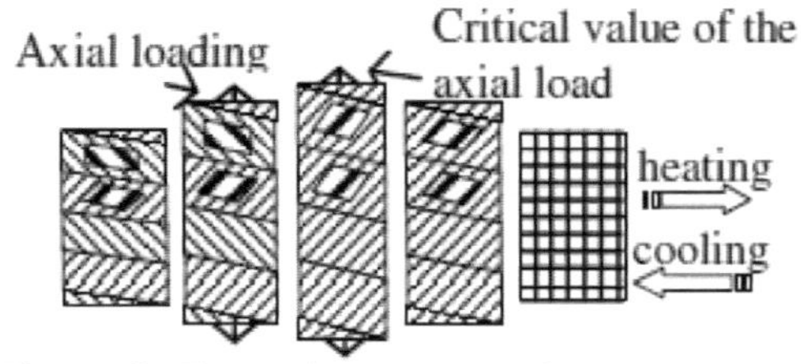

Figure 2. The various stages of the martensitic transformation and the shape memory effect [9]

For a representative elementary volume (REV), internal variables denoted by x_A, x_+ and x_- are associated and denote the volume fractions of the corresponding phases "A", "M+" and "M-".

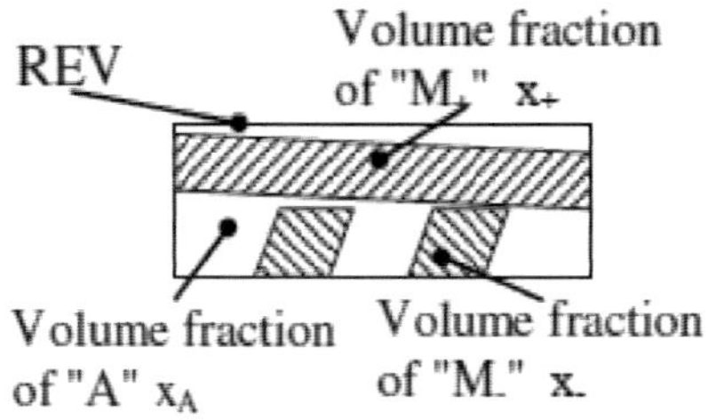

Figure 3. The REV and the volume fractions ofthe phases

If we denote by Δ_A , Δ_- and Δ_+ the expectation values of the length changes in the phases calculated from statistical thermodynamics [18] (Heintze *et al*, 2008), we have:

$$\begin{cases} \Delta = x_A\, \Delta_A + x_-\, \Delta_- + x_+\, \Delta_+ \\ x_A + x_- + x_+ = 1 \end{cases} \tag{1}$$

The rate of change of the phase fractions is given by the following equations:

$$\begin{cases} \dot{x}_+ = -x_+ P^{+A} + x_A P^{A+} \\ \dot{x}_- = -x_- P^{-A} + x_A P^{A-} \end{cases} \tag{2}$$

The quantities P_{ab} are the transition probabilities from phase "a" to phase "b" , which also can be calculated from statistical thermodynarnics. They're functions of the parameters x_A, x_+, x_- and T. For stress σ and strain ε, it is clear that we have:

$$\begin{cases} \sigma = \partial\Psi/\partial\varepsilon \\ \partial\sigma/\partial\varepsilon = E \end{cases} \tag{3}$$

where Ψ represents the considered effective potential energy and E the elasticity modulus.

When incorporating the possibility of electric heating, which is crucial for an application as actuator, the energetic balance leads to:

$$\rho c\dot{T} = -\alpha_c\, S_v\, (T - T_0) + j(t) - (h_{M+} - h_A)\, x_+(t) - (h_{M-} - h_A)\, x_-(t) \tag{4}$$

the above equation summarizes the different heat transfer phenomena affecting the SMA sample temperature T.

T_0: the environment at temperature, ρ and c are respectively the density and the specific heat of the SMA sample, S_v the sample heat exchange surface. h_{M+}, h_{M-} and h_A are the specific enthalpies of the different phases. (h_{M+} -h_A) and (h_{M-} - h_A) represent the transformation latent heats of the two variants of martensite. $j(t)$ the Joule heating ($-\alpha_c S_v$ $(T-T_0)$) convection heat exchange and α_c the thermal conductivity coefficient

The equations (2) and (4) together with (1) constitute a stiff system of nonlinearly coupled ODEs:

$$\frac{d}{dt}\begin{Bmatrix} x_+ \\ x_- \\ \rho c T \end{Bmatrix} = \begin{bmatrix} -p^{+A} & 0 & +p^{A+} & 0 \\ 0 & -p^{-A} & +p^{A-} & 0 \\ -(h_{M+}-h_A) & -(h_{M-}-h_A) & 0 & -\alpha_c S_v \end{bmatrix} \begin{Bmatrix} x_+ \\ x_- \\ x_A \\ T \end{Bmatrix} + \begin{Bmatrix} 0 \\ 0 \\ 0 \\ j(t)+\alpha_c S_v T_0 \end{Bmatrix} \qquad (5)$$

4. Stiff system algorithm resolution

By considering the notations:

$$\frac{d}{dt}\langle x_+ \;\; x_- \;\; \rho c T\rangle^T = \frac{d}{dt}\{Y\}^T \qquad (6)$$

$$[A(Y)] = \begin{bmatrix} -p^{+A} & 0 & +p^{A+} & 0 \\ 0 & -p^{-A} & +p^{A-} & 0 \\ -(h_{M+}-h_A) & -(h_{M-}-h_A) & 0 & -\alpha_c S_v \end{bmatrix} \qquad (7)$$

$$\{B\}^T = \langle 0 \;\; 0 \;\; 0 \;\; j(t)+\alpha_c S_v T_0\rangle \qquad (8)$$

And by introducing the initial conditions, we obtain:

$$\begin{cases} \frac{d}{dt}\{Y\} = [A(Y)]\{Y\} + \{B\} \\ \{Y(0)\} = \{Y_0\} \end{cases} \qquad (9)$$

(9) is a nonlinear stiff initial value problem. Stiffness in this case comes from the difference in scale between different dynamic phenomena evolving simultaneously: volume fractions of the phases, latent heat and temperature T. These conditions make it difficult to use conventional methods of resolution due to the very high number of integration steps and the associated convergence problems [11]. For the resolution we have developed a highly efficient algorithmic allowing us to apply the MAS model to the study of the pseudo-elastic response of SMA.

5. Results and Discussion

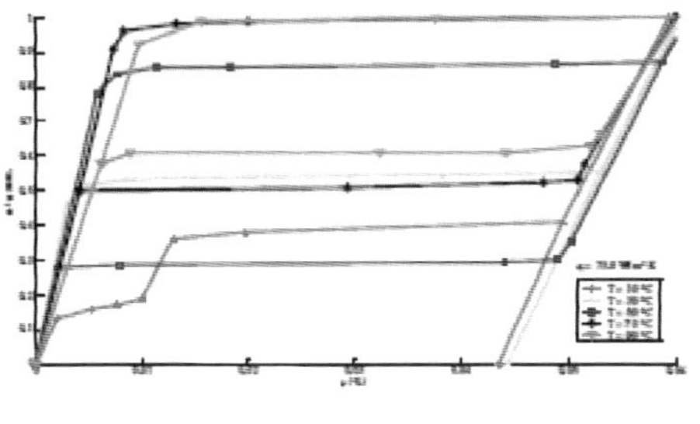

Figure 4. Influence of the operating temperature on the hysteresis loop

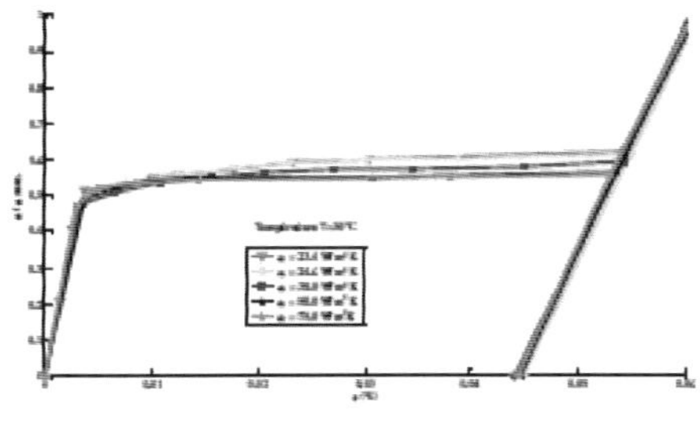

Figure 5. Influence of the heat exchange factor on the hysteresis loop

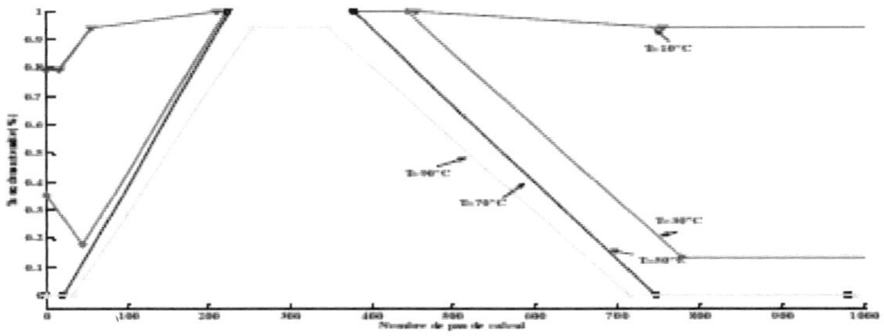

Figure 6. Influence of operating temperature on the overall rate of martensite

Figure 4 shows the clear influence of the operating temperature on the pseudo-elastic response of the SMA sample. Figure 5 shows that the SMA sample is not sensitive to the heat transfer mode. By analyzing the Figure 6, at low temperature (T = 10°C), the SMA sample consists mainly of the martensitic phase (80%), during the martensitic transformation this rate increases to the value of 95% and then stabilize around 100%.

6. Conclusion

The presented model was developed on the basis of the "MAS" model. It allows a very efficient way for analysing the pseudo-elastic response of a SMA sample for the manufacture of actuators. The mathematical formalism has allowed the development of a numerical simulation code. Obtained results are very significant and can reproduce the SMA pseudoelastic behaviour.

References

[1] Chemisky, Y. 2009 thesis of the Paul Verlaine university Metz Frannce

[2] Khandelwal, A.; Buravalla, V. 2009 *International Journal of Structural Changes in Solids – mechanics and applications* **1** 111

[3] Duval, A. 2009 thesis of the Henri Poincaré university Nancy 1 France

[4] Peultier, B.; Ben Zineb, T.; Patoor, E. 2008 *Materials Science and Engineering* A **384**

[5] Seelecke, S.; Müller, I. 2004 *Appl. Mech. Rev.* **57(1)** 23

[6] Li, Q. 2006 Ph.D. Thesis North Carolina State University USA

[7] Achenbach, M.; Müller, I. 1982 *International Conference on Martensitic Transformations* C4-**163**

[8] Heintze, O. 2004 Ph.D. Thesis North Carolina State University USA

[9] Heintze, O.; Seelecke, S. 2008 *Mat. Sci. Eng.* A **339**

[10] McDonald, A. D.; Walker, S. J. I. 2010 Proceedings of the ASME Conference on Smart Materials, Adaptive Structures and Intelligent Systems

[11] Thohura, S.; Rahman, A. 2013 *Global Journal of Science Frontier Research Mathematics and Decision Sciences* **13 Issue 6**

Substantiation of Reference Method For Determining Concrete's Freeze-Thaw Resistance

S Nikolskiy, O Pertseva

Peter the Great St. Petersburg Polytechnic University, St. Petersburg, Russia

E-mail: bagira@cef.spbstu.ru, olya_perceva@mail.ru

Abstract. It has been analytically proved that using concrete's rate of set ε as a measure of damage, instead of decreasing of tensile strength R, increases freeze-thaw resistance's accuracy of estimation a lot under otherwise equal conditions by the time of freeze-thaw cycling. Also it has been experimentally shown that ratio of relative decreasing R to ε in direction, perpendicular to compression, is assumed to be independent on values R and ε for a given concrete and on the ways of achieving them during mechanical or freeze-thaw cycling. Taking this into account patented methods for estimation of concrete's freeze-thaw resistance as per values R and ε received after freezing and thawing cycles of some specimens and their postliminary failure by linear compression was substantiated.

1. Introduction

Freeze-thaw resistance of concrete is an ability of water-saturated concrete specimen to maintain repeated standard thermo cycles without noticeable damage. Different types of water pressure cause concrete's freeze-thaw deterioration, such as hydraulic and osmotic pressure [1], capillary pressure [2] and other types of water influence according to existing freeze-thaw resistance theory. Decreasing strenght of construction material is associated with the processes of its water freezing (for example, rock [3]). Water gets into structure of porous bodies, separates particles and breaks coupling between them [4]. Porosity of material is determinative factor for frost resistance and durability subsequently [5,6]. So, strength of concrete could be pictured as a porosity function [7]. In order to determine the concrete mix composition, it is necessary to take into account freeze-thaw resistance. Worldwide experience offers some test methods to determine durability of concrete by freeze-thaw damage, such as Slab test [8], CDF [9], CIF-Test [10] and Cube-Test [11]. But they differ in terms of their procedures and conditions and are referenced to particular conditions.

There are two different standard types of methods of determining the freeze-thaw resistance of concrete- basic [12] and reference [13] in the Russian Federation. Considerable spread of values of strength for concrete specimens (variation coefficient $\rho \approx 17\%$) in constant conditions of preparing and testing determines random spread of choosing average values $\overline{R}$ in the range of $\Delta R / R = \pm 3\rho\sqrt{n}$, where n is the volume of bath of specimens. In this case for proving the significance of relative decreasing $\overline{R}$ on 0,05…0,15 it is necessary to test more that 50 specimens and support $\Delta\overline{R} / \overline{R} = 0.068\overline{R}$, when confidence figure is equal 0,95. Therefore, main disadvantages of the basic method are high labor input and small operability by virtue of the fact that duration of the basic thermo cycle is equal 4,5 at least and $F >> 50$.

One of the existing reference methods is a Dilatometric rapid method of determining the freeze-thaw resistance of concrete [13]. This method is a prototype for the method, which has been offered by me. In this method concrete's freeze-thaw resistance is determined by the maximum relative difference of volume deformations of the tested concrete and standard specimens in accordance with tables provided in standard specification [12] taking into account concrete's type, its form and the size of specimens.

However, the results from the tables provided in state standard specification are acceptable only for Portland cement concrete and slag Portland cement concrete without surface-active additives (PEAHENS), such concretes are used extremely seldom now. Now lot of new concretes are investigated, tested and used, for example, nanomodificated concrete [14], high-strength concrete [15], concrete on the basis of fine-grained dry powder mixes [16], concrete with using recycling concrete aggregates [17] etc. In order to obtain new tables long labour-consuming experiences, which imply using, basic methods are needed [18].

2. Project's objectives

The aims of this paper are to obtain the dependence between relative decreasing in strength by rate of strain and substantiate the new method for determination of concrete frost resistance. Using of measure of damage D by the time of freeze-thaw cycling is offered for removal of disadvantages. In accordance with experimental data [19, 20] dependence $D = Ac^q$ is well approximated, where: $A = D_0$ (when c=1) – value of D after first thermo cycle; c – number of thermo cycles; q – constant of material.

In accordance with this statement, evaluation of maximal measurement errors (ΔF and $\Delta F / F$) by the time of determination F is possible with help of dependence:

$$\delta R = A \cdot c^q, \tag{1}$$

where $A = \delta R_1$ (when c=1) – decreasing of strength after first thermo cycle and δR – decreasing of original strength R. The first consequence from (1) covers to constant of material q:

$$q = \{\ln(\delta R_c / \delta R_1)\}/(\ln c) \tag{2}$$

After differentiation (3):

$$q\frac{\Delta F}{F} + \{\ln(F/c)\}\Delta q = \frac{\Delta R}{R} + \frac{\Delta \delta R_c}{\delta R_c} \Rightarrow \pm\frac{\Delta F}{F} - \frac{1}{q}\left\{\left|\frac{\Delta R}{R}\right| + \left|\frac{\Delta \delta R_c}{\delta R_c}\right|\right\} + \{\ln(F/c)\}\left|\frac{\Delta q}{q}\right| = 0, \tag{3}$$

where $\Delta R/R$, $\Delta \delta R/\delta R_c$ >> $\Delta R/R$ and $\Delta q/q$ are maximal relative measurement errors of determination original value R, decreasing R after c thermo cycles or values q correspondingly. By the time of using non-destructive method [21] $\Delta R/R$ =0,03 or $\Delta R/R$=0,068, when we used bathes from 25 specimens. So, $\Delta \delta R / 2\Delta R$.

Therefore, $\Delta F/F$ depends on the measurement errors of strength $\Delta R/R + \Delta \delta R_c / \delta R_c$ with accumulation factor $1/q$ and measurement error of $\Delta q/q$ with accumulation factor $\ln(F/c)$. If $q = 1$ ($\Delta q = 0$), that we will get measurement error of calculation F by $\delta R / \delta R_{1c}$ from (3):

$$\pm\frac{\Delta F}{F} = \left|\frac{\Delta R}{R}\right| + \left|\frac{\Delta \delta R_c}{\delta R_c}\right|, \tag{4}$$

where $\Delta R/R \approx 0{,}03$; $\Delta \delta R_c = 2\Delta R \approx 0{,}06R$; $\Delta R_c \le [\delta R/R] = (0{,}05...0{,}15)R$.

After differentiation (2):

$$\frac{\Delta q}{q} = \frac{1}{\ln(\delta R_c / \delta R_1)} \left\{ \frac{\Delta \delta R_c}{\delta R_c} + \frac{\Delta \delta R_1}{\delta R_1} \right\}, \tag{5}$$

where $\Delta \delta R_c = \Delta \delta R_1 = 2\Delta R;\ \delta R_c = \delta R_1 c^q; \Delta \delta R_1 / \delta R_1 >> \Delta R / R.$ After plugging (5) in (3):

$$\frac{\Delta F}{F} \approx \frac{1}{q} \frac{\Delta \delta R_c}{\delta R_c} \left\{ 1 + \left(\ln \frac{F}{C} \right) \left(\frac{1 + c^q}{\ln c} \right) \right\}. \tag{6}$$

Accumulation factor $\frac{1 + c^q}{\ln c}$ (when $q = 1$) has a minimum when $c = 4$ and is numerically equal 3,6.

When F = 40 and $c = 4 \ln(F / c) = 2{,}3$ = 4; $\left(\ln \frac{F}{C} \right) \left(\frac{1 + c^q}{\ln c} \right) \approx 8{,}3$, and $\frac{\Delta F}{F} \approx \frac{9{,}3}{q} \frac{\Delta \delta R_c}{\delta R_c}$. While $\Delta \delta R_c = \delta Rc$ (when c is too small), it is $\Delta F / F \approx 9{,}3 / q$. The result proved that contribution of measurement error of q in $\Delta F / F$ is more in 8 times than contribution of measurement error of R. It's the case, when measure of damage is a decreasing of strength. To decide this problem using of rate of strain $\varepsilon = \delta l / l$, where $l \approx 100mm$ – base length, $\delta l = l_c - l$ – the base deformation and l_c – base length after "c" thermo cycles, is offered. But for that it is necessary to determine dependence between decreasing in strength and rate of strain. So, this value is defined as $z = (\delta R / R) / \varepsilon$.

3. Formatting the text

To determine this dependence concrete mix was tested. Mix contained a portlandt cement (12%) of brand 400, sand (25%), granite crushed stone 5 … 20 mm (56%) and water (7%). 108 samples cubes with an edge of 150 mm were prepared. There points of intersection of diagonals of opposite sides were spaced far apart from each other (to 1,5 mm). Samples hardened 28 days in water at the room temperature, and then 60 days in damp sand at 18 … 26°C. From these 108 samples, 8 samples were cycled alternately thermally and mechanically, and 100 samples were used for a frost resistance assessment by a basic method. At realization of a basic method l distance between points of each sample was measured by means of a tool microscope before tests for frost resistance or for durability. Rate of strain was counted as. Changing of distance between points at realization of a method [22] was carried out on 20±2 °C before and after cycling by means of a bracket with variable base and a measuring head of our type (the price of division of 1 micron). Threshold loading of a water-saturated sample was determined by a way [23], registering the acoustic issue (AI) by means of the AF-15 device at cyclic loading and unloading of a sample to zero. In the first experience load of L was brought to 11 t; in the absence of acoustic issue in the course of the end of unloading value of L was increased by 5% and so until at the end of unloading there was no AE. For accepted an average of L two last cycles. The limit of long durability was found as the relation to the average area of two loaded sides. Dependence z was determined as $(L_1 - L_2) / L_1 \varepsilon$, where L_1 is a maximal value L in the first loading and L_2 – in the second loading (fig. 1). Results of this experiment are given in the Table 1. Dependence z is near to linear.

Table 1. Results of experimental determination dependence between R and ε after 20, 42, 84 and 105 cycles (thermocycles and mechanical cycles)

№ of specimen		1	2	3	4	5	6	7	8	Average
Origin value R_0, [MPa]		9,45	16,8	22,7	23,4	24,3	24,3	25,0	30,1	22,0
After 20 cycles	R_0, [MPa]	9,18	16,2	22,2	22,7	23,5	23,9	24,1	29,4	
	$\delta R/R \cdot 10^5$	2857	3571	2203	2991	3292	1646	3600	2326	2811
	$\varepsilon \cdot 10^3$	216	210	204	206	204	200	202	190	204
	z	13,23	17,00	10,80	14,92	16,14	8,23	17,82	12,24	13,8
After 42 cycles	R_0, [MPa]	9,03	16,15	21,9	22,3	23,4	22,9	23,9	20,4	
	$\delta R/R \cdot 10^5$	4482	3869	3524	4701	3704	561	4440	3222	4213
	$\varepsilon \cdot 10^3$	406	452	404	408	394	396	410	394	408
	z	11,04	8,56	8,72	11,52	9,40	14,55	10,73	8,18	10,32
After 84 cycles	R_0, [MPa]	8,10	14,7	19,1	21,2	21,0	23,0	22,6	26,7	
	$\delta R/R \cdot 10^5$	1429	893	1586	940	1244	1230	1253	1130	1213
	$\varepsilon \cdot 10^3$	1206	1217	1188	1196	1201	1207	1196	991	1175
	z	11,85	5,31	13,35	7,86	10,36	9,31	7,36	11,04	9,55
After 105 cycles	R_0, [MPa]	7,75	14,2	18,6	20,1	21,1	20,9	21,4	26,4	18,8
	$\delta R/R \cdot 10^5$	1799	1547	1806	1410	1317	1399	1440	1229	1488
	$\varepsilon \cdot 10^3$	1582	1440	1435	1534	1561	1459	1297	1196	1439
	z	11,37	10,74	12,59	9,19	8,43	9,59	11,10	10,28	10,40
$\bar{z}$ for specimen		11,40	10,76	11,04	10,73	10,21	10,30	11,05	10,34	$\bar{\bar{z}}$=10,73
R- average value of the long-term strength in conditions of compression, δR decreasing R after specimen cycling,								S^2	S	ρ
ε - rate of strain in the direction, perpendicular compression; $z=(\delta R/R)/\varepsilon$;								1,94	1,394	0,134
z_k-value z after 105 cycles; $\bar{z}$ - average value z for specimen; $\bar{\bar{z}}$ - average value for $\bar{z}$; S^2 –sampling variance; S-mean square deviation; ρ- variation coefficient								0,183	0,428	0,04

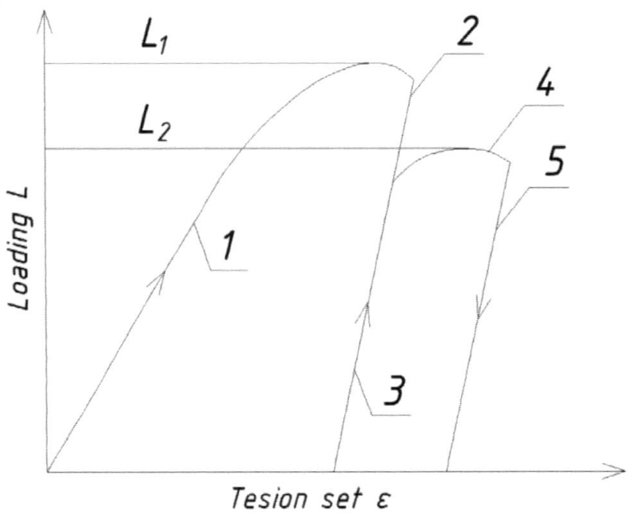

Figure 1. This is a shedule of specimen loading in conditions of monoaxial compression after thermocycles: 1 – line of first loading; 2 – line of first unloading; 3 – line of second loadind; 4 - line of second loading; 5 - line of second unloading; ε – relative longitudinal tension set; L_1 – extreme loading of the first loading; L_2 – extreme loading of the second loading.

Also results by non-destructive method were compared with results by standard method (Table 2)

Table 2. Comparison results by non-destructive method and standard method.

Average value of:	Non-destructive method	Basic method
δR/R	0,1488	0,161
ε	0,01439	0,01448
z	10,4	11,1

Results

Results by non-destructive method differ 6,3% from results by basic method. Also this dependence is not delicate to changing some thermo cycles to mechanical cycles. So, value of z is constant and dependence of relative decreasing in strength is linear. After mathematical manipuleations it becomes clear, that using ε as a measure of damage is more accurate than using δR, in approximately in 35 times.

After taking into account theese results new method for determination of concrete frost resistance was offered by application for a patent [24].

References

[1] Rønning, T.F. 2001 *The Norwegian Institute of Technology Press: Trondheim*. **416**
[2] Davie, C.T., Pearce, C.J., Bićanić, N. 2005 *Proc. of the 13 ACME conference: Sheffield*. **46-49**
[3] Kurilko, A.S., Novopashin, M. D. 2005 *Journal of Mining Science*. **41** 119-122.
[4] Cherepanov, V.I., Nekrasova, E.V., Chernyh, N.A., Panchenko, Ju.F. 2013 *Construction Materials* **15** pp. 10-12. (rus)
[5] Shashank, B. 2004 *University of Tokyo Press: Tokyo*. **94**
[6] Shanshan, J., Jinxi, Z., Baoshan, H. 2013 *Construction and Building Material*. **47** 126–130.
[7] Nesvetaev, G.V., Kardumjan, G.S. 2013 *Beton i zhelezobeton* **45** 6-8 (rus)
[8] SS 137244. Concrete testing – Hardened Concrete-Frost Resistance.
[9] RILEM TC 117-FDC 2001 *RILEM Technical Committee*
[10] RILEM TC 176 2004 *RILEM Technical Committee*
[11] Bunke, N. 1991 *Schriftenreihe des Deutschen Ausschusses für Stahlbeton: Berlin*, **422**
[12] GOST 10060.1-95 (rus)
[13] GOST 10060.3-95 (rus)
[14] Ponomarev, A.N. 2007 *Construction and Building Materials* **24** 22-25. (rus)
[15] Barabanshchikov, Yu.G., Semenov, K.V. 2007 *Power technology and engineering* **36** 197-200
[16] Kalashnikov, V.I., Tarakanov, O.V., Kusnetsov, Y.S., Volodin, V.M., Belyakova, E.A. 2012 *Magazine of Civil Engineering* **34** 47-53.
[17] Gokce, A., Nagataki, C., Saeki, T., Hisada, M. 2010 *Construction and building material* **25** 2426-2431.
[18] Dikun, A.D., at al. 2005 *Construction Materials* **52** 55-56 (rus)
[19] Kuncevich, O.V. 1983 *Stroyizdat: Leningrad*. **132** (rus)
[20] Nikolskiy, S.G. Nikolskaya, T.S. 2009 *Proc. of the 3rd IC Populjarnoe betonovedenie* **35-43** (rus)
[21] Nikolskiy, S.G., at al. 2010 RU Patent 2,380,681 (rus)
[22] Nikolskiy, S.G., at al. 2009 RU Patent 2,305,281 (rus)
[23] Vorontsova, E.A. Nikolskiy, S.G., 2011 *Proc. of the XL Week of Science: St. Petersburg*, pp. **344-430** (rus)
[24] Pertseva, O.N., Nikolskiy, S.G. 2014 RU Application for a patient 2,014,116,713 from 24, May (rus)

IOP Conf. Series: Materials Science and Engineering **123** (2016) 012048 doi:10.1088/1757-899X/123/1/012048

Kinetic study of non-isothermal decomposition of a composite diasporic-boehmitic bauxite

M Samouhos[1]*, P Angelopoulos[1], G Pilatos[2], M Taxiarchou1, S Papageorgiou[2]

[1]Laboratory of Metallurgy, School of Mining and Metallurgical Engineering, National Technical University of Athens, Greece
[2]Demokritos National Center of scientific research, Agia Paraskevi, Attiki, Athens, Greece

Email: msamouhos@metal.ntua.gr

Abstract. In the current study, the kinetic of the thermal decomposition of a composite diasporic-boehmitic bauxite is investigated under non-isothermal heating conditions by means of thermogravimetric analysis. The calculation of activation energy (*Ea*) has been performed by various methods and the attained values range from 185000 to 190000 J·mol-1 for conversion rate of 0.5. The first-order kinetic model can adequately be used to describe thermal decomposition of bauxite composite, while the pre-exponential factor value was found to be 2.088*1011 min-1.The calculated activation energy value together with pre-exponential factor and the solid state kinetic model factors enable the simulation of the thermal decomposition under various heating conditions by the implementation of an ordinary differential equation. The calculated values are in satisfactory agreement with the experimental ones.

1. Introduction

Bauxite is an aluminum ore which mainly consists of the minerals gibbsite [$Al(OH)_3$], bayerite [a-$Al(OH)_3$], boehmite [γ-AlO(OH)], diaspore [α-AlO(OH)] and minor phases of iron oxides, kaolinite and anatase. Bauxite consist the major primary source for the alumina (Al_2O_3) and metallic aluminum. Non metallurgical applications of rough bauxite ore include the production of hydraulic fracturing proppants and low cost abrasives [1]. Recently, a heightened interest is presented concerning the production of low cost refractory cements and dense ceramics using a concrete rich in bauxite material [2]. The above applications require the calcination of bauxite which practically equivalents with the dehydroxylation of its aluminum phases. Although the thermal decomposition kinetic of gibbsitic and diasporic bauxites has been examined [3, 4], in the current study, for first time in the literature, the thermal decomposition of composite diasporic-boehmitic bauxite is kinetically studied in detail.

2. Experimental

The diasporic-boehmitic bauxite ore used in the present study originates from an open pit of the Greek aluminum producer "Aluminum of Greece" located in Parnassos Mountain, Greece. Its granulometric distribution is: +80-120 μm (22.4% w/w), +120-160 μm (41.6%) and +160-200 μm (36%). The bulk chemical analysis of the bauxite ore is presented in Table 1, while its mineralogical analysis can see in Figure 1. The thermogravimetric analysis experiments that were used in the current kinetic analysis were carried out on a Labsys DTA/DSC Setaram. Approximately 80 mg of dried bauxite composite samples were placed in platinum crucibles and heated in helium atmosphere with a flow rate of 30 mL·min^{-1}. In Figure 2 the TG, DTG and DSC diagrams versus sample temperature are presented at four constant heating rates; 2, 5, 10 and 20°C min.

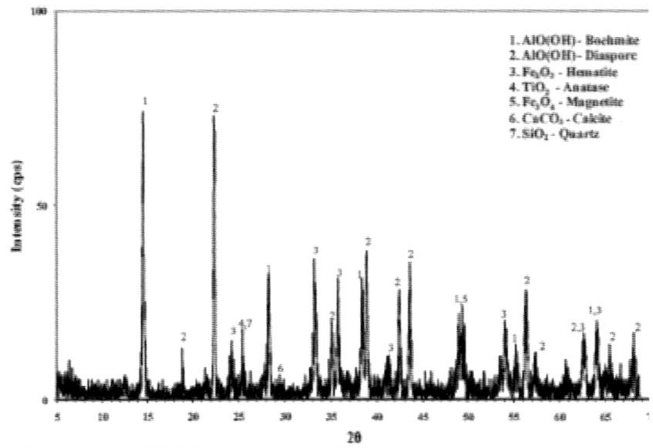

Figure 1. XRD pattern of the rough bauxite ore

Table 1. XRF analysis of the bauxite ore

Species	Mass(w/w.%)
Al_2O_3	58.68
Fe_2O_3	22.53
SiO_2	3.36
TiO_2	2.37
ZrO_2	0.88
CaO	0.43
Cr_2O_3	0.151
Na_2O	0.056
L.O.I	11.04

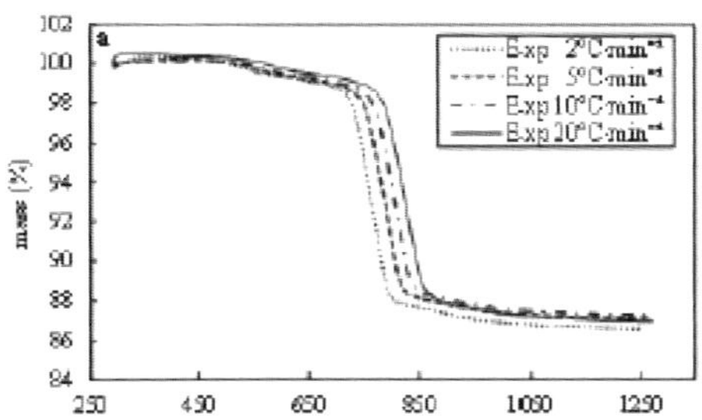

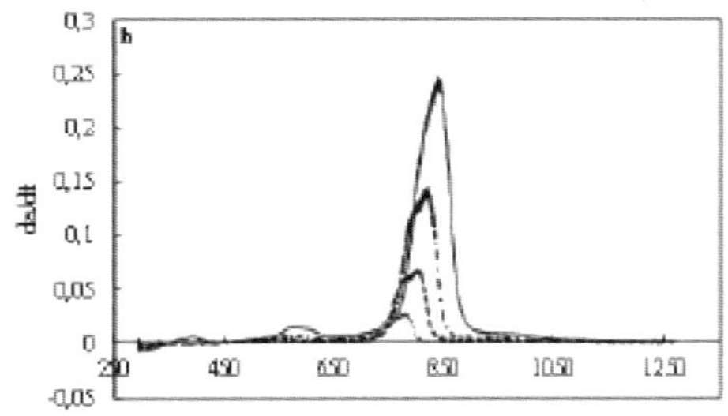

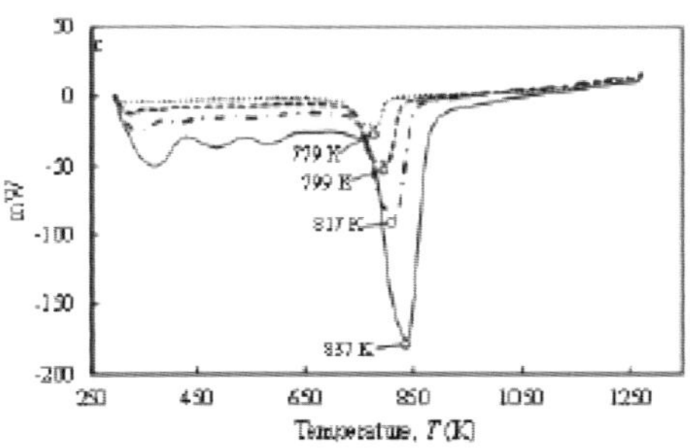

Figure 2. **(a)** TG, **(b)**DTG and **(c)** DSC plot of diaspore-boehmite composite bauxite heated under various heating rates

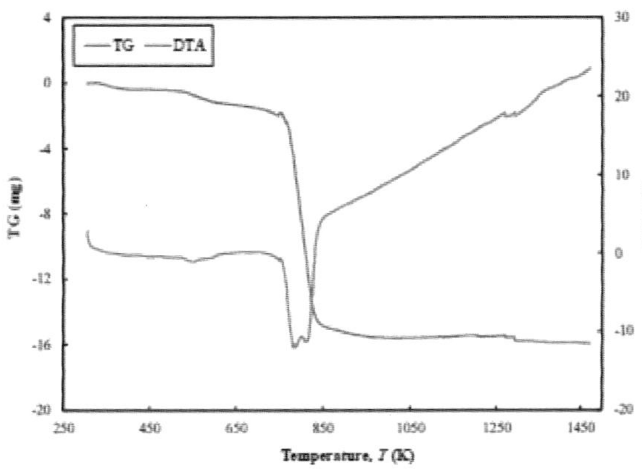

Figure 3. Thermogravimetric analysis of the rough bauxite ore

Figure 3 presents the TG-DTA curves of the bauxite ore in the temperature range between 300 and 1473K with a 10°C·min-1 heating rate. The data curve shows two main endothermic peaks at 778.3 and 805.5 K that correspond to the dehydroxylation of the boehmite and the diaspore respectively and their transformation to Al_2O_3. The weight loss between 523 and 723K can be attributed to the dehydroxylation of minor amorphous, undetectable by X-ray diffractometry, aluminum phases. The slight weight loss that observed around 1273K can be attributed to the partially decomposition of hematite to magnetite (Fe_3O_4) and metastable wustite (FeO).

3. Kinetic analysis

The governing equation for kinetic analysis of solid-state decomposition under isobaric conditions is described by the following equation [5]:

$$\frac{da}{dt} = k(T)f(a) \tag{1}$$

Where, T is the temperature, a is the extension of conversion and, k is the reaction rate constant. Typically, the dependence of the decomposition evolution by the temperature is introduced in Eq. 2 by replacing *k(T)* with the Arrhenius equation, which gives the following form:

$$\frac{da}{dt} = A \cdot exp\left(\frac{-E_a}{R \cdot T}\right) f(a) \tag{2}$$

Where, (*A*) denotes the pre-exponential factor, (E_a) is the activation energy and (*R*) the gas constant. The integral form, with respect to temperature, of the equation (2) can be written as:

$$g(a) = \frac{A}{\beta}\int_0^T \exp\left(\frac{-E}{RT}\right) dT \tag{3}$$

The objective of the kinetic analysis constitutes the calculation of the three basic parameters, namely, the activation energy, the pre-exponential factor and the kinetic model.

Three model free methods (Table 2) where implemented for the calculation of activation energy distribution throughout the conversion range allowing the estimation of the apparent activation energy at progressive degrees of conversion [6-10].

Table 2. Model free methods used for the calculation of the activation energy

Methods	Equation	Calculation
Friedman method	$\ln\left[\beta_i\left(\frac{da}{dT}\right)_{a,i}\right] = \ln[A_a f(a)] - \left(\frac{E_{a,a}}{RT_a}\right)$	$ln[\beta_i(da/dT)_{a,i}]$ versus $1/T_a$
Flynn, Wall, Ozawa method	$\frac{d\log\beta}{d1/T_a} \approx 0.457\frac{E_a}{R}$	$\log\beta$ versus $1/T$
Flynn, Wall, Ozawa method with DSC data	$\frac{d\log\beta}{d1/T_a} \approx 0.457\frac{E_a}{R}$	log β versus $1/T_m$ (T_m *when DSC max peak)*
Lyon method	$E_{a,a} = -R\left[\frac{d\ln\beta}{d1/T_a} + 2T_a\right]$	$ln\beta$ versus $1/T_\alpha$

The determination of the appropriate kinetic model requires the introduction of the function *z(a)* which is derived by multiplying the integral and the differential form of the reaction kinetic models [20]:

$$z(a) = f(a) \cdot g(a) = \left(\frac{da}{dt}\right)_a T_a^2 \left[\frac{\pi(x)}{x}\right] \tag{4}$$

Where π(x), is a function and x=E/RT. It has been established [11] that the term in the brackets of Eq. 4 has a negligible effect on the shape of *z(a)* function. Thus, the values of *z(a)* can be determined by multiplying the experimental values $(d\alpha/dt)_\alpha$ and $T_{\alpha 2}$. The resulting experimental values of $z(\alpha)$ are plotted as a function of α and compared against theoretical $z(\alpha)$ master plots of several kinetic models. A suitable model is identified as the best match between the experimental and the theoretical $z(\alpha)$ master plots using the Criado method [12]. The adequacy of each model to simulate the non-isothermal decomposition of bauxite was tested by two error factors (*RSS*) and (*RSQ*). Finally, the pre-exponential factor (*A*) is calculated by the following equation [13]:

$$A = \frac{-\beta Ea}{RT_{max}^2 f'(a_{max})} \exp\left(\frac{E_a}{RT_{max}}\right) \quad (5)$$

Where, the subscript a_{max} in the aforementioned equation denotes the conversion value where the weight loss rate maximizes.

4. Results and discussion

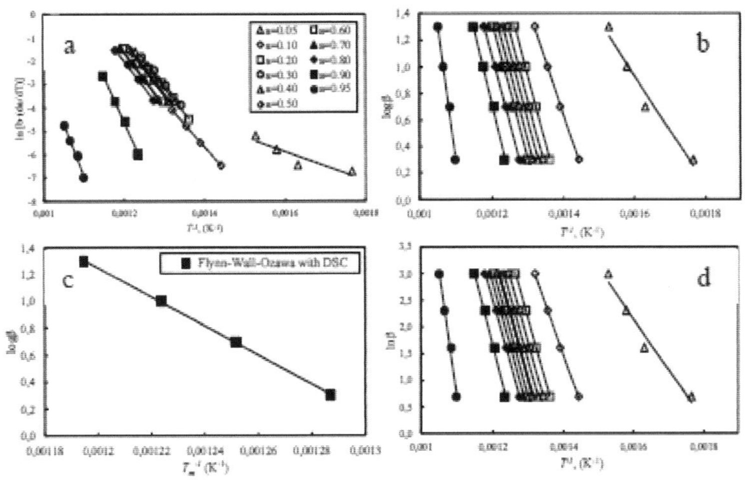

Figure 4. Plotting of diagrams used in the calculation of activation energy through: Friedman (a), Flynn-Wall-Ozawa (b), Flynn-Wall-Ozawa with DSC data (c) and Lyon methods (d)

The methodology of activation energy calculation via various methods is presented in Figure 4, while its values throughout the whole conversion range are presented in the diagram of Figure 5. The results show that the activation energy calculated by the aforementioned methods is almost constant for a wide (a) range from about 0.1 to 0.9.

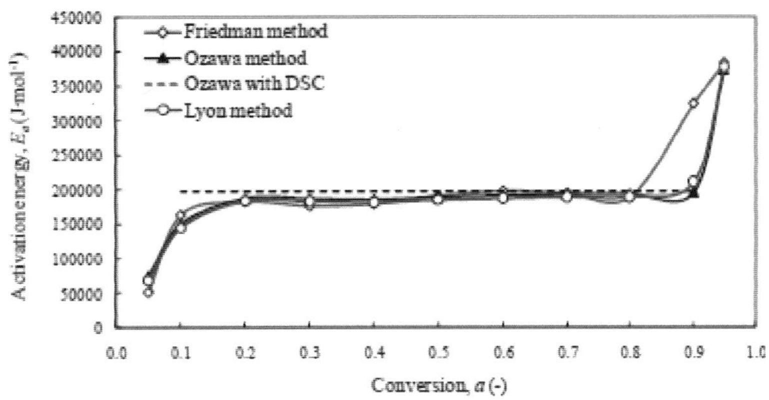

Figure 5. Activation energy upon conversion rate as calculated by various isoconversional methods

For the determination of the kinetic model that describes the thermal decomposition of bauxite composite, the Criado diagram has been created and presented in Figure 6. For each theoretical model, $Z(a)\cdot Z(0.5)^{-1}$ values throughout the whole conversion range accrued by solving Eq. 4 for *g(a)* and *f(a)* values of specific kinetic models. The determination of the deviation of each theoretical model from experimental values that are presented with black and white symbols, was performed by calculating the error factors *RSS* and *RSQ*. A value of *RSQ* close to unity implies that the model is consistent with the experimental data, while, as far as the *RSS* factor is concerned, the lower its value, the lower is the divergence between model and experimental values.

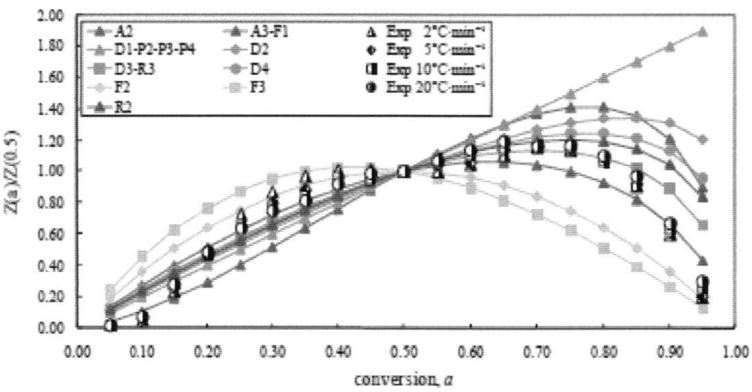

Figure 6. $Z(a)\cdot Z(0.5)^{-1}$ diagram of theoretical models and experimental values throughout the whole conversion range

It was found that the models A_3 and F_1 can adequately describe the decomposition of bauxite composite under non-isothermal, constant heating rate conditions. The mean *RSQ* value is 0,965 while the mean RSS 0.237 implying adequate accuracy. In order to find which is the most appropriate model between A_3 and F_1, a new formula is introduced, namely *y(a)* [14].

$$y(a) = \left(\frac{da}{dT}\right)_a exp\left(\frac{E_o}{RT_a}\right) \tag{6}$$

The value of the conversion (a_m) when maximum value of *y(a)* occurs is characteristic for each model. As far as the experimental data of bauxite thermal decomposition are concerned and regardless of the heating rate, the maximum *y(a)* values occur at the initial stage, namely for $a_m= 0$. In case of A_3 model, a_m value equals to 0,283, while for the F_1 model, the value of a_m equals 0 [14]. It's thus clear, that the thermal decomposition of bauxite, can be described by F_1 model, namely a first order reaction model. The pre-exponential factor calculated by Eq. 5 was found to be $2.088{*}10^{11}min^{-1}$. After calculation of the kinetic triplet, namely the activation energy, pre-exponential factor as well as the kinetic model, Eq. 4 takes the following form:

$$\frac{da}{dT} = \frac{2.088\cdot 10^{11}}{\beta}\cdot \exp\left(\frac{-183000}{8.314\cdot T}\right)\cdot (1-\alpha) \tag{7}$$

In Figure 7, experimental data from thermogravimetry are compared to the modeling data that have been accrued by Eq. 7. In fact, Eq. 7 fails in calculating the decomposition progress at the initial and final stages, namely at temperature below 550K and above 900 K due to the realization of minor reactions , as already has been mentioned, at these temperature ranges. In order to identify whether Eq. 7 is capable to simulate the weight loss of bauxite composite due to dehydroxylation of boehmite and diaspore, the conversion was assumed to be 0 at 650K while it was considered that the decomposition has been completed at 950 K, where *a*=1. By applying Eq. 7, and under the aforementioned assumption, it can be seen that the theoretically calculated conversion progress is in good agreement with the experimental data.

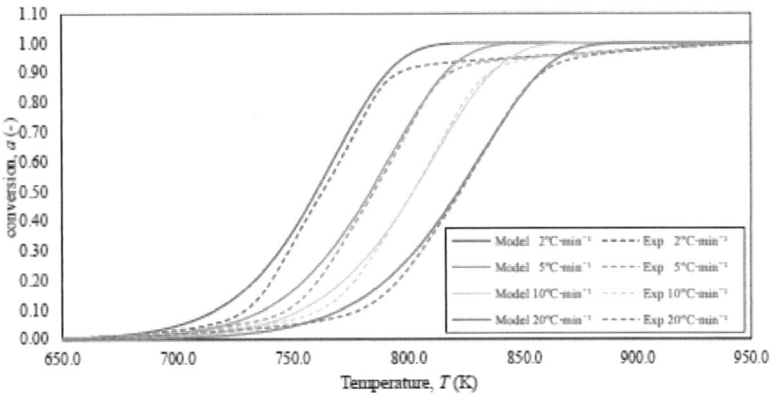

Figure 7. Experimental and model (a) values of thermal decomposition of bauxite under non-isothermal, constant heating rate conditions and temperature range between 650 K to 950 K.

5. Conclusion

In the current study, the kinetic of thermal decomposition of a diasporic-boehmitic bauxite ore were investigated at the temperature range between 300 and 1273 K. The kinetic study under non isothermal conditions is based on the determination of kinetic triplet values, namely E_a, A and $f(a)$.The value of E_a ranges depending on the method used, between 184994 and 190357 J·mol$_{-1}$ for a 0.5 conversion rate. The dependence of decomposition progress by the conversion was found to obey the F_1. Although decomposition of diaspore and boehmite occur in slightly different temperature, distinction of reactions is impossible and thus it is inevitable the assumption of a single step reaction. Thus, the decomposition of the composite bauxite ore kinetically follows a single process pathway due to the almost simultaneous decomposition reactions of boehmite and diaspore to Al_2O_3.

References

[1] Russell A, Dickson D, Harben P W, Keegan N 1999 *Bauxite & alumina: a guide to non metallurgical uses and markets* (Worcester Prk: Industrial Minerals Information Ltd.)
[2] Dong Y, Feng X, Feng X, Ding Y, Liu X, Meng G 2008 *Journal of Alloys and Compounds* **460** 599
[3] Li H, Shao T, Lia D, Chen D 2005 e *Thermochimica Acta* **427** 9
[4] Ray D, Parya T K 2013 *International Ceramic Review* **62** 120
[5] Brown M E 2002 *Introduction to thermal Analysis: Techniques and Applications* (Springer: 2nd edition)
[6] Friedman 1964 *Journal of Polymer Science Part* C **6** 183
[7] Flynn H J, Wall L A 1966 *Journal of Polymer Science Part B: Polymer Letters* **4** 323
[8] Doyle C 1961 *Journal of Applied Polymer Science* **5** 285
[9] Musanic S M, Houra I F, Suceska M 2010 *Central European Journal of Energetic Materials* **7** 233
[10] Lyon R E *Thermochimica Acta* **297** 117
[11] J. Malek 1995*Thermochim. Acta* **267** 61
[12] Criado J M 1978 *Thermochimica Acta* **24** 186
[13] Malek J 1989 *Thermochimica Acta* **138** 337
[14] Malek J 1992 *Thermochimica Acta* **200** 257

Novel applications of microwaves in the metallurgical processing of a nickel-ferrous laterite ore and an aluminum industry waste

M Samouhos[1], M Taxiarchou[1] and E Kouvelos[2]

[1]Laboratory of Metallurgy, School of Mining and Metallurgical Engineering, National Technical University of Athens, Zografou Campus, Athens, Greece
[2]Demokritos National Center of scientific research, Agia Paraskevi, Attiki, Athens, Greece

Email: msamouhos@metal.ntua.gr

Abstract. Microwave radiation is a relatively new source of energy in the pyrometallurgical process. In this study, the application of microwaves in the carbothermic reductive roasting of a nickeliferous hematitic laterite ore and a red mud are investigated. The effective microwave heating (at temperatures above 900 oC) of the aforementioned materials is attainable due to the relatively high values of their imaginary permittivity (ε΄΄). In both two cases, the reduction of the included hematite was attempted, and the reduction degree was calculated as a function of: (a) the heating time and (b) the supplied microwave power. The mechanism of Fe+3 to Fe0 conversion was investigated using Mössbaurer spectroscopy revealing the formation of magnetite, fayalite and nano-structured metallic iron. Finally, it should be noted that that the gaseous products of the microwave pyrometallurgical process were analyzed using a mass spectroscopic technique, which is an important novelty.

1. Introduction

Microwaves are electromagnetic radiation covering the frequency range between 300 MHz and 300 GHz. Microwave radiation interacts with microwave absorptive materials (dielectrics), generating heat instantaneously in their core. It is converted to heat via two main mechanisms; (a) hindered permanent dipole rotation and (b) electron/ion resistivity. The interaction between the microwave radiation and a non magnetic media is determined by the temperature-depended complex permittivity of the concrete media [1]:

$$\varepsilon^*(T) = i \cdot \varepsilon_r'(\mathrm{T}) - j \cdot \varepsilon_r''(T) \quad (1)$$

Where, $\varepsilon^*(T)$, is the complex relative permittivity, $\varepsilon_r'(T)$ is the real relative permittivity, and $\varepsilon_r''(T)$ is the imaginary relative permittivity. The real part represents the degree of the microwave absorbance, while the imaginary part represents the degree of the microwave power conversion to heat. The average power (PL) absorbed per unit volume is estimated by [1]:

$$P_L = \frac{1}{2}\omega\varepsilon_0\varepsilon_r''(T)\left|E_{\text{int}}\right|^2 \quad \text{(watts/cubic metre)} \tag{2}$$

Where: ω, is the angular frequency (2πf, in radians –s^{-1}), E_{int} is the internal electric field in the sample (Vm^{-1}) and ε_0 is the permittivity of free space.

This study discuses a novel reductive roasting processing of a laterite ore and a red mud under microwave heating. Both materials are of great industrial significance; laterite is the main nickel source, while the processing of red mud, which is the major waste of the aluminum industry, presents worldwide growing interest due to its high environmental impact [2, 3].

2. Experimental setup

Bulk chemical analysis of the rough and the reduced samples was performed using a Xepos Spectro Bench XRF spectrometer while their mineralogical characterization was carried out using a Bruker D8 Focus analyzer. The determination of the iron phases that appeared in the reduced laterite samples was realized by Mössbauer spectroscopy. The reduction degree of laterite was calculated following the E 1071-85 ASTM method. The metallic iron (Fem) content of the reduced red mud samples was determined following the method of Zhiyong Xu et al. [4].

The reductive roasting experiments of the laterite ore and the red mud residue were performed using an appropriately modified Ceralink ThermWave 1.3 microwave oven working at 2.45 GHz (Figure 1a). A second microwave oven setup was designed for the chemical analysis of the off gases that were produced during the reduction of the laterite ore (Figure 1b). The sample of laterite-lignite mixture was placed into an airtight cylindrical fused silica sample holder. An air mixture of O_2-He_2 (21:79 % w/w) was channeled by an aperture in the top of the sample holder, while the off gases were issued, and guided to a Pfeiffer (OmniStarTM) spectrometer, by a second lateral aperture. The total O_2-He_2 flow rate was 160 mL/min.

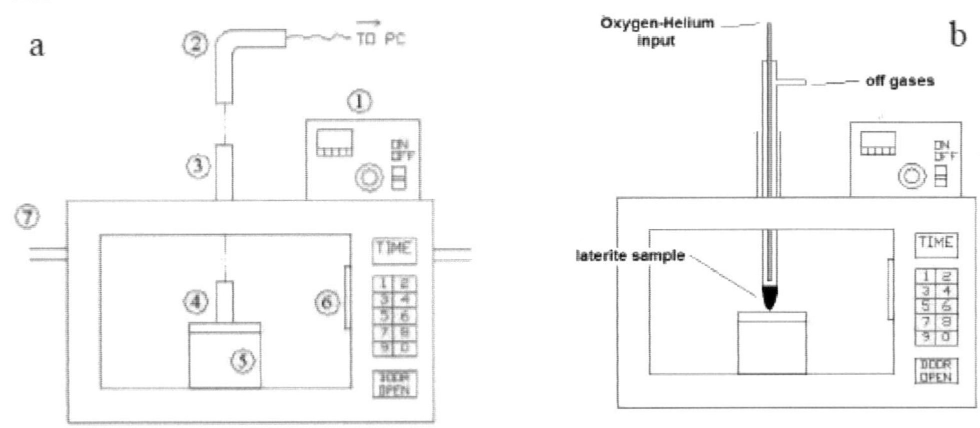

Figure 1. (a) Schematic diagram of the microwave oven used in the microwave reduction of the ores (1) power controller, (2) optical pyrometer, (3) 2.45 GHz chock tube, (4) alumina crucible, (5) alumina latform, (6) waveguide position, (7) water cooling system [5], **(b)** Positioning of the sample holder for the chemical analysis of the laterite reduction off gases

The hematitic nickel-ferrous laterite ore that was used in the current study originates from the main open pit mine of the Greek FeNi producer LARCO in Lokrida, Greece, while the respective red mud sample was provided by the Greek aluminum industry "Aluminum of Greece S.A". The bulk XRF chemical analyses of the laterite and the red mud samples are presented in Table 1. X-ray diffractometry showed that the iron content in both materials is almost exclusively contained in hematite phase (Fe^{+3}). The reducing agent used at the reduction experiments was lignite (30.15% Cfix).

Table 1. Bulk XRF chemical analyses of the laterite ore and the red mud residue

Species (w/w %)	Fe_2O_3	SiO_2	Al_2O_3	MgO	CaO	TiO_2	Cr_2O_3	NiO	K_2O	Na2O	Zr_2O	L.O.I
Laterite	40.7	30.55	4.80	4.52	3.20	-	4.00	1.33	0.75	-	-	8.0
RM	43.19	7.37	14.63	0.31	12.03	6.05	0.28	-	0.07	2.14	1.12	12.19

3. Results and discussion

The heating behavior of the laterite-lignite mixture powder under the microwave irradiation is presented in Figure 2, where the top-surface temperature of the sample is displayed as a function of the heating time, the power supply and the amount of carbon content.

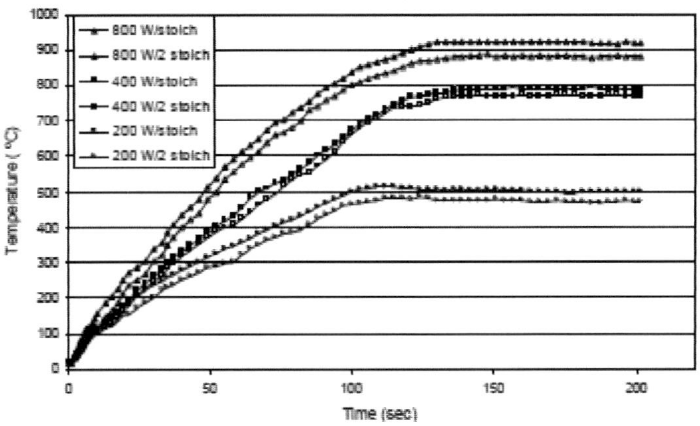

Figure 2. Top surface temperature of laterite-lignite mixtures as a function of the output power supply (200, 400 and 800 W) and the lignite content [5]

It is evident that the mixture is susceptible in microwave absorption and can be rapidly heated over 900°C with an average heating rate of about 6.4°C/s at 800 W. Figure 3 presents the Mössbauer spectra of four reduced laterite samples received after their reduction for; (a) 2, (b) 4, (c) 6 and (d) 8 min. The conditions used were 800 W of output power supply and a twice stoichiometric carbon addition. "stoichiometric", is defined as the required carbon amount for the complete reduction of the initial contained hematite to metallic iron according to the following reaction:

$$Fe_2O_3 \quad + \quad 1.5\,C \quad \rightarrow \quad 2\,Fe \quad + \quad 1.5\,CO_2 \qquad \Delta H = 232\text{ kJ/mol} \qquad (3)$$

The relative amounts of the iron-containing phases are directly calculated from their corresponding sub-structural areas. Results show the gradual decreasing of the initial Fe^{+3} content and its conversion to Fe^{+3}/Fe^{+2} (magnetite), Fe^{+2} (fayalite, ferrosillite) and Fe^0 (metallic iron) phases. The Fe^{+2} content (initially zero in the rough laterite) is already high (42%) after 2 min of roasting while the metallic iron [under bcc Fe(Ni) form] content is 11%. A significant part of the formed iron-nickel alloy is present in a non-magnetic nanoscale phase. The reduction degree of the laterite ore was parametrically examined as a function of (a) the roasting time, and (b) the output power.

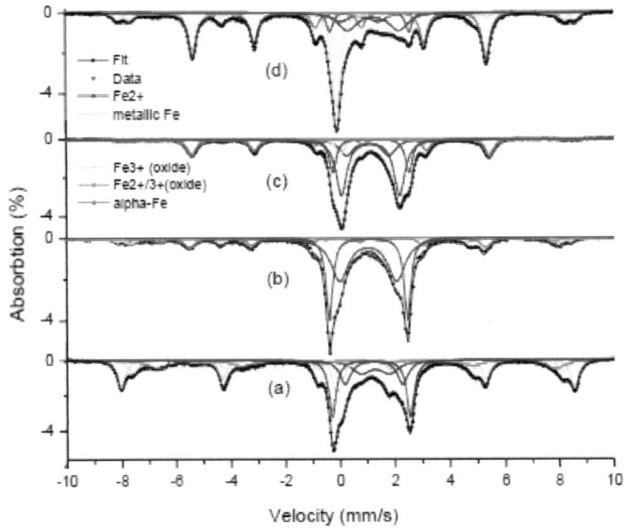

Phases	samples			
	a	b	c	d
Fe^{+3}	31	9	4	13
Fe^{+2}	42	72	69	29
Fe^{+2}/Fe^{+3}	16	5	2	7
nmFe(Ni)	7	4	6	18
a-Fe(Ni)	4	10	19	34

Figure 3. The Mössbauer spectra of reduced laterite samples after 2 (a), 4 (b), 6 (c) and 8 (d) min of roasting and their respective phase contents (wt.%) [5]

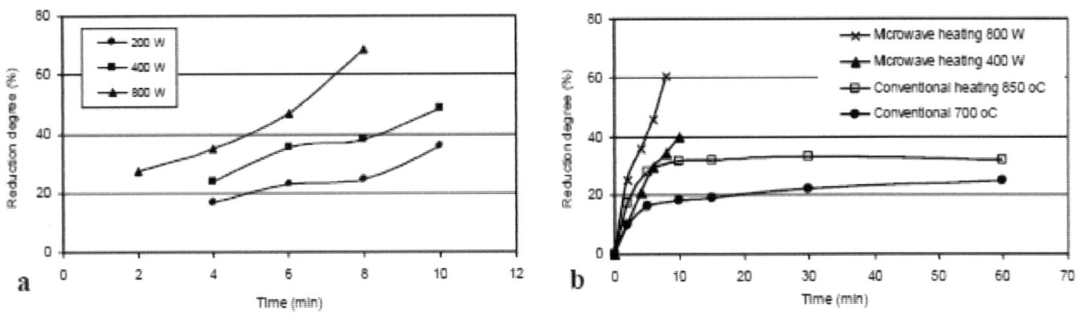

Figure 4. (a) Reduction degree of laterite as a function of roasting time and output power supply [5], **(b)** comparative reduction degree rates of microwave and conventionally heated laterite-lignite mixtures [6]

Figures 4(a) present the reduction degree of the laterite ore at three output power levels using a twice stoichiometric carbon addition. It is clear that the reduction reaction progress is significantly affected by the output microwave power. A rough comparison of the reduction degree rates between a conventionally and a microwave heated hematitic laterite (both from Lokrida open pit mines) is attempted in Figure 4(b). The conventionally roasting experiments were performed using a tube furnace [6]. Despite the differentiation of the experimental conditions (the conventional reductions were done under isothermal conditions, while the microwave reductions under non isothermal conditions), the results indicate the increase of the reduction reaction rate in case of the microwave heated laterite ore. The mass spectra of the off gases emitted by the laterite reduction in relation to the heating time are presented in Figure 5.

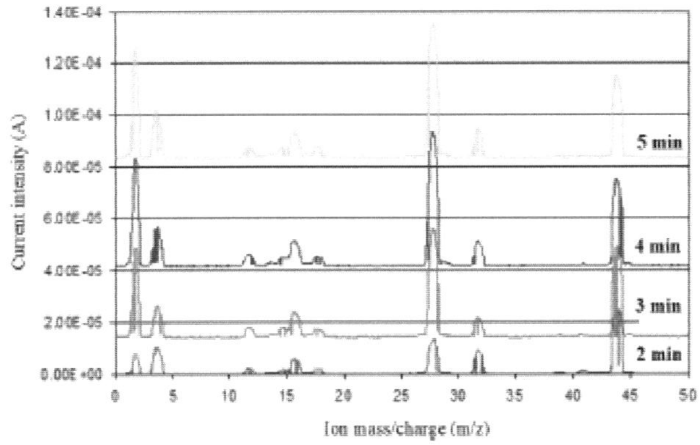

Figure 5. The mass spectra of the off gases emitted by the carbothermic reduction of the laterite ore in relation to the roasting time

The results indicate the emission of H2 (m/z=2), CO (m/z=28 and m/z=12) and CO_2 (m/z=44) after 2 min of microwave heating. The production of CO can be attributed to the Boudouard reaction while H_2 is produced via the following reaction after 700°C [7, 8]:

$$C(s) + H_2O(g) \rightarrow H_2(g) + CO(g) \qquad \Delta H = 131.3\ KJ/mol \qquad (4)$$

It should be noted that the sample's temperature at this moment exceeds 800 oC. Comparing the progress of the laterite ore's reduction degree (Figure 9) with the evolution of the off gases emissions, it can be concluded that initially (up to 2 min) the reduction follows a solid-solid reaction mechanism while after 2 min the reducing agents H2 and CO accelerates the reduction process.

Red mud residue presents a similar to laterite microwave heating behavior. As it can be seen in Figure 6, the heating of a red mud-lignite mixture is sufficient at 800 W, while the addition of 1wt.% (per total mixture's mass) graphite, increases the heating rate and the maximum developed temperature.

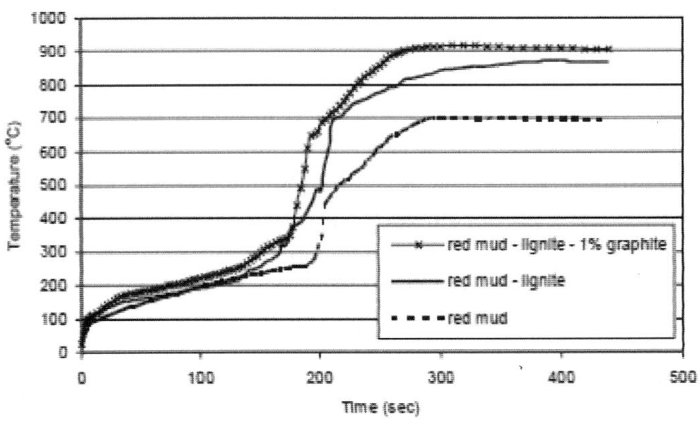

Figure 6. Top surface temperature of red mud, red mud-lignite and red mud-lignite- graphite samples heated at 800 W (samples mass: 22 g, lignite/red mud ratio: 0.3 wt.%)

The reducibility of the red mud was examined as a function of; (a) the roasting time, and (b) the amount of microwave power. Figure 7(a) shows the effect of roasting time on the metallization degree of a 22 g red mud sample using 600 W of power supply and a 0.3 (w/w) ratio of lignite/red mud.
The effect of microwave power supply on the metallization degree is shown in Figure 7(b). The

roasting time was 12 min while a 2.5 times stoichiometric addition of carbon was used. The roasted concentrates were milled at the granulometry of -200 Mesh and subsequently they were magnetically separated into a Carpco (model MWL 3465) separator applying a 0.3 A current intensity. The magnetic product that was taken is enriched 10% in metallic iron [Figure 7(b)]. Although the high metallization degree (69%) of the magnetic concentrate, its content in total iron is relatively low.

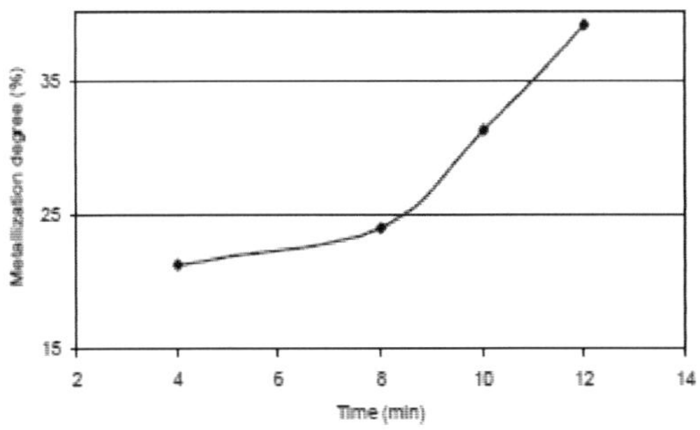

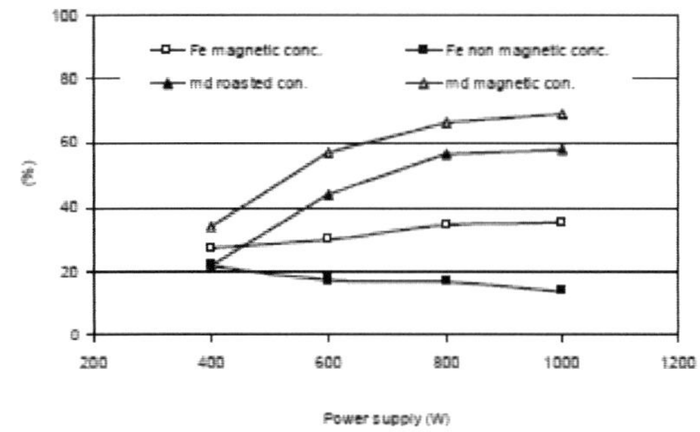

Figure 7. (a) Metallization degree of red mud heated at 600 W as a function of time, **(b)** Metallization degree for roasted red mud samples (roasted con.) and its respective magnetic fractions (magnetic con.), Fetot in the magnetic and non magnetic fractions as function of power supply [9]

4. Conclusions and future work

Laterite-lignite and red mud-lignite mixtures are susceptible to microwave heating presenting an initial heating rate > 6°C/s and a temperature stabilization over 900°C. The maximum achieved reduction degree of the laterite ore was 69% after 8 min of heating at 800 W. Similar results were achieved in the case of red mud reduction (metallization degree: 59% after 12 min at 1000 W). Mössbauer spectroscopy showed the formation of metallic iron, magnetite and iron-silicate (fayalite) phases during laterite reduction. Mass spectroscopic analysis of the off gases that were produced during laterite's reductive roasting, showed the formation of CO and H_2.

Although reductive roasting of red mud proceeds rapidly and a high metallization degree is achieved, the magnetic separation of the roasted concentrate is ineffective due to the enclosing of Fem particles into an Al-Si-Ca amorphous matrix. As a future research goal, the reductive roasting of the red mud under gentle heating conditions, to Tmax=700°C, is proposed. Under these conditions: (a) the vetrification will be avoided, while (b) hematite (Fe^{+3}) will be reduced to magnetite (Fe^{+3}, Fe^{+2}). The last phase disposes a high magnetic susceptibility (χ) and will be effectively separated in a magnetic fraction.

References

[1] Metaxas C A, Meredith JMR 1983 *Industrial Microwave Heating* (London: Peter Peregrinus)
[2] http://redmud.org/home.html
[3] Hind A R, Bhargava S K, Grocott S C 1999 *Physicochemical Engineering Aspects* **146** 359
[4] Zhiyong X, Hwang J, Greenlund R, Huang X, Luo J, Anschuetz S 2003 *Journal of Minerals and Materials Characterization and Engineering* **2** 65
[5] Samouhos M, Taxiarchou M, Hutcheon E, Devlin E 2012 *Minerals Engineering* **34** 19
[6] Zevgolis E, Zografidis C, Halikia I 2010 *Mineral Processing and Extractive Metallurgy* **1** 9
[7] Li Y L, Guo L J, Zhang X M, Jin H, Lu Y J 2010 *International Journal of Hydrogen Energy* **35** 751
[8] Zhang R, Jiiang W, Cheng L, Sun B, Bi J 2010 *International Journal of Hydrogen Energy*, **35** 11810
[9] Samouhos M, Taxiarchou M, Tsakiridis E, Potiriadis K 2013 *Journal of Hazardous Materials* **254-255** 193

Crack Formation in Cement-Based Composites

A Sprince[1], L Pakrastinsh[2] and N Vatin[3]
[1, 2]Department of Structural Engineering, Riga Technical University, Riga, Latvia
[3]Institute of Civil Engineering, Saint-Petersburg State Polytechnical University, St Petersburg, Russia

E-mail: andina.sprince@rtu.lv

Abstract. The cracking properties in cement-based composites widely influences mechanical behavior of construction structures. The challenge of present investigation is to evaluate the crack propagation near the crack tip. During experiments the tension strength and crack mouth opening displacement of several types of concrete compositions was determined. For each composition the Compact Tension (CT) specimens were prepared with dimensions 150x150x12mm. Specimens were subjected to a tensile load. Deformations and crack mouth opening displacement were measured with extensometers. Cracks initiation and propagation were analyzed using a digital image analysis technique. The formation and propagation of the tensile cracks was traced on the surface of the specimens using a high resolution digital camera with 60 mm focal length. Images were captured during testing with a time interval of one second. The obtained experimental curve shows the stages of crack development.

1. Introduction

The development of new types of concrete focuses on High Performance Fiber-Reinforced Cement Composites (HPFRCC). Fiber-Reinforced Concrete (FRC) is a composite material in which certain characteristics are developed for a particular application and environment. These characteristics are not only strength, but improved durability, increased resistance to various external agents, high rate of hardening, better aspect, etc. This concrete is characterized by an enhanced postcracking tensile residual strength, also defined as toughness in the following, due to the fiber reinforcement mechanisms provided by fibers bridging the crack surfaces. Fibers in concrete provide improved mechanical and physical properties of the material [1, 2].

The durability and deformation characteristics of a material are just as important as its strength properties. Therefore, designers and engineers need to know those properties of concrete and must be able to take them into account in the structure analysis [2, 3].

This paper reports on an experimental investigation of the cracking resistance of several types of HPFRC using the Compact Tension test.

2. Experimental program

The experimental work included the preparation of two HPFRC compositions with polyvinyl alcohol (PVA) fibers constituting 2% of the total amount of cement with and without nanosilica. The HPFRC mixtures consisted of cement, water, quartz sand (grain size 0.3 - 1 mm), silica fume, nanosilica, microsilika, plasticizer Sikament and two types of PVA fibers. PVA fiber properties are listed in Table 1. For the purposes of this paper the batches containing microsilica were designated MS PVA and the ones containing microsilika and nanosilika – NS PVA [4].

Table 1. Properties of PVA fibres

Fiber type	Ø [μm]	L [mm]	f_t [GPa]	E [GPa]
MC 40/8	40	8	1.6	42
MC 200/12	200	12	1.0	30

The testing procedure developed for the present work consisted of applying a tensile load to a single-edge notched specimen. The selected specimen shape resembles the one used for the evaluation of crack propagation behavior in metals (ASTM-E647 2005) [5], the so-called Compact Tension (CT) specimen. With the purpose of maximizing the stress intensity at the tip of the notch, the notch thickness was reduced to 0.5 mm using a small diamond cutting disc. The depth of the specimen was also reduced, promoting the plain stress state. Summarizing, the dimensions adopted for the specimens were 150 mm by 150 mm (perpendicular and parallel to the notch) and 12 mm (thickness). The available path for progress of the initiated crack parallel to the notch was 30 mm. The distance between the loading axis and the tip of the notch was also 50 mm (see Figure 1).

The testing sequence consisted of subjecting the specimens to a tensile load at a constant displacement rate, transmitted by two rods with a diameter of 20 mm (see Figure 2). The use of the two rods allowed the transmission of the displacement while keeping free the rotation of the specimen with respect to the rods. The adopted displacement rate was 1 μm/s [6].

The formation and propagation of the tensile crack was traced on the surface of the specimens using a high resolution digital camera positioned 70 mm away from the specimen. The lens with a 60 mm focal length allowed the observation of a 36 mm by 24 mm area on the surface of the specimen (see Figure 2). Shooting settings – lens aperture F11, shutter speed 1/30, ISO 200. Images with a 24 megapixel resolution were captured during testing at an interval of one second. These images were subsequently used for conditions for the strain field interpolation were met with the need of applying a speckle pattern on the surface of the specimen.

To experimentally measure the deformation during the CT tests the ARAMIS software was used.

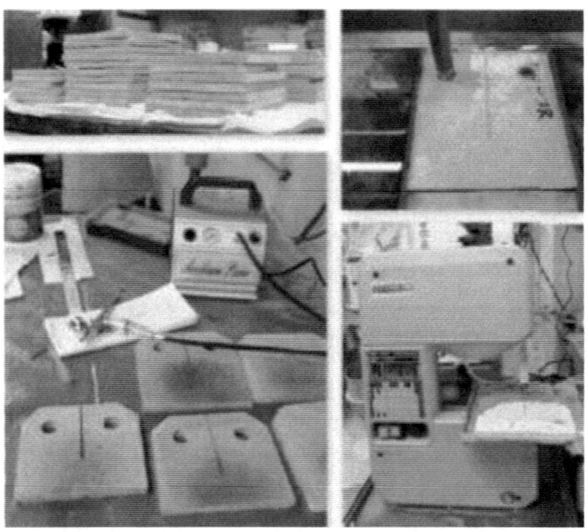

Figure 1. Preparation of Compact Tension specimen

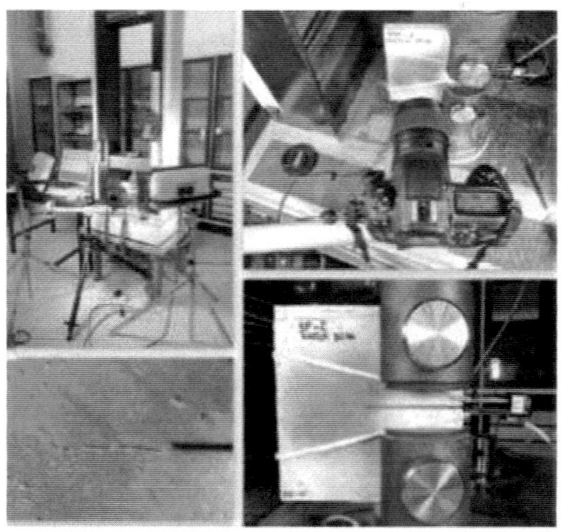

Figure 2. Compact Tension test setup

In the present case, fluorescent lights were used and positioned in such a way that the intensity of the light reflected by the specimen surface was even. The images of the surface of the specimens were analyzed prior to testing and sufficient image correlation was obtained. Each facet was composed of 15x15 pixels. Each pixel covered a real area of 6x6 μm2. The total area of 36x36 mm2 was modeled by a facet mesh overlay composed of approximately 400x260 facets [6].

Before testing, to measure crack mouth opening displacement (CMOD), the extensometer with precision +/-2.5 mm was centrally and symmetrically positioned at the edge of the test plate's notch.

3. Results and Discussion

The "cocktail" of two types of PVA fibers exhibited different behavior in HPFRCC. The main role of short dispersed fibers was to control the opening and propagation of small cracks, whereas long fibers control the large cracks and serve as a bridge between the two sides of the crack [1]. Figure 3 shows the behavior of short and long fibers and the experimental load – crack mouth opening displacement (CMOD) curves are presented.

Several stages of HPFRCC behavior can be identified. The first stage 0A (from zero to about 60-80% of peak load) corresponds to the ascending linear elastic portion of the curve where the section is not cracked. The second stage AB (from about 60-80% of peak load to peak load) corresponds to the portion of the curve where cracking starts and during which crack formation stabilizes and the specimen reaches its peak load. A peak point – B where the maximum resistance is attained. The third stage BC (from peak load to about 60% of peak load) correspond to the portions of the curve where strain softening starts and where the load–strain curve response is almost linear elastic and crack widths increase and the fourth stage (from C to the end) is a non-linear portion where either or both materials are in their non-linear range and crack widths continue to increase (according to Naaman [7]).

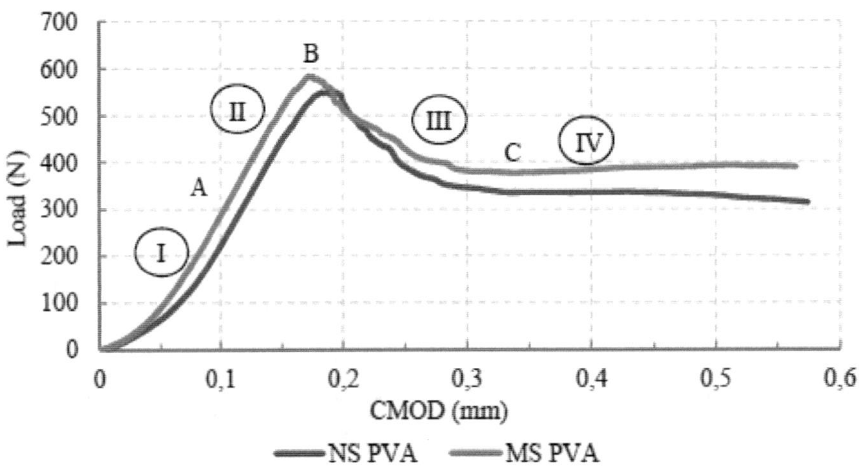

Figure 3. Load – CMOD curves measured during testing

The experimental results (see Figure 3) are quite similar and they indicate that nanosilica does not have significant influence on the tensile load and CMOD properties of HPFRC. The average tensile load for NS PVA specimens was 570 N and CMOD was 0.57 mm respectively. The average tensile load for MS PVA specimens was 613 N and CMOD was 0.59 mm.

The Figure 4 shows the crack visibility to the human eye and by the system "Aramis". By human eye crack is visible when specimen is already collapsed but system "Aramis" allows the observation of how the cracks initiate and propagate in specimen. The crack becomes clearly visible, although their openings are very small. Detailed information about those experiments look for [4].

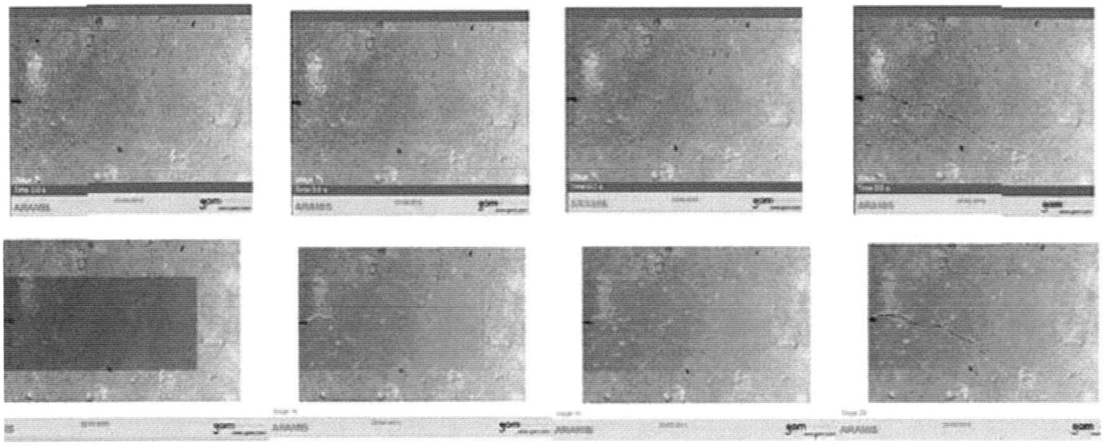

Figure 4. Crack visibility to the human eye and by system "Aramis"

4. Conclusion

Two High Performance Fiber-Reinforced Cement Coomposite (HPFRCC) mixes with polyvinyl alcohol (PVA) fibers containing 2% of the total amount of cement by weight with and without nanosilica were prepared for a laboratory examination. The tensile load and the cracks mouth opening displacement (CMOD) were determined.

The experimental study indicates that nanosilica does not have significant influence on the HPFRCC tensile strength and on the CMOD properties.

Wider use of this material permit the construction of sustainable next generation structures with thin walls and large spans that cannot be built using the traditional concrete.

Acknowledgments

This work has been supported by the European Social Fund within the scope of the project "Support for the Implementation of Doctoral Studies at Riga Technical University".

References

[1] Brandt A.M 2008 *Composite Structures* **86** 3
[2] Prisco M, Plizzari G, Vandewalle L 2009 *Materials and Structures* **42** 1261
[3] Benin AV, Semenov AS, Semenov SG 2014 *Advanced Materials Research* **831** 364
[4] Sprince A, Fischer G, Pakrastinsh L, Korjakins A 2014 *Advanced Materials Research* **842** 470
[5] ASTM-E647 2005 Standard test method for measurement of fatigue crack growth rates ASTM Standard vol.03.01 (US: ASTM Standard)
[6] Pereira EB, Fischer G, Barros JAO, Lepech M 2010 Crack formation and tensile stress-crack opening behavior of fiber reinforced cementitious composites (FRCC) *Proc. of FA MCoS-7* 1638
[7] Naaman AE, Reinhardt HW 1996 Characterization of High Performance Fiber Reinforced Cement Composites *Proc. of 2nd Int. Workshop on HPFRCC* (London: E&FN Spon) chapter 41 1-41

Gas pressure atmosphere annealing: A novel method for the preparation of SiC nanowires

X Zhang[1,3,4,*], **B Zhong**[2], **L Liu**[1], **X Huang**[1], **G Wen**[1,2], **Y Huang**[3], **and J Bollmann**[1,4]

[1] School of Materials Science and Engineering, Harbin Institute of Technology, Harbin, China
[2] School of Materials Science and Engineering, Harbin Institute of Technology, Weihai, China
[3] School of Chemical Engineering and Technology, Harbin Institute of Technology, Harbin, China
[4] Institute of Electronic and Sensor Materials, TU Bergakademie Freiberg, Freiberg, Germany

E-mail: zhangxiaodong@hit.edu.cn

Abstract. Silicon carbide nanowires were fabricated by gas pressure annealing of SiOC nanocomposite powders, which were synthesized by pyrolysis of a SiO_2 - sucrose gel. The reaction was carried out in an atmosphere sintering furnace without any additives. The nanowires have pronounced homogenous diameters smaller than 100 nm and lengths of up to several millimetres. The X-ray diffraction pattern indicates the formation of the β-SiC phase and transmission electron microscopy analysis show the monocrystalline structure of the nanowires.

1. Introduction

In the last years one-dimensional quantum wires found considerable interest in low-dimensional physics and materials research due to their promising applications in nanoelectronics, micromechanics, field emission, and nanocomposites [1–3]. Among one-dimensional quantum wires, silicon carbide (SiC) nanowires attracted increasing interest during the past decade [4]. Recently, elaborated measurements, combining atomic force microscopy and lithography techniques, indicates that the strength value of SiC nanowires is about two times higher compared to the value of SiC whiskers with micrometre diameter [5, 6]. Furthermore, SiC is an important wide band-gap semiconductor with outstanding properties, such as high disruptive strength, high thermal conductivity, high saturation drift velocity, and excellent physical and chemical stability [7]. The combination of these distinctive physical and chemical properties make SiC nanowires a promising candidate for the fabrication of various nanoscaled devices [8-10].

However, most syntheses of SiC nanowires are quite complicated, difficult to control, and catalytic metal additives are usually essential. Therefore, the application of SiC nanowires is noticeable

[*] To whom any correspondence should be addressed

restricted so far. In this paper, a novel method for the synthesis of SiC nanowires is proposed and the preparation and structural characterization of such SiC nanowires are presented.

2. Experimental

2.1 Synthesis of SiOC nanocomposite

Sucrose and silica sol where used as starting materials to synthesize SiOC nanocomposite in the following way: At first, sucrose was intermixed with silica sol by magnetic stirring. Next, the resulting sol was converted into dry gel at 90°C. Finally, a SiOC nano-composite is obtained by pyrolysis of the dry gel at 900°C for 30 min in a quartz tube under flowing nitrogen. The final powders are black in colour, amorphous in structure and nano-scaled in particle size.

2.2 Growth of SiC nanowires

The SiOC nanocomposite was annealed in a graphite crucible for 60 min under argon cover atmosphere in a gas pressured furnace at 1500°C. The initial argon pressure was 0.5 MPa. The nanowires grew at the surfaces of the powder and could be easily collected by brushing them from the sintered body.

3. Results and Discussion

A high-resolution scanning electron microscopy (SEM) image, shown in Figure 1, reveals that the SiOC nanoparticles are quite homogeneous in size. The diameter varies from 20 to 50nm. As a result, the SiOC powders have a very large specific surface and will exhibit a high reaction activity. Figure 2 shows the XRD pattern of SiOC powders, indicating that the powders are amorphous. Analysis of element yields that the molar ratio of Si:O:C is 1.101:2.032:2.907 whereas the percentage of impurities is less than 0.7wt%.

Figure 1. SEM of SiOC nano-powders

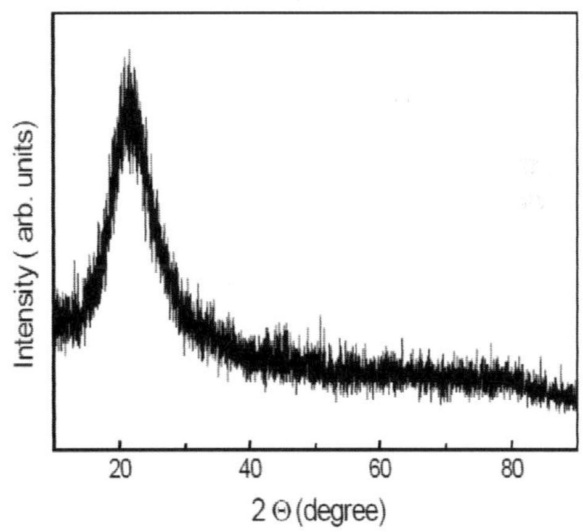

Figure 2. XRD pattern of SiOC nano-powders

The SEM morphologies and EDX patterns of the β-SiC nanowires are shown in Figure 3. Obviously, most of the products are wire-like shaped with diameters ranging from 80 to 120 nm and lengths up to several millimetres. The EDX patterns (see insert of Fig.3a) of the SiC nanowires confirm that the wires consist of Si, C and a small amount of O as well.

XRD measurements of bulk nanowires (on the sintered composite substrate) are presented in Figure 4 to identify the overall structure. The diffraction peaks can be indexed to a cubic structure with a lattice constant of 0.4358nm, in well agreement with reference values for β-SiC (JCPDS Card. No. 29–1129). The observed peaks at 35.8°, 60.1° and 71.8° are diffractions from (111)- , (220)- and 311)-lattice planes of SiC crystal, respectively.

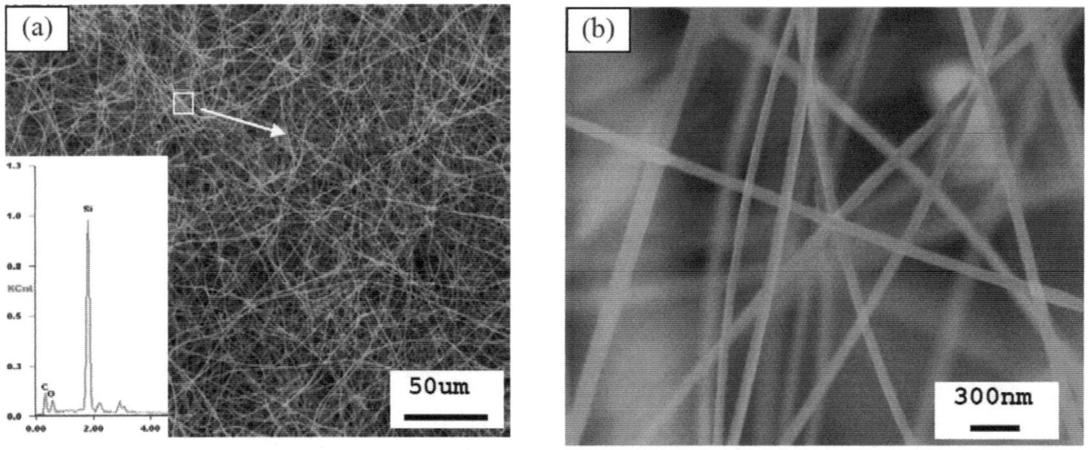

Figure 3. SEM morphologies of the β-SiC nanowires

The high intensities of the diffraction peaks relative to the background signal indicate the high purity cubic β-SiC phase of the wires. However, the peak at 21.6° is the diffraction of the (101) planes of SiO2 (JCPDS Card. No. 83–1403), which probably originates from a fractional crystallization of the SiOC powder.

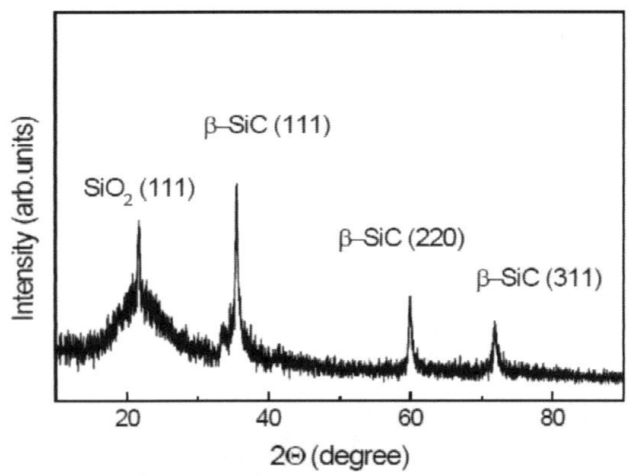

Figure 4. (a) XRD pattern of the SiC nanowires with substrate

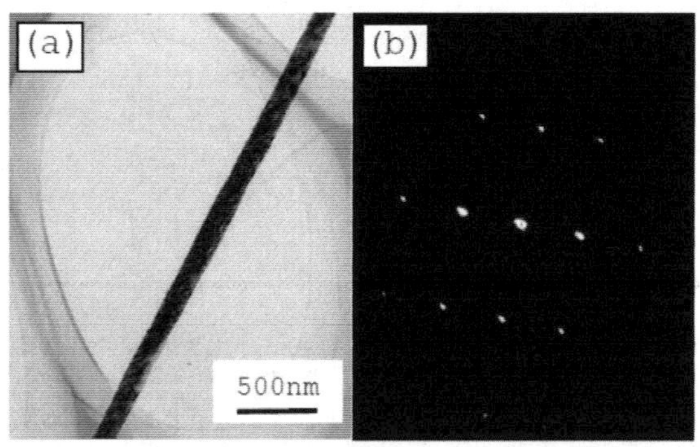

Figure 5. (a) TEM morphology of the SiC nano-wire and (b) the corresponding SAED

To characterize the structure of the wires in more detail, transmission electron microscopy (TEM) images and selected area electron diffraction (SAED) pattern were taken. Figure 3a shows a typical TEM image of the SiC nanowires, revealing that the nanowire is very uniform in diameter. The surface of the SiC nanowire is very clean, without any coating of amorphous materials and is very smooth. The corresponding selected-area electron diffraction pattern (Fig.3b) verifies that the nanowire is in a single crystalline phase. This spot pattern is due to the cubic β-SiC structure and indicates that the nanowires are perfectly crystallized.

Thermal analysis (TG) of the SiOC nano-powder was performed under argon ambient to gain more insight into the growth mechanism of SiC nanowires. DSC-TG curves (see Fig.6) show a significant weight loss above 1315°C and (endothermic) minima at 1245°C, 1315°C, 1412°C, and 1426°C, which are typically for the occurring reactions. We found initial reaction temperatures much lower than reported data in ref. [14] (see Tab.1).

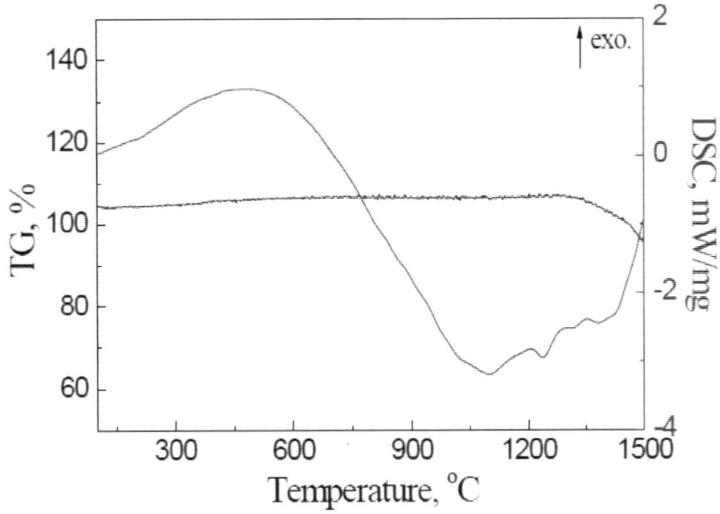

Figure 6. DSC-TG curves of the SiOC nano-powders

From our experimental results the mechanism for the SiC nanowire growth will be the following one. The formation of SiC is an epitaxial growth starting on some random distributed spots at the surfaces of the SiOC powder matrix. During the carbo-thermal reduction, reaction (1) may start at 1493K [14], which introduces the nucleation of SiC. With increasing temperature, silicon monoxide (SiO) gas and carbon monoxide (CO) gas is generated by reaction (2) to (4). The SiO(g) and CO(g) now reacts directly at some nucleation sites (presumably β-SiC crystallites at the surface of the powder) to produce β-SiC nanowires via the vapor reaction (5). Due to thermodynamic principle, the nanowires grew along the [111] direction of β-SiC which is the preferential growth direction to minimize Gibbs free energy.

$$SiO2(s) + 3C(s) = SiC(s) + 2CO(g) \quad (1)$$
$$SiO2(s) + 2C(s) = Si(l) + 2CO(g) \quad (2)$$
$$SiO2(s) + 2C(s) = SiO(g) + CO(g) \quad (3)$$
$$CO2(g) + C(s) = CO(g) \quad (4)$$
$$SiO(g) + CO(g) = SiC(s) + CO2(g), \quad (5)$$

where (s), (l) and (g) refer to the solid, the liquid and the gaseous state, respectively.

Equation (1) is either a purely solid-solid or a liquid-solid reaction (above 1450°C, quartz melts).

In our study, both SiO2 and carbon exist in nano-sized and amorphous state so that the melting point of SiO2 and the initial reaction temperature might be lowered by about 200°C [14-16]. Moreover, the high argon gas pressure used here ensures the saturation of SiO gas and CO gas.

Table 1. Gibbs free energy change of silica due to graphite reactions [14]

Reaction equation	ΔG (J/mol)	Reaction temperature (K)
SiO2+2C = Si+2CO	151865-8T	1814
SiO2+3C = SiC+2CO	116300-76.95T	1493
SiO2+SiC = 3Si+2CO	258870-106.63T	2240
SiO2(l)+C = SiO+CO	144050-69.36T	2070
SiO2(l)+Si(l) = 2SiO(g)	133420-53.25T	2505
SiO2(l)+Si(g) = 2SiO(g)	58950-35.66T	1653

3. Conclusions

In summary, large quantities of high-purity crystalline β-SiC nanowires could be synthesized in an atmosphere sintering furnace without any catalytic additives. The results show that the nanowires with diameter below 100 nm are oriented along the [111] direction and are in the single-crystalline β-SiC phase without any wrapping of amorphous material. The demonstrated manufacturing process has inherent advantages, including a simple production procedure, low-cost raw materials, lower synthesis temperature and a high yield, and can be easily scaled-up.

Acknowledgements

The work was supported by the National Natural Science Foundation of China (NSFC, Grant number 51021002, 51172050, 51102063, 51372052), the Natural Scientific Research Innovation Foundation in Harbin Institute of Technology (HIT. NSRIF. 2011109, HIT. NSRIF. 2010121, MH20120757), and the Fundamental Research Funds for the Central Universities (HIT. ICRST. 2010009), and Yunshan Carbon Industry Co., Ltd.

References

[1] Alivisatos A P *Science* 1996 **271** 933
[2] Duan X, Huang Y, Cui Y, Wang J and Lieber C M *Nature* 2001 **409** 66
[3] de Heer W A, Chatelain A and Ugarte D *Science* 1995 **270** 1179
[4] Dai H J, Wong E W, Yuan Z, Lu, Fan S and Lieber C M *Nature* 1995 **375** 769
[5] Wong E W, Sheehan P E and Lieber C M *Science* 1997 **277** 1971
[6] Kim P and C. M. Lieber. Science, 1999, **286** 2148
[7] Wong K W, Zhou X T, Frederick C K A, Lai H L, Lee C S and Lee S.T. *Appl. Phys. Lett.* 1999 **75** 2918
[8] Pan Z W, Lai H L, Frederick C K A , Duan X F, Zhou W, Shi WS, Wang N, Lee C S, Wong N B, Lee S T and Me S S *Adv. Mater.* 2000 **12** 1186
[9] Seong H K and Choi H J. *Applied Physics Letters* 2004 **85**(7)1256
[10] Seeger T, Rediich P K and Ruffhle M *Adv. Mater.* 2000 **12** 279-282
[11] Hu J Q, Lu Q Y, Tang K B, Deng B, Jiang R R, Qian Y T, Yu W C, Zhou G E, Liu X M and Wu J X *J. Phys. Chem. B.* 2000 **104** 5251
[12] Zhang Y W and Wang N L *Solid State Communications* 2001 **118** 595
[13] Morkoç H, Strite S, Gao G B, Lin M E, Sverdlov B and BurnsM *Journal of Applied Physics* 1994 **76** 1363
[14] Youbao C and Qibao W. *Chinese Journal of Chemical Physics* 1998 **6**(3) 256
[15] Zhang X, Huang X, Wen G, Geng X, Zhu J, Zhang T and Bai H *Nanotechnology* 2010 **21** 385601
[16] Liu X, Antonietti M and Giordano C, *Chem. Mater.* 2013 **25** 2021

Preparation of porous carbons from polymeric precursors modified with acrylated kraft lignin

M Sobiesiak

Department of Polymer Chemistry, Faculty of Chemistry, Maria Curie Sklodowska University, Lublin, Poland

E-mail: magdalena.sobiesiak@umcs.pl

Abstract. The presented studies concern the preparation of porous carbons from a BPA.DA-St polymer containing acrylated kraft lignin as a monomer. The porous polymeric precursor in the form of microspheres was synthesized in suspension polymerization process. Next samples of the polymer were impregnated with acetic acid or aqueous solution of acetates (potassium or ammonia), dried and carbonised in nitrogen atmosphere at 450°C. After carbonization microspherical shape of the materials was remained, that is desired feature for potential application in chromatography or SPE technique. Chemical and textural properties of the porous carbon adsorbents were characterized using infrared spectroscopy (ATR-FTIR), thermogravimetry analyses with mass spectrometry of released gases (TG-MS) and nitrogen sorption experiments. The presented studies revealed the impregnation is useful method for development of porous structure of carbonaceous materials. The highest values of porous structure parameters were obtained when acetic acid and ammonium acetate were used as impregnating substances. On the surface of the materials oxygen functional groups are present that is important for specific interactions during sorption processes. The highest contents of functionalities were observed for carbon BPA.DA-St-LA-C-AcNH$_4$.

1. Introduction

Porous carbon sorbents are widely used for water purification purposes for removing both organic and inorganic pollutants. The main features deciding on applicability of carbon adsorbents for sorption are porous structure and surface chemistry. Both of them are formed during thermal treatment and activation process, and strongly depend on pyrolysis conditions and the nature of precursor. [1-3]

Currently the interest in production of low-cost adsorbents is growing. This aim can be realized by utilization of low values by-products or residual materials as carbonaceous precursors.

Lignin is a substance that fulfil above requirements. Commercially, it is by-product of paper industry, but from chemical point of view possesses a great potential. Polyaromatic macromolecules of lignin with numerous functional groups can be equally precursors of carbon adsorbents as well as reagents in chemical syntheses of new polymeric materials [3, 4].

The presented studies concern the preparation of porous carbons from a polymeric composite containing acrylated kraft lignin as a monomer. The porous polymeric precursor in the form of microspheres was synthesized in suspension polymerization process. Next samples of the polymer were impregnated with aqueous solution of activating substance (acetic acid, potassium and ammonia acetates) and then carbonised at 450°C. Chemical and textural properties of the porous carbon adsorbents were characterized using nitrogen sorption experiments, infrared spectroscopy and thermogravimetric analyses coupled with mass spectrometry of evolved gases.

Porous structure and chemical properties of the prepared carbons were different in relation to substance used for activation. The results of ATR-FTIR and MS analyses revealed the presence of oxygen functional groups on the surface of the prepared carbon materials that is important for specific interactions during sorption processes. Additionally, after carbonization the materials remained their microspherical shape, what is desired feature for potential application in chromatography or SPE technique.

2. Experimental

The composite (BPA.DA-St-LA), used as polymeric precursor for preparation of the carbons, was synthesised from diacrylated bisphenol A (BPA.DA), styrene (St) and acrylated kraft lignin (LA), according to the scheme presented in Figure 1. In order to obtain polymer in the form of microspheres suspension polymerization process was applied. Details of this process were published in the previous paper [5].

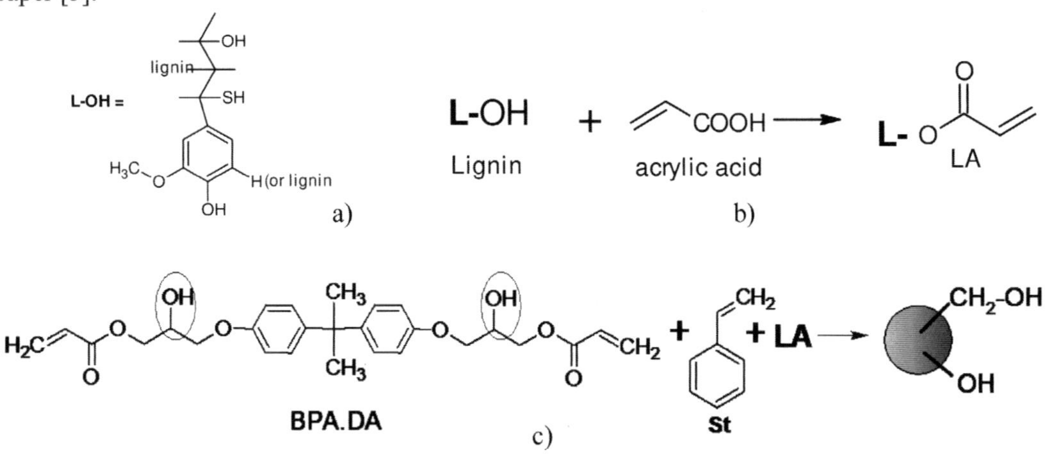

Figure 1. a) Fragment of chemical structure of lignin, b) Scheme of reaction of lignin with acrylic acid, c) Scheme of synthesis of the polymeric composite - BPA.DA-St-LA, used as the precursor

2.1. Carbonisation process of polymeric microspheres

Samples of the polymeric precursor (BPA.DA-St-LA) were impregnated with aqueous solution of acetic acid (BPA.DA-St-LA-**C-AcA**) or acetates (potassium - BPA.DA-St-LA-**C-AcK** or ammonia - BPA.DA-St-LA-**C-AcNH$_4$**). Next, they were dried in the air. Carbonization process was carried out in tubular furnace under nitrogen atmosphere (100mL/min.). The samples were heated from room temperature to 450°C with the rate 10°C/min. The final temperature was maintained for 30 min. The prepared porous carbons were washed with hot water and dried. Sample carbonized in the same conditions, but without impregnation was used as a reference material (BPA.DA-St-LA-**C**). After carbonization all prepared carbons maintained spherical shape except for BPA.DA-St-LA-**C-AcK**.

2.2. Apparatus

Carbonisation processes were performed in tubular furnace (Czylok, Poland) with quartz reactor tube.

The porous structure of prepared materials were characterised by nitrogen adsorption at 77K using an adsorption analyser ASAP 2405 (Micrometrics Inc., USA). Before analysis the samples were outgassed at 140°C for 2 h. Specific surface areas were calculated by the BET method, assuming that the area of a single nitrogen molecule in adsorbed state is 16.2 Å^2. Pore volumes and pore size distributions were determined by the BJH method.

ATR-FTIR (Attenuated Total Reflectance) spectra were obtained using a Bruker FTIR spectrophotometer TENSOR 27. The spectra were acquired in the spectral region 400-4000 cm^{-1}. Resolution was set as 4 cm^{-1} and number of scans per spectrum was 32.

TG-MS (Thermogravimetry coupled with Mass Spectrometry of evolved gasses) analyses were performed on the thermal analyzer STA 449 F1 Jupiter (Netzsch, Selb, Germany). Measurements were carried out at the heating rate of 10°C/min in the temperature range 30-1000°C under helium atmosphere (40 mL/min). Gaseous products evolved of the sample at the time of heating were analyzed by on line coupled QMS 403C Aeolos (Netzsch, Germany).

3. Results and Discussion

Characteristics of the carbonised materials were performed in comparison to the reference carbon material (BPA.DA-St-LA-**C**) and the polymeric precursor (BPA.DA-St-LA).

Table 1 summarise porous structure parameters of the studied materials. The highest values of specific surface areas (S_{BET}) and total pore volumes (V_{TOT}) were obtained for the carbons impregnated with acetic acid and ammonia acetate. Development of porosity (S_{BET}) in these two cases was ca. 50 and 80 times larger in comparison to that of the polymeric precursor. Similar tendency was observed for values of total pore volumes. Although, porous structure of the presented materials is quite different. The polymeric precursor in its structure possesses pores of width in the range of 3 to 15nm. Carbonization without any impregnation agent resulted in shrinking of pores, and consequently porous structure of BPA.DA-St-LA-**C** was dominated by small mesopores (2.8nm). In the case of carbons activated with acetic acid and ammonia acetate, pore size distribution possesses bimodal character. However, pores of BPA.DA-St-LA-C-**AcA** had a little larger widths then those of BPA.DA-St-LA-C-**AcNH₄**. Mean pore size of BPA.DA-St-LA-C-**AcK** was ca. 33nm.

Table 1. Porous structure parameters of the studied materials

material	S_{BET} (m^2g^{-1})	V_{TOT} (cm^3g^{-1})	Pore size (nm)
BPA.DA-St-LA	3	0.025	3-15
BPA.DA-St-LA-**C**	5	0.018	2.8
BPA.DA-St-LA-**C-AcA**	155	0.090	13.5; 37.7
BPA.DA-St-LA-**C-AcNH$_4$**	244	0.117	4.5; 28.7
BPA.DA-St-LA-**C-AcK**	9	0.074	33

Chemical structures of the prepared materials were evaluated on the basis of ATR-FTIR analyses (Fig. 2) and results of mass spectrometry analyses of evolved gases (Fig. 3 b)-d)).

In the ATR –FTIR spectra of polymeric precursor numerous bands coming from functional groups are present. Signals of ester groups:1724, 1285, 1100 cm^{-1}; carboxylic: 1650, 1360 cm^{-1}; ether: 1222, 1031 cm^{-1}; hydroxyl:1200-1050 cm^{-1}; aromatic rings:1606, 1508, 827, 754, 700 cm^{-1}; methyl and methylene groups: 1448, 754 cm^{-1} are observed. After carbonization quantity of these species decreased, but carboxylic, ester and hydroxyl groups were still present in polyaromatic structure of the prepared carbons.

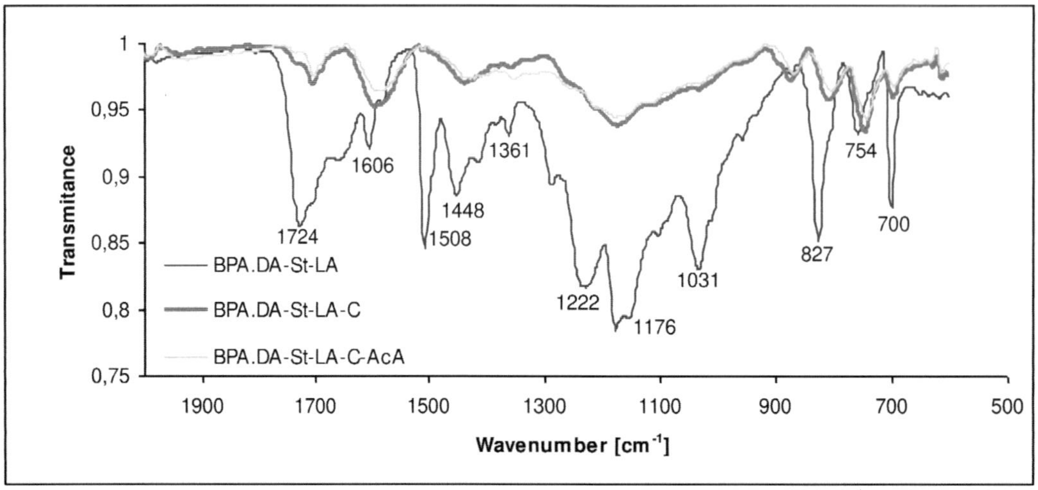

Figure 2. ATR-FTIR spectrograms of chosen materials

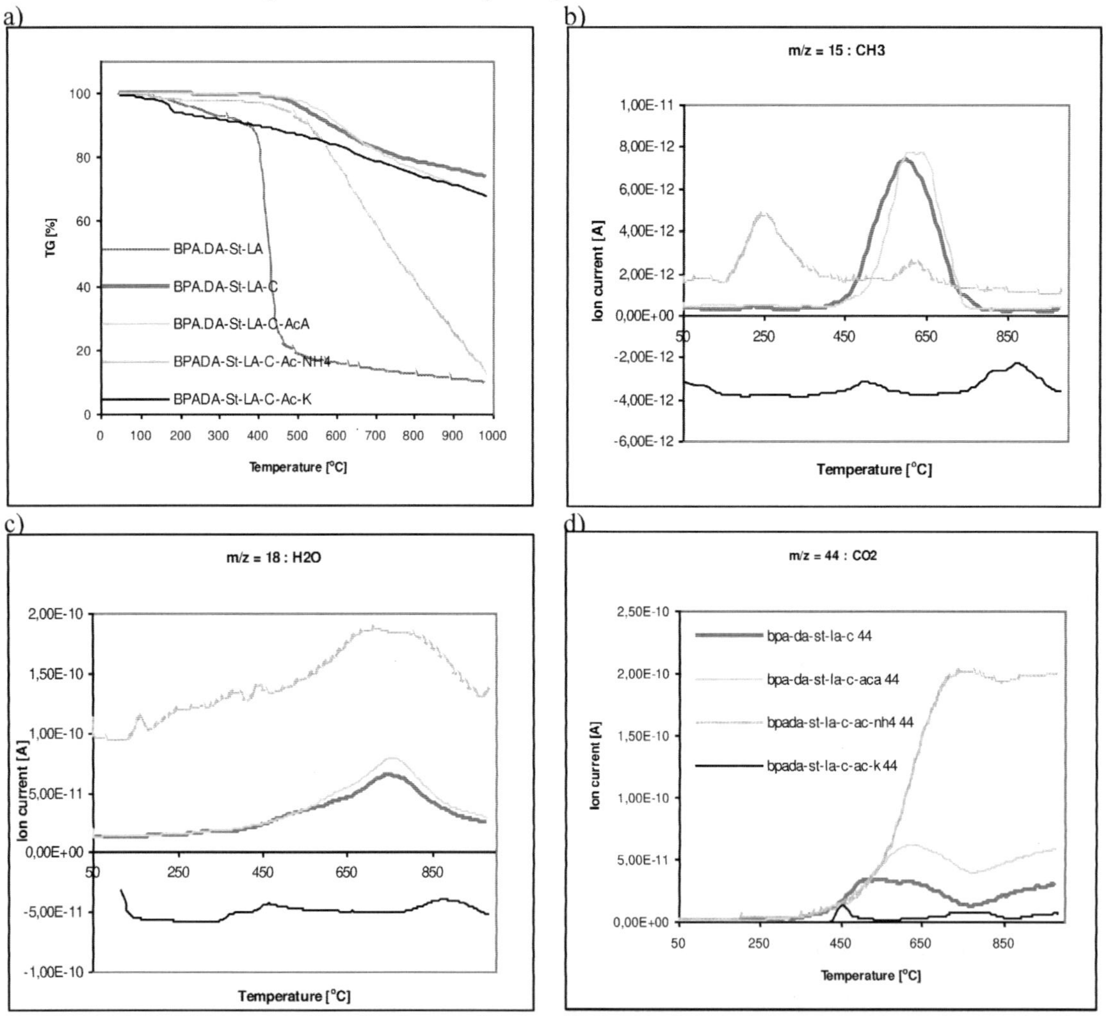

Figure 3. a) Thermogravimetric curves of the studied materials, mass spectrometry curves of the evolved gases for b) m/z = 15, c) m/z = 18, d) m/z = 44, respectively

These data were also confirmed by TG-MS analyses. Different functionalities decompose in specific range of temperatures evolving molecules of simple substances such as CO_2, CO, H_2O, hydrocarbons, etc observed as peaks on MS-profiles [6]. Typical decomposition temperature of carbonyl groups is in range of 700 to 900°C, ethers at ca. 700°C, carboxylic anhydrides about 400 and 600°C, carboxylic acids 100-400°C, and hydroxyl groups 600-700°C. The strongest signals of CO_2 are in the range of 350-700°C, and for H_2O in range of 300-900°C. These data prove the presence of hydroxyl, ester, carboxyl and carbonyl groups. Signals of simple hydrocarbons are observed at decomposition temperature of materials, that for the carbons activated with acid and the reference one is 450-700°C, for the carbons activated with ammonia acetate is 250 and 600°C, whereas for activated with potassium acetate - 500 and 800°C.

The highest contents of functional groups possessed BPA.DA-St-LA-**C-AcNH$_4$**.

4. Conclusions

The presented studies revealed the impregnation is useful method for development of porous structure of carbonaceous materials. The specific surface areas were ca. 30 and 50 times larger than that of the reference carbon, if acetic acid and ammonium acetate were used as activators.

All prepared carbons were mesoporous. Mean pore diameters of the carbons impregnated with acetic acid were13.5 and 37.7 nm, and with ammonium acetate – 4.5 and 28.7nm.

TG-MS profiles and FTIR-ATR spectra proved the presence of: ester, acid anhydride, carboxylic, hyroxyl (both aromatic and aliphatic) functional groups. The highest contents of functionalities were observed for carbon BPA.DA-St-LA-C-AcNH$_4$.

5. Acknowledgments

The research leading to these results has received funding from the People Programme (Marie Curie Actions) of the European Union's Seventh Framework Programme FP7/2007-2013/ under REA grant agreement No PIRSES-GA-2013-612484

References

[1] Sobiesiak M 2012 *Adsorption* **19** 349
[2] Puziy A M, Poddubnaya O I, Sobiesiak M, and Gawdzik B 2013 *Adsorption* **19** 717
[3] Myglovets M, Poddubnaya O I, Sevastyanova O, Lindström M E, Gawdzik B, Sobiesiak M, Tryba M M and Puziy A M 2014 *Proceedings of 5th NWBC, March 25th-27th, 2014*. Stockholm, Sweden
[4] Stewart D 2008 Industrial Crops and Products **27** 202
[5] Gawdzik B, Sobiesiak M, Podkoscielna B and Sevastyanova O 2014 *COST Action FP1105 "Understanding Wood Cell Wall Structure, Biopolymer Interaction and Composition: Implications for Current Products and New Material Innovation,"* 52
[6] Stein A, Wang Z and Fierke M A 2009 *Advanced Materials* **21** 265

Acoustical and thermo physical properties of metal-ceramics composites in dependence on few volume concentration of metal

A Abramovich

St. Petersburg State Technological University of Plant Polymers and St. Petersburg State University, Physical Department St. Petersburg, Russia

E-mail: andrew@ns2740.spb.edu

Abstract Metal-ceramics composites (cermets) are modern construction material used in different industry branches. Their strength and heat resistance depend on elastic and thermos physical properties. In this work cermets based on corundum and stainless steel (sintered in high vacuum at temperatures 1500 – 1600oC) are investigated. The volume steel concentration in the samples varies up 2 to 20 vol %. The elastic modules were measured by ultrasonic method at room temperature, measuring of thermo conductivity coefficient were carried out at temperatures 100, 200oC by method of continued heating in adiabatic calorimeter. We founded appearance of two extremes on dependences of elastic modules (E, G) on stainless steel concentrations, nature of which is unknown, modules values change in range: E = 110 - 310, G = 60 - 130GPa (for different temperatures of sintering). Similar dependence is observed for thermo conductivity coefficient which values varies up 10 to 40 W/(m.K). There is presented also discussion of results based on structure cermet model as multiphase micro heterogeneous media with isotropic physical properties in the work.

1. Introduction

Composite materials on the base of ceramics and metal are known in different industry branches. These materials were created for industrial application to produce constructions possessing a high strength, thermal stability and resistance to aggressive media. Investigation results of physical technical properties of the materials are published in periodical issues and monographs [1, 2]. It should be noted that thermo-shock resistance composites have to possess high thermo conductivity and mechanical stability which are determined by elastic modules. The measurement of mechanical stability requires special preparation of many samples to provide reliable results, whereas for measurement of dynamic elastic modules one sample is required. Now there is no uniform understanding about connection of elastic and thermo physical properties for sintered dispersed materials. In this work we explore this connection for composite corundum-metal.

2. Experimental technique

We investigated samples of cermets based on α-Al_2O_3 in combination with the commercial quality stainless steel (18 - Cr, 9 - Ni, 1 - Ti, 72 wt% Fe). To fabricate a cermet, the initial fine grained mixture was prepared by milling α -Al_2O_3 powder in a ball mill in the presence of balls 1–2 cm in diameter made from the stainless steel. The milling was terminated when the steel content in the α-Al_2O_3 powder became equal to 2.2, 4.0, 5.5, 11.0, 21.0vol %. Then, the mixture obtained was doped with a plasticizer and subjected to dry compaction under a pressure of 100MPa and then sintered in a vacuum at 1500, 1600$_o$C. Finally, the samples were cooled in the furnace at an average rate 100$_o$C/h and no further treatment was made. The cermets thus prepared ranged in volume porosity from 3 to 7% and had the steel content from 2.2 to 21.0vol%. The cermet samples (two series for two temperatures of sintering) were then cut into smaller parts of the required dimensions (10 mm – diameter, 8-15 mm - length), which were further ground and polished, depending on the measuring method. To investigate the elastic properties of the cermets, we measured their density by the hydrostatic method and the velocity of longitudinal and transverse ultrasound waves of a frequency of 2MHz. We used the pulsed phase-interferometer method [3], which made it possible to determine the velocity of ultrasound within an accuracy of 0.1–0.2% and 5% for the elastic modules. With the data obtained, we calculated Young's modulus E and the shear modulus G from the well-known formulas of the elasticity theory for an isotropic medium [3]. The thermo conductivity coefficient λ was measured by adiabatic calorimeter NT-λ-400 (Thermometer company). The measuring λ was carried out by method of comparison with the standard sample. The measurement accuracy of λ is 2-3% for all samples. Examination of the surface structure (cleaved facet) of the cermet samples was made with a JSM-840 scanning electron microscope (the Jeol company).

3. Results and discussion

In fig.1 typical microstructure of studied cermets obtained by scanning electron microscopy (SEM) are presented. It is seen there are three phases in samples researched: corundum grains, metal drops and pores. The metal drops with in size 2-3 μm are disposed usually on joint of corundum grains, more small are disposed on the grain faces. It should be noted that inter phase and inter grain boundaries are not observed and size of corundum grains don't change after sintering.

Figure 1. Microstructure of the cermet sample sintered in vacuum at 1500°C by SEM method

Fig.2 shows concentration dependences of elastic moduli E and G for two series of samples (5 pieces in each) sintered at 1500 and 1600$_o$C temperatures. We researched the samples with concentration up 2.2.to 21.0vol%. (at less values of metal volume share sintering is not possible). As it is seen from fig.2 these curves have complicated view for both serious of samples: moduli E and G decrease and increase at concentration change, (moreover, besides) having maxima at11.0 vol. %. Minima are observed for curves 1, 1', 2'at concentration 2.2-5.5vol %.

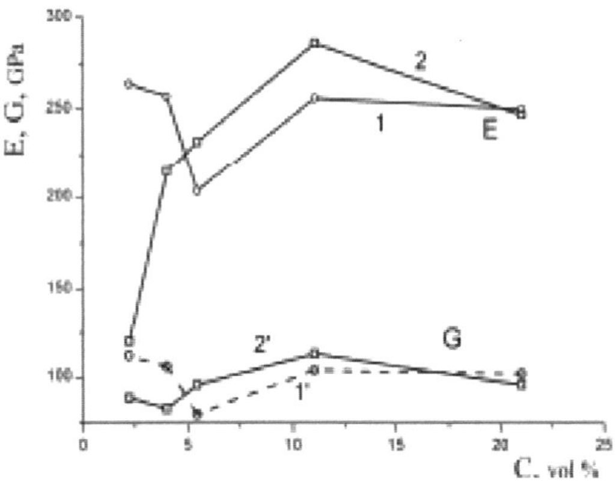

Figure 2. Concentration dependences of elastic modules E, G for cermets, sintered at various temperatures: 1500°C – curves 1, 1'; 1600°C – curves 2, 2'

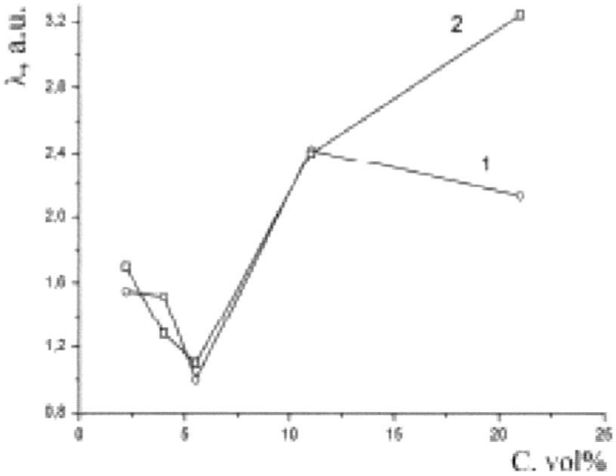

Figure 3. Concentration dependences of thermo-conductivity coefficient (arbitrary units) in cermets, sintered at temperatures: 1 – 1500, 2 – 1600°C (λ = 1 corresponds to the standard sample value)

In work [1] the similar concentration dependences of strength for composites metal-ceramic are presented. They have two maxima at 15 and 85vol %. Authors [1] suggest the qualitative explanation of effect based on influence new inter grain phase appeared in sintering process. This phase as result thermo chemical reaction possesses new elastic properties. In the estimation authors [4] the volume share of new phase can average tens percents. Authors [1] consider formation of this phase as important factor for produce composite with optimal properties. If no components interaction exists the composite will be unusable "mechanical mixture". However this mixture is used as simple model for estimation composite properties. In [5] this model allowed to determine elastic modules of inter grain phase for no porosity Cu powder and ceramic MgO sintered in vacuum. It should be noted that SEM and XRES methods rarely determine presence of inter grain phase (for example, she absent in fig.1). In studied cermets the above phase can exist as one of kind spinels [1], one of them we in [6] and authors [7] founded in boundary medium and identified as $FeAl_2O_4$ Therefore it is of interest to compare the results of studying elastic and thermo physical properties observing the formation the inter grain phase. The decision of this problem is the aim of this work.

Fig.3 shows concentration dependences of thermo conductivity coefficient λ in arbitrary units for the same samples. We see that the values λ for both series of samples decrease at 2.2 - 5.5vol% and increase for second series at 5.5 - 21vol% concentrations as for first series values λ decrease in this range. Thus minimum elastic moduli (fig.2) correspond to minimum thermo conductivity coefficient (fig.3). This effect is valid for both series samples sintered at various temperatures and in our view may be connect with the formation the inter grain phase described in [1, 2, 5, 7]. Indeed at small metal concentration (~1-2 vol%) the boundaries of inter grain phase can't be formed, the layer is crumbly therefore the composite has low elastic moduli and thermo conductivity coefficient (fig.2,3). At concentration ~5.5vol% formation of the boundaries are completing, the inter grain layer possesses perfect structure, in this connection the composite gains high properties. If the metal concentration increases to 11vol% elastic moduli value decrease because $E_{ceramics} > E_{steel}$ therefore value of thermos conductivity coefficient increases because $\lambda_{steel} > \lambda_{ceramics}$. Further increasing concentration to 21vol % lead to noticeable decrease in values E,G for all samples and value λ of one of them (Fig.3).The increase in porosity of cermets from 3 to7% may be reason of above effect. By our opinion the gain of thermo conductivity coefficient for second series will be continued at further increasing volume part of steel because $\lambda_{steel} > \lambda_{ceramics}$. At steel concentration 20 – 30vol % the composite structure approximates to the model "mechanical mixture" [1].

The reason of above considered model of inter grain phase formation may be phonon theory heat transmission in crystals. The theory has been used to explain experimental results of "heat pulses" method in sintered composites [6, 7]. This method studies the time dependence of propagation of no equilibrium phonons (~10_{12} Hz) at liquid-helium temperatures in ceramics [6, 7]. In our work [6] the "heat pulses" method first have been used for cermets. It has been found that appearance of inter grain phase leads: a) to decrease of phonons diffusion coefficient (~10_3 times) as compared with value for the basic corundum ceramic, b) to decrease of composite thermo conductivity at the few metal concentrations. In addition in the work [7] phonons diffusion coefficient in nano disperse corundum-iron composites was investigated at low temperatures. It is shown that noticeable electron-phonon effect is observed if size of metal particles is compared with phonon wave length therefore "phonon capture" takes place. This effect decreases the composite thermo conductivity.

4. Conclusions

Thus in the work: 1) at few volume steel concentration the decrees effects of elastic moduli and thermo conductivity coefficient are establish; 2) complex concentration dependences of this parameters at all researched range are founded; 3) the value of optimal steel concentration for making industrial composite are determined. The employment of acoustical and thermo physical methods and also the of phonons spectroscopy results obtained earlier make possible to propose quality model of inter grain phase formation and to understand her influence on strength and thermo shock protection.

References

[1] Kislyi P.S., Bodnaruk M.S., Borovikova M.S. 1985 *Cermets*. Kiev: Naukova dumka (in Russian)
[2] Garshin A.P., Gropyanov V.M., Zaitsev G.P., and Semenov S.S. 1997 *Ceramics fot Mechanical Engineering*. S.-Petersburg St.Technol.Univ. (in Russian)
[3] Truell R., Elbaum C., and Chick B. B. 1972 Ultrasonic Methods in Solid State Physics. *Academic Press*, New York
[4] Gusev A.I., Rempel A.I. 2001 Nanocrystalline materials. *Moscow: PHYSMATHLIT* (in Russian)
[5] Slepnev, A.G. 2007 J. Pis'ma v *JTF*. **33** 85 (in Russian)
[6] Ivanov, S.N.; Khazanov, E.N.; Taranov, A.V.; Mikhailova, I. S.; Gropyanov, V.M. and Abramovich A.A. 2001 *J. Physics of the Solid State*. **43** 665 (in Russian)
[7] Karban, O.V.; Salamatov, E.I.; Taranov, A.V.; Khazanov,E.N.; Khasanov,O.L. 2009. *Journal of Experimental and Theoretical Physics*. **135** 758 (in Russian)

IOP Conf. Series: Materials Science and Engineering **123** (2016) 012054 doi:10.1088/1757-899X/123/1/012054

Improving the connection between wood and cement using LBL nanocoating to create a lightweight, eco-friendly structural material

L Bejo, B Major, L Csoka, Z Hantos, Zs Karacsonyi

University of West-Hungary, Simony Karoly Faculty of Engineering, Wood Sciences and Applied Arts

E-mail: laszlo.bejo@skk.nyme.hu

Abstract. Structural elements made out of cement bonded wood may be an excellent alternative to flammable organic bonded composite beams, and CO2 intensive, heavy and nonrenewable reinforced concrete. Unfortunately, preliminary studies showed that a sufficient load-bearing performance is difficult to achieve. Improving the compatibility of cement and wood by LbL nanocoating may be a significant step towards creating viable cement bonded wood load bearing elements.

The study involved creating multi layer nanocoating on the surface of poplar veneer using various polyelectrolyte combinations and numbers of treatment cycles, and testing the withdrawal resistance of the samples from a cement matrix. PDDA-PSS treatment was found to form increasingly uniform coating on the surface of wood, while the results were less straightforward for PAH-PSS. Both types and all levels of treatment caused dramatic improvement in load withdrawal resistance. The best result – a more than tenfold improvement – was achieved by at least 10 cycles of PDDA-PSS treatment. PAH-PSS treatment yielded a somewhat more modest improvement, which was already evident after five treatment cycles. The results point to the excellent potential of LbL nanocoating for creating cement bonded structural wood based composite materials.

1. Introduction

In many ways, wood based materials are ideal: they are lightweight, renewable and recyclable, and can be tailored to suit many applications. Among its many uses, wood based panels and beams have been used for structural purposes. Structural wood based beams, like LVL, PSL and LSL (Laminated Veneer Lumber, Parallel Strand Lumber, and Laminated Strand Lumber) in particular, are becoming increasingly widespread in the USA, and are also gaining popularity in Europe [1].

One significant drawback of these products is their limited fire performance. Along with solid wood, they are generally classified as flammable, with normal combustibility. For this reason, some regions enforce strict building codes that severely limit their application, despite their excellent specific strength and environment friendly production technology.

Steel reinforced concrete construction elements are perhaps the most prevalent in Europe, especially for large residential and commercial construction. While these materials are „tried and true", they are very heavy, made of non-renewable materials, require much energy and emit much carbon dioxide during their production.

A combination of the concepts of wood based structural beams and reinforced concrete could be the ideal solution for this problem. However, existing experience with cement bonded wood based materials, as well as some preliminary experiments showed that high-strength cement bonded wood based materials are difficult to produce. Their strength is much lower than required for structural applications [2]. One of the reasons identified was the chemical incompatibility of wood and cement. LbL nanotechnology is a relatively new technique developed in the 1990s. The idea behind this process is changing the surface chemistry of materials by the alternate application of cationic and anionic polyelectrolytes. The materials in the polyelectrolyte solutions bond to the surface, and form a nano-layer, to which the next layer of nanomaterial can bond. After the first two layers, the process can be repeated several times to form a multi-layer coating. Under ideal conditions, the layers become increasingly uniform, and the surface charge becomes homogeneous [3].

LbL nanotechnology is a very simple, cheap and efficient process. Several different kinds of polyelectrolytes or even other materials (e.g. nanomineral colloids) can be applied to a range of different substrates. de Villiers et al. [4] wrote a comprehensive review of the different materials methods and applications used in various studies. LVL nanocoatings are especially helpful in medical technology, where change of surface chemistry can bring dramatic improvements, e.g. in the delivery of certain drugs. It is also useful in many industrial applications, e.g. for creating special composites, improving paints and surface treatment, or for better adhesion, etc.

The earliest studies to create LBL nanocoating for wood surfaces were carried out around the turn of the century [5]. At first, these experiments typically concentrated on creating nanocoating on individual wood fibres, mainly to improve the characteristics of paper, or improved bonding in composite materials ([6],[7],[8],[9],[10],[11]). Lingström et al. [12] provided a thorough overview of many studies, and gives a systematic overview of the application of polyelectrolyte nanocoatings on lignocellulosic fibres.

There are few studies available that deal with treating larger particles or even large wood surfaces. Renaccer and Zhou [13] established that nanoscale films on wood can be deposited without changing the microscopic and macroscopic texture. Zhou [14] used various polyelectrolytes and coating techniques to create nanocoatings on pine veneer that can be used as adhesives.

LbL nanotechnology may also be used to improve the compatibility of wood and cement. Applying just one layer of montmorillionite nanomineral and PDDA to hybrid poplar wood wool, Alpar et al [15] managed to improve the bending strength of cement bonded wood-wool by 20%. These encouraging results will hopefully inspire further research in this area. The purpose of this study is to examine the potential of polyelectrolyte treatment to improve the connection between wood and cement. The objectives include:

- Treating wood particles with various types of nano-polyelectrolytes and using different numbers of treatment to change its surface charge,
- Monitoring the uptake of the nanomaterials in poplar wood, and
- Examining the effect of the treatments on the wood-cement interface by mechanical testing.

2. Materials and methods

The basic substrate chosen for the treatment was 2.5 mm thick peeled veneer made of hybrid poplar (Populus × euramericana cv. 'I-214'). This poplar hybrid is very popular in Hungary, due to its relatively high strength and durability. It is also advantageous because it has a relatively low sugar content that would inhibit cement curing. The veneer was cut into 25 mm wide strips, at least 250 mm in length, and stored under normal conditions (21 °C, 65 % RH) for several days to reach a moisture content of approx. 12 %. The specimens were chosen for treatments in a random manner (CRD). 0.1 concentration (m/m) aqueous solutions were prepared of the following polyelectrolytes, using distilled water:

- Poly(Diallyl Dymethil Ammonium Chloride (PDDA) (Mw: 400-500 kDa)
- Poly(Allylamin Hydrochloride) (PAH) (Mw: 58 kDa)
- Poly(Sodium 4-Styrenesulfonate) (PSS) (Mw: 70 kDa)

The above materials are frequently used to create LbL nanocoatings. PDDA and PAH are strong polycations, and PSS is a polyanion. The following treatments were applied to the veneer strips:

- PDDA + PSS, repeated 5, 10, and 15 times,
- PAH + PSS, repeated 5 and 10 times,
- distilled water only (repeated 30 times), as control.

In each case, three veneer specimens were treated by dipping an 80 mm section at the end of the veneer strips into the solution for 10 minutes. The amount of solution was determined based on the mass of the treated sample. After each treatment, a sample of the solution was analysed using a WPA lightwave, Diode-array, S2000 UV/Vis spectrophotometer. Difference absorbance spectra were created using the corresponding measurements on the control samples, to verify how much material remained in solution after the treatment.

After completing the LbL treatments, the specimens – including the control – were encased in a cement matrix to a depth of 80 mm (see figure 1). Commercial Portland cement, type CEM I 42.5 based on EN 197-1, was used to create the paste, with a water to cement ratio of 7:20. The encased samples were left to cure for 28 days before testing.

After curing, pullout tests were performed on the specimens, as indicated in figure 2. The tests were carried out under displacement control, at a rate of 0.5 mm/min. After testing, the adhesion of cement to the surface of the extracted veneer strips was evaluated visually.

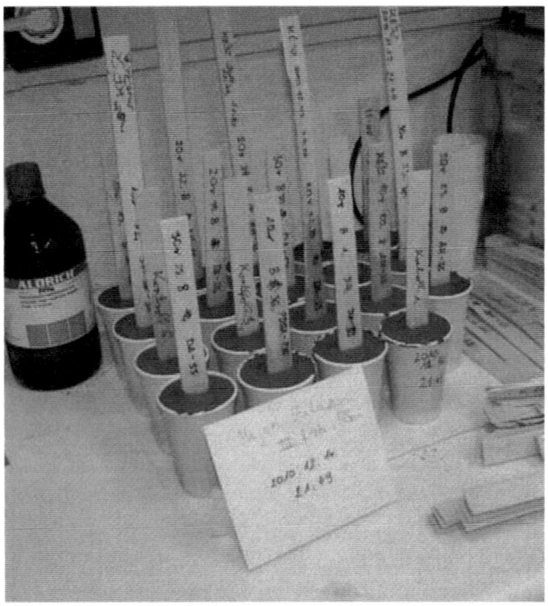

Figure 1. LbL treated poplar veneer samples encased in cement matrix

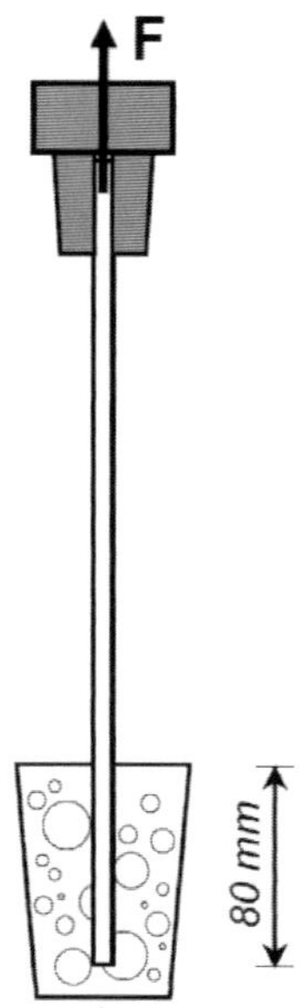

Figure 2. Schematic of the pullout test

3. Results and discussion

PDDA-PSS treatment proceeded as expected. Figures 3a and 3b show the difference absorbance spectra of PDDA and PSS, respectively (15 repetition samples). These diagrams display more-or-less decreasing values, i.e. decreasing residual concentrations with increasing number of treatments. This is consistent with our expectations; as the number of nanolayers increases, more and more uniform coverage is achieved, which progressively facilitates the bonding of the next layer. After about 5 PDDA-PSS treatment cycles, the residual concentration approaches zero, i.e. there is practically no PDDA or PSS remaining in the solution. (N.B.: slightly negative values calculated after several treatment cycles reflect experimental error, rather than physical reality.)

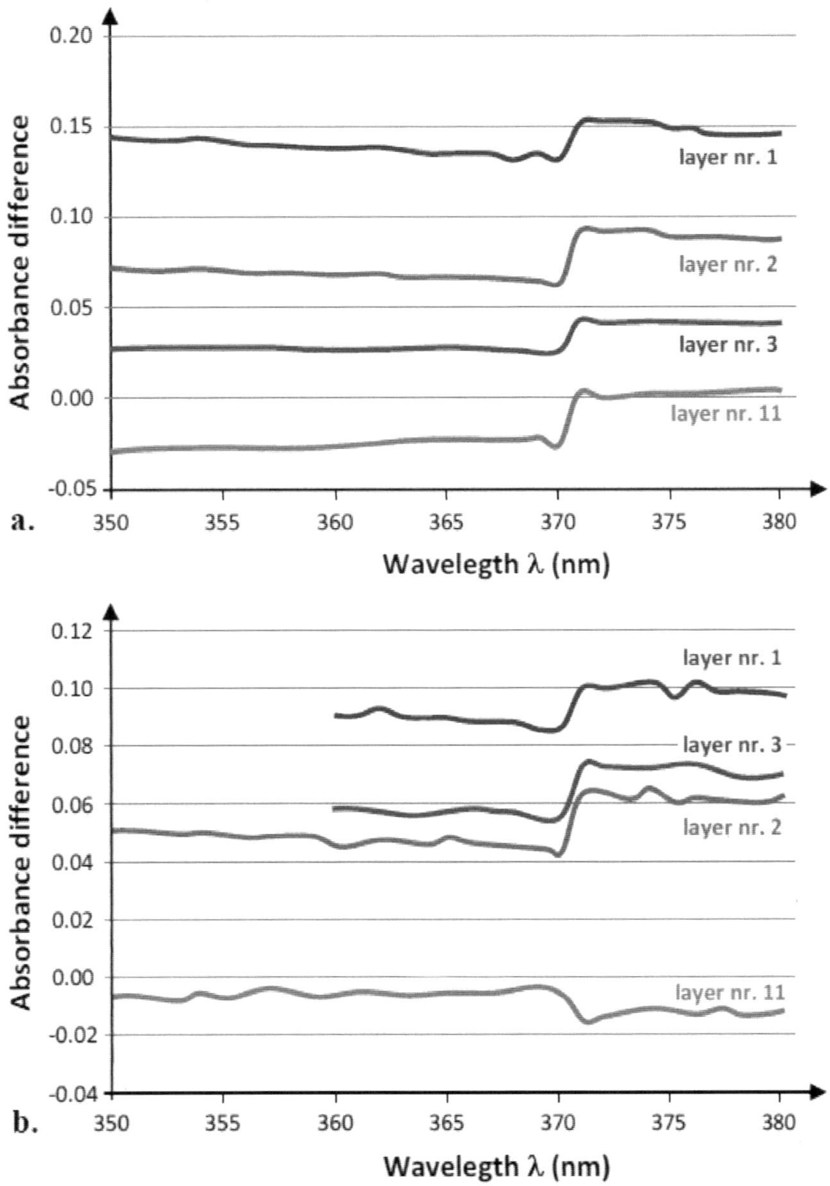

Figure 3. Difference absorbance spectra of the PDDA (a) and PSS (b) solutions after the adsorption of the polyelectrolytes on the surface, measured in various stages of the LbL treatment

PAH-PSS treatment yielded less straightforward results. Even though there was a general decreasing trend, residual concentrations changed somewhat more erratically. One possible explanation lies in the molecular weight of PAH, which is lower than that of PDDA. The small molecular dimensions may have prevented the formation of continuous nanolayers in some cases, and deteriorated the next layer's ability to bond. Nevertheless, by the end of the 5th cycle, fairly continuous nanolayer formation likely occurred. The experiences show that higher concentrations of PAH solution may be beneficial in similar future experiments.

Figure 4 shows the results of the pullout tests. In general, all of the treatments led to dramatic improvement in pullout force. 5 cycles and 10 cycles of PDDA-PSS treatment increased the load withdrawal resistance more than fivefold and tenfold, respectively. After 10 cycles, further improvement was not observed. Similarly, 5 cycles of PAH-PSS treatment caused an almost tenfold improvement, after which no significant further increase occurred. Visual observation confirmed the excellent cement adhesion achieved after the LBL treatment, as opposed to the untreated samples.

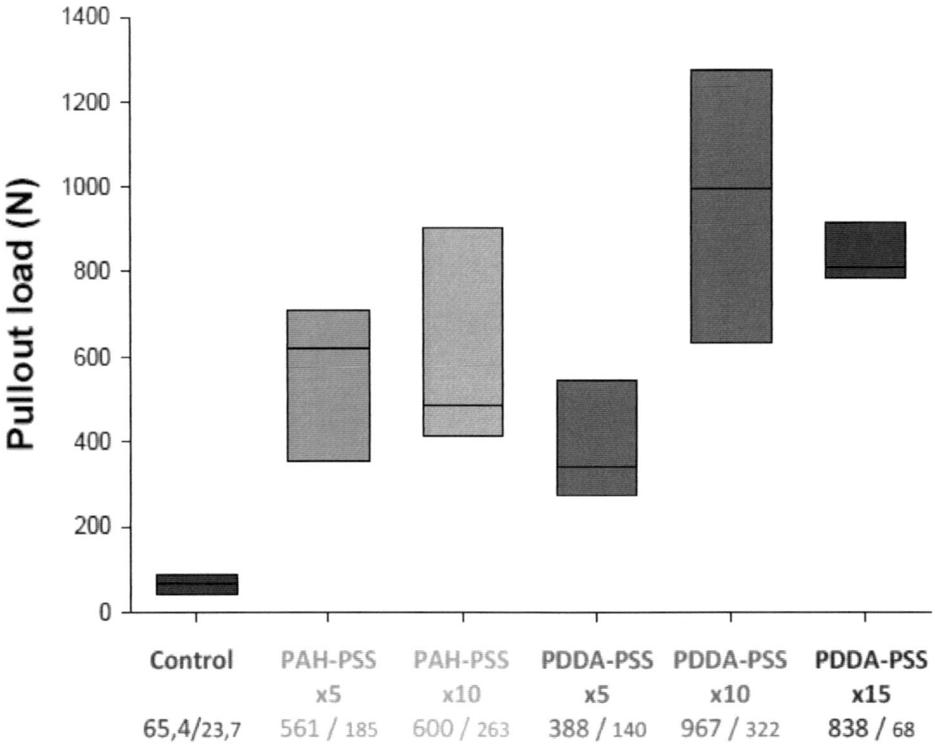

Figure 4. Pullout force measurement results of veneer samples treated with various numbers of polyelectrolyte layers. The average and standard deviation values are indicated on the horizontal axis

4. Summary and conclusions

The research described in this paper, concerning the use of LbL polyelectrolyte treatment of hybrid poplar veneer to improve its compatibility with cement, yielded the following conclusions:

- PDDA-PSS treatment was found to form nanolayers of increasing uniformity, as evidenced by UV-VIS difference spectroscopy results. After 5 treatment cycles, the residual concentration in the solutions was close to zero.
- PAH-PSS treatment performed reasonably well, but the results were less straightforward.
- Higher PAH concentrations may be required for uniform layer formation.
- All of the treatments caused dramatic improvement in load withdrawal resistance, with a more than tenfold ultimate increase in withdrawal load in case of PDDA-PSS treatment. No further improvement was observed in withdrawal resistance after 5 and 10 treatment cycles, in case of PDDA-PSS and PAH-PSS treatment, respectively.

The results indicate the excellent potential of LbL nanotechnology to improve the strength of cement bonded wood based materials. Further experiments are required, however, to establish the applicability of this technique to create structural load bearing cement bonded wood composites.

Acknowledgements
This study was supported by the Environment conscious energy efficient building TAMOP-4.2.2.A-11/1/KONV-2012-0068 project sponsored by the EU and European Social Foundation.

References

[1] Forest Products Laboratory 2010 *Wood handbook* – Wood as an engineering material, General Technical Report FPL-GTR-190. 508
[2] Bejo, L., P. Takats, N. Vass. 2005 *Acta Silvatica et Lignaria Hungarica* **1** 111
[3] Decher, G., Y.M. Lvov, J. Schmitt 1994 *Thin Solid Films* **244** 771. DOI: 10.1016/0040-6090(94)90569-X
[4] de Villiers, M.M., D.P. Otto, S.J. Strydom, Y.M. Lvov 2011 *Advanced Drug Delivery Reviews* **63** 701 DOI: 10.1016/j.addr.2011.05.011.
[5] Forschberg, S, L. Wågberg 2000 Production of Particles or Fibres having a Coating of Polyelectrolytes Interacting with Each Other and Paper or Nonwove Products with Improved Opacity Therefrom. *SCA Hygene Products* AB Sweden, 19
[6] Agarwal, M., Y.M. Lvov, K. Varahramyan 2006 *Nanotechnology* **17** (21) 5319. doi:10.1088/0957-4484/17/21/006
[7] Zheng, Z., J. McDonald, R. Khillan, Y. Su, T. Shutava, G. Grozdits, Y.M. Lvov 2006 *Journal of Nanoscience and Nanotechnology* **6**(3) 624. DOI: 10.1166/jnn.2006.081
[8] Lu, Z., S. Eadula, Z. Zheng, K. Xu, G. Grozdits, Y. Lvov 2007. *Colloids and Surfaces* A: Physicochemical and Engineering Aspects **292** (1) 56
[9] Peng, C.Q., Y.S. Thio, R. A. Gerhardt. 2008. *Nanotechnology* **19** (50) 505603. doi: 10.1088/0957-4484/19/50/505603
[10] Lin, Z., S. Renneckar 2011. Composites Part A: *Applied Science and Manufacturing* **42** (1) 84. DOI: 10.1016/j.compositesa.2010.10.011
[11] Lee, J, J. Ryu, H.J. Youn. 2012. *Cellulose* **19** (6) 2153. DOI: 10.1007/s10570-012-9781-6
[12] Lingström, R., E. Johansson, L. Wågberg 2009 Polyelectrolyte Multilayers for Fibre Engineering. In: L.A. Lucia and O.J. Rojas eds. *The Nanoscience and Technology of Renewable Biomaterials* Wiley and Sons, Chichester, UK. 123.
[13] Renneckar, S., Y. Zhou 2009. *ACS Appl Mater Interfaces* **1**(3 559. doi: 10.1021/am800119q.
[14] Zhou, Y 2008 Nanoscale surface modification of wood veneers for adhesion. MS Thesis, Virginia Polytechnic Institute and State University, Blacksburg, VA, USA. 80 pp.
[15] Alpar, T.L., A.A. Pavlekovics, L. Csoka, L. Horvath 2011 Wood wool cement boards produced with nano minerals. In: *Proceedings 3rd Int. Scientific Conference on Hardwood Processing* (ISCHP3 2011) I: Peer reviewed contributions. October 16-18, 2011. Blacksburg, VA, USA. pp. 75-82. ISBN 978-0-9837700-0-8.

Performance evaluation of high modulus asphalt concrete mixes

V Haritonovs, J Tihonovs, M Zaumanis

Riga Technical University, Faculty of Building and Civil Engineering, The Institute of Transportation, Department of Roads and Bridges, Riga, Latvia Russia

E-mail: viktors.haritonovs@rtu.lv

Abstract. Dolomite is one of the most available sedimentary rocks in the territory of Latvia. Dolomite quarries contain about 1000 million tons of this material. However, according to Latvian Road Specifications, this dolomite cannot be used for average and high intensity roads because of its low quality (mainly, LA index). Therefore, mostly imported magmatic rocks (granite, diabase, gabbro, basalt) or imported dolomite are used which makes asphalt expensive. However, practical experience shows that even with these high quality materials roads exhibit rutting, fatigue and thermal cracks. The aim of the research is to develop a high performance asphalt concrete for base and binder courses using only locally available aggregates. In order to achieve resistance against deformations at a high ambient temperature, a hard grade binder was used. Workability, fatigue and thermal cracking resistance, as well as sufficient water resistance is achieved by low porosity (3-5%) and higher binder content compared to traditional asphalt mixtures. The design of the asphalt includes a combination of empirical and performance based tests, which in laboratory circumstances allow simulating traffic and environmental loads. High performance AC 16 base asphalt concrete was created using local dolomite aggregate with polymer modified (PMB 10/40-65) and hard grade (B20/30) bitumen. The mixtures were specified based on fundamental properties in accordance to EN 13108-1 standard.

1. Introduction

If the local material does not fulfill requirements, then one should seek the way for the improvement of its properties. If this is not possible, then one should seek the technological solution which will allow application of the weaker material [1-2]. One of a proper solution might be the use of dolomite as a component of High Modulus Asphalt Concrete (HMAC). Knowing that the binder courses, situated between 5 and 12 cm below the road surface, are subject to the highest stresses, high stiffness is probably the most important requirements for HMAC [3-4]. HMAC is a mixture of asphalt concrete designed for use in base and binder course of asphalt pavement. It has closed structure with comparatively large content of bitumen. Hard road bitumen grades are applied, mainly 10/20, 15/25, 20/30 and polymer modified bitumen. Hard bitumen assures the mixtures resistance to rutting. However large content of bitumen assure workability, fatigue durability and water resistance [5-6]. This type of an asphalt mixture is designed not only by empirical properties but also by performance based properties (rut test, stiffness modulus and fatigue tests) [7-8].

France was also one of the first countries in which mechanistic asphalt pavement design was introduced into the general practice [9]. In France, it is known under the acronym EM. In Poland, the acronym is AC WMS. Possible application of weaker mineral aggregate is one of the advantages of EM (in English HMAC). Application of High Modulus Asphalt Concrete allowed for saving on asphalt pavement's thickness thanks to higher stiffness modulus which reduces tension strains in asphalt base layer.

The aim of the paper is to develop high performance asphalt concrete for base and binder courses with using only locally available aggregates – crushed dolomite. In order to achieve resistance to deformations at high ambient temperature hard grade binder was used.

2. Materials

The basic materials used in this study are fractionated crushed dolomite aggregate, unmodified hard grade bitumen B20/30 and SBS modified bitumen PMB 10/40-65 (B70/100 for reference mixture). Crushed dolomite aggregate were obtained from Pļaviņu DM ltd. (Latvia), and hard grade bitumen B20/30 from Grupa LOTOS S.A (Poland).

2.1. Bitumen characteristics

The binder properties have been tested by means of conventional binder tests: needle penetration, softening point, aging and Fraas breaking point. The test results are listed in Table 1.

Table 1. Typical characteristics of the bitumens.

	Bitumen			
Parameter	B 20/30	PMB 10/40-65	B70/100	Standard
Penetration at 25°C, dmm	25.3	35.0	80.0	LVS EN 1426
Softening point, °C	62.6	64.2	46.0	LVS EN 1427
Fraas temperature, °C	- 13.0	- 17.0	- 21.0	LVS EN 12593
Kinematic viscosity, mm_2/s	1460	-	346	LVS EN 12595
Dynamic viscosity, Pa·s	3277	-	160	LVS EN 12596
Elastic recovery, %	-	85.0	-	LVS EN 13398
Ageing characteristics of bitumen under the influence of heat and air (RTFOT method)				
Loss in mass, %	-0.02	0.01	0.05	LVS EN 12607-1
Retained penetration, %	75.9	69.7	76.0	LVS EN 1426
Increase of a softening point, °C	6.9	5.5	5.4	LVS EN 1427
Fraas breaking point after aging, °C	-11.0	- 16.0	- 17.0	LVS EN 12593
Retained elastic recovery,%	-	> 50	-	LVS EN 13398

2.2. Properties of dolomite aggregate

The test results of dolomite main properties show very low flakiness index – 5, high frost resistance with average MS value of 7 and low fines content – 0,6%. However LA value is only 33. These aggregates are suitable for use as a component of High Modulus Asphalt Concrete, where permitted LA value up to 40 (SPENS). The properties of dolomite aggregate are shown in Table 2.

Table 2. Physical and mechanical characteristics of the dolomite aggregate

Physical and mechanical properties	Results	Standard
Los Angeles coefficient (LA), %	33	LVS EN 1097-2
Resistance to wear. Nordic test (A_N), %	21	LVS EN 1097-9
Flakiness Index (FI), %	5	LVS EN 933-3
Water absorption, %	2	LVS EN 1097-6
Grain density, Mg/m3	2.80	LVS EN 1097-6
Fine content, %	1.0	LVS EN 933-1
Freeze/thawing (MS), %	7	LVS EN 1367-2

3. Max design

The basic materials used in this study are fractionated crushed dolomite aggregate, unmodified hard grade bitumenHMAC-16 asphalt concrete mixtures have been designed by using unconventional (bitumen - B20/30, PMB 10/40-65 and dolomite aggregate LA > 30) raw materials (Fig. 1). The basic idea of HMAC is to design a mix with hard grade bitumen at high binder content [10-11]. The Marshall mix design procedure was used for the determination of the optimal bitumen content for the reference mixture, considering the mixture test results for Marshall stability and flow, as well as the volumetric values: air voids (V), voids in mineral aggregate (VMA) and voids filled with bitumen (VFB) [12]. Test specimens for Marshall Test were prepared in the laboratory by impact compactor according to LVS EN 12697-30 with 2×50 blows of hammer 150°C temperature.

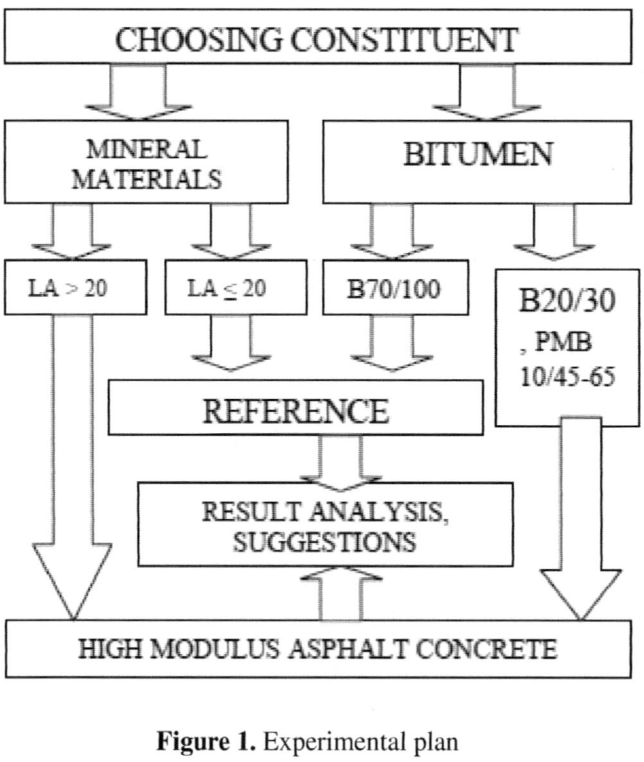

Figure 1. Experimental plan

4. Results

4.1. Volumetric properties

Analysis of volumetric parameters of the different asphalt mixtures at different binder contents was performed. The results are presented in Table 3. The binder content was optimized according to HMAC requirements developed in the SPENS programme (Sustainable Pavement for European New Member States).

4.2. Marshall test

Table 4 contains Marshall test results at different binder contents for the mixtures that passed the requirement of having less than 5% air voids. The results show that HMAC mixtures have higher Marshall Stability compared to the reference mixture.

Table 3. Volumetric properties of HMAC mixtures

Parameter	Mixtures					
	PMB 10/45 -65		B20/30			B70/100
	HMAC-1/1	HMAC-1/2	HMAC-2/1	HMAC-2/2	HMAC-2/3	Reference
Bulk density, kg/m3	2455	2457	2430	2455	2457	2550
Maximum density, kg/m3	2555	2551	2586	2555	2551	2680
Voids content, %	3,9	3,7	6.0	3.9	3.7	4.85
VMA	17,8	18	18.3	17.8	18	17.6
VFB	78,2	79,6	67.2	78.2	79.6	72.4
Bitumen content, %	5,67	5,83	5.06	5.67	5.83	5.0

Table 4. Marshall test results

Parameter	Mixtures					
	PMB 10/45 -65		B20/30			B70/100
	HMAC-1/1	HMAC-1/2	HMAC-2/1	HMAC-2/2	HMAC-2/3	Reference
Bitumen content, %	5,67	5,83	5.06	5.67	5.83	5.0
Specimen height, mm	64.2	64.0	Not tested voids content > 5%	63.6	62.9	63.3
Stability at 60°C (kN)	16.5	15.8		16.6	15.4	12.0
Flow at 60°C (mm)	2.8	3.4		3.8	5.9	4.2

4.3. Wheel Tracking Test

A wheel tracking apparatus was used to simulate the effect of traffic and to measure the plastic deformations of the asphalt concrete samples. Tests were performed according to standard EN 12697-22 method B (wheel tracking test with small size device in air). This test method is designed to repeat the stress conditions observed in the field and therefore can be categorized as simulative. The resistance of asphalt mixture to permanent deformation is assessed by measuring the rut depth and its increments caused by repetitive cycles (26.5 cycles per minute) under constant temperature at 60°C. The rut depths are monitored by means of two linear variable displacement transducers (LVDTs), which measure the vertical displacements of each of the two wheel axles independently as rutting progresses. The obtained results after 20 000 wheel passes demonstrate that the highest rut depth of 5.7mm for the HMAC mixture with 5.83% bitumen content. The reference mixture demonstrates similar result having 5.3mm rut depth. HMAC mixture having 5.67% bitumen content shows the highest rutting resistance with 3.8mm. Figure 2 summarizes the wheel tracking test results.

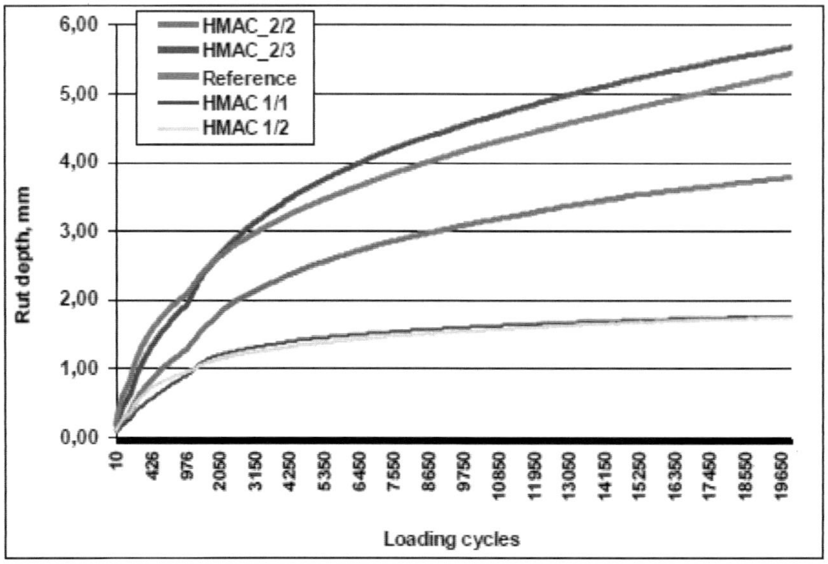

Figure 2. Wheel tracking test results

4.4. Fatigue

To determine the fatigue life of the prepared asphalt concrete mixes, a four point bending beam fatigue test was conducted. The test was run at 10°C, using 10Hz frequency at 130 μm/m strain level. The beams were compacted in the laboratory using roller compactor. They were saw cut to the required dimensions of 50mm width, 50mm height and 400mm length. The failure criterion used in the study is the traditional 50% reduction from initial stiffness. The obtained results indicate that both HMAC mixtures have high resistance to fatigue, compared to the results of reference mixture made with conventional aggregates and bitumen. HMAC mixes fatigue resistance corresponds to standard category ε6-130. However reference mixture did not pass the fatigue test. HMAC mixes compliance with the SPENS requirements are given in Table 5.

Table 5. Compliance with SPENS requirements

Parameter	Mixtures				Requirement
	PMB 10/45 -65		B20/30		
	HMAC-1/1	HMAC-1/2	HMAC-2/2	HMAC-2/3	
Voids content, %	3,9	3,7	3.9	3.7	3.0 – 5.0
Rut resistance, mm/1000ciyles	0.04	0.04	0.14	0.22	0.03 – 0.25
Stiffness (10°C, 10Hz), MPa	16700	16100	17100	17900	Smin 14000
Fatigue (10°C, 10Hz, 130μmm/mm)	ε6-130	ε6-130	ε6-130	ε6-130	ε6-130 (≤ 50%)
Water sensitivity, ITSR, %	100	100	98	94	ITSR 80

5. Conclusion

Use of relatively weak dolomite aggregate in High Modulus Asphalt Concrete was evaluated. This mixture was designed to have less than 5% air voids using Marshall compactor, and had high hard (B20/30) and polymer modified (PMB 10/45-65) bitumen content. Testing was performed to compare this mix with $AC16_{bin}$ reference mixture that was produced using conventional bitumen B70/100 and granite aggregate. Test results demonstrated that with optimum mix design HMAC mixture can provide high rut und fatigue resistance. However mixtures with hard grade bitumen B20/30 showed a little lower resistance to rutting. The reference mixture while having high rut resistance, proved that lower binder content results in shorter fatigue life.

These results provide confidence that the weak Latvian dolomite may be applied in High Modulus Asphalt Concrete for base and binder courses. HMAC mixtures fulfill the HMAC asphalt concrete requirements in accordance with SPENS project recommendations (SPENS).

Acknowledgements

This research is co-financed by the ESF within the project "New "Intelligent" Nano-Composite Materials for Roads, Bridges, Buildings and Transport Vehicle", agreement No. 2013/0025/1DP/1.1.1.2.0/13/APIA/VIAA/019

References

[1] Sybilski, D.; Bankowski, W.; Krajewski, M. 2010. *International Journal of Pavement Research and Technology* **3**, No. 2. 96

[2] Haritonovs, V.; Zaumanis, M.; Brencis, G.; Smirnovs, J.; 2013. *The Baltic Jaurnal of Road and Bridge Engineering* **8**, No 4 91

[3] Backer, C.; Visscher, J.; Glorie, L.; Vanelstraete, A.; Vansteenkiste, S.; Heleven, L. 2008. A comparative high – modulus experiment in Belgium. Proceedings of Transport Research Arena Europe 2008 (TRA 2008) International Conference. 21-24 April 2008, Ljubljana, Slovenia.

[4] Brosseaud, Y.; Farcas, F.; Mouillet, V. 2012. High modulus asphalt mixes with high rate of RA: what does it happen? International Symposium on Heavy Duty Asphalt Pavements and Bridge Deck Pavements (ISAP2012), May 23-25, 2012. Nanjing, China.

[5] Sybilski, D.; Bankowski, W.; Krajewski, M. 2010. *International Journal of Pavement Research and Technology* **3**, No. 2 96

[6] Sivapatham, P.; Beckedahl, H.J.; Jannsen, S. 2010. High Stable Asphalt for Heavy loaded Bus Test Lane Sections. Proceedings of the 11_{th} Internationa Conference on Asphalt Pavements ISAP 2010. 1-6 August 2010, Nagoya, Aichi, Japan.

[7] Tuminiene, F; Laurinavičius, A. 2013. *The Baltic Jaurnal of Road and Bridge Engineering* **8**, No 2. 142

[8] Sustainable Pavements for European New Member States (SPENS) Document No. D8. [online]. [accessed on 10.10.2014.]. Available: http://www.spens.fehrl.org

[9] AFNOR – Association Franēaise de Normalisation. "Enrobēs Hydrocarbonēs: Couches d'assises: enrobēs a module ēlevē (EME): NF P 98-140" Paris, 1999. (in French).

[10] Sivilevičius, H.; Podvezko, V.; Vakrinienė, S. 2011. *Construction and Building Materials* **25**, No 1. 115

[11] Rohde, L.; Ceratti, J. A. P.; Nunez P.V.; Vitorello T. 2008. Using APT and Laboratory Testing to Evaluate the Performance of High Modulus Asphalt Concrete for Base Courses in Brazil. In Proceedings of the Third International Conference on Accelerated Pavement Testing

[12] Sivilevičus, H.; Vislavičius, K. 2008. *Construction and Building Materials* **22**, No 9. 2007

IOP Conf. Series: Materials Science and Engineering **123** (2016) 012056 doi:10.1088/1757-899X/123/1/012056

Fracture Behaviour of Glass Columns Experimental Study of Axial Loaded Glass Columns

A Jakab[1], K Nehme[2], S G Nehme[3]

[1]Ph.D. Student, Department of Construction Materials and Engineering Geology, BME, Budapest, Hungary
[2]Associate Professor, Department of Civil Engineering, University of Debrecen, Debrecen, Hungary
[3]Associate Professor, Head of the Laboratory, Department of Construction Materials and Engineering Geology, BME, Budapest, Hungary

E-mail: jakab.andras@epito.bme.hu

Abstract. Nowadays supporting structures can be transparent due to the development of glass strengthening procedures. The building glass as a versatile building material enables the efforts of the architects due to its transparency. This paper focuses on glass columns in the topic of load-bearing glasses and also on the design and load bearing capacity of fins and stability issues. Laboratory experiments were carried out at the BME, Department of Building Materials and Engineering Geology on the fracture behaviour of centrally compressed glass columns. More than 120 specimens where loaded until fracture. The load and deformations were measured. Based on the experimental results the critical force was determined and with force-deflection diagrams were illustrated the fracture and stability processes. Authors are going to compare the results of the laboratory experiments and theoretical calculations.

1. Introduction

Glass is called also the material of the third millennium, although glass is a brittle material and for a long time its brittleness was a well-known property besides its transparency [1]. This paper focuses on the buckling behaviour of the load bearing glass elements, especially on the glass columns. Laboratory tests were carried on axially compressed glasses in the BME, Department of Construction Material and Engineering Geology. Previous studies were demonstrated by the Authors of present paper about the topic: effect of non-standardised loads on glass columns which preferred to take into account beside of the standardised loads [2]. Based on the laboratory experimental results were presented e.g. effect of the rate of loading, heat strengthening, effect of the different height of the specimen and the effect of the lamination with total thickness of 12 mm (single layer, and laminated – two, and three layers – glasses) on the behaviour of glass columns. Three different stages were determined and introduced in the buckling behaviour of glass columns [2]. The critical buckling force was studied based on the international theoretical results and the laboratory tests [3]. Continuing the previous laboratory research, additional tests were done to clean earlier issues and to draw attention to the design problems of the glass columns. Present paper introduces the effect of the support on the buckling behaviour of glass columns e.g. influence of the damping material, and the fracture pattern of glass columns were studied as well. The fracture of this structural element can cause stability problems in the whole building, therefore the fracture process of glasses used in load bearing elements, must be extensively studied. The widespread application of glass justifies the economic efficiency of the construction beside of the safety.

2. Phenomenon of flexural buckling

Axial loaded columns start to deform without horizontal displacement at the beginning of the loading process. Damping material (rubber min. SHORE 60) is recommended to apply between the glass and the supporting steel surface. Therefore, the vertical displacement contains the deformation of the glass and the damping material as well. When the compression load reaches a critical value, the buckling of the column begins. This force is called (N_{cr}) critical buckling force. In the first Stage of the buckling process the loaded element can be unloaded without visible residual deformations. Until the loading force is increased up to the critical buckling force it is mainly a stability problem (SLS- Service Limit State of the columns). After reaching the critical buckling force post-critical Stages follow. In the case of further increase of the loading force, the column reaches the Ultimate Limit State (ULS), where the risk of the whole construction failure is significant. Significant displacements can be observed between SLS and ULS which serves as a reserve of the glass column in the post-critical Stages. The critical buckling force calculated with the application of the Euler formula was studied as follows:

$$N_{CR} = \frac{\pi^2 EI}{(\upsilon L)^2} \tag{1}$$

This formula includes physical properties of the glass element. However, the critical buckling force is sensitive of varying in the effective length factor. This factor can have different values depending on the shape of the buckled elements. Figure 1 introduces general buckled shape with the value of the effective length factor.

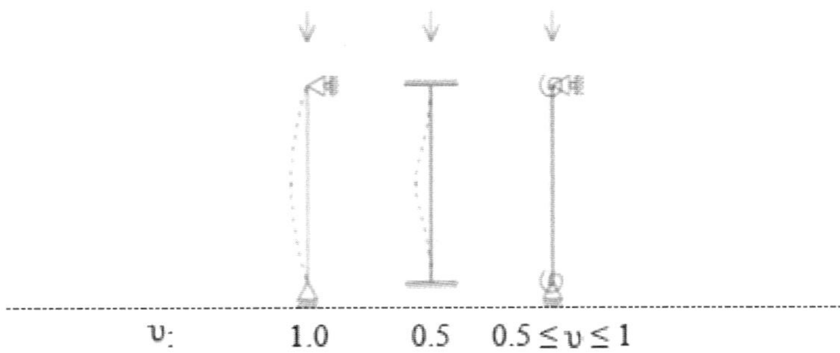

Figure 1. Critical buckling shape and the effective length factor

The effective length factor in the reality of a glass columns is between 0.5 and 1. Although the damping material causes displacement in the supports, - and functions like a spring in the fixings - the real effective length factor cannot be more than the value 1. The displacement of the damping material is not relevant compared to the overall column. The effective length factor varies during the loading process. Until the SLS the effective length factor approaches the value 1 (it functions rather like a pinned support) and after the SLS it reduces. The buckling of the specimen causes more fixed state in the supports (functions rather like a fixed support).

3. Laboratory experiments

3.1. Test parameters

Laboratory experiments were carried out to study the buckling behaviour of single and laminated glass columns. The specimens were tested with use of Instron 5989 testing machine. Single layer float and heat-strengthened glass and laminated glass consisted of both float and heat-strengthened glass layers were tested. The scales of the geometry of specimens (height, thickness, width) were selected on the basis of existing glass columns from international and Hungarian realized projects. Test parameters of glass specimens were the following: Constants: test arrangement, the type of support; width of glass (80 mm); interlayer material (EVA foil with thickness of 0.38 mm); edgework; temperature (+23 ± 5°C). Variables: type of glass layers: HSG/ non heat-treated Float; height of specimens: 1000 mm; 920 mm; 840 mm; number of glass layers and the thickness of specimens: single layer: 8 mm; 12 mm, laminated: 4.4 mm; 6.6 mm; 8.4 mm, laminated: 4.4.4 mm; The rate of loading: 0.5 mm/min; 1 mm/min. Support: Height of fixing: 95 mm; rubber plate (Shore A 80) was used between the steel supports and the glass. Simplified designation is used to distinguish the studied specimens, these are e.g. H_2(4.4)_2_920_0.5: ~ H, F: Type of glass: H – HSG; F – non heat-treated float glass; 2(4.4): Number of glass layers ex.: 4.4 mm laminated glass; 2: The number of specimen; 920: Nominate height of specimen [mm]; 0.5: Rate of loading [mm/min]. Abbreviations are used for the float laminated glass VG and for heat-strengthened laminated glass VSG [2-4].

3.2. Experimental test set-up

The load and vertical displacement of the upper cross-head of the Instron 5989 universal testing machine were continuously measured. At three different heights the buckling displacement (horizontal displacement) of all specimens were continuously measured with HBM displacement transducers during the tests. Strains at centre point on the surface of the glass panels were measured with HBM LY11-10/120 strain gauges. The tests were carried out at a room temperature (+23 ± 5°C). At least three specimens were tested at each testing combination. Laminated specimens were loaded until all glass layers were fractured. In total, 120 specimens were tested. The specimens were mounted as shown in figure 2 [2-4].

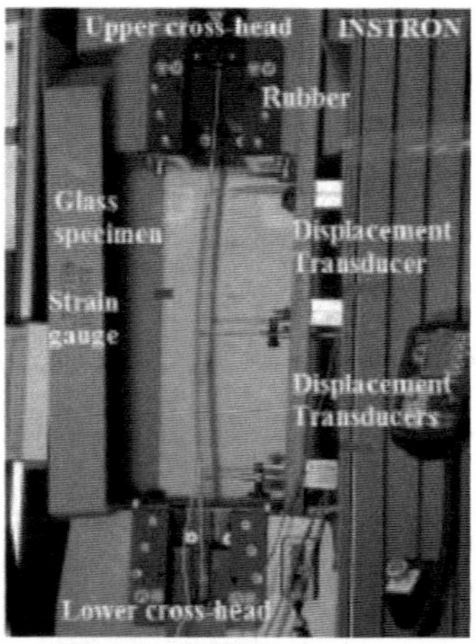

Figure 2. Test-set up

3.3. *New experimental results*

As mentioned in section 1 three different stages can be distinguished (figure 3 where number and arrows demonstrate the stages). In the 1st Stage, the elastic deformation of the damping materials (rubber plates) influences the vertical and horizontal displacements and no buckling occur (first stable stage). The allowed buckling load during structural design calculations is suggested to be the maximal load of the 1st Stage (stable state) reduced with safety factors. The 2nd Stage is a short term stage which indicates a geometrical unstable condition (the direction of horizontal displacement can be shown) and the specimen loses its former stability (bound phenomenon, instability). The 2nd Stage in the buckling behaviour is mainly influenced by the type of the supporting structure (fixed/pinned) and the stiffness of the glass columns. In the 3rd Stage, both the vertical and horizontal displacement increase until fracture of the glass (second stable stage). Considering the experimental results in certain cases the 2nd Stage was absent during the loading procedure. A specific value to the critical buckling load cannot be observed and the beginning of the post-critical Stage is difficult to determine. Strain gauges were placed in the centre of both outer surfaces of the glass columns. The figure 4 indicates the loading force vs. strains on the glass surface. At the beginning both outer surface of the glass column is in compression after that, the compression starts to decrease at one outer glass surface and tensile stresses develop. The buckling process starts during this phenomenon.

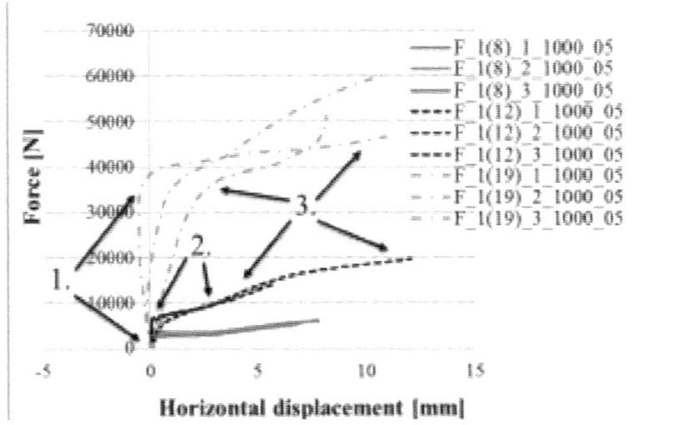

Figure 3. Stages of the bucking behavior

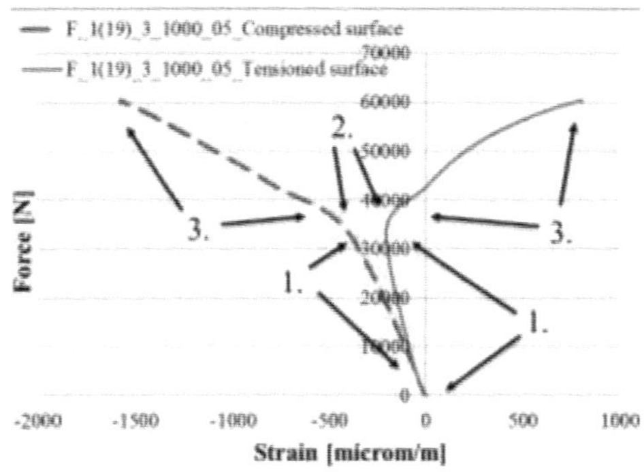

Figure 4. Strain vs. Force

Mean stresses calculated at the moment of buckling are shown in table 1. The stresses increase with the increase of the cross-sectional area of the glass. The stresses increase with the decrease of the height of the specimens in the case of laminated glasses. No significant difference can be observed between the stresses at the moment of buckling of float or HSG glasses with same thicknesses and with the height of 1000 mm, although this difference increases with the decrease of the height.

Table 1. Mean stresses at the moment of buckling

Number of layers	Thickness [mm]	Height [mm]	Stress of float glasses [N/mm^2]	Stress of HSG glasses [N/mm^2]
	8	1000	6.26	3.99
	8	920	3.79	5.81
1	8	840	7.58	5.12
	12	1000	9.33	10.72
	19	1000	25.32	22.57
	4.4	1000	4.10	3.98
	4.4	920	6.13	4.73
	4.4	840	8.69	5.11
2	6.6	1000	7.90	11.13
	8.8	1000	13.29	12.37
	8.4	1000	-	7.36
	10.10	1000	14.95	17.17
3	4.4.4	1000	7.50	8.22

3.4. Fracture pattern

The glass specimen fractures when the tensile strength of glass is exceeded by the stresses developed on the tensile glass surface. Actually the cracks start from the local edge defects (which were initiated e.g. by edgework). Two inflection points develop during the buckling process. The locations of the inflection points are affected by the height of the glass columns, the stiffness and the supporting conditions as well. Dense crack pattern can be observed in the region of the inflections. Two inflection regions developed, which are near to the supports [4].

Local maximal tensile stresses develop in the inflection regions. In the 3rd Stage the maximal tensile stress developed in the mid of the height of the specimen, opposite to the tensile glass layer at the inflection region. The buckling length of the glass column reduces in the case of fixed support compared to pinned, therefore the location of the local and maximal tensile stresses are influenced by the supporting conditions. In float glasses the cracks with critical thickness can propagate in the location of the local stress regions as well. In the case of heat-strengthened glasses the distribution of cracks influences the strength of the glass, therefore the strength of a glass column is mainly influenced by the area of glass layers in tension.

In the case of glasses consisted of glass layers of 4 mm thicknesses, the fracture originates at the mid of the height of the specimens. With the increase of the thickness of glass layers from 4 mm up to 6 mm the local stresses increase in the inflection regions, therefore in these regions cracks develop. In the case of thickness of glass layers the normal stresses increased in the vertical direction and the cracks propagate also in the vertical direction in the glass layers. In the case of unsymmetrical laminated glasses consisted of thicknesses of 8 and 4 mm glass layers the thicker glass layer fractured at first. More cracks were developed in the thicker (8 mm) glass layer than in the thinner (4 mm) glass layer. Vertical cracks formed in the glass layer with thickness of 8 mm as well (Figure 5-6).

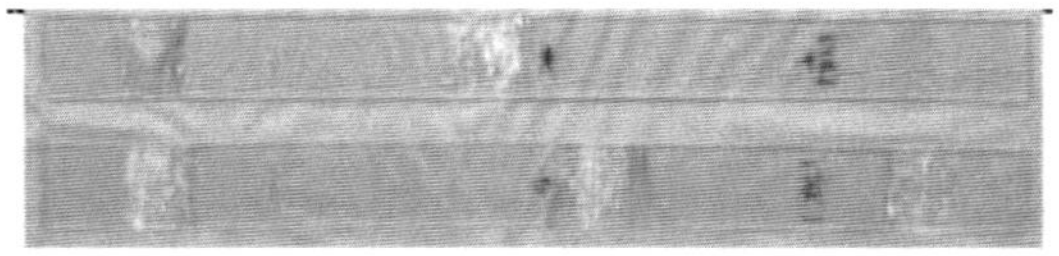

Figure 5. Cracks types and crack regions. Float/VG consisted of 6.6 mm

Figure 6. Cracks types and crack regions. HSG/VSG consisted of 8.4 mm

3.4.1. Studies based on the international experiments.

Laboratory experiments were carried out by Italian researches [5] on laminated specimens consisted of glass layers with thicknesses of 4 mm, and with use of PVB interlayer material. They applied constant width of 300 mm and the heights of the specimens were varied (400, 500 and 600 mm). The specimens were loaded with axial compression. All of the glass layers were fractured, and three regions of cracks were observed in the glass layers. The glass layers fractured at the mid of them and near to the supports. The fragmentation pattern is connected with the more rigid behaviour of these specimens compared to that of used in experiments (e.g. VG 4.4 with height of 1000 mm). In our experiments the scales of the specimens were determined based on the size of the realized glass columns in the literature review, which are commonly used in practice. The ratio height/ width (H/B) varied from 1000/80 to 840/ 80 and the ratio width/ thickness varied from 80/12 to 80/4. The Italian researchers applied in their calculations n=0.5 (buckling length factor) value to evaluate the critical buckling load, although rubber as a damping material was used in the supports. Rubber as a damping material allows deformations (e.g. rotations of the glass specimen in the supports) due to its elastic behaviour.

4. Calculation results. Slenderness and the critical buckling force

4.1. Single layer float glasses

To compare the experimental results a common effective length factor must be determined. The theoretical limits of the effective length factors are comparable in the case of single layer glasses and laminated glasses based on the experimental results [2]. Figure 7 indicates the slenderness of the tested glass columns vs. critical buckling force. The mean results of specimens are above of the effective length factor value 1, which means that the supports are rather pinned. Above of the curves are numbers which show the thickness of the specimens.

4.2. Laminated float glasses

In the case of laminated glasses significant difference can be observed in the buckling behaviour. The laboratory experimental results of laminated glasses consisted of two glass layers, - thickness of 4.4; 6.6; 8.8; 10.10 mm - indicated a more fixed supporting state. Although it means the supports are stiffer, the test set-up did not vary. The changes are justified by the effects of the interlayer material. When calculating the border curves of the critical buckling load, the effect of the interlayer material cannot taken into account. In the laboratory experiments short term loading was applied (loading rate of 0.5 mm/min). In the future it is suggested to study the stress sharing effects in the case of long term loading (Figure 8).

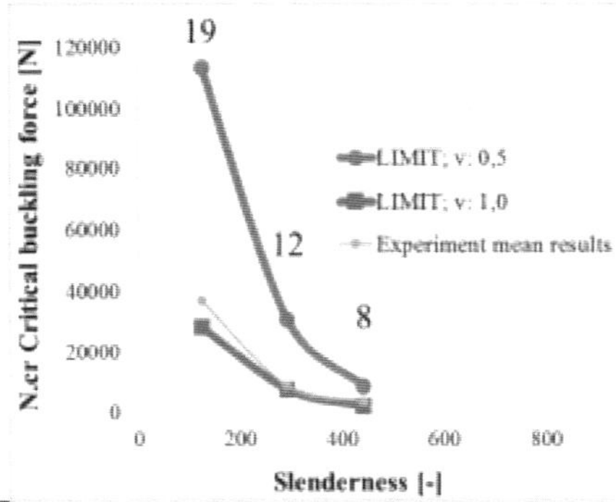

Figure 7. Slenderness vs. Critical buckling force in case of single layer glasses

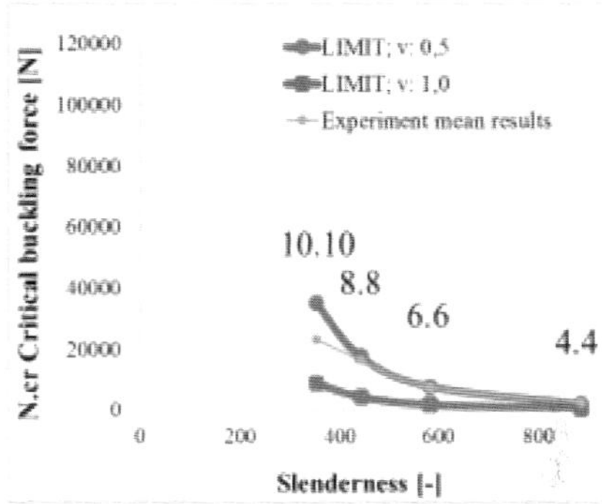

Figure 8. Slenderness vs. Critical buckling force in case of laminated glasses

5. Conclusion

Three different stages can be distinguished in the buckling behaviour of glass columns. However, the 2nd Stage may be missing depending on the initial shape and stiffness of the supporting system. Dense crack pattern can be observed in the region of the inflections. The width of fractured region increases with the increase of thickness of the specimen. The fracture pattern of HSG glasses is denser compared to that of float glasses and contains vertical cracks, as well. No significant difference can be observed between the stresses at the moment of buckling of float or HSG glasses with same thicknesses and with the height of 1000 mm, although this difference increases with the decrease of the height. The range of the effective length factor in the reality of a glass columns varies between 0.5 and 1. Although the damping material causes displacement in the supports, the real effective length factor cannot be more than 1. The supports of single layer glass are rather pinned until the results of laminated glasses are more closed to the fixed supporting condition based on the experiments and calculation.

Acknowledgement

Authors express their gratitude to Rákosy Glass Ltd. for providing the specimens. Authors are thankful to the Department of Construction Materials and Engineering Geology, BME and Mr. A Eipl (Struktúra Ltd.) and Mr. P Molnár (Struktúra Ltd.) for their technical support.

References

[1] Pankhardt K 2012 Load Bearing Glasses. *Testing of Constuction Glasses* (Saarbrücken: Lap Lambert, ISBN: 978 3 8473 2191 0

[2] Nehme K, Jakab A, Nehme S G 2013 Experiments on the Buckling Behaviour of Glass Columns Part 1. (Budapest: *Építőanyag* **65/3** http.//dx.doi.org/10.14382/epitoanyag-jsbcm.2013.13)

[3] Nehme K, Jakab A, Nehme S G 2013 Experiments on the Buckling Behaviour of Glass Columns Part 2. (Budapest: *Építőanyag* **65/4** http.//dx.doi.org/10.14382/epitoanyag-jsbcm.2013.21)

[4] Jakab A, Nehme K, Nehme S G 2014 Fracture Behaviour of Glass columns (Düsseldorf: *Glasstec Engineered Transparency*, ISBN:978 3 86780 402 8)

[5] Aiello S, Campione G, Minafo G, Scibilia N 2011 Compressive Behaviour of Laminated Structural Glass Members (Palermo: *Elsevier*, DOI:10.1016/J.ENGSTRUCT.2011.07.004)

IOP Conf. Series: Materials Science and Engineering **123** (2016) 012057 doi:10.1088/1757-899X/123/1/012057

Performance Characteristics of Waste Glass Powder Substituting Portland Cement in Mortar Mixtures

P Kara[1], L J Csetényi[2] and A Borosnyói[3]

[1]Department of Building Materials and Products, Institute of Materials and Structures, Riga Technical University, Riga, Latvia
[2]Concrete Technology Unit, School of Engineering, Physics and Mathematics, University of Dundee, Dundee, United Kingdom
[3]Department of Construction Materials and Engineering Geology, Budapest University of Technology and Economics, Hungary

E-mail: patricija.kara@rtu.lv

Abstract. In the present work, soda-lime glass cullet (flint, amber, green) and special glass cullet (soda-alkaline earth-silicate glass coming from low pressure mercury-discharge lamp cullet and incandescent light bulb borosilicate glass waste cullet) were ground into fine powders in a laboratory planetary ball mill for 30 minutes. CEM I 42.5N Portland cement was applied in mortar mixtures, substituted with waste glass powder at levels of 20% and 30%. Characterisation and testing of waste glass powders included fineness by laser diffraction particle size analysis, specific surface area by nitrogen adsorption technique, particle density by pycnometry and chemical analysis by X-ray fluorescence spectrophotometry. Compressive strength, early age shrinkage cracking and drying shrinkage tests, heat of hydration of mortars, temperature of hydration, X-ray diffraction analysis and volume stability tests were performed to observe the influence of waste glass powder substitution for Portland cement on physical and engineering properties of mortar mixtures.

1. Introduction

The concrete industry is improving its sustainability performance and as well as meeting this through cement technology is increasingly moving towards the use of recycled aggregates [1]. Waste glass is one of the least recycled materials in the majority of countries as it requires relatively large amounts of energy to be consumed to melt the cullet [2]. The quantity of treated waste glass has risen by about 25% within the last 10 years due to an improved waste glass collecting system in Latvia, notwithstanding the glass waste recycling infrastructure still suffers from the lack of an adequate network of local recycling companies [3] and alternative solutions are required to solve the problem of the stockpiled waste glass. Being amorphous and having relatively high silicon and calcium contents, glass is pozzolanic or even cementitious especially when the fineness of glass powder is much greater than that of Portland cement. Caijun Shi has stated in [4] that ground glass possesses pozzolanic reactivity, but cannot be used as a cement substitution in conventional concrete because of potential alkali-aggregate reaction. But when the particle size of glass powder is less than 75 μm, silica may dissolve relatively quickly, reacting with free portlandite ($Ca(OH)_2$) and acting as a pozzolanic material [5], giving possible control over the alkali-silica reaction in concrete, therefore making this material a possible substitute for Portland cement [6].

2. Materials and Methods

2.1. Materials

The constituent materials used in the laboratory to produce mortar mixtures comprised:

(i) *Portland cement (PC):* CEM I 42.5N with specific surface area of 388 m^2/kg (Blaine); soundness of 1.0 mm (Le Chatelier); setting time of 122/220 min (initial/final); compressive strength at 28 days of 55 MPa; mineral composition (mass - %) C_3S 51.7%, C_2S 19.3%, C_3A 5.4%, C_4AF 15.90%; density of 3150 kg/m^3; particle size distribution is shown in figure 1.

(ii) *Waste glass powders:* soda-lime glass cullet (flint (F), amber (A), green (G)) and special glass cullet (soda-alkaline earth-silicate glass coming from low pressure mercury-discharge lamp cullet (L) and incandescent light bulb borosilicate glass waste cullet (D)) were subjected to grinding using a planetary ball mill (Retsch PM 400, equipped with 4 steel jars of 500 ml capacity each), operating at 300 rev · min^{-1} for 30 minutes. Hardened steel spheres with 17 mm diameter and 430g mass per steel jar were used as grinding media. Characterisation included fineness by laser diffraction particle size analysis (see figure 1), specific surface area by nitrogen adsorption technique, density by specific gravity bottle/pycnometer and chemical composition by X-ray fluorescence spectrophotometric analysis (shown in the table 1).

(iii) *Water* used within the mixtures was simple tap water, clean and free from impurities with conductivity ranging between 80 and 85 µS/cm and pH being 7.5□}1.

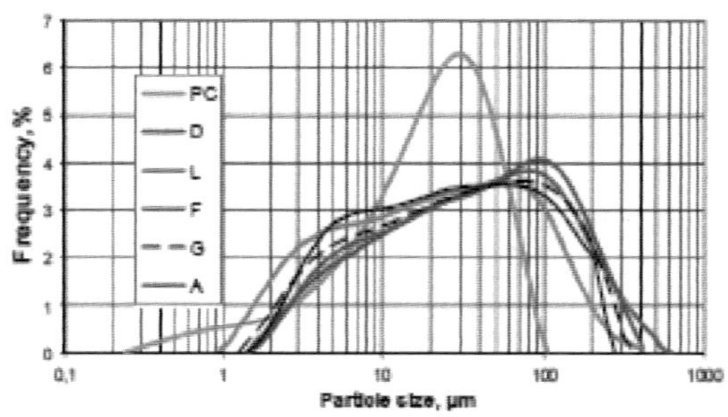

Figure 1. Particle size distribution of Portland cement and waste glass powders

Table 1. Chemical composition, fineness and density of Portland cement and waste glass powders

	Bulk oxide, w/w %											Fineness	Density,
	SiO_2	Al_2O_3	Fe_2O_3	CaO	Na_2O	K_2O	MnO	MgO	TiO_2	SO_3	P_2O_5	m^2/kg	kg/m^3
PC	17.93	5.22	2.97	62.03	0.20	1.76	0.140	4.06	0.360	3.76	0.440	388	3150
D	71.14	2.60	0.17	1.32	3.30	1.70	0.006	0.62	0.006	0.00	0.023	608	3688
L	65.52	1.22	0.11	5.11	12.35	1.88	0.010	2.95	0.027	0.14	0.038	542	2767
F	69.61	1.34	0.08	11.30	11.21	0.56	0.008	0.46	0.028	0.13	0.021	502	2741
G	67.80	1.53	0.36	10.93	11.09	0.55	0.018	1.58	0.062	0.03	0.023	463	2723
A	69.26	1.42	0.43	8.89	11.43	0.51	0.018	3.08	0.052	0.01	0.026	542	2843

2.2. Mixture proportions

The control mortar mixture and the various mortar mixtures with waste glass powders (WGP) as Portland cement substitute (at 20% and 30%) were mixed in a laboratory mortar mixer according to EN 196-1:2005. The water/cement ratio was selected to be w/c = 0.29. Where cement substitution by glass powder was applied, the water/binder ratio was changed to w/b = 0.34 (20% WGP) and w/b = 0.37 (30% WGP), however, the water/cement ratio was kept constant at w/c=0.29.

2.3. Mechanical and physical properties

Cubic (50 mm × 50 mm × 50 mm) specimens were prepared in order to determine compressive strength. Mortars were cast into the moulds; the exposed surface of the specimens was covered with polyethylene film to prevent moisture evaporation. Demoulding was carried out 24 hours after production and the specimens were cured in water at a temperature of +20±2°C for 28 days. Compressive strength tests were conducted on a Controls (model 50-C56G2) semi-automatic compression machine (3000 kN). The compressive strength of mortar specimens was determined at the ages of 7, 28, 84, 118, 270 and 660 days in accordance with LVS EN 12390-3:2009/AC:2011 and using triplicate specimens at each age. Physical properties of hardened mortar samples such as density (by the hydrostatic weighing method LVS EN 12390-7:2009) and water absorption were also included.

2.4. Early age shrinkage cracking and drying shrinkage [7]
The test method according to the Hungarian standard MSZ 523-5:1975 was used for the drying shrinkage tests. Three 40 mm × 40 mm × 160 mm prismatic specimens were prepared for each series to the drying shrinkage tests. The ring test method was performed for the early age shrinkage cracking. The substitution of 20% and 30% per mass of Portland cement seemed to be applicable in view of drying shrinkage with total deformation up to 2.5 ‰ in the period of 592 days. However, the increase in waste glass powder addition led to an increase in drying shrinkage. The WGP addition contributed to a slowdown in the rate of hydration of the cement mortar, so the early age shrinkage cracking tendency became more prominent, which could be seen in the longer cracking time result during the ring tests. In the point of view of early age shrinkage cracking tendency, the soda-lime WGPs needed from 40 to 47 hours for the appearance of the first early age crack, whereas the special WGPs required some longer time, from 44 to 53 hours, which in both cases was about 35% longer compared to the time for the reference Portland cement mortar.

2.5. Heat of hydration
Isothermal conduction calorimetric tests were carried out on mortar samples with cement and waste glass powder (as Portland cement substitute, at levels of 20% and 30%) using an I-Cal 4000 Isothermal Calorimeter in compliance with ASTM C1679, applying curing temperatures of less than 50°C and total test times of not less than 72 hours.

2.6. Temperature rise under semi-adiabatic conditions
The temperature rise under semi-adiabatic conditions inside the concrete mixtures was measured by an apparatus consisting of a plywood box with a 50 mm insulation layer of Finn foam FI – 300 (λ of 0.033 W/(mK)) and embedded four demountable 50 mm × 50 mm × 50 mm plywood moulds with thermocouples located in the center of one face of each mould and connected to a data transmitting device. Sample temperature was monitored until it reached its highest value and whilst maintaining an ambient air temperature of 20±°C outside of the box [8].

2.7. X-ray diffraction (XRD) analysis
The hardened cement mortars were used for XRD analysis. The XRD measurements were performed with a Hilton-brooks HG62 diffractometer of using CuK_{α} radiation (k = 1.540562 A). The angular range was from 3 to 60° 2-Theta with a step width of 0.1° and a measuring time of 3 s/step. For XRD quantitative phase analysis based on Rietveld refinement, the samples were mixed with 5 wt.% Al_2O_3 as internal standard (corundum, a standard material widely used in XRD analysis due to having only a few, but tall and narrow peaks). The Rietveld fitting procedure also permits the estimation of the amount of non-crystalline phases.

2.8. Volume stability
Mortar bar specimens were prepared following the procedure described in RILEM TC 106 AAR-2 "Detection of potential alkali-reactivity of aggregates – the ultra-accelerated mortar-bar test" [9]. There were 33 mortar bar specimens prepared for cement and five types of WGPs: 3 reference mortar bar specimens, 15 mortar bar specimens with cement substitution by WGPs at 20 wt. % and again 15

mortar bar specimens with cement substitution by WGPs at 30 wt. %. The mortar bar specimens were cured at 20±2°C and 95% relative humidity in the moulds, demoulded after 24±2 h when their initial length was measured. The specimens were then placed in water, transferred to an oven at temperature 80±2°C for 24 h, removed from the water and the length measured immediately before the temperature could drop substantially. Thereafter the specimens were placed in 1 M NaOH solution preheated to a temperature of 80±2°C, the containers were sealed and placed in an oven at 80±2°C (for 14 more days). Length measurements were taken only on the 7th and 14th days by means of a digital length comparator (±0.001 mm accuracy).

3. Results and discussions

The development of the compressive strength is indicated in figures 2 to 4 for the five WGPs in comparison to that of neat cement mortar. A rapid development of the compressive strength is visible between the ages of 28 to 118 days, being faster than that of the hardened neat cement mortar that is attributed to a supposed very active hydration of the WGPs during that period of time. The application f special waste glass cullet at 20% substitution resulted in higher compressive strengths than the application of soda-lime cullet and also compared to cases with 30% substitution.

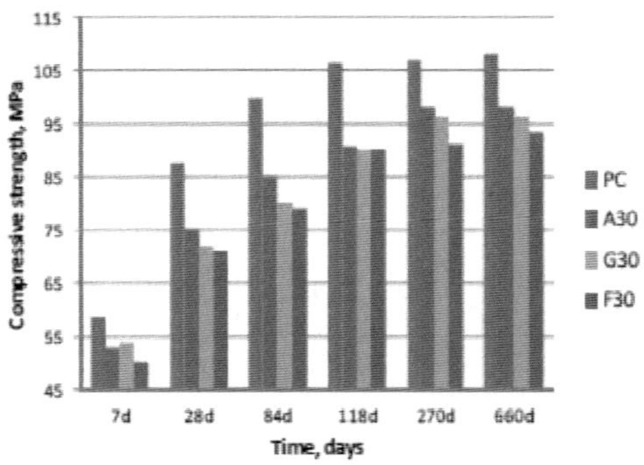

Figure 2. The evolution of compressive strength of mortar specimens, at 30% soda-lime WGPs substitution for PC

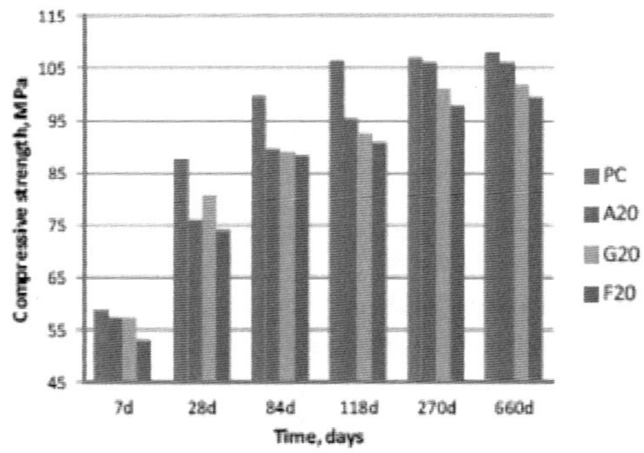

Figure 3. The evolution of compressive strength of mortar specimens, at 20% soda-lime WGPs substitution for PC

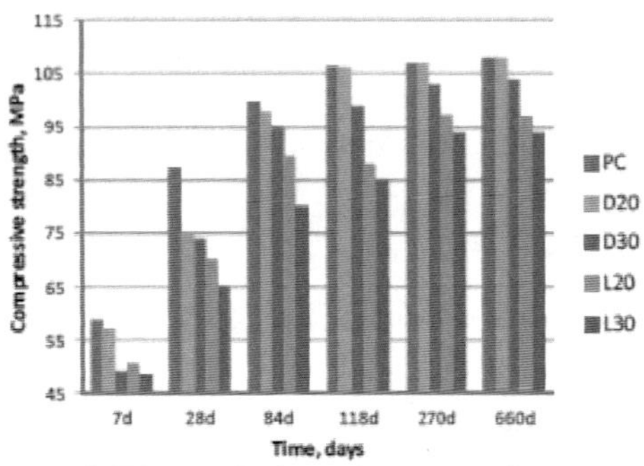

Figure 4. The evolution of compressive strength of mortar specimens, at 20% and 30% special WGPs substitution for PC

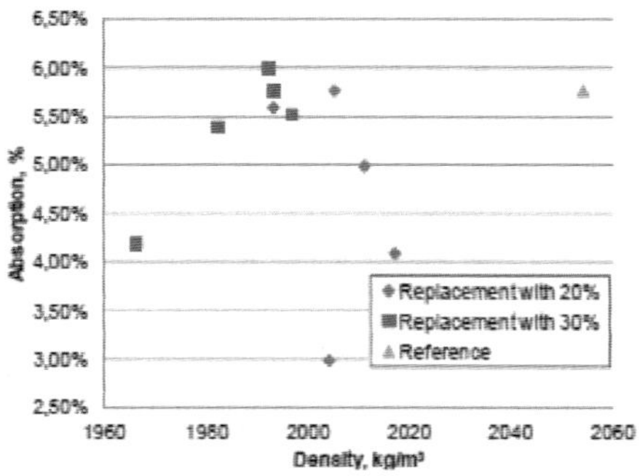

Figure 5. Absorption and density of mortar specimens, at 20% and 30% WGPs substitution for PC

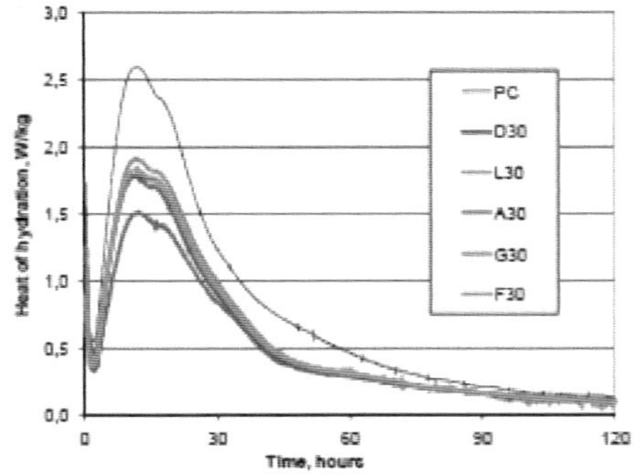

Figure 6. The evolution of the heat rate of a cement mortar (CEM I, w/b=0.29), at 30% WGPs substitution for PC

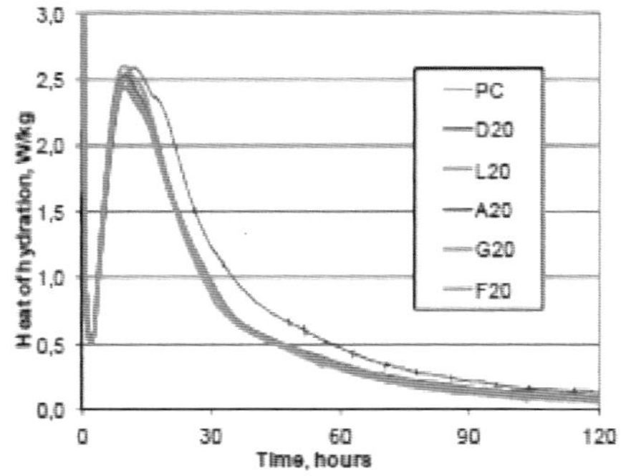

Figure 7. The evolution of the heat rate of a cement mortar (CEM I, w/b=0.29), at 20% WGPs substitution for PC

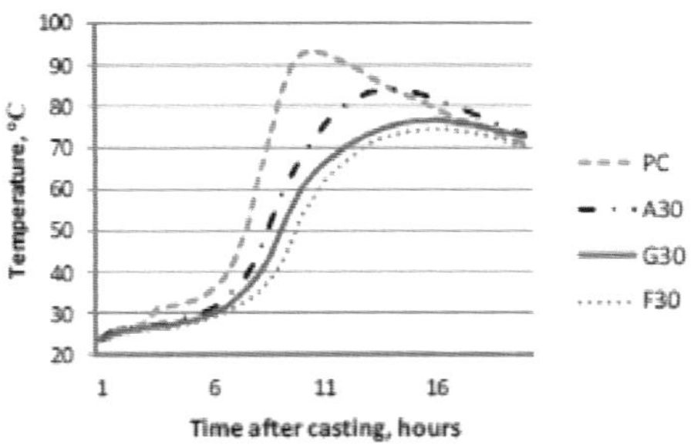

Figure 8. The temperature rise of a cement mortar (CEM I, w/b=0.29), at 30% soda-lime WGPs substitution for PC

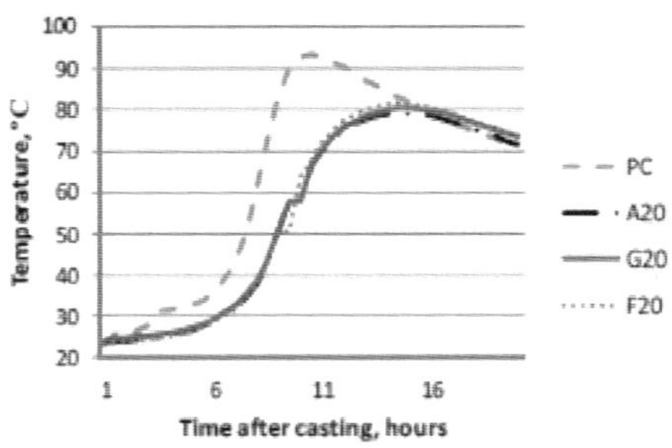

Figure 9. The temperature rise of a cement mortar (CEM I,w/b=0.29), at 20% soda-lime WGPs substitution for PC

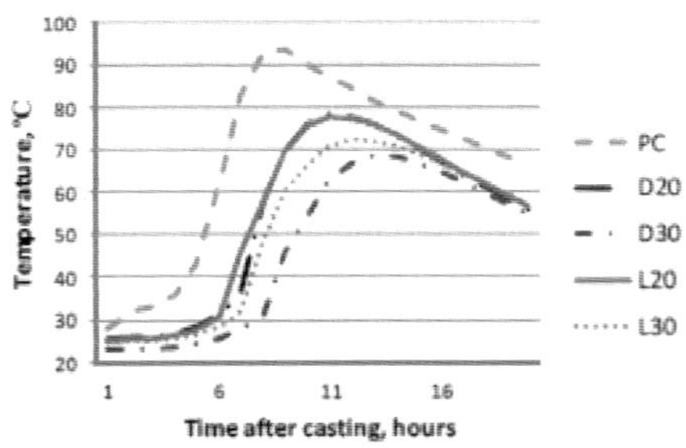

Figure 10. The temperature rise of a cement mortar (CEM I, w/b=0.29), at 20% and 30% special WGPs substitution for PC

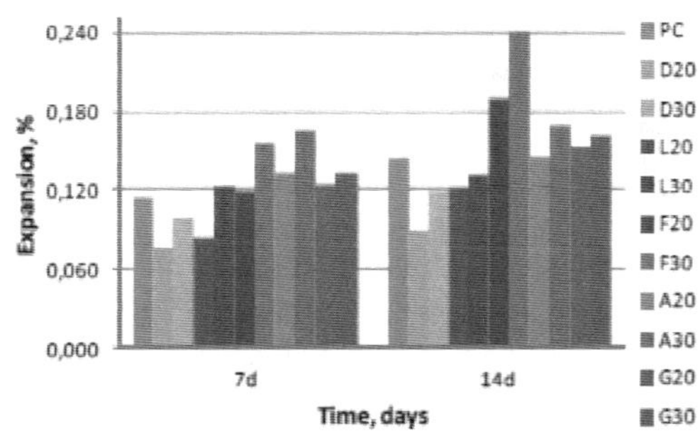

Figure 11. Volume stability of mortar specimens, at 20%and 30% WGPs substitution for PC

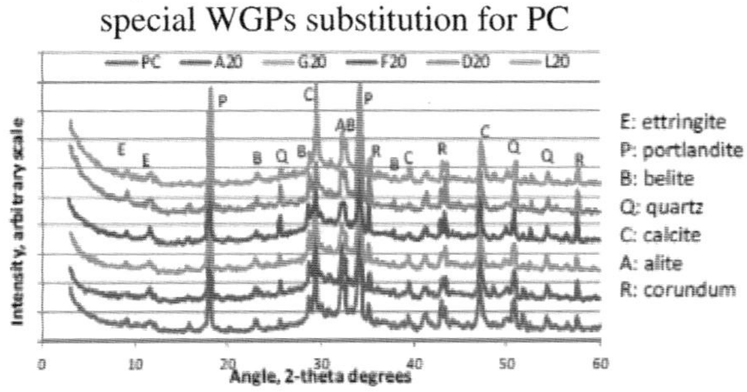

Figure 12. X-ray diffractograms of hydrated mortars, at 20% WGPs substitution for PC

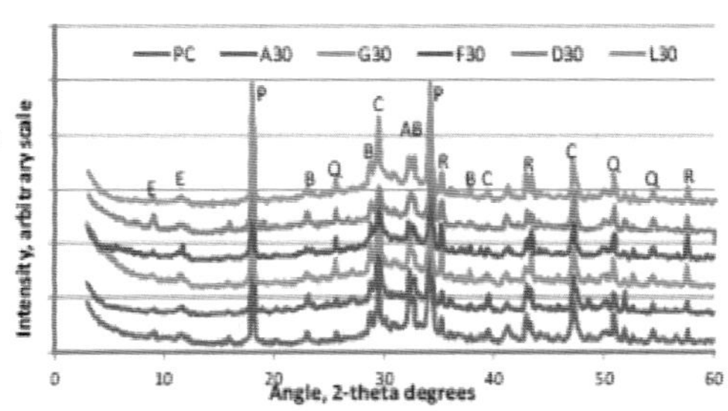

Figure 13. X-ray diffractograms of hydrated mortars, at 30% WGPs substitution for PC

The significant effect on the compressive strength development is due to the fineness and chemical composition of WGPs. The cement substitution with D20 WGP and A20 WGP gives equal compressive strength to Portland cement at the later age (see figures 3 and 4). The absorption and density of mortar specimens depending on the substitution level is shown in figure 5. It can be seen that a higher substitution level gives lower densities and increases the rate of absorption of the specimens; this may be attributed to the higher fineness of WGPs in comparison to Portland Cement (see table 1). The heat rate evolution of cement mortars represented in the two diagrams in figures 6 and 7 confirms the effect of a decreased heat rate when substituting cement mortar with waste glass at higher substitution levels, especially when using D30 WGP (see figure 6). The temperature rise curves are shown in figures 8 to 10. The peak value of the temperature within the reference mortar specimen was clearly defined and took place approximately after 10 h (93°C). In contrast, for the specimens with WGPs this was retarded up to 5 hours later with temperatures ranging from 68°C to 84°C. The ASR mortar expansion is demonstrated in figure 11. Experimental results have shown that both particle size and chemical composition influence expansion during the alkali-silica reaction; inasmuch as finer glass particles exhibit considerably lower expansion; whilst pozzolanic activity increases as fineness increases, also Na_2O_{eq} of Portland cement has strong impact [6]. Fine particles of waste glass powder also tend to perform a relatively rapid pozzolanic reaction with Portland cement on the contrary to the much slower alkali-silica reaction. The mortar expansion rate decreases with the substitution of cement by finely ground waste glass powders, in particular for D20 WGP. The X-ray diffraction (XRD) analysis (see figures 12 and 13) indicates that inclusion of WGPs does affect chemical reactions in the mixtures. Less unhydrated material (e.g. combined residual alite and belite contents) was detected in the samples made with WGPs than the level of substitution for Portland cement may have caused. Rietveld curve fitting shows unhydrated material being 11.4% in the PC sample, whereas, with removal of the dilution effect of substitution, 11.2, 10.8, 10.7, 10.2 and 10.1% in A20, D20, F20, L20 and G20, as well as 9.6, 9.3, 8.9, 8.7 and 8.3% in A30, F30, L30, G30 and D30, respectively. However, these sequences may not directly reflect any order of reactivity, as there are other important factors, such as fineness and packing, to be considered when assessing the mechanical performance of the mixtures.

4. Conclusions

Considering the performance of different waste glass types as finely ground waste glass powders which can partially substitute Portland cement, it can be concluded that this by-product can be applied in concrete technology. The study has identified important effects on mechanical and durability properties when special glass cullet (low pressure mercury-discharge lamp cullet and incandescent light bulb borosilicate glass waste cullet) was used in comparison to soda-lime cullet, for example, increased compressive strength up to 10% at later ages due to the up to 30% higher obtained fineness by the grinding process; lower mortar expansion (< 0.12%) during alkali-silica reaction; up to 20% lower heat rate and up to 10% lower peak hydration temperature and more favorable early age shrinkage cracking tendency, and indicated that finely ground waste glass powder is a valuable Portland cement substitute. The use of waste glass powder in concrete industry may moderate the problem of dumped waste glass and reduce CO_2 emissions into the atmosphere by decreasing the proportion of Portland cement in unit volume of concrete produced.

References

[1] McCarthy M J, Jones M R, Csetenyi L J, Halliday J E 2009 *Concrete* **32** August

[2] Poutos K H, Alani A M, Walden P J, Sangha C M 2008 *Construction and Building Materials* **22** 557 DOI: 10.1016/j.conbuildmat.2006.11.018

[3] Kara P 2014 *Journal of Sustainable Architecture and Civil Engineering,* **1(6)** 42 DOI: 10.5755/j01.sace.6.1.6127

[4] Shi C 2002 Composition of materials for use in cellular lightweight concrete and methods thereof *Patent US 6488762 B1*

[5] Dhir R K, Dyer T D, Tang M C 2009 *Materials and Structures*, **42**. 1451, DOI: 10.1617/s11527-008-9465-8

[6] Kara P 2014 Challenge of ASR in the use of waste glass slurry *Alkali Aggregate Reactions (AAR) in Concrete: Workshop Proceeding No.11 from a Nordic-Baltic Miniseminar, The Nordic Concrete Federation*, Latvia, Riga, 21-22 November 2013 (Oslo: Helli-Visuell kommunikasjon, ISBN:978-82-8208-039-2), 77

[7] Kara P, Borosnyoi A, Fenyvesi O 2014 *Építőanyag - Journal of Silicate Based and Composite Materials,* **66 (1)** 18 DOI: 10.14382/epitoanyag-jsbcm.2014.4

[8] Jones M R, McCarthy A 2006 *Cement and Concrete Research*, **36(6)**, 1032

[9] RILEM 2000 TC 106-AAR: alkali aggregate reaction, *Materials and Structures*, **33(5)** 283 DOI: 10.1007/BF02479697

Influence of raw materials composition on firing shrinkage, porosity, heat conductivity and microstructure of ceramic tiles

E Kurovics[1], **A Y Buzimov**[2] **and L A Gömze**[1]

[1]Intsitute of Ceramics and Polymer Engineering, University of Miskolc, Hungary
[2]National research Tomsk State University, Tomsk, Russia

E-mail: mse205@gmail.com and buzimov92@gmail.com

Abstract. In this work some new raw material compositions from alumina, conventional brick-clays and sawdust were mixed, compacted and heat treated by the authors. Depending on raw material compositions and firing temperatures the specimens were examined on shrinkage, water absorption, heat conductivity and microstructures. The real raised experiments have shown the important role of firing temperature and raw material composition on color, heat conductivity and microstructure of the final product.

1. Introduction

In the last years in the convectional ceramic industry several works have shown the important roles of bio-original pore-forming additives both in technological processes and quality of final products [1-6]. The microstructure of the convectional brick-clays was examined and detailed described in the above works under paining the importance of the gaps between the solid particles of the used materials. To improved the thermal insulation of convectional brick-products waste materials were used by [7]. To predict the thermal performance of lightened fired clay-bricks organic additive materials were used by [8, 9]. Using sawdust and other bio-origin waste materials the relations between the microstructures and drying properties of ceramic-bricks and roof tiles were examined in [10, 11]. Their experiments have shown that the drying temperature i.e. intensity of drying influence essentially on dried and fired bricks and ceramic roof tiles.

The aims of this work are the development of porous ceramic tiles with increased thermal insulating properties using mixtures of convectional Hungarian brick-clays, alumina powders and sawdust.

2. Materials and experiments

For the laboratory experiments the basic materials was convectional grey brick-clay from Leier Mátraderecske in Hungary. The other ceramic raw material was MARTOXID KMS-94 alumina powder from Martinswerk in 0, 20, 40, 60, 80, 100 m%. To alumina-clay mixtures sawdust pore-forming additives of fraction 0-2 mm were added in 0, 2, 4, 6, 8 m%. Before the experiments the ceramic raw materials were dried at 110 °C during 24 hours. After the drying the powders of brick-clays and alumina were measured in proportion as it is sown in Table 1 and mixed. Thereafter0, 3, 6, 9 and 12 grams of sawdust were added and mixed again. From the so prepared batches of 150, 153, 156, 159 and 162 grams powder mixes ceramic disc of 20 mm diameters were pressed at 220 MPa uniaxial compression equipment and method. Before firing each discs have 5 grams weights.

Table 1. The raw material mixtures

<table>
<tr><th colspan="3">Quantity [g]</th></tr>
<tr><th>Clay</th><th>Al_2O_3</th><th>Sawdust</th></tr>
<tr><td>150</td><td>0</td><td rowspan="6">0
3
6
9
12</td></tr>
<tr><td>120</td><td>30</td></tr>
<tr><td>90</td><td>60</td></tr>
<tr><td>60</td><td>90</td></tr>
<tr><td>30</td><td>120</td></tr>
<tr><td>0</td><td>150</td></tr>
</table>

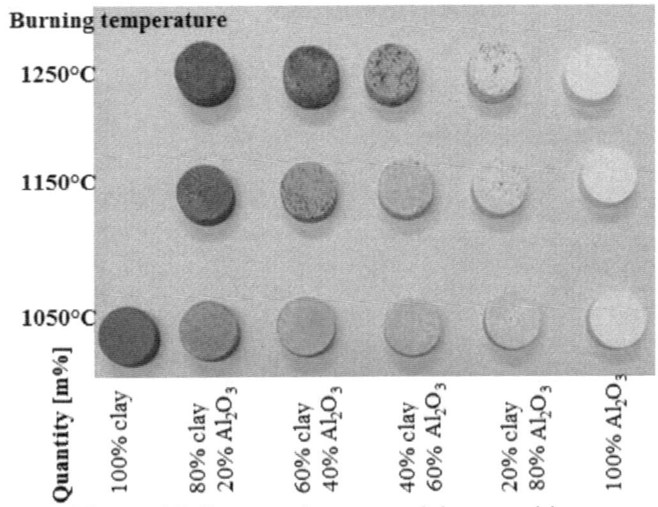

Figure 1. Influence of raw material composition and firing temperature on color of specimens

After compacting the ceramic disc specimens were divided to 3 groups by 10 pieces in each and heat treated (fired) at temperature of 1050 °C, 1150 °C and 1250 °C in laboratory chamber kiln. In Figure 1 are shown how the raw material composition and firing temperature are influenced on color of prepared for testing specimens.

The diameters and weights of each specimen were measured before and after firing. The water absorption and firing shrinkage were also measured and determined on each specimen.

3. Results and Discussion

The firing shrinkages of specimens as function of alumina proportion and sawdust containments at different firing temperatures are shown in Figure 2, Figure 3 and Figure 4 and the water absorptions are shown in Figures 5-7. The water absorption of fired at different temperature specimens determined by Archimedes method.

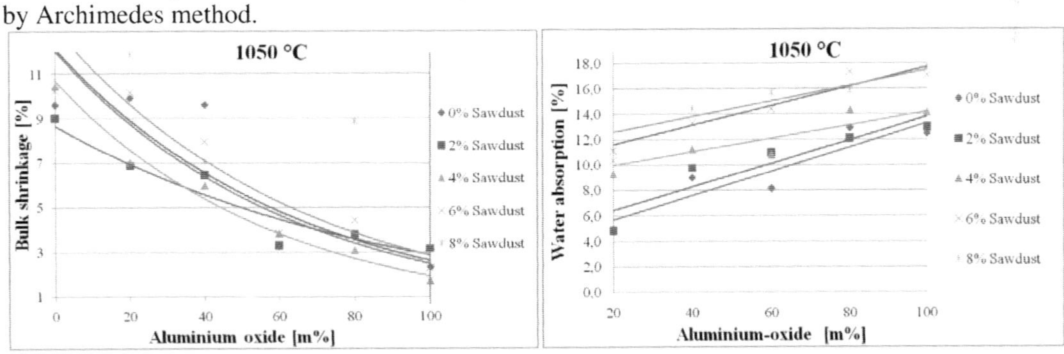

Figure 2. The bulk shrinkage and the water absorption of the test specimens burnt at 1050 °C

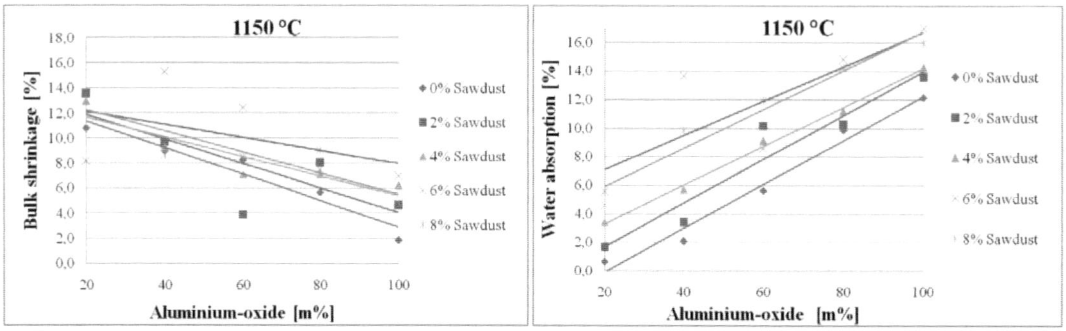

Figure 3. The bulk shrinkage and the water absorption of the test specimens burnt at 1150 °C

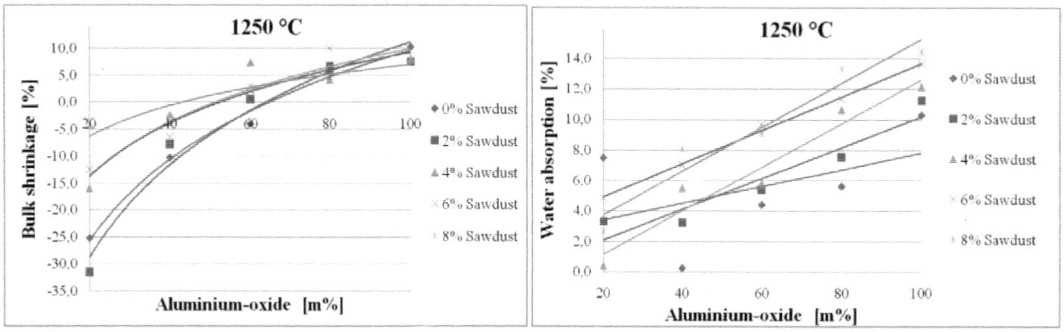

Figure 4. The bulk shrinkage and the water absorption of the test specimens burnt at 1250 °C

Generally the fired ceramics [12-14] convection brick clays have porous microstructures. These microstructures very strong depend on quality and quantity of the used pore forming additives light sawdust, rice husks, sunflower husks and others. From the works [1] and [15] it is well known that the thermal conductivity of fired ceramic bricks strong depends on the quantity and geometrical sizes of mixed in pore forming additives. The thermal conductivities of fired specimens were measured with instrument of C-THERM TCi and are shown in Figures 4.

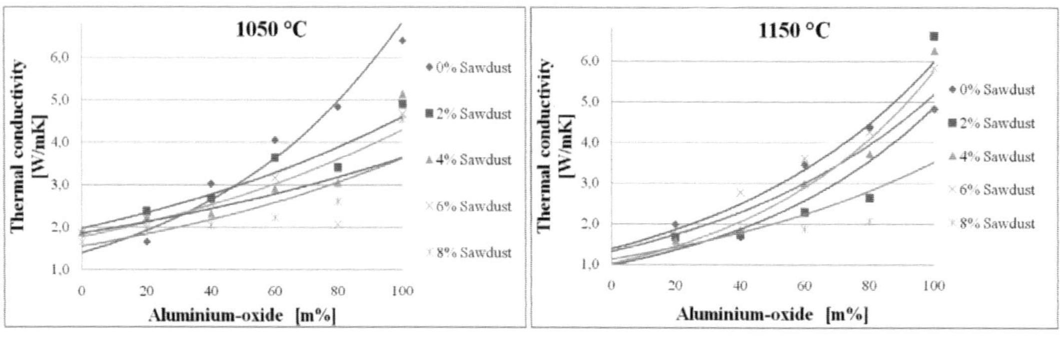

Figure 4. The heat conductivity of the test specimens burnt at 1050 °C and 1150 °C

4. Conclusions

The raw material composition and firing temperature are influencing very strong on microstructure, firing shrinkage, water absorption and heat conductivity of fired ceramic tiles. The realized experiments have shown that the water absorption, therefore the porosity are very strong depended on the volume of alumina in the raw materials mixtures. The thermal conductivities of prepared ceramic someplace have increased considerable with increasing the containment of alumina in the raw materials mixtures.

References

[1] Bánhidi V, Gömze A L 2008 *Materials Science Forum* **589** 1

[2] Kristály F, Gömze A L 2008 *Építőanyag-JSBCM* **60** 34 http://dx.doi.org/10.14382/epitoanyag-jsbcm.2008.7

[3] Bories C, Borredon M E, Vedrenne E 2014 *Journal of Nvironmentalmanagement* **143** 186

[4] Bories C, Aouba L, Vedrenne E, Vilarem G 2015 *Construction and Building Materials* **91** 158

[5] Rubia-García MD, Yerba-Rodríguez Á, Eliche-Quesada D, Corpas-Inglesias F A, López-Galindo A 2012 *Construction and Building Materials* **36** 495

[6] Géber R, Kocserha I, Orosz V, Simon A, Paróczai C 2010 *Materials Science Forum* **659** 477

[7] Kadir A A, Sarani N A, Leman AM 2014 *Materials Science Forum* **803** 330

[8] Aouba L, Coutand M, Perrin B, Lemercier H 2015 *Construction and Building Materials* **91** 59

[9] Sarani N A, Kadir A A 2014 *Applied Mechanics and Materials* **465** 872

[10] Gömze A L, Gömze N L 2008 *Építőanyag-JSBCM* **60** 102 http://dx.doi.org/10.14382/epitoanyag-jsbcm.2008.16

[11] Kadir A A, Hinta H, Sarani N A 2015 *ARPN Journal of Engineering and Applied Science* **10** (15) 6289

[12] Kalatur E S, Buyakova S P, Kulkov S N, Gotman I, Kocserha I 2014 *Építőanyag-JSBCM* **66** 31 http://dx.doi.org/10.14382/epitoanyag-jsbcm.2014.6

[13] Kulkov S N, Dedova E S, Pedraza F, Erdélyi J 2014 *Építőanyag-JSBCM* **66** 35 http://dx.doi.org/10.14382/epitoanyag-jsbcm.2014.7

[14] Shavchenko N L, Sablina T Y, Sevostyanova I N, Buyakova S P, Kulkov S N 2015 *Russian Physics Journal* **58** (11) 1

[15] Sarani N A, Kadir A A 2013 *Advanced Materials Research* **690-693** 919

IOP Conf. Series: Materials Science and Engineering **123** (2016) 012059 doi:10.1088/1757-899X/123/1/012059

Cross-laminated timber made of Hungarian raw materials

G Marko, L Bejo, P Takats

University of West-Hungary, Simony Karoly Faculty of Engineering, Wood Sciences and Applied Arts

E-mail: laszlo.bejo@skk.nyme.hu

Abstract. Cross-laminated timber (CLT), generally made out of softwood, enjoys increasing popularity throughout Europe. This material offers a versatile, eco-friendly technology to create strong, lightweight and energy-efficient buildings. Unfortunately, the sites and climatic conditions in Hungary are not suitable for growing high-quality coniferous trees. Transporting raw materials from other countries (sometimes thousands of kilometres away) negates the environmental advantages of wood-based construction. Local options are definitely preferable from an ecological aspect.

Poplar wood (populus spp.) is of great economic importance in Hungary. There are several relatively high density, high strength varieties growing in large quantities in Hungary, that may be used as alternatives to softwood, with comparable properties. There is an increasing interest in using poplar as a construction material, especially in regions were there is a shortage of traditional construction timber.

This paper presents the results of a preliminary investigation to create CLT using poplar lumber. Laboratory-scale CLT specimens were created in a hot press, and tested for their loadbearing capacity. The MOR values of poplar CLT are comparable to, albeit somewhat lowerthan those of softwood CLT. Further investigations are required to establish the economic viability and technological conditions for the commercial production of poplar CLT.

1. Introduction

Lightweight residential construction is gaining in popularity all over Europe, especially due to the increasingly stringent regulations that is pushing for near-zero emission buildings. Lightframe houses are especially suitable for this purpose, due to their superior thermal insulation, and low embodied energy. In the meantime, they have been criticized because of their low heat storage capacity [1], and also regarding the potentially unhealthy living environment that may arise depending on the type of ventilation system used [2]. Solid wood wall structures (e.g. log houses) present a healthier alternative, with improved heat storage, but also pose certain technical challenges.

A new type of wood-based construction material, Cross-Laminated Timber (CLT) was first proposed in the mid-1990s in Austria. It was developed at the Graz University of Technology, Austria [3]. The basic idea behind the new material is creating a strong, dimensionally stable, solid wood material, similar to plywood in structure (i.e. composed of layers of alternating, perpendicular grain direction), but made out of lumber, rather than wood veneer. The resulting panels (typically composed of 3-9 layers, but may be as many as 17 layers) are strong, dimensionally stable, versatile, easily tailored to create any kind of wall, ceiling, or roof surfaces. They have a relatively large mass to facilitate heat storage, provide a healthy living environment if left exposed (due to wood's excellent humidity control capacity), and pleasant in appearance.

There are several facilities that produce CLT in Central-Europe. The raw material is typically coniferous lumber, with a strength grade of C16 or higher (according to EN 14081-1, [4]). Recently it also become increasingly popular in the US [5], and in Canada [6]. However, in regions without significant softwood forests (e.g. in Hungary), the production, and also the utilisation of CLT is limited or nonexistent.

Even though CLT was first developed in Europe, there is currently no European standard governing the production and requirements of CLT. A draft standard was published in 2011, but is not in effect to date. In North America, APA – the Engineered Wood Association published a standard for performance-rated CLT panels [8], which contains reference values for the required performance of CLT.

Poplar wood (Populus spp.) has relatively low density. It is typically used for manufacturing wood veneer, and various simple wood products like matches, crates, pellets, etc. Poplar's mechanical characteristics are similar to those of some coniferous species, and are, in fact, classified together with softwoods according to the European Standards [4]. There are some successful examples of using poplar in structural products like LVL [9], and even in glulam [10]. However, it has not been considered for manufacturing CLT until recently, when our study, and a parallel study by Kramer [3] both focused on this potential raw material.

The purpose of the research project described in this study was to manufacture Cross-Laminated Timber made of poplar wood, and assess its properties to compare it to softwood CLT, and to existing industrial standards.

2. Materials and methods

The raw material chosen for the experiments was a hybrid poplar (Populus × euramericana cv. 'I-214'). This poplar hybrid is very popular in Hungary, due to it relatively fast growth rate and uniform structure. The resin was a structural grade polyurethane adhesive (Jowapur 686.60).

Several 3 m long logs were obtained, and cut up in a sawmill into lumber of varying width, 45 mm in thickness. The logs were green at the time of sawmilling (63% MC). After sawmilling, the lumber was cross-cut into lengths of 2.03 m and 0.97 m, and edged using a bandsaw. After this, the lumber was kiln-dried using a gentle drying schedule for two weeks, until it reached a moisture content of 12±2 %. After drying, the lumber was left to cool and equalize for one week in an indoor climate. The dried lumber was planed down to 33 mm in thickness, and various widths.

Before creating the layup for the CLT panel, the lamellae were stress-graded nondestructively, according to the standard EN 338-2003 [11]. The measurement is based on measuring the density and the longitudinal vibration frequency of the lamellae. The measurement has been carried out by the Portable Lumber Grader equipment that has been certified for grading lumber according to the European Norm [12]. All of the lamellae were graded. Low grade long lamellae were cut up into shorter pieces to be used in the core, while higher grade material was used in the face layer. The moisture content of the lamellae was also verified using a resistance moisture meter.

After preparing the lamellae, one 3-layer CLT panel (2m×1m×0.1m) was created according to the following procedure:

- Laying out two face layers and one core layer using the lamellae. The lamellae were laid out in such a manner that any curvature was facing outwards from the centre (see figure 1).
- Clamping each layer laterally, so as to eliminate any gaps between the lamellae.
- Adhesive application by pouring the amount of resin specified in the technical guidelines of the adhesive onto the layup, and distributing it using serrated spatulas.
- Placing the layers on top of one another, and transferring the final layup into the press (while maintaining lateral clamping, see figure 2)
- Closing the press, applying 0.4 N/mm^2 pressure, for 12 hrs (no heat application.)
- Opening the press and removing the CLT panel.

After removing, the panel was cut up along its length into 3 pieces of 300 mm wide bending specimens, and tested on a large scale structural testing machine in 4-point bending, according to EN

408:2010 [13], using a video-extensometer to measure the centerpoint deflection (figure 3). Load and deflection data were recorded using a data acquisition frequency of 10/s. The beams were tested to failure, and the failure mode observed.

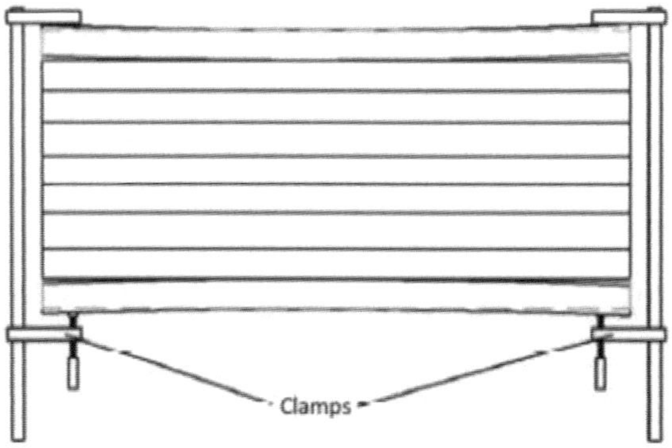

Figure 1. Laying up and clamping the lamellae so that they curve outward

Figure 2. The final layup placed in the press while maintaining lateral pressure

Figure 3. Testing the CLT specimen to failure using a 4-point bending setup. Deflection is measured using a video-extensometer

IOP Conf. Series: Materials Science and Engineering **123** (2016) 012059 doi:10.1088/1757-899X/123/1/012059

3. Results and discussion

Figure 4 shows the lamella strength classes determined by nondestructive testing. Generally, the strength classes were higher than expected. This allowed all of the face layers to contain lamellea of strength classes C18 or above. This is close to fulfilling the technical requirements set forth by the German Institute of Building Technology [14], which stipulates that at least 90% and 30% of the individual boards running parallel and perpendicular to the longitudinal direction of the element, respectively, shall correspond to strength class C24 or higher, and the rest should be at least C16, according to EN 338.

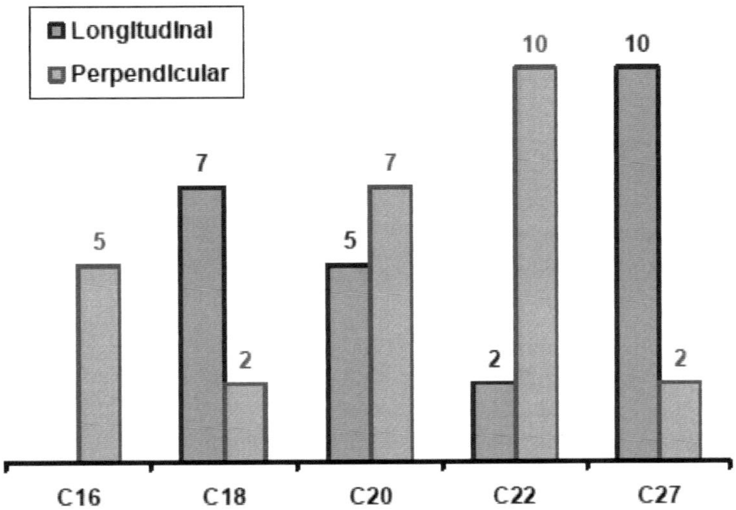

Figure 4. Non-destructively measured Strength classes of the face and core layer (longitudinal and perpendicular) poplar lamellae

Figure 5 displays the failure modes of the three tested specimens. The failure was typically a combination of bending and rolling shear, the latter usually occurring close to the glueline. Glueline failure occurring in specimen nr. 1 is probably due to the imperfection of the manual adhesive application process, which could be remedied by using industrial glue applicators. The combined failure mode is similar to that found in a parallel study [3].

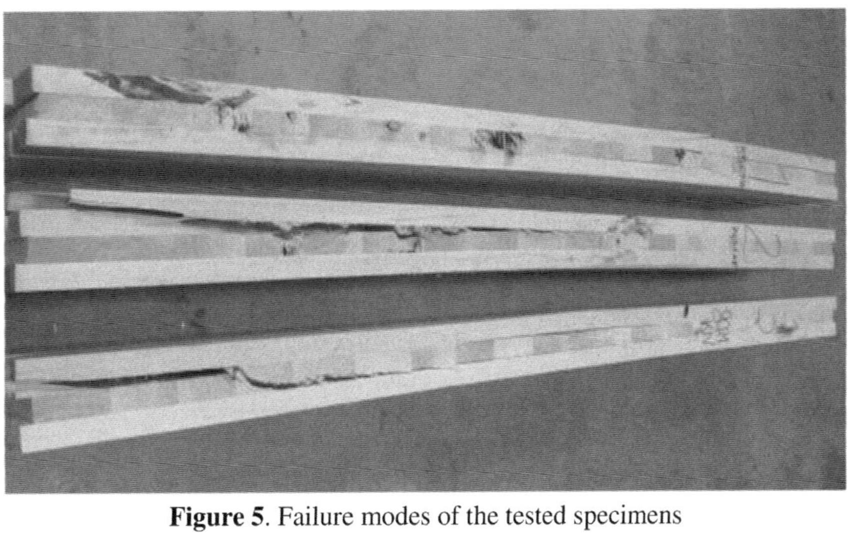

Figure 5. Failure modes of the tested specimens

Table 1. Measured MOR and MOE values, and design values based on ANSI/APA PRG 320 [8]

	MOR (N/mm^2)	MOE (N/mm^2)
Specimen #1	34.0	7,552
Specimen #2	46.9	8,354
Specimen #3	48.8	7,787
Average	**43.2**	**7,898**
Design values [8]	*12.7 ... 28.2*	*8,270 ...11,720*

Table 1 contains the measured MOR and MOE values of the three specimens. The table also shows the design values belonging to various CLT grades, according to the American standard APA PRG 320. Our study was not extensive enough to calculate design values for poplar CLT, but if measured values are indicative of the typical MOR values to be expected of this product, poplar LVL can probably fulfill the requirements of at least the lower grade CLT. On the other hand, MOE values were markedly lower than those specified in the American standard. These findings are consistent with a parallel, larger scale study undertaken at the Oregon State University [3], which also found poplar LVL sufficiently strong, but somewhat falling behind requirements in terms of MOE.

4. Summary and conclusions

A small-scale study of the viability of poplar wood to produce Cross-Laminated Timber elements led to the following results and conclusions:

- Hybrid poplar, as a raw material, can yield lumber of sufficiently high strength grade to produce commercially viable CLT.
- The CLT specimens manufactured and tested in this study yielded relatively high bending strength values, but low MOE. These findings are consistent with a parallel study by Kramer [3].
- Despite the relatively low MOE values, poplar may be a viable raw material for CLT, either by selecting and using high-grade lumber only (through non-destructive testing), or using poplar as a mix-in species along with softwood lumber.

The above results show that poplar lumber, while not automatically suitable for this purpose has the potential to be used in CLT manufacturing, either by choosing premium grade material, or as a mix-in species, in conjunction with softwood material.

Acknowledgements

This study was supported by the Environment conscious energy efficient building TAMOP.4.2.2.A-11/1/KONV-2012-0068 project sponsored by the EU and European Social Foundation.

References

[1] Hacker, J.N., T.P. De Saulles, A.J. Minson, M.J.Homes. 2008. *Energy and Buildings* **40**:375.

[2] Hens, H. 2012. Passive Houses: *ASHRAE Transactions* **118**(1):1077

[3] Kramer, A. 2014. Cross-Laminated Timber Engineering: Improvement and Application. MS Thesis, Oregon State University, OR, USA. 74 pp.

[4] EN 14081-1:2005 Timeber structures – Strength graded structural timber with rectangular crosssection – Part 1: General requirements. Euroean Committee for Standardisation, Brussels, Belgium.

[5] Gagnon, S. and C. Pirvu (eds). 2011. *CLT Handbook*: Cross-Laminated Timber. FP Innovations Special Publication SP-528E. Pointe-Claire, QC, Canada.

[6] Karacabeyli, E. and B. Douglas (eds). 2013. CLT Handbook – US ed. FP Innovations Special Publication SP-529E. Pointe-Claire, QC, Canada.
[7] prEN 16351:2011 Timber structures – Cross laminated timber – Requirements. Euroean Committee for Standardisation, Brussels, Belgium.
[8] ANSI/APA PRG 320-2012. Standard for Performance-Rated Cross-Laminated Timber. APA – The Engineered Wood Association, Tacoma, WA, USA. 29 pp.
[9] Molnar, S. and M. Bariska. 2002. Wood species of Hungary. Szaktudás Publishing House, Budapest, Hungary. 210 pp.
[10] Schlosser, M., N. Horvath, L. Bejo. 2012. Glulam beams made of Hungarian raw materials. In: R. Nemeth, A. Teischinger (eds.) Proc. 5th Conference on Hardwood Research and Utilisation in Europe. Sopron, Hungary, 2011.09.10-11. pp. 383-392.
[11] EN 338:2009. Structural timber. Strength classes. Euroean Committee for Standardisation, Brussels, Belgium.
[12] FAKOPP Enterprise. 2012. PORTABLE LUMBER GRADER. Software and Hardware Guide, Version 2.0. <www.fakopp.com/site/download/PLG_Guide.pdf> Accessed: 25.11.2014
[13] EN 408:2010. Timber Structures. Structural timber and glued laminated timber. Determination of some physical and mechanical properties. Euroean Committee for Standardisation, Brussels, Belgium.
[14] Deutsches Institut für Bautechnik 2011. European technical approval ETA-06/0009

IOP Conf. Series: Materials Science and Engineering **123** (2016) 012060 doi:10.1088/1757-899X/123/1/012060

Experimental investigations and numerical simulations of notch effect in cellular plastic materials

L Marsavina[1], E Linul[1], T Voiconi[1] and R Negru[1]

[1] Department of Mechanics and Strength of Materials, POLITEHNICA University Timişoara, Blvd. Mihai Viteazu, No. 1, 300222 Timişoara, Romania

E-mail: liviu.marsavina@upt.ro

Abstract. Cellular plastics are light weight structures with many applications in civil, aeronautical, automotive and mechanical engineering. Properties of cellular materials depend on the properties of the solid material, on the shape and dimensions of the cellular structure and on the relative density of the cellular material. Most of cellular plastic materials are crushing in compression and have a brittle behavior in tension. The effect of notches represents an important issue in such materials, taking into account that for packing applications for example, notches/holes should be introduced in the cellular material. This paper investigates the effect of notches in compression for three different densities 100, 145 and 300 kg/m^3 polyurethane (PUR) foams. Experimental investigations were performed on rectangular blocks of 100x100x25 mm with 16, 28 and 40 mm central holes. The mechanism of damage was monitored with an IR camera FLIR A40M. Purpose of the numerical simulations was to calibrate a material model, based on compression test for un-notched specimens using the CRUSHABLE FOAM models implemented in ABAQUS SIMULIA. Then the material models were used to simulate the experimental tests on notched blocks. Good agreement was obtained for the load - displacement curves obtained experimentally and from simulation. Also the plastic deformation patterns observed experimentally by IR thermograpghy were obtained numerically using the CRUSHABLE FOAM material model.

1. Introduction

Cellular plastics are light weight structures with many applications in civil, aeronautical, automotive and mechanical engineering. Properties of cellular materials depend on the properties of the solid material, on the shape and dimensions of the cellular structure and on the relative density of the cellular material, [1, 2]. Most of cellular plastic materials are crushing in compression and have a brittle behavior in tension, [3]. Components made of cellular or porous materials may have micro-structural defects like cracks, filled cells or missing walls holes induced by manufacturing process, [4]. On the other hand macro-structural notches or holes may be introduced in design of components. The influence of notches on the strength of the structures is an important issue, which should be considered on the design stage. Up to now there are only few studies investigating notch effect on foam and porous materials mainly regarding metallic foams [5-7] and polymeric foams [5, 8, 9]. The notch effect was experimentally investigated [5, 8, 9], but also computational studies [6, 7, 10] were performed.

Two approaches were considered for the characterization of notch effect in compression of PUR foams experimental investigations and finite element analysis. The experimental investigations in compression were performed on a three different densities (100, 145 and 300 kg/m3), closed cell rigid

PUR foam. IR thermography was used to identify the plastic deformations. A numerical assessment of CRUSHBLE FOAM model was carried out in order to represent the behavior of notched specimens.

2. Experimental investigations

Polyurethane (PUR) materials of three different densities (100, 145 and 300 kg/m^3) manufactured by Necumer GmbH, Germany under commercial designation Necuron (100, 160 and 301) were investigated. Microscopic investigations of these materials show a closed cell structure [11].

Square specimens (W = 80 mm) having a thickness *b* of 25 mm with central hole of different diameters (D = 16, 28 and 40 mm), were used, figure 1. One face of the specimens was sprayed with matt black paint in order to have a constant emissivity for thermographic measurements. The experimental tests were performed using a universal testing machine LBG 100 kN on displacement control (v=2 mm/min) and at room temperature, figure 1.

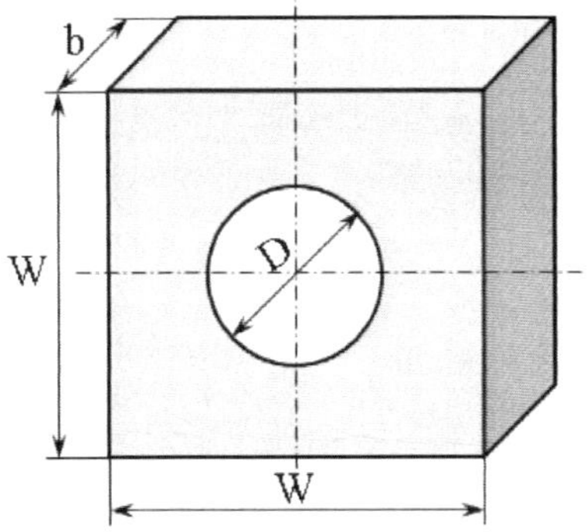

Figure 1. Specimen with holes

Figure 2. Experimental set-up

Typical load - displacement curves for the three foam densities are shown in figure 3.a for 16 mm hole diameter. An increase of supported load with increase of density was observed. A drop of load occurs after the plateau stress is reached, at this point the ultimate tensile stress is reached on the hole upper and bottom edges, where tensile occurs and a crack initiates. The load carrying capacity of holed foams decreases with increasing the hole diameter, figure 3.b plotting results for foam of 100 kg/m^3 density.

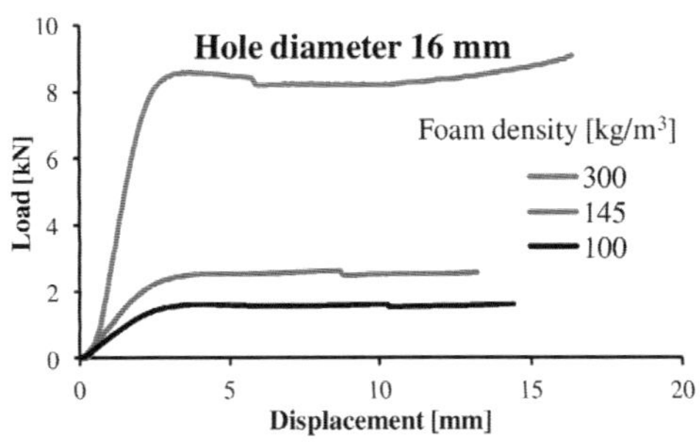

a) influence of density

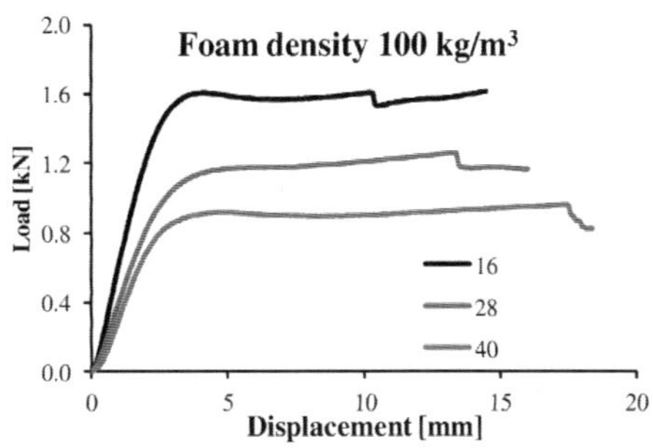

b) influence of hole diameter

Figure 3. Load-displacement curves

Figure 4 presents the ratio between the maximum net stress of notched specimen σ_{max} and the ultimate tensile strength of the foam σ_{UTS} versus ratio between hole diameter D and specimen width W. For all three investigated foams a notch insensitive response in compression was observed, which could be explained by the ability of foams to crush at a constant plateau stress σ_{pl}.

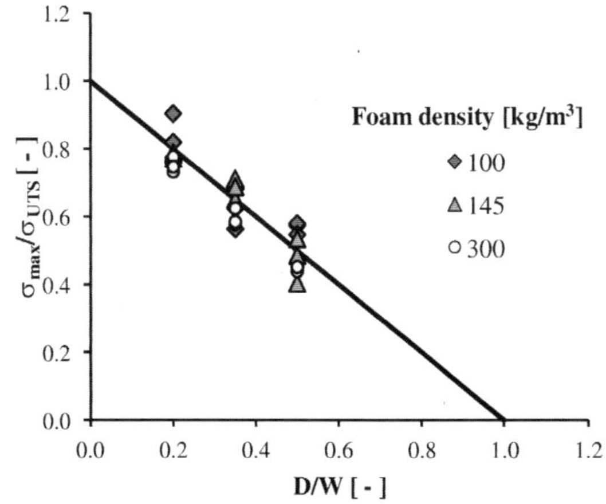

Figure 4. The effect of hole diameter on the compressive strength of PUR foams blocks with central holes

Thermography was used in order to identify the damage mechanism. A FLIR A40M infrared camera was used to measure emitted infrared radiation from the specimen which increases with plastic deformations occurred in the foam specimens due to loading. For example in figure 6 are presented different stages of temperature distribution, corresponding to different load stages (displacements 0, 4.2, 8.5, 10.5 and 11.7 mm), from the compression test of foam density of 145 kg/m^3, with a central hole of 16 mm. After a short linear elastic part, the temperature starts to increase in the vicinity of the hole due to plastic deformations (figure 5. b, c), than a crack initiates and propagates from the top and bottom surfaces of the hole (figure 5. d, e). The increase of temperature could be also seen plotting the temperature variations together with load-displacement curve, figure 6.

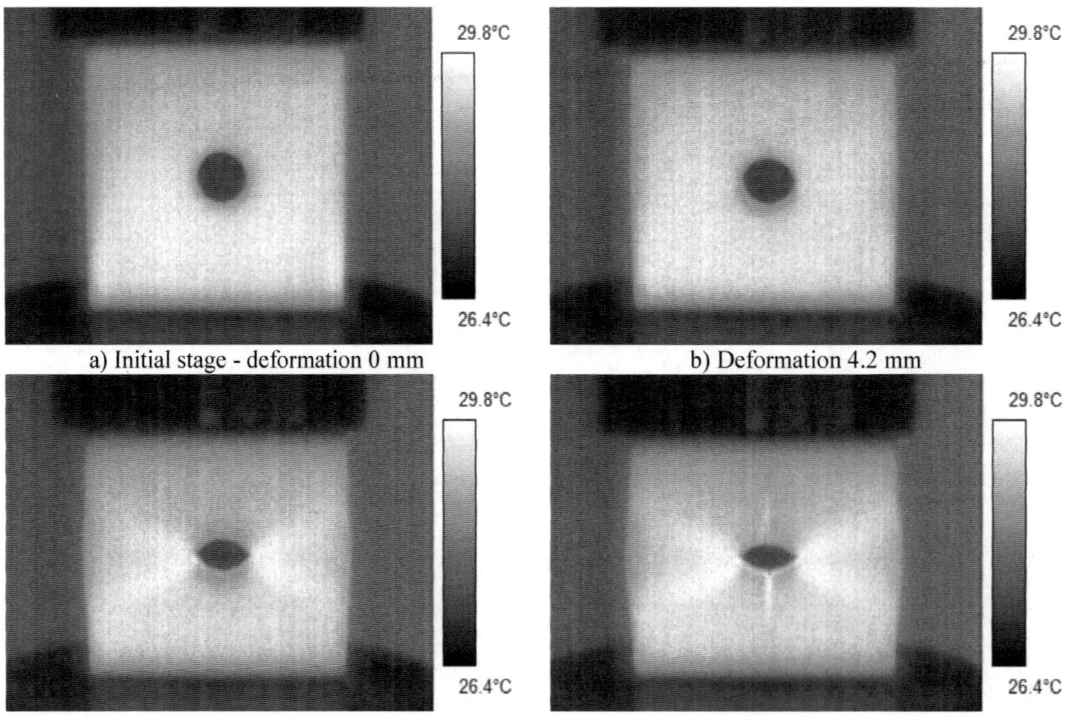

a) Initial stage - deformation 0 mm b) Deformation 4.2 mm

c) Deformation 8.5 mm d) Deformation 10.5 mm

e) Deformation 11.7 mm

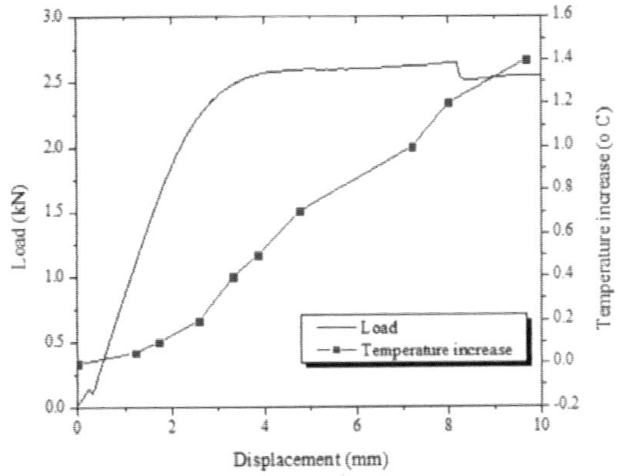

Figure 5. Temperature distributions at different load stages from compression testing of 145 kg/m^3 foam with a hole of 16 mm.

Figure 6. Load - displacement curve for foam density block of 145 kg/m^3 with hole with 16 mm diameter and the temperature increase.

From the thermographic measurements the angle of maximum temperature (figure 7), which corresponds to bands of deformation of cellular structure of the foams and the maximum temperature increase on these directions (figure 9) were determined. It could be observed that angle of the bands of deformation increases with increasing hole diameter, but is not influenced by the foam density. In contrary the maximum temperature increases with increasing density from 0.8°C for 100 kg/m^3 density to 1.6°C for 300 kg/m^3 density.

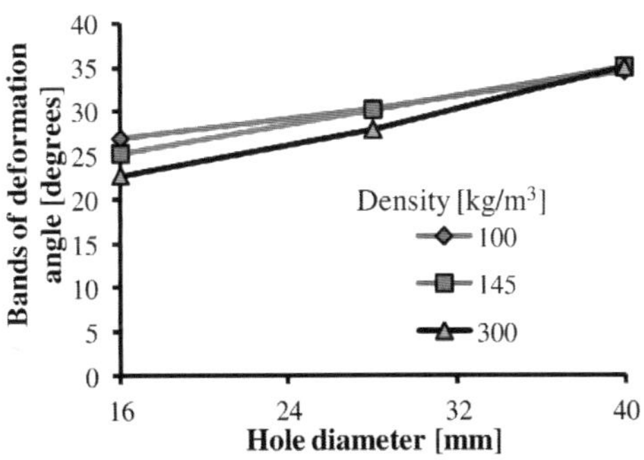

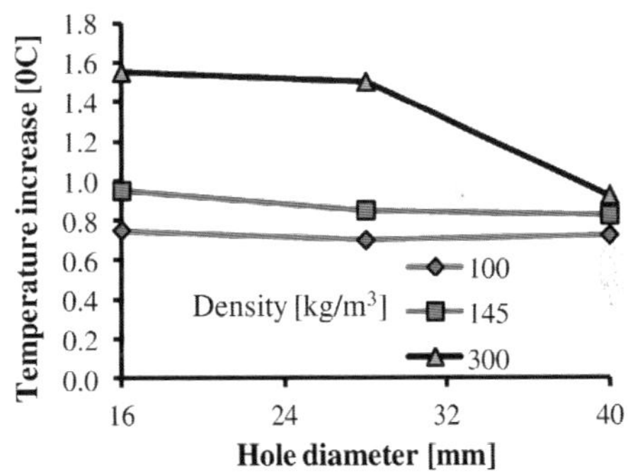

Figure 7. Bands of deformation angles

Figure 8. Maximum temperature increase on the band of deformation direction

3. Numerical simulation

The CRUSHABLE FOAM plasticity models were proposed [12] and implemented [13] in order to analysis the behavior in compression of crushable materials (like foams, balsa wood). These models are able to describe the damage in compression characterized by cell wall buckling processes. Two types of crushable foams models are implemented in ABAQUS: volumetric hardening and isotropic hardening. The isotropic hardening model uses for the yield surface an ellipse centered at the origin in the $\sigma_e - \sigma_m$ stress plane. The yield surface evolves in a self-similar manner, and the evolution is governed by the equivalent plastic stress [13]:

$$\sigma_e^2 + \alpha^2 \sigma_m^2 = B^2 \tag{1}$$

where σ_e represents is the von Mises stress and σ_m is the mean stress, α defines the shape of the yield surface in relation with the relative magnitude of the axes. This parameter is computed based on the ratio between the initial yield stress in uniaxial compression σ_C^0 and initial yield stress in hydrostatic compression p_C^0: $k = \sigma_C^0 / p_C^0$:

$$\alpha = \frac{3k}{\sqrt{9-k^2}} \tag{2}$$

The size of the of the yield ellipse on σ_e - axis is:

$$B = \sigma_C \sqrt{1+\left(\frac{\alpha}{3}\right)^2} \tag{3}$$

with σ_C the absolute value of the yield stress in uniaxial compression.

The specimens (80x80x25 mm) with holes (diameter 16, 28, 40 mm) used in the experiments were modeled in ABAQUS software. 3D 20 node quadratic solid (C3D20) elements were used, with a refined mesh near the hole, resulting 11850 elements connected in 53483 elements. A convergence study was carried out resulting the present mesh topology. The boundary conditions represent the experimental setup: 0 displacements of vertical direction were imposed at the bottom side of the specimen, while 15 mm displacements were applied on the top side.

The uniaxial compression experimental stress - strain data (20 points for each foam density) were used to define the CRUSHBLE FOAM material model. The isotropic hardening feature was implemented by k=1 and the plastic Poisson ratio ν_p=0. Self contact on hole edges was defined.

A typical result from simulation, vertical reaction force from bottom edge versus applied displacement is shown in figure 9, together with experimental load - displacement curve for foam density of 100 kg/m^3 and a 28 mm hole diameter. It could be observed that the numerical result fall between the experimental curves indicating a good agreement. The CRUSHBLE FOAM model allows to indentify the plastic equivalent strain, figure 10. The shape and path of plastic equivalent strain is similar with the termographic distribution of temperature increase due to plastic deformation.

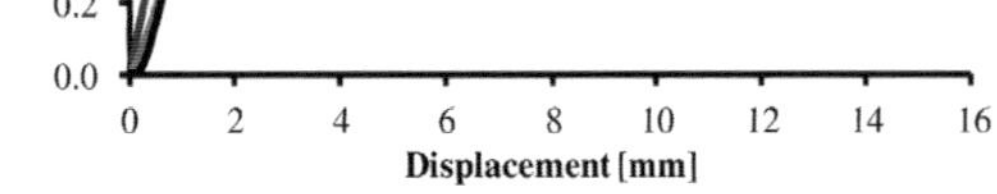

Figure 9. Comparison between experimental and numerical simulation results for 100 kg/m3 foam density and a hole of 28 mm0 diameter

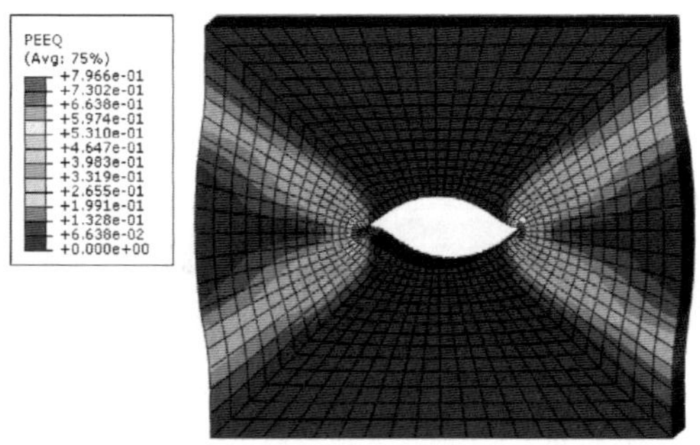

Figure 10. The effective plastic strain for modeling of 145 kg/m3 foam density with a 16 mm hole diameter

4. Conclusions

The paper presents two methods for investigation of notch effect in PUR foams under compression loading. The experimental procedure consists in compression tests performed on PUR foam (densities 100, 145 and 300 kg/m^3) blocks (80x80x25 mm) with central holes of different diameters (16, 28 and 40 mm).

From the practical point of view thermography, as a full filed temperature measurement, could be employed to investigate the damage of components or structures made of PUR foams.

On the other hand CRUSHABLE FOAM model is able to reproduce the behaviour of PUR foams in compression and could be used for designing of PUR foam components and structures above to the yielding point.

Acknowledgement
This work was partially supported by the strategic grant POSDRU/159/1.5/S/137070 (2014) of the Ministry of National Education, Romania, co-financed by the European Social Fund – Investing in People, within the Sectoral Operational Programme Human Resources Development 2007-2013.

The experimental work was carried out in the framework of the grant from the Romanian National Authority for Scientific Research, CNCS – UEFISCDI, project PN-II-ID-PCE-2011-3-0456, contract number 172/2011.

References

[1] Gibson LJ and Ashby MF 1997 *Cellular solids, Structure and properties, Second edition*, Cambridge University Press
[2] Ashby MF 2005 Cellular Solids - Scaling of Proprieties. In Scheffler, M Colombo P (Eds), *Cellular Ceramics, Structure, Manufacturing, Properties and Applications*, Wiley-VCH Verlag Gmbh & Co pp 3-17.
[3] Marsavina L 2010 *Fracture Mechanics of Cellular Solids* in Altenbach H and Ochsner A (Eds.) Cellular and porous materials in structures and processes, Springer, Wien pp 1-46
[4] Andrews EW and Gibson LJ 2001 *Acta Mater.* **49** 2975
[5] Fleck NA, Olurin OB, Chen C and Ashby MF *J. Mech. Phys. Solids* **49** 2015
[6] Combaz E, Rossoll A and Mortensen A 2011 *Acta Mater.* **59** 572
[7] Mangipudi KR and Onck PR 2001 *Acta Mater.* **49** 7356
[8] Legan MA, Kolodezev VE and Sheremet AS 2005 *Strength, Fracture and Complexity* **3** 217
[9] Voiconi T, Negru R, Linul E, Marsavina L, Filipescu H 2014 *Frattura ed Integrita Strutturale* **30** 101
[10] Srivastava V, Srivastava R 2014 *J Mater Sci* **49** 2681
[11] Marsavina L, Constantinescu DM, Linul E, Apostol DA, Voiconi T and Sadowski T 2014 *Eng Fract Mech* **129** 54
[12] Deshpande VS, Fleck NA 2000 *J Mech Phys Solids* **48** 1253
[13] ABAQUS Anaysis Users Manual 2010 Dassault Systèmes, Providence

IOP Conf. Series: Materials Science and Engineering **123** (2016) 012061 doi:10.1088/1757-899X/123/1/012061

Production of durable expanded perlite microspheres in a Vertical Electrical Furnace

M Panagiotis, P Angelopoulos, M Taxiarchou, I Paspaliaris

School of Mining and Metallurgical Engineering, National Technical University of Athens, Athens, Greece

E-mail: pangelopoulos@metal.ntua.gr

Abstract. Expanded perlite constitutes one of the most competitive insulating materials that is widely used in construction and manufacturing industry due to its unique properties combination; it is white, natural, lightweight, chemically inert, and exhibits superior insulating properties (thermal and acoustic) and fire resistance. Conventionally, perlite expansion is performed in vertical gas-fired furnaces; the conventional perlite expansion process has certain disadvantages which affect expanded products quality, thus limiting their performance and range of applications. In order to overcome the drawbacks of the conventional expansion technique, a new perlite expansion process has been designed based on a vertical electrical furnace (VEF). In the current study, fine perlite samples (-150 μm) from Milos Island, Greece, were expansed in the novel VEF and a conventional gas-fired furnace with the aim to evaluate and compare the main physical properties of the expanded products. The novel expanded perlite particles were characterised by superior properties, namely increased compression strength, competitive water and oil absorption capability, size homogeneity, spherical shape and decreased surface porosity in comparison to conventionally expanded samples.

1. Introduction

Perlite is a natural occurring volcanic rock that mainly consists of amorphous silica (70-76% wt SiO2) and low amounts of other metal oxides (Al2O3, K2O, Na2O, Fe2O3, CaO, MgO). Due to the presence of 2-6% chemically bound water in its structure, perlite can be expanded up to 20 times its original volume when heated rapidly at a temperature that exceeds its softening point (in the range between 700 and 1260 °C) [1,2], as the contained water vaporises and pushes its way out of the grain. Expanded perlite particles are characterised by an extensive network of air bubbles that has beneficial effect on their properties. Expanded perlite is a lightweight material (bulk density lower than 150 kg·m-3), inorganic, chemically inert, fire resistant, with superior thermal and sound insulating properties, while its properties are invariant in time.

More than 70% of the perlite world production is intended for construction industry, where perlite is used to reduce construction weight and improve thermal and sound insulation and fire resistance. Moreover, it is used in various other industrial sectors, like horticulture applications, food and chemical industry. The main physical properties of expanded perlite, like loose bulk density, thermal and acoustical properties, water retention, grain durability and stability, are mainly affected by the expansion conditions [2,3].

On industrial scale the perlite expansion is mainly conducted in vertical gas-fired furnaces, consisting of a vertical expansion chamber with a direct flame at its bottom-end directing a forced air flow upwardly [4,5]. Although this technique is in use for more than 50 years, it still remains an empirical industrial process, despite the rapid increase and development of expanded perlite applications over the last decades. One of the major drawbacks of this technique is the lack of detailed knowledge and control of the expansion conditions (temperature and air velocity) inside the heating chamber, which have an adverse impact on the expanded particles properties and characteristics. Conventionally expanded perlite grains have irregular shape, extended network of open pores and obvious signs of structural destructions, resulting in lack of durability or stability [6]. These properties limit significantly the range of its applications in the construction and chemical industry. In order to meet the technical targets for innovative applications, there is need for expanded perlites with improved properties in respect to their mechanical and surface characteristics. [6,7].

In order to overcome these disadvantages and produce high quality expanded perlite, a new expansion process based on the use of a Vertical Electrical Furnace has been designed and constructed by the Laboratory of Metallurgy of the NTUA [6,8-10]. The new furnace is characterised by a) precise control and the flexibility of the expansion conditions, depending on the raw material characteristics and the required properties of the expanded product; b) broader range of grain sizes that can be sufficiently operated in the furnace, and c) the use of milder temperature profile [8,10].

In this work, the physical and structural properties of ultrafine perlite samples expanded in the new VEF and in a conventional gas-fired furnace under various expansion conditions were studied and compared. In detail, the effect of both expansion techniques on the loose bulk density (LBD), water repellency, oil absorption, compressive strength, surface morphology and particle size distribution was investigated.

2. Materials and methods

2.1. Materials

Two different raw perlite samples (CH and TR) were used in the present study, originating from two different quarries of S&B in Milos Island, Greece, and supplied by S&B Industrial Minerals S.A. The chemical composition of the two perlites are presented in Table 1. The chemical analysis was performed using Energy Dispersive X- Ray Fluorescence (EDXRF) instrument Xepos (SPECTRO A.I. GmbH Company), and the Al2O3 content was defined by flame Atomic Absorption Spectroscopy (FAAS, Perkin Elmer 2100 spectrophotometer) after dissolution in a HF/HClO4/HNO3/HCl mixture. Loss of Ignition (L.O.I.) was determined by heating the dried, pulverized samples up to 1050 °C for 1 hour. Although both samples have a typical perlitic composition, TR is characterised by higher L.O.I., attributed to the bound water and volatiles [(CO, CO2, H2, I2, Cl2, F2) or (H2, CO2, CO, SO2, O2)] [11,12], and potassium (K2O) content. These elements play a significant role in the expandability of perlite, as they affect the melt dynamic viscosity [13]. The mineralogical composition was determined by X-Ray diffraction (XRD) on a SIEMENS D5000 diffractometer with Cu-Kα1 (Ni filtered) radiation, in the 2theta range from 5 to 75° and 0.02°/sec step, and indicated the dominance of amorphous silica, but also the presence of the crystalline phases quartz, alkali and plagioclase feldspars, and biotite (Fig. 1). The granulometry of the raw materials studied was below 150 μm, measured on a Malvern Laser Particle Analyser.

For the characterisation of the expanded materials Loose Bulk Density (LBD), water repellency, oil absorption and compressive strength were determined. The particles morphology was also studied through Scanning Electron Microscopy using a JEM 1600 analyser. The measurement of oil absorption was based on an ISO standard method [14], while the water repellency (determination of the absorbed water volume in the perlite sample after a saturation cycle) was measured using an internal test procedure that has been developed by S&B Industrial Minerals S.A. [15]. The compression strength measurement was based on the calculation through a hydraulic ram, of the pressure established at the sample surface when the sample height was reduced by 3 cm [16]. For the calculation of the LBD the total volume of the expanded material, including the solid volume (skeletal volume), the volume of closed and open voids, and the volume of the inter-particle space was taken into account. The expanded perlite particles size distribution was measured using a Malvern Laser Particle Analyzer for diameters below 500 μm and through sieving for the coarser particles.

Table 1. Raw perlite chemical composition

Oxides	CH (%)	TR (%)
SiO_2	72.57	73.03
Al_2O_3	13.73	12.70
Na_2O	3.99	3.43
K_2O	3.62	5.39
CaO	1.39	0.84
Fe_2O_3	1.25	0.91
MgO	0.37	0.30
TiO_2	0.15	0.08
L.O.I.	2.92	3.32

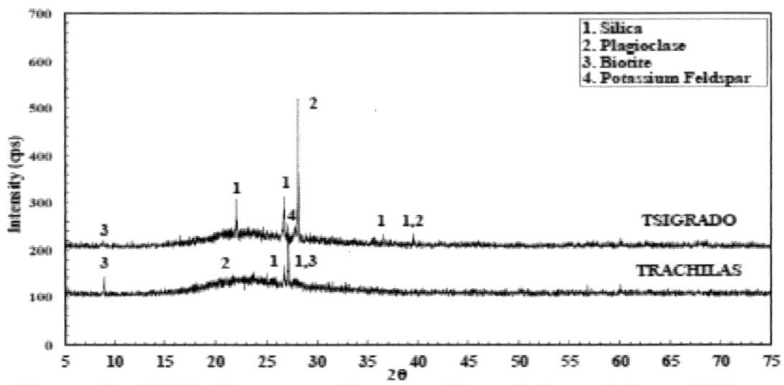

Figure 2. XRD diagrams of raw perlite originated from different ore mines.

2.2. Methods

In this study, two different expansion techniques are considered, the one based on the use of the new VEF and the other on the use of the conventional vertical gas-fired furnace.

The VEF includes the: a. heating chamber consisting of a cylindrical tube with length 3m and diameter 13.4 cm made of Kanthal APM FeCrAl alloy [17]; b. heating sources consisting of 6 pairs of electrical Kanthal resistances (total heating capacity 24 kW, maximum temperature 1250°C), located around and along the tube, defining six heating zones, the setting points of which can be adjusted independently; c. raw material feeding system, d. the air supply system and e. particle collector [6]. The raw perlite particles are fed directly at the top of the heating chamber and as they are moving downwards through regions of progressively increasing temperature, they are expanded, and are collected at the bottom of the chamber. An air stream can be injected at the direction of the gravity force in order to better control the particle residence time in the chamber. Temperature measurements within each thermal zone are obtained via ceramic sheathed N-type thermocouple (Ni-Cr-Si/Ni-Si) placed at each zone centre. The particle heating rate and processing time can be controled by proper adjustment of the heating resistances temperature and air feeding rate.

The conventional expansion furnace consists of a vertical heating chamber, the cyclone and the aspiration system. The chamber contains a direct flame at its bottom-end directing a forced air flow upwardly. The perlite particles are introduced through side ports to the hottest region of the expansion chamber near the flame at 1450oC, where they expand almost instantaneously [4,7]. The lightweight particles are carried away by the hot off gases stream; the air-solid stream exits the heating chamber by means of the aspiration system. The average residence time of perlite particles in the expansion chamber is less than one second, except for the coarsest particles [5].

5 perlite samples were subjected to expansion both in the new *VEF* and in the conventional vertical gas-fired furnace, under various temperature and air feeding rate conditions, summarised in Table 2.

Table 2. Studied expansion conditions in the *VEF*

Parameter	Value range
Furnace temperature (°C)	600-1200 (gradient in the 6 zones)
Air feeding rate ($L \cdot min^{-1}$)	15-100
Perlite type	TR, CH
Feed granulometry	-150μm

3. Results and Discussion

The studied properties of the expanded materials produced by both expansion techniques are shown in Fig. 2. The samples are compared in pairs. In each pair the same raw perlite sample has been expanded with the two difference techniques to produce two expanded samples of similar LBD. Regardless of the applied expansion technique, the lighter the material becomes the lower its compression strength gets. In all cases, the compression strength of the samples expanded in the VEF is 75 to 460% higher compared to the corresponding conventionally expanded ones. The percentage of improvement depends on the perlite type and it is much higher in the case of TR perlite (pairs 1, 2 and 3).

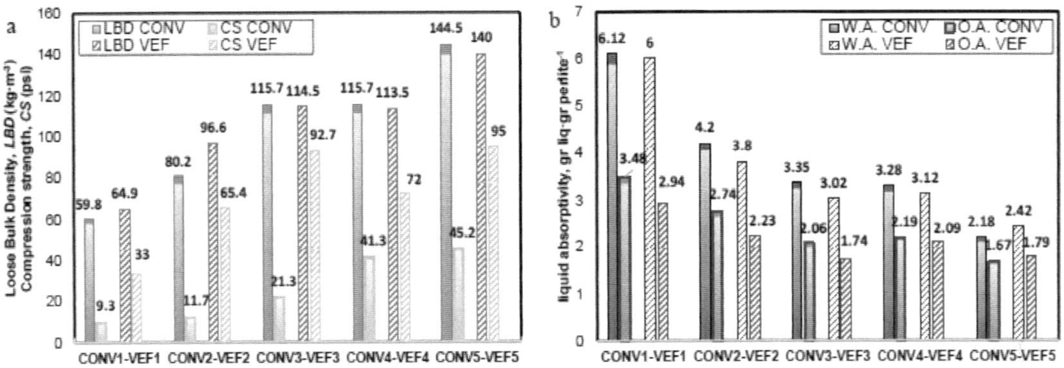

Figure 2. LBD, compression strength, water repellency and oil absorption of the various expanded samples

Contrary to compressive strength, the water repellency and oil absorption of the samples expanded in the VEF was not significantly improved comparted to the corresponding conventionally expanded materials. Depending on the perlite type and LBD of the expanded material, the water repellency is up to 10% improved compared to the corresponding conventionally expanded materials, while the oil absorption is up to 20% improved. Generally, liquid absorption increases as LBD decreases, indicating that more extended porosity is created at higher expansion degree (lower LBD). For the same LBD, the improvement of oil absorption is quite higher than the improvement of water repellency, signifying a difference in the pore diameter of the VEF expanded materials and the corresponding expanded ones. These results are in agreement with the particles morphology as shown in the SEM photos of perlite samples expanded by both expansion techniques, presented in Fig. 3. Significant morphological differences are observed depending on the applied expansion technique, which corroborate the values of the measured physical properties; conventionally expanded samples consist of irregular particles with extended network of open pores on their external surface. On the other hand, perlite samples that have been expanded in *VEF* consist of spherical particles of smooth surface and limited number of pores or cracks on their surface.

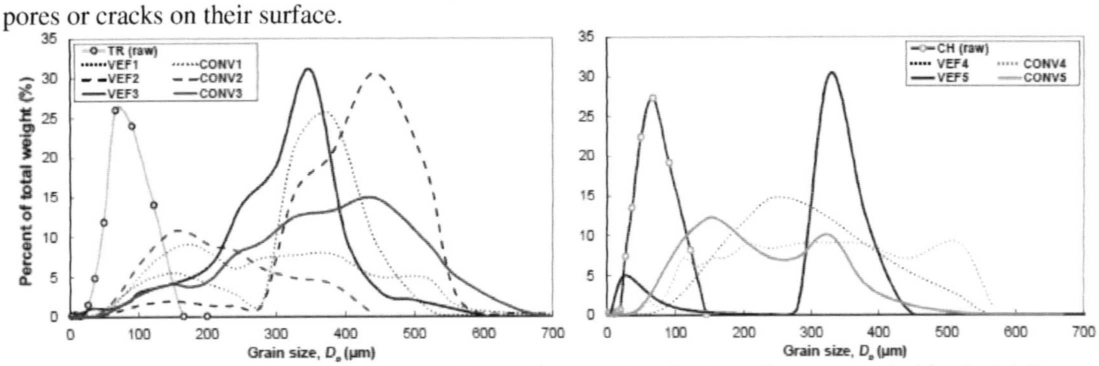

Figure 4. Granulometry of the expandes perlite samples that have been expanded int he VEF (blacklines) and in conventional gas-fired furnace (coloured lines)

Evidently, the applied expansion technique affects the morphology and consequently the physical properties of the expanded products. The extensive surface porosity of the conventionally expanded particles is attributed to the violent heating conditions during expansion, leading to particle splitting, breakage or even explosion. On the other hand, the open porosity is limited in the cases where expansion has been performed in the *VEF*, where the particle heating is milder, resulting in a more gentle expansion performance. The spherical shape and closed surface of the particles expanded in theVEF is related to the increased mechanical strength of these materials.

The particle size distribution of all the expanded samples together with that of both raw materials (CH, TR) are presented in Fig. 4. The granulometry of the materials expanded in the VEF was narrower and their mean size was larger compared to conventionally expanded ones of similar LBD. These results indicated that VEF particles were better expanded and had a more uniform and spherical shape, as also shown in SEM photos of Fig. 5. Moreover, the high content of conventional expanded samples in fine particles can be attributed to their splitting during the violent expansion conditions.

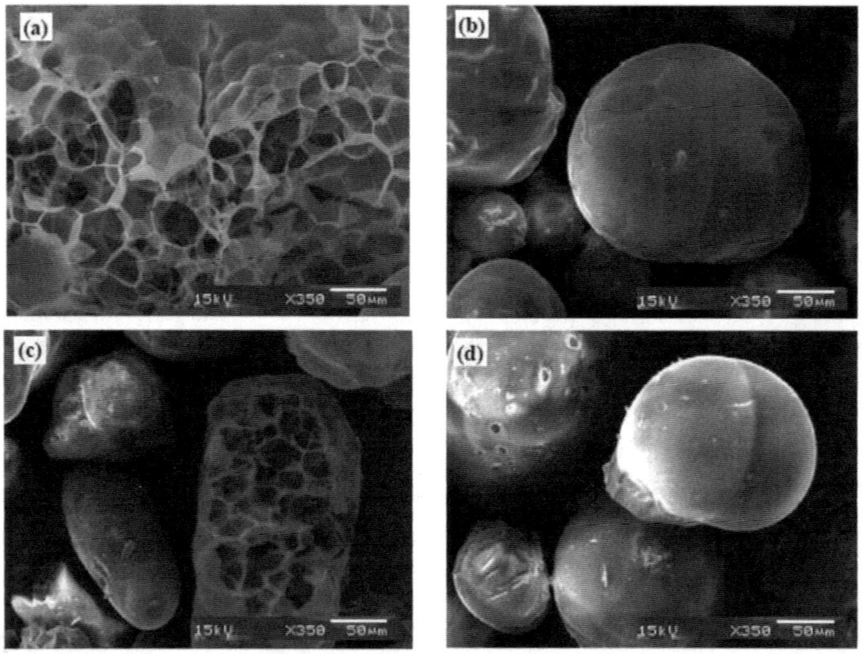

Figure 3. SEM micrographs of sample CONV1 (a), VEF1 (b), CONV5 (c) and VEF5 (d)

4. Conclusions

In this study, the physical and structural properties of two different perlites originating from Milos Island, Greece, expanded with the application of a new expansion process based on the use of a Vertical Electrical Furnace (VEF) were investigated, and compared to those of conventionally expanded samples of similar density and perlite type.

The experimental results confirmed that the novel expansion process can produce well expanded perlite microspheres with superior mechanical properties and marginally improved liquid absorption compared to conventionally expanded perlite, as a result of the milder expansion conditions and the precise control of the expansion process that the furnace can provide. The perlite particles expanded in the VEF are characterised by 75 to 460% increased compression strength, up to 10% improved water repellency and up to 20% improved oil absorption. Moreover, the samples treated in the VEF have more homogeneous particle size distribution and larger mean size. Finally, the expanded particles have an almost spherical shape with limited surface porosity or cracks in contrast to conventionally expanded particles, which are characterised by extensive surface porosity and irregular shape.

Acknowledgement
This research work has been financed by the European Union FP7 under Grant Agreement Number CP-IP 228697-2 ("Efficient exploitation of EU perlite resources for the development of a new generation of innovative and high added value micro-perlite based materials for the chemical, construction and manufacturing industry", EXPERL). Dr. P.M. Angelopoulos would also like to gratefully acknowledge the financial support of the Greek State Scholarships Foundation (IKY).

References
[1] Chatterjee K 2008 Uses of Industrial Minerals, Rocks and Freshwater *Nova Science Publisher*, New York
[2] Barker JM, Santini K 2006 Perlite, *Industrial Minerals and Rocks* (Kogel, J.E., Trivedi, N.C., Barker J.M. and. Krukowski, S.T. editors) SME, 685-702
[3] Karalis T, Taxiarchou M 2006 Current applications and future trends of perlite worldwide, *Proceedings of the 2nd International Conference on Advances in Mineral Resources Management and Environmental Geotechnology AMIREG* (Chania, Greece). 227-234.
[4] Papanastasiou DJ 1979 Perlite expansion in a vertical furnace – a simplified theoretical analysis, *Perlite Institute Annual Proceedings* (Dubrovik). 67-71.
[5] Sodeyama K, Sakka Y, Kamino Y, Seki H, *J Mater Sci.* **34**, 2461-2468.
[6] Peppas A, Taxiarchou M, Koffa E, Karalis T, Amanatidis A 2006 Development of closed porous microcellular products from perlite *Proceedings of the 2nd International Conference on Advances in Mineral Resources Management and Environmental Geotechnology AMIREG* (Chania, Greece). 235-240.
[7] Klipfel A Founti M Zahringer K Martin J and Petit J 1998 *Flow Turbul. Combust.* **60** 283-300
[8] Angelopoulos P Gerogiorgis DI and Paspaliaris I 2014 *Appl. Math. Model.* **38** 1799-1822
[9] Angelopoulos PM Gerogiorgis DI and Paspaliaris I 2013 *Ind. Eng. Chem. Res.* **52** 17953-17975
[10] Angelopoulos P Kapralou C and Taxiarchou M 2011 CFD modeling of vertical electrical furnace for perlite expansion: study of the air temperature and velocity profiles. *In Proc of the 7th GRACM* (Athens, Greece, June 30- July 2 2011) CD edition
[11] Dalakishvili AI 2005 *Glass Phys Chem.* **31**, 820-822.
[12] Shackley D, 1988 Characterization and expansion of perlite. *PhD Dissertation*, Nottingham University
[13] Zähringer K, Martin J, Petit J 2001 *J Mater Sci.* **36**, 2691–2705
[14] ISO 787-5 1980 General methods of test for pigments and extenders - *Part 5 Determination of oil absorption value ISO Standards Handbook: Paints and varnishes Vol 3: Raw materials* (2002) Ed.3
[15] S&B Industrial Minerals S.A., 2003. Method for water repellency test of expanded perlite. Internal Test Procedure (ITP) Documentation Greece
[16] S&B Industrial Minerals S.A., 2003 Method for durability measurement of 3-centimetres expanded perlite compression strength. Internal Test Procedure (ITP) Documentation Greece
[17] Kanthal Company Website http://www.kanthal.com/en/products/materialdatasheets/tube/kanthal-apm. Access date: Feb 2014.

Development of foamed Inorganic Polymeric Materials based on Perlite

G-M Tsaousi[1], I Douni, M Taxiarchou, D Panias and I Paspaliaris

Department of Mining and Metallurgical Engineering, National Technical University of Athens, Zografos, Athens, Greece

E-mail: margitsaousi@metal.ntua.gr

Abstract. This work deals with the development of lightweight geopolymeric boards for use in construction sector utilizing a solid perlitic waste as the main raw material. Hydrogen peroxide (H_2O_2) was used for the foaming of geopolymeric pastes and the production of porous and lightweight inorganic polymeric materials. The effect of geopolymeric synthesis parameters, such as the composition of activator and the curing conditions, on paste's properties that affect the foaming process, such as setting time and viscosity, were studied in detailed. Finally, the effects of H_2O_2 concentration on the properties (apparent density and % cell volume) and the microstructure of foamed boards were also studied. The produced porous boards have effective densities in-between 540 – 900 Kg/m^3 and the thermal conductivity of the optimum product is 0.08 W/m.K. Based on their properties, the developed lightweight geopolymeric boards have high potential to be used as building elements in construction industry.

1. Introduction

Lightweight porous materials offer some significant advantages to the construction sector such as lighter high rise buildings, less structural steel and concrete consumptions, improved thermal efficiency and superior fire resistance as well as seismic performance. The market of lightweight porous materials is exclusively occupied by the autoclaved aerated concrete (AAC). AAC is produced by foaming OPC concrete through hydrogen gas that is produced when aluminum powder reacts with the alkaline water of concrete [1]. The hydrogen gas after the end of the foaming process escapes to the environment and is replaced by air. The AAC after foaming is very soft and in order to acquire satisfactory mechanical strength is placed in an autoclaved chamber at 180°C-200°C and 8-12 bars steam pressure [2, 3]. The AAC production suffers from some disadvantages such as increased plant precautions due to the explosive nature of hydrogen gas, difficulties in the production control and high energy consumption [4, 5].

The development of lightweight foamed inorganic polymeric materials based on perlitic wastes as alternatives of the autoclaved aerated concrete seems to be very attractive. Geopolymers can replace OPC with significant environmental (80-90% reduction in CO_2 emissions) and technical (improved resistance to aggressive chemicals and fire) benefits while at the same time have identical mechanical properties [6]. In addition, geopolymers are normally produced utilizing waste materials such as fly ashes, metallurgical slags and mine wastes [7-10] strengthening thus their environmental character and contributing towards a zero waste circular economy. In Europe the annual production of perlite was estimated about 1.835.000 tons for 2012 [11] while the perlite fine wastes were about 275.250 tons which equals almost 15% of the raw perlite. Perlite wastes have not been extensively studied as potential aluminosilicate source for the synthesis of inorganic polymers through the geopolymerization technology. The geopolymerization process is based on an exothermic heterogeneous chemical reaction between a solid aluminosilicate raw material and an alkali metal silicate solution under atmospheric conditions and temperatures up to 100°C. The materials synthesized by this reaction present mainly an amorphous or semi-crystalline Si-O-Al and/or Si-O-Si backbone. Extended research activity in the field of geopolymeric synthesis the last decades has proven that geopolymers possess excellent mechanical and physicochemical properties [12-14] that render them ideal alternatives for many industrial applications, mainly in the construction sector.

Foaming process aims ideally at the formation of a large number of small individual voids (closed cells) or interconnected series of voids (open cells) inside a viscous material (paste or polymer). The control of the nature, size and distribution of voids is the most critical step in production of foamed materials, as the voids determine their density and strength [15]. Voids can be produced by two main methods: 1) by endogenous gas generation, which can be achieved by mixing gas-releasing agents such as H_2O_2 or fine aluminium or zinc powders in cement paste or mortar, or 2) by introducing a very large volume fraction of air bubbles, usually through the use of an organic foaming agent. Hydrogen peroxide (H_2O_2) is a widely known foaming agent which is thermodynamically unstable and therefore can be easily decomposed to water and oxygen gas (chemical equation (1)) with the latter playing the role of the blowing agent [16].

$$2H_2O_2 \longrightarrow 2H_2O+O_2 \quad (1)$$

This paper deals with the development of lightweight foamed geopolymeric boards as alternatives to the marketable autoclaved aerated concrete. The geopolymeric pastes were prepared utilizing milling plants perlitic wastes as raw material and were foamed by using hydrogen peroxide as blowing agent. The effect of main synthesis parameters on the properties of geopolymeric pastes (setting time and viscosity) were studied in extent. In addition, the effect of the amount of added foaming agent on the final foamed board's characteristics (apparent density and % cell volume) and microstructure was also studied.

2. Experimental

2.1 Materials

The raw materials used for the synthesis of geopolymeric paste were perlitic waste, sodium hydroxide and hydrogen peroxide. The perlitic waste originated from the milling plant of S&B Industrial Minerals S.A. in Greek island Milos. The chemical analysis in dry basis, determined by XEPOS X-ray Fluorescence diffractometer utilizing X-LAB software, is given in Table 1. The main chemical components of perlitic waste are silicon, aluminum, sodium and potassium while magnesium, calcium, titanium and manganese are minor constituents.

Table 1. Chemical Analysis of perlite waste

Oxides (%) w/w								
Na_2O	MgO	Al_2O_3	SiO_2	K_2O	CaO	TiO_2	Fe_2O_3	MnO
4.00	0.64	11.80	72.29	5.26	0.90	0.12	1.41	0.13

The raw perlitic waste is extremely fine with mean particle size d_{50} = 6.50 μm measured on a MALVERN Laser Particle Size Analyzer. The mineralogical analysis confirmed that perlitic waste is an amorphous volcanic alumino-silicate glass which contains some crystalline phases as illite, feldspars (albite and anorthite) and quartz.

2.2 Experimental procedure

The geopolymeric viscous paste was prepared by mixing perlite with the alkaline activating solution for 5 minutes until a homogeneous mixture was obtained. The NaOH concentration in the alkaline activating solution was varied in-between 2-12M while the solid/liquid ratio was kept constant at 1.2g/mL. The resulted paste was cured at four different temperatures (50°C, 70°C, 90°C, 100°C) and its setting time was determined by a MATEST Vicat apparatus according to the EN 196-3:2005. The viscosity of the geopolymeric pastes was measured immediately after their preparation at ambient temperature by Brookfield Viscometer LV+.

The geopolymeric foams were prepared by mixing the geopolymeric viscous paste (solid/liquid ratio =1,2 g/mL, CNaOH 8M) with hydrogen peroxide in a mechanical mixer for 5 minutes. The amount of added H_2O_2 was varied in-between 0.25-1.25%w/w. Then the mixture was molded in 15 x 15 x 4 mm molds and cured in three consecutive stages (stage 1: 90°C for 24h, stage 2: 50°C for 72 h, stage 3: 35°C for 24h). After curing the foamed boards were demolded and kept in a dry atmosphere for further evaluation. The % cell volume of the foamed boards was estimated through the equation (2) by measuring their apparent (ρ) and skeletal (d) densities [16]. The apparent density of foamed boards was determined by measuring the weight of cured boards with a precision balance as well as their dimensions with a precision electronic caliper. The skeletal density was determined by QUANTACHROME stereopycnometer.

$$C = \left(1 - \frac{\rho}{d}\right) x100 \qquad (2)$$

The microstructure of foamed boards was observed with SEM microscopy using JEOL 2100 HR Scanning Electron Microscope. Finally, the thermal conductivity of foamed boards was measured with the Netzsch HFM 436/3 Lamda, Heat flow meter.

3. Results and discussion

3.1 Synthesis and evaluation of the geopolymeric paste

The effect of initial NaOH concentration in the activating phase on the properties of geopolymeric pastes was studied in the region from 2M to 12M. The solid/liquid ratio was kept constant at 1.2 g/mL while the pastes cured at 50°C, 70°C, 90°C and 100°C. The results are shown in Figure 1 where the setting time of synthesized geopolymeric pastes at different temperatures is plotted versus the initial concentration of NaOH solution. The results showed a strong trend towards increase of setting time as a function of NaOH concentration. At 50°C and as the NaOH concentration rises from 2 to 6 M, the setting time rises from 20 to 120 hours indicating an increase of almost 6-fold, while for NaOH concentration higher than 6 M the setting time is invariable. It is observed that at 70°C the setting time is generally faster than at 50°C. The setting time is invariable in the region of NaOH concentration in-between 2-5M but the setting process is strongly decelerated when the NaOH concentration becomes higher than 5M reaching again a plateau after a NaOH concentration higher than 8M. The setting of the geopolymeric pastes is substantially accelerated at 90°C and 100°C. The results revealed that the geopolymeric perlitic pastes solidify faster under lower NaOH concentrations in the alkaline activating phase as well as at higher curing temperatures. The former of the above observations seems opposing to the common perception in the geopolymerization technology according to which geopolymerization is effective at high NaOH concentrations. Indeed as is seen in Figure 2, Si dissolution from perlitic waste after 24 hours retention time at 50°C increases as the initial NaOH concentration increases but the 10-fold increase of NaOH concentration (from 1M to 10M) causes only 2,5-fold increase on the silicon dissolution (from 10% to 25%) leaving, according to the dissolution reaction (equation 3), a substantial excess amount of free NaOH and establishing lower SiO_2/Na_2O mass ratios in the aqueous phase. The lower SiO_2/Na_2O mass ratios favor the formation of monomer and oligomer Si species (Figure3) [17] and therefore inhibit the polycondensation phenomena prolonging the setting time of geopolymeric perlitic pastes.

$$SiO_2 + NaOH + H_2O\ SiO(OH)_3^- + Na^+ \tag{3}$$

The increase of temperature improves the dissolution reaction kinetics and therefore, decreases the excess amount of free NaOH and increases the SiO_2/Na_2O mass ratios in the aqueous phase accelerating the polycondensation phenomena (Figure 3) and making the geopolymeric paste setting faster (Figure 1).

In Figure 4, the effect of NaOH concentration on the viscosity of perlitic pastes is shown. The perlitic pastes are non-Newtonian shear thinning fluids that achieve substantially low viscosities at high shear stresses. Not significant dependence of NaOH concentration on viscosity was revealed, indicating that the early viscosity is more related to the density of suspended perlite particles in the paste rather than to the physical properties of the aqueous phase itself. An exception is the high viscosity values from 800.000 to 1.400.000 centipoise measured at spindle speed lower than 1 rpm for the paste composed with 2M NaOH solution as activator. This observation proves once more the faster polymerization and therefore solidification at 2M NaOH indicating that the polycondensation phenomena in the perlitic pastes are optimized under low NaOH concentration.

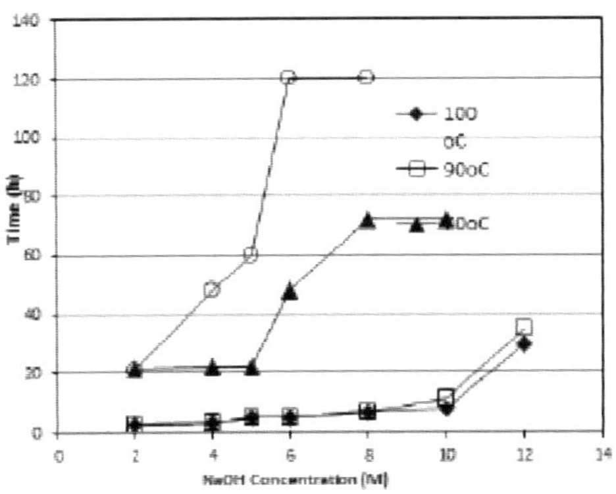

Figure 1. Setting time as a function of CNaOH.

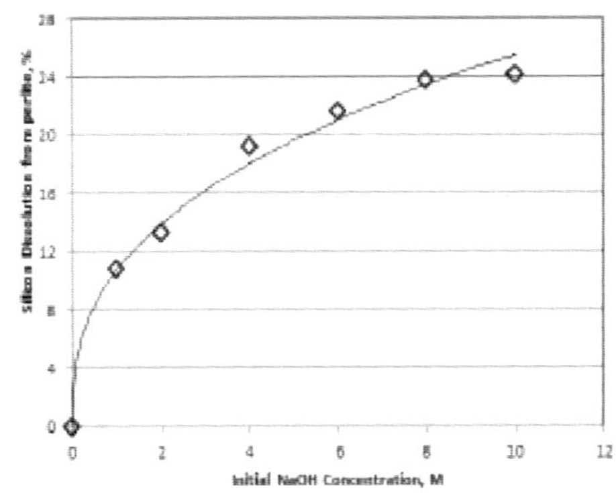

Figure 2. Silicon dissolution from perlitic waste in several NaOH solutions at 50°C, 24h and pulp density 16g/L

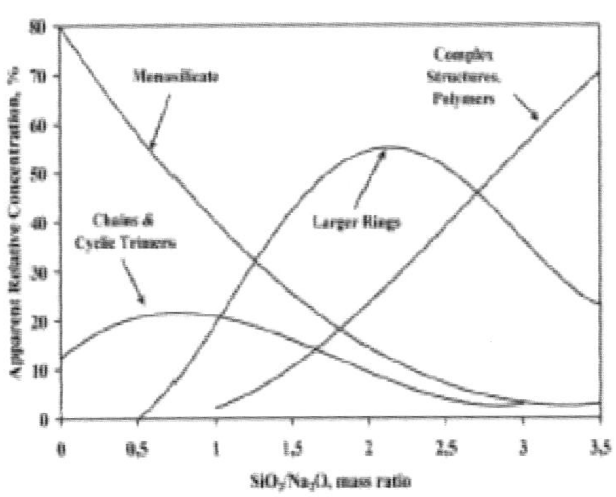

Figure 3. Qualitative interpretation of soluble silicate species equilibria in 1m aqueous silicate solution

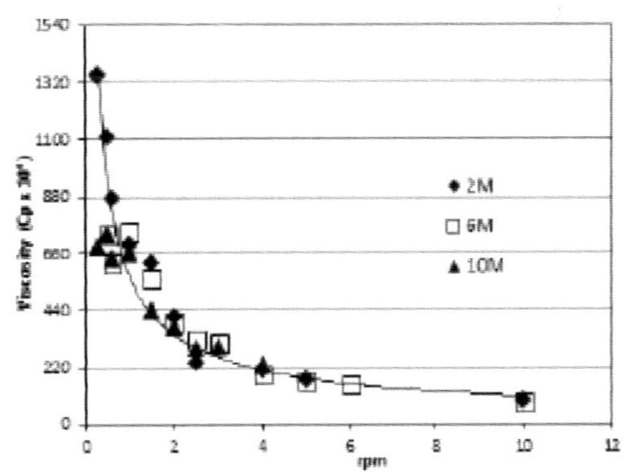

Figure 4. Viscosity as a function of CNaOH

3.2 Synthesis and evaluation of foamed boards

In order for a large number of relatively small individual gas bubbles to be dispersed inside the geopolymeric paste, two main processes have to be appropriately controlled: a) bubble swelling and b) bubble upward movement. The former is mainly affected by the amount of blowing agent per volume of paste while the latter is affected by the viscosity of paste [18]. The highest the viscosity, the lowest the bubble rise terminal velocity. Because the initial viscosity of the geopolymeric perlitic pastes under ambient temperature (where the foaming is taking place) seems to be almost unaffected by the composition of the aqueous phase, the most important parameter that determines the effective entrapment of bubbles inside the paste is its setting time which is shorten substantially at temperatures as high as 90°C. Therefore, the curing of perlitic pastes was decided to take place initially at 90°C for 24h for fast setting and then to lower temperatures (50°C for 72h and 35°C for 24h) for avoiding the surface cracking.

The apparent density of the foamed boards as a function of the % H_2O_2 addition in the paste is shown in Figure 5. The apparent density decreases substantially as the % H_2O_2 increases up to an optimum value of 0.5% reaching a value which is slightly lower than 600 kg/m3. Further addition of H_2O_2 seems to have detrimental effect on the apparent density of foamed boards increasing it to a value around 700 kg/m^3. The increase of the % H_2O_2 up to the optimum value of 0.5% increases the % cell volume in the boards (Figure 6) as well as increases the size of cells and makes thinner their walls (Figure 7a & b). Further increase of % H_2O_2 leads to the perforation of the very thin cell walls causing coalescence of neighboring cells (Figure 7c, d & e) and collapse of the foamed structure which has as a consequence the decrease of % cell volume (Figure 6) and therefore the increase of the apparent density (Figure 5).

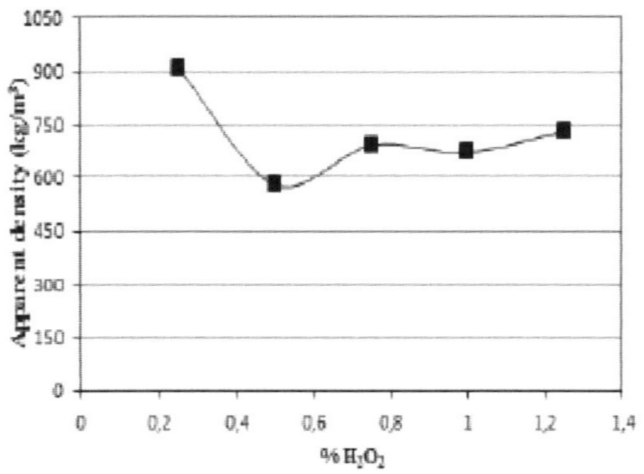

Figure 5. Apparent Density of foamed boards in function of H_2O_2content in the paste

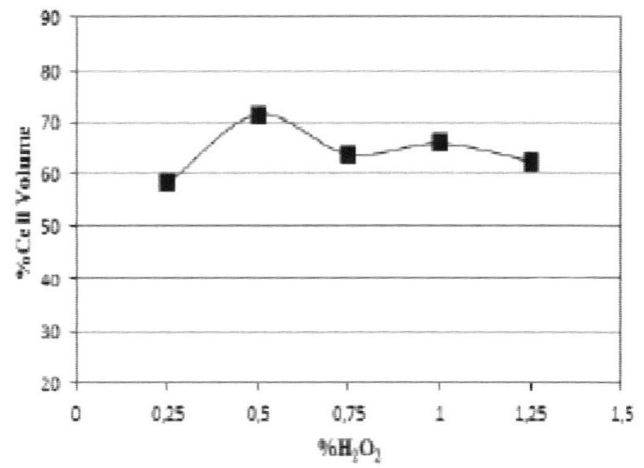

Figure 6. %Cell Volume of foamed boards in function of H_2O_2 content in the paste

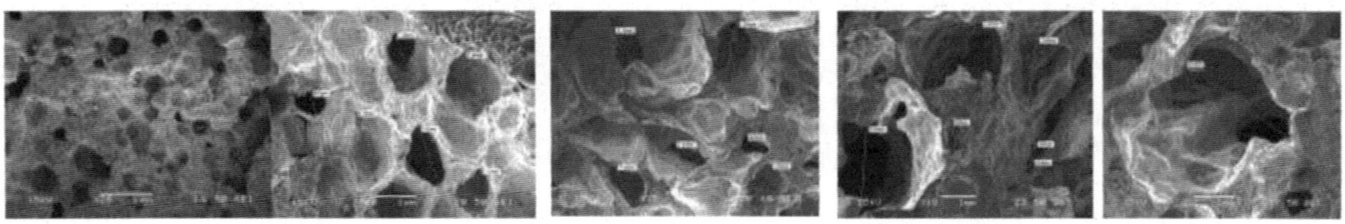

(a) (b) (c) (d) (e)

Figure 7. Scanning Electron Microscopy pictures of geopolymeris foamed boards synthesized with different H_2O_2content. a) 0.25%, b) 0.5%, c) 0.75%, d) 1% and e) 1.25% H_2O_2.

A comparison of the most important properties of aerated concrete, autoclaved aerated concrete and developed geopolymer foams is given in Table 2. The developed geopolymer foamed boards have directly comparable thermal conductivity and apparent density with the aerated concrete and autoclaved aerated concrete and therefore have high potential to be used as lightweight construction materials.

Table 2. Conventional Foamy Materials vs Geopolymeric Foams with H_2O_2

Material	Aerated Concrete	Autoclaved Aerated Concrete(AAC)	Foamed Geopolymer Boards
d(Kg/m^3)	400-1600	300-1800	540-900
λ(W/mK)	0.06-1.4	0.07-0.21	0.08-0.1

4. Conclusions

In this work, the development of lightweight geopolymeric boards was investigated. The evaluations of the geopolymeric paste's properties that are important for the foaming process were performed. The concentration of sodium hydroxide in the liquid phase and the curing temperature of the pastes were proved as important synthesis parameters affecting substantially the geopolymeric paste's setting time. Under the optimum synthesis conditions defined in this work, the perlite based inorganic polymeric pastes had setting time shorter than 2h. The produced pastes are viscoelastic fluids with viscosity values up to 1400 cp x 10^3 while their initial value is more determined by the density of suspended perlite particles in the paste rather than the physical properties of the aqueous phase itself. Hydrogen peroxide (H_2O_2) proved to be an effective foaming agent for this type of geopolymeric pastes. The foamed materials that came out were lightweight and porous and the foaming process proved to be more efficient at 0,5%(w/w) H_2O_2 addition in the geoplolymeric paste. The best foamed boards produced had apparent density 540 Kg/m^3, cell volume 70% and thermal conductivity 0.08 W/mK. In comparison with the other conventional lightweight materials (mainly AAC), the foamed boards from perlite have similar thermal properties and densities that render them potential lightweight materials for the construction sector.

References

[1] Song Y, Li B, Yang E, Liu Y and Ding T 2014 *Cement and Concrete Composites* **56** 51
[2] Holt E and Raivio P 2004 *Cement and Concrete Research* **35** 796
[3] Jerman M, Keppert M, Vyborny J and Cerny R 2013 *Construction and Building Mater*ials **41** 352
[4] Svanholm G 1990 Pouring into molds with removable walls, stiffening, autoclaving *US4902211 A*
[5] Mostafa NY 2005 *Cement and Concrete Res*earch **35** 1349
[6] Giannopoulou I, Dimas D, Maragos I and Panias D 2009 *Global NEST* Journal **11** 127
[7] Xu H and Van Deventer J.S.J 2000 *International Journal of Minerals Processing* **59** 247
[8] Palomo A, Grutzeck M.W and Blanco M.T 1999 *Cement and Concrete Research* **29** 1323